RICHARD OF WALLINGFORD

RICHARD
OF
WALLINGFORD

An edition of his writings
with introductions, English translation
and commentary by

J. D. NORTH

I

TEXTS AND TRANSLATIONS

OXFORD
AT THE CLARENDON PRESS

This book has been printed digitally and produced in a standard specification in order to ensure its continuing availability

OXFORD
UNIVERSITY PRESS

Great Clarendon Street, Oxford OX2 6DP

Oxford University Press is a department of the University of Oxford. It furthers the University's objective of excellence in research, scholarship, and education by publishing worldwide in

Oxford New York

Auckland Cape Town Dar es Salaam Hong Kong Karachi Kuala Lumpur Madrid Melbourne Mexico City Nairobi New Delhi Shanghai Taipei Toronto

With offices in

Argentina Austria Brazil Chile Czech Republic France Greece Guatemala Hungary Italy Japan South Korea Poland Portugal Singapore Switzerland Thailand Turkey Ukraine Vietnam

Oxford is a registered trade mark of Oxford University Press in the UK and in certain other countries

Published in the United States by Oxford University Press Inc., New York

Reprinted 2006

ISBN 978-0-19-922744-0 Volume I
ISBN 978-0-19-922745-7 Volume II
ISBN 978-0-19-922746-4 Volume III

Printed and bound by CPI Antony Rowe, Eastbourne

TO MARION

PREFACE

THROUGHOUT this collection of the writings of Richard of Wallingford, abbot and scholar of the fourteenth century, I have tried to temper an editor's ambition for completeness with the demands of historical relevance. Completeness is not the easiest of things to defend, even at the best of times (which these are not); but the need for it in such a work as this should be as evident as the need to include uninteresting people in telephone directories. No doubt historical testimony is least welcome when it seems to offer nothing analogous to present experience, but there is something presumptuous about an editor who undertakes to filter away all elements of fourteenth-century thought which might appear irrelevant to the twentieth century. Richard of Wallingford was no Archimedes, nor did he propose a new system to rival Ptolemy's, and by these crude tokens there are some who will consider him to be hardly worth reading. The more discerning will be content that he was not only a great abbot, but far and away the most original English mathematical astronomer of the late Middle Ages, and a man who plainly influenced European astronomy during the fifteenth and sixteenth centuries. He has been often enough the victim of brief and empty praise, and yet none of his principal writings has been published before in anything approaching its entirety, and none has been correctly interpreted in any printed work of which I am aware. The best textual work done hitherto was that by the late Reverend H. E. Salter, of Magdalen College, whose versions of the short treatise *Rectangulus* and of a part of *Albion* deserved to be better understood than they ever were.

In collecting together Richard of Wallingford's writings, astronomical, mathematical, ecclesiastical, and devotional, I have added to the usual list, deleted from it, and discussed works to which doubt attaches. The most notable addition is that of a draft treatise on the St. Albans astronomical clock, known now in some detail from a solitary Bodleian manuscript which I first recognized in 1965. (This account of Richard's astronomical clock, the oldest purely mechanical clock of which the details are known, entails a drastic revision of the accepted view of fourteenth-century technological potential.) I also claim for Richard of Wallingford works which are usually given to John Maudith; here, partly because doubts as to authorship remain, the works are not printed in full. I have naturally included the astrological works, written by one of those whom Bouché-Leclercq calls 'les disciples récalcitrants des géomètres'. One of the two treatises is of doubtful authorship, and did not seem to me to merit a translation, while the other, *Exafrenon*, is an exception to the general rule that authentic texts are provided with a translation into modern English. Since a Middle English translation of a substantial part of *Exafrenon* survives, that has been given instead, supplemented by a modern translation only where necessary.

The three volumes into which this work is divided are so arranged that the text (vol. i), the translation (facing the text), the commentary (vol. ii), and the appropriate diagrams (vol. iii, numbered to correspond to the text), may all be opened simultaneously. Commentary of a more general kind is included in the Appendices (vol. iii), while vol. ii opens with a short biography.

A translator offers hostages to fortune, but surely not inadvisedly in writings such as these, where, for example, '*latitudo*' might correspond to our modern 'declination', and where in the same paragraph '*declinatio*' might mean 'latitude' or correspond to yet a third idea. Of course, as Erasmus found when he vainly argued that five words with understanding are better than ten thousand in an unknown tongue, there will always be those who prefer the austerity of unrelieved obscurity. It is certainly less expensive. I should like to believe that more people are interested in medieval history than read medieval Latin fluently. The depth of my commentaries is not, perhaps, commensurate with their length, but they are written with the thought that no one is ever likely to read these volumes consecutively, and that the commentary on each work should therefore be more or less self-contained. Each of the principal works is provided with an essay (in Vol. II) which is meant to set it in a historical context, and any such chapter could be read in isolation from the edition as a whole.

I hope that I make my debts on specific matters clear as the occasions arise. The greater part of the work was done between 1965 and 1968, when the Nuffield Foundation, and also the University of Frankfurt, made it possible for me to give to it my undivided attention. The work was complete by the end of 1971, and a short list of important supplementary material which has come to light since then will be found in Vol. III, before the bibliography. My wife typed the work from the first page to the last, and its dedication to her is small recompense for her dedication to it. Without the promptings and help of Dr. A. C. Crombie, Dr. R. W. Hunt, Mr. F. R. Maddision, and Mr. T. I. M. Beardsworth, I should never have begun the task. Dr. Crombie in particular deserves my thanks for having kept the history of scientific thought alive as an academic discipline in Oxford. I owe debts of a general kind to many other people, among whom I must mention Professor Willy Hartner and his colleagues at Frankfurt, Monsieur Emmanuel Poulle, Dr. Lorenzo Minio-Paluello, Professor Olaf Pedersen and his colleagues at Århus, Dr. R. P. Lorch, and Mr. Francis Steer.

I should like to thank those many librarians, seen and unseen, who have gone out of their way to help; and especially those in Duke Humphrey's Library, Oxford, and Mr. Peter Brown and Mr. William O'Sullivan of Trinity College, Dublin. My enormous debt to the officers of the Clarendon Press will be immediately apparent, but I must here single out Mr. Bernard Dod, who not only acted as the book's guardian spirit, but has saved me from many solecisms. To the Curators of the Bodleian Library, the Syndics of the Cambridge University Library, the Trustees of the British Library, the Österreichische Nationalbibliothek, the Museum of the History of Science, Oxford, and the President and Fellows of Corpus Christi College, Oxford, I offer acknowledgement and thanks for permission to reproduce illustrations from manuscripts in their keeping.

Finally I wish to place on record my gratitude to the Fellows of the Royal Society and the Radcliffe–Skinner Trustees, for their very liberal financial assistance towards the cost of publication. Tycho Brahe once called on the women of Denmark to send their linen to the paper-mills so that their national history—as recorded by Saxo Grammaticus—could be preserved in print at a time of economic crisis. That I was not driven to follow Tycho's example was in large measure due to the generosity of the Royal Society and the Radcliffe–Skinner Trust.

Oxford
October 1975

J. D. N.

CONTENTS

VOLUME I

EDITORIAL PRINCIPLES AND ABBREVIATIONS

THE LATIN WORKS

Our chief purpose is to restore the text of Richard of Wallingford's extant writings to a form as close as possible to that which he intended. There are, however, occasions—as in the proof of theorems, and in the presentation of arguments generally—when his achievements must take precedence over what his intentions should ideally have been. For the most part, it is not difficult to determine the substance of the original, but variant readings are recorded where there is doubt, or where they throw light on the filiation of manuscripts, or where they are merely of some philological interest or importance. Each text is based on a small number of manuscripts, and these are discussed, together with all other known copies of the work in question, in the Introductions in vol. ii. Each text must be regarded as self-contained, for each presents problems peculiar to itself. There are nevertheless several principles followed throughout. Scribal contractions have almost always been resolved, proper names (including *Sol*, *Luna*, and *Terra*) capitalized, and punctuation supplied. No attempt has been made to be faithful in every detail, and at all costs, to a single manuscript. (Diplomatic editions are presumably so called in deference to Wilde's remark that to be intelligible is to be found out.)

'There can be no doubt', wrote Salter in his *Registrum Annalium Collegii Mertonensis* (Oxford Historical Society, 1923, p. xxxvi), 'that the superior person of the fifteenth century considered that he showed his breeding by spelling words as they were spelt by no one else.' The fourteenth century—like the fifteenth—did have an orthography of sorts, and I have tried to judge this on the strength of the two manuscripts closest to Richard in time and provenance (MSS. Ashmole 1796 and Corpus Christi College, Oxford, (D) 144). I cannot claim to have been completely ruthless, and have tended to allow the text to follow the best manuscript in each individual case. Trivial differences in spelling, unnecessary and idiosyncratic addition of the word *et*, unimportant inversions of word order, the substitution of certain close synonyms (*quia* for *quod*, *igitur* for *ergo*, *ista* for *illa*, and so forth) are all ignored; and, more controversially, no regard is paid to the shape which any scribe happens to give to the letters (and what *is* a letter?) *u* and *i*. Here the letter-forms *u*, *v*, *i*, and occasionally *j*, are used for much the same reasons as that the book is printed in the form of a codex rather than a roll: the modern reader finds it convenient, especially when it is a question of distinguishing between the vocalic *u* and the often consonantal *v*. One is saved from mistaking *volvit* for *voluit*, for instance; and who prefers *uuluula* to *vulvula*? Numerals are here given in modern form, unpointed; and geometrical points are denoted by capital letters, rather than by the lower-case letters of the manuscripts.

We may illustrate the problem of orthography sufficiently well by reference to a single text, *Albion*. The manuscripts collated are close in time, and the differences in spelling are few. We give below a list of variants which are not recorded in footnotes to the text—and it should be added that many common variants resulting from an interchange of *c* and *t*, *y* and *i*, *d* and *t* (in words like *sed*), are not listed there or here. We give here the spelling generally preferred, usually but not invariably following MS. C, together with preferred, common, or interesting spellings from manuscripts which differ:

albion: C *albyon*; L *albeon*
almanac: L *almanak*
almucantarath: C *almuchantharather*, *almucantarat*, *almicanterath*, *almuchantarath*; A *almicantaraz*; L *almicantarader*
azimuth: C *asmuch*
eclipsis: C *eclipcis*
epiciclus: A *epiciculus*
geusahar: A *geuzar*; and see our glossary
intrinsecum: C *intrincecum*
membrana: CAHH′ *membramina*,
resecemus: CA *resesemus*.

Changes in the tense or mood of verbs are generally recorded. In phrases like *per primam [partem] huius quinte*, referring to the treatise itself, MS. A almost invariablg makes the first adjective agree with a masculine noun, presumably *liber*. The scores of variant readings which result are not recorded. There is a great variety of usages in recording numerical data in degrees, minutes, etc. Most commonly the names of the units are put in the genitive case thus: 3 graduum 56 minutorum 14 secundorum. Generally speaking this convention is accepted here, although in the Latin text we frequently abbreviate thus: 3 gra. 56 min. 14 sec.

Different scribes were obviously ill at ease with the construction

arcus qui/que est distancia inter,

being uncertain as to whether the relative pronoun should agree with the first or second noun. Here the first (and usual classical) alternative is accepted. They were also uncertain as to how to regard the frequently encountered word *diameter* (Gr. *διάμετρος*) both as regards gender and declension. Here, on the basis of classical usage and manuscript evidence, it is taken as a feminine noun of the second declension, *diameter*, *-tri*. The copyist of the fragments which make up the treatise *Horologium* invariably accepts *diametrum* for the nominative case, and this was probably the author's own usage.

Problems of a different kind present themselves when we come to the Middle English translation of *Exafrenon*, and it is desirable to discuss them separately.

The discussion of more general textual matters is put in footnotes to the translation, in order to keep the text as free from interference as possible. When square brackets occur in the translation, they do not necessarily carry an implication of textual omission. They are to allow of elaboration and clarification, and occasionally the Latin will be such that the sense of the added words would have been understood by the reader. Editorial principles used in connection with tables are *ad hoc*, and are explained as the occasion arises.

When we are dealing with a Latin translation, from an original in Arabic, which is almost always the case, we use the Latinized form of the author's name. The principle is difficult to apply when we discuss both the original and the translation together, but it saves repeated caveats about the probability of there being interpolated material, and even about the possibly mistaken identity of the author. Likewise in our own translation, we shall treat many Latin words as though they were English, in order to avoid confusion with their strict modern equivalents. Examples are: 'motus', 'sinus', 'extremum', and 'semicirculus' (when it refers, that is, to the pallets of the clock escapement; but I have used 'semicircle' in the translation, where there could be no confusion, as a reminder that as a technical term this was a very homely one, to a geometer, at least). 'Kardaga', 'cata', and 'iomyn' are other examples of words I treat as English, in the absence of near English equivalents.

ON THE TRANSCRIPTION OF THE MIDDLE ENGLISH TRANSLATION

> And fayn wolde I satysfy every man... And certanely, our language now used varyeth ferre from that whiche was used and spoken whan I was borne. For we englysshemen ben borne under the domynacyon of the mone, which is never stedfaste but ever waverynge, wexynge one season, and waneth and dycreaseth another season.
>
> Caxton, Prol. to transl. of Virgil's *Eneydos* (1490)

In the fourteenth and fifteenth centuries it was customary to add a flourish to a final consonant (especially *n*, *m*, *h*, and *r*), which (i) occasionally indicates a final *e*, and (ii) with *n* is often now transcribed as *un*, a move which in some cases at least stems from reading the *n* as *u*, and treating the flourish as a typical abbreviation for *n*. Thus *questioñ* has on occasion been taken by editors as *question*, or as *questione*, more usually *questioun*, but even as *questionn*. English orthography at this time having been much more fluid than that of Latin (a fact which tends to support the view that the significance of final flourishes is very largely decorative), some editors of English texts have cautiously printed the flourishes as such, while others, advocating an idiosyncratic approach, have given them all the significance of omitted letters. The first attitude is perhaps preferable to the second; but here where the sense of the texts and therefore ease of reading is of first importance, a third approach is called for. Since a given scribe will often spell a word on two occasions differently, even within a single sentence, it is no argument to claim that final flourishes are meaningless because they are often omitted; but it is at least a proof that in these cases it is not incorrect, in the light of late medieval practice, to ignore them. This is the apology offered for choosing to ignore doubtful flourishes. Thus the preposition is often spelt *for*, rather than *fore*; *saithe* is preferred to *saithee*, *men* to *menn*, *when* to *whenn*, *upon* to *uponn*, *him* to *himm*, and so on. More controversial are words like *governa*(*u*)*nce*, *exaltacio*(*u*)*n*, and *opposicio*(*u*)*n*, which are here presented without the *u* unless the evidence of the script is clearly to the contrary. The Middle English *Exafrenon* comes from a period of transition, and when doubt attaches to a reading, the later form is preferred. The final flourish indicating an *e* of the plural form of an adjective is, of course, not ignored, although the final *e* was often omitted by the scribe. In general, the spelling is never arbitrarily standardized.

The translations are from a time when dialectal peculiarities had ceased to be well defined amongst educated men. Those found in our translation are as likely to originate with the scribe as with the author. The argument against standardization when two or more manuscripts are being collated is that there is a nucleus of would-be technical words not widely represented in other texts, and it would be wrong to prefer one word to another (in attempting to establish the source translation) on the slender statistical evidence of a handful of copies. Furthermore, that this is a period of transition is also evident from the very inconsistencies of any given copy. Thus our scribes seem to have been only faintly conscious of a rule of preferring an infinitive ending in *-en* (rather than *-e*) before a word beginning with a vowel, and conversely before a consonant.

Needless to say, none of our rules of transcription is based on any presumption about fourteenth- or fifteenth-century pronunciation, although partly on the grounds of modern pronunciation, the letters thorn (þ) and yogh (ȝ—the Old English *g*) are treated differently. The former, written by then in a way indistinguishable from *y*, was used quite randomly in MS. D as an alternative to *th*, exactly as it was to be commonly used well into the eighteenth century. In MS. T we scarcely ever find *th*. In our translation, we treat þ as a scribal contraction, and expand it accordingly. The letter ȝ was used with greater deliberation, however, in MS. D. The time was yet to come when it would be generally dropped in favour of *y* (as in *year* or *yif*), or in favour of *gh* (as in *night*). One must suppose that long after the substitution the pronunciations given to ȝ were guttural (that in *year*, for instance, being palatalized, especially in the south, and perhaps aspirated as *hyear*). The substitution was perhaps forced on English in due course partly by the exigencies of printing in continental founts lacking the letters þ and ȝ—and this meant a greater loss to the language in the latter case. Our rule, then, is as follows: we adopt an italic *y* for the palatalized forms, and an italic *g* for those which have come down to us with a hard g, as against an italic *gh* for the rest. By this means, legibility is combined with an opportunity for the reader to restore the original. Again, in the conflation of *u* and *v*, and *i* and *j*, our adoption of modern usage does not prevent a person so minded from easily restoring the original. As for punctuation, we have nothing to add to what was said of the Latin texts.

Words which are not likely to be easily understood by the modern equivalent of Chaucer's 'every maner man, That English understandė kan' are included in a glossary (in vol. iii).

EDITORIAL ABBREVIATIONS

Abbreviations are used to indicate three types of variant: addition, omission, and alteration. Thus

(i) '^{n} ACH +videlicet spera' is to be read 'manuscripts with sigla A, C, and H, add the words "videlicet spera"'.

(ii) '^{n} DF λ' means 'manuscripts D and F lack the word flagged'. In addition, the footnote '^{n} ACE 7λ' means 'the manuscripts A, C, and E omit seven consecutive words of the printed text, the first being the word flagged'.

(iii) '^{n} BL moventur' means simply that the flagged word is rendered 'moventur' in the manuscripts named; while, combining the symbols of (i) and (ii), we denote alterations of greater extent thus: '^{n} BCD 4λ+oculi', which means that four consecutive words, beginning with the word flagged, are rendered by the one word 'oculi'.

Semicolons separate footnotes applied to the same word. Where a vertical stroke is used in the text, it indicates a page change in the manuscript. It is of especial significance in *Horologium*, where actual folio references are easily found from a conspectus at p. 320, vol. ii.

Square brackets, [ ], enclose words for which there is no justification in any early manuscript. Such brackets are occasionally used to restore a passage which has been lost from a damaged manuscript, or to improve the sense of a rather literal translation, for example. Angle brackets are occasionally used elsewhere; for instance, to indicate conscious editorial additions which are thought to be required by the text without being present in the original. Here, if a footnote shows that all manuscripts collated differ from the printed text, then the reading is obviously the editor's.

Asterisked numerals (e.g. 135*) are those which are in error, but which are beyond doubt the work of the author; which is to say that they cannot be corrected, for internal reasons. On those occasions when the author is given the benefit of the doubt, the fact is amply recorded in the manner explained.

The sigla assigned to manuscripts apply only within a given section, and are listed at the beginning of each treatise. An index of manuscripts will be found in vol. iii.

A siglum with a prime (as H′) denotes not a new manuscript but the work of a glossator or one who has in some way altered the manuscript (in this example, H).

MATHEMATICAL AND ASTRONOMICAL NOTATIONS

Proofs

In the translation of *Quadripartitum* (beginning with I. 1) the justification of each stage in the demonstration is given by reference to earlier propositions, and earlier (numbered) steps in the proof.

Superscript signs

p	part, generally a sixtieth part of a standard radius (*R*).
°, ′, ″, etc.	degree, minute, second, etc. On the convention which makes a standard radius 150 parts, this is written 150^p, and then $60' = 1^p$, in the accepted manner.
k	The kardaga unit (15 degrees). See vol. ii, pp. 46–7.
*	'is known to me'. This and other uses of the asterisk are explained when the occasion arises.
‾ (macron)	is used with a term denoting a quantity (e.g. mean motus) which changes in time at a constant rate.

Suffixes

l *longior* } see Appendix 29.
p *propior* }
o mean }

Trigonometrical notation

▱	parallelogram (which will often be a rectangle).
Δ*ABC*/*DEF*	triangle *ABC* with transversal *DEF*, *D* being on *BC*, etc. See Appendix 20.
R	standard radius (regardless of the number of parts into which it is supposedly divided).
G	standard gnomon.

Astronomical symbols

The symbols for the signs of the zodiac are given in Appendix 29 (see vol. iii, p. 169).

The cardinal points of the division into the astrological houses are:

H ascendent, or horoscope; the point of the ecliptic which is rising over the horizon.

M *medium celum*, mid-heaven (elsewhere often denoted by *MC*); the ecliptic degree which is crossing the meridian.

D descendent; the degree of the ecliptic which is setting.

I *imum medium celum*, lower mid-heaven; the second point of the ecliptic crossing the meridian, diametrically opposite *M*. (*I* was often said to be on the 'midnight line'.) Elsewhere often denoted by *IMC*.

Symbols generally used in the theory of planetary longitude, and explained more fully in Appendix 29, are:

T Earth.

C centre of the deferent circle.

E equant point. (In the case of the Sun, *C* behaves as an equant, and in the case of the Moon, *T*.)

E′ pole of the mean apogee (Moon only).

A aux of Sun, Moon, or planet.

O centre of epicycle.

S Sun.

M Moon.

P planet.

V First Point of Aries, or vernal point.

r epicycle radius, here taken (as is e below) in ratio to the deferent radius.

e see Appendix 29. e ($= CT$) may be called the 'eccentricity' of the planet's deferent, but different writers use the term differently. With Mercury, e is equal to the closest approach of C to T.

λ true longitude, or true place of planet.

$\bar{\mu}$ mean motus, mean longitude of the centre of the epicycle (may be called 'mean argument' in the case of the Sun).

μ true motus, true longitude of centre of the epicycle.

$\bar{\alpha}$ mean argument, or anomaly.

α true argument, or anomaly.

ω longitude of aux (*aux propria*).

$\bar{\gamma}$ mean centre (i.e. the elongation of the centre of the epicycle from the line of aux). Corresponds to $2\bar{\eta}$ (q.v.), in the case of the Moon.

$\bar{\eta}$ elongation of mean Sun from mean Moon and line of aux ($2\bar{\eta}$, the elongation of the mean Moon from the line of aux, is known as the 'double elongation').

ξ equation of centre. The angle subtended at the centre of the epicycle by the line joining T to the equant or quasi-equant, i.e. by TC (Sun), $E'T$ (Moon), or ET (Mercury, Venus, Mars, Jupiter, or Saturn). For the Sun, ξ may be called simply 'equation', or 'equation of argument'.

ζ equation of the argument, prosthaphairesis for anomaly.

d distance of the centre of the epicycle (O) from the Earth (T).

θ angular diameter of the epicycle as seen from T.

Π proportional minutes.

Δ *diversitas diametri*. (Π and Δ may denote several analogous but different functions, according to context. See Appendix 29.)

δ proportional part (*pars proportionalis*; equal by definition to $\Delta . \Pi/60$).

t epoch.

Notations for trains of gears

Each simple train is represented by a product of ratios written thus:

$$\frac{(T_1)}{a} \cdot \frac{b}{c} \cdot \frac{d}{e} \cdots \frac{p}{q} \cdot \frac{r}{(T_2)},$$

which denotes a wheel of a teeth, and period of revolution T_1, engaging a wheel of b teeth, on the arbor of which is a wheel of c, engaging in turn a wheel of d,..., and so on, the last wheel of r having a period of T_2. We do not distinguish between a pair of co-axial wheels with equal numbers of teeth (n), and an idle wheel. Both may be expressed as a ratio n/n.

The advantage of the notation is that the product of the ratios must be unity, and that if we wish to find the period of some intermediate wheel, we may do so at once by an obvious procedure. Thus the period of the wheel of e is simply

$$\frac{(T_1)}{a} \cdot \frac{b}{c} \cdot d.$$

It will be necessary to extend the notation to cover the case of a transported train of gears.

It will be supposed that an arm rotating about a point A, and co-axial with a wheel of a teeth which is either fixed or moving with respect to some reference frame, carries a train of wheels which may be expressed:

$$\frac{\ }{a}\frac{b}{c} \cdot \frac{d}{e} \cdots \frac{p}{q}.$$

The last of the train on the arm, having q teeth, engages a wheel which is co-axial with the wheel of a, but not fixed to it. T_0 is the period of the arm with respect to the wheel of a. (In a relatively simple arrangement, the wheel of a will be fixed to the frame of the instrument, and then T_0 will be the period with respect to that frame.)

We shall now write:

with respect to the arm $$\left\{ \frac{(T_0)}{a} \cdot \frac{b}{c} \cdot \frac{d}{e} \cdots \frac{p}{q} \cdot \frac{r}{(T_1)} \;\Bigg\|\; \left(\frac{r}{\dfrac{1}{\dfrac{1}{T_0} \pm \dfrac{1}{T_1}}} \right) \right\}$$ with respect to the wheel of a.

The product of the left will be unity, as before. The choice of algebraic sign on the right is decided as the sign of $(-1)^n$, where n is the number of ratios on the left. The great advantage of this type of train, which Richard of Wallingford used for his astronomical clock, is that a good approximation (T_0) to some desired periodic time is usually a simple matter to obtain. The transported train may then be so chosen as to improve the approximation, in a completely separate calculation.

BIBLIOGRAPHICAL ABBREVIATIONS

We follow medieval practice in referring to Richard of Wallingford's works by their short titles, *Albion*, *Rectangulus*, and so on. There is no title in the manuscript for the treatise on the astronomical clock, which we call *Tractatus horologii astronomici*, by analogy, shortening this to *Horologium*.

References in the notes and commentaries are usually given in full. When the first citation cannot be found, the bibliography (in vol. iii) should serve to identify the work. There are these few titles which have been used sufficiently often to merit abbreviation:

Albategni	*Opus astronomicum* . . . (al-Battānī), ed. and tr. into Latin by C. A. Nallino, vols. i and ii (Pubblicazioni del Reale Osservatorio di Brera in Milano, no. 40), in three parts, Milan, 1899, 1903, 1907. When other editions are used, the fact is made clear.
Almagest	*Claudii Ptolemaei opera*, I: *Syntaxis mathematica*, ed. J. L. Heiberg, 2 vols., Leipzig, 1898, 1905. A German translation made from this is: *Claudius Ptolemäus: Handbuch der Astronomie*, tr. Karl Manitius, 2 vols., Leipzig, 1912, 1913 (repr. 1963).
Curtze	'Urkunden zur Geschichte der Trigonometrie im christlichen Mittelalter', *Bibliotheca Mathematica* (3), i (1900), pp. 321–416.
Emden	*A Biographical Register of the University of Oxford to* A.D. *1500*, 3 vols., Oxford, 1957, 1958, 1959 (paginated consecutively).
Euclid	*Elementa*, ed. J. L. Heiberg, with tr. into Latin, 4 vols., Leipzig, 1883–8. Translated into English from this edition, and with much additional material with reference to historically important editions, is *The Thirteen Books of Euclid's Elements*, tr. T. L. Heath, 3 vols., second ed., Cambridge, 1926.
Gebir	*De astronomia libri IX* . . . (issued with *Petri Apiani Instrumentum primi mobilis*), Joh. Petreius, Nuremberg, 1534.
Gesta	*Gesta abbatum monasterii Sancti Albani a Thoma Walsingham . . . compilata*, vol. i (A.D. 793–1290), 1867; vol. ii (1290–1349), 1867. Rolls Series, London.
Knowles	*The Religious Orders in England*, vols. i and ii, Cambridge, 1948, 1955.
MLWL, MLWL 2	*Medieval Latin Word-list from British and Irish Sources*, by J. H. Baxter and C. Johnson, London, 1934. *Revised Medieval Latin Word-list* etc., by R. E. Latham, London, 1965.
Pantin	W. A. Pantin, *Documents illustrating the Activities of the General and Provincial Chapters of the English Black Monks, 1215–1540*, 3 vols., Camden Third Series, vols. xlv, xlvii, liv, London, 1931, 1933, 1937.
T.K.	L. Thorndike and P. Kibre, *A Catalogue of Mediaeval Scientific Writings in Latin*, revised ed., Mediaeval Academy of America, London, England, 1963.

I

CANONS TO THE TABLES OF JOHN MAUDITH

CANONS TO THE TABLES OF JOHN MAUDITH

The evidence for Richard of Wallingford's authorship

IT will be suggested that Richard of Wallingford was the author of two different, but closely related treatises, both of which may be regarded as canons to the tables of John Maudith, an Oxford astronomer somewhat older than Richard. Since doubt must inevitably attach to this conjecture, and since the complexity of the textual problems they present is out of all proportion to their value, the treatises are not printed in full, but parallel synopses are offered in their place. For the sake of convenience, we may refer to these two works as *Tractatus de quatuor tabulis* and *Canones super quatuor tabulis*, or, briefly, *Tractatus* and *Canones*. (Neither title is used in the manuscripts of the works in question.) The *Canones* are found in reasonably similar form in two different manuscripts:

(i) Cambridge University, MS. 1572 (= Gg. VI. 3), ff. 45^r–48^r;

(ii) Melk, Stiftsbibl. MS. 51, ff. 193^r–194^r.

In the first manuscript, but not in the second, they are associated with the tables ('super tabulis mirabiliter inventis') of John Maudith,[1] tables which are to be found without canons in

(iii) Bodleian Library, MS. Laud misc. 674, ff. 67^r–72^v.[2]

The *Tractatus* apparently survives in only one manuscript, of the latter part of the fourteenth century:

(iv) British Museum, Royal MS. 12.C.18, ff. 22^r–25^v.

The anonymous treatise is incomplete, owing to the loss of perhaps two folios at the end, and no tables are now to be found in association with it.

Before considering the contents of the two works, we should turn to the second chapter of *Exafrenon*. There we find three allusions to Maudith's tables, two of them indicating that he drew up tables of ascensions for Oxford, and one containing this very explicit statement: 'You will find this canon, explaining how to find the equinoctial ascension from the rising or setting of the Sun, among the canons which I wrote for you over Maudith's tables . . .'[3] The phrase 'canones quas tibi

[1] The tables are at ff. 37^r–40^r in (i).

[2] M. R. James in *Fasciculus Jo. Willis Clark dictatus*, p. 56, notes that a 'Tabula Jo. Maudith' was once in the library of the Austin Friars, York.

[3] See p. 192, below.

scripsi super tabulas Maudith' is unambiguous. It is worth comparing with the rubric to the Cambridge version of the *Canones*: 'Hic incipit quidam parvus tractatus editus a magistro Johanne Mauduth super quatuor tabulis mirabiliter inventis.'[1] The latter apparently provides the main reason for supposing that Maudith rather than Richard wrote what I have called the *Canones* to Maudith's tables. Certainly every writer to have listed the work has followed the lead of this title.[2] Clearly it is not enough to know that Richard (for we here suppose it established that he wrote *Exafrenon*) wrote canons to the tables, since there might have been two sets. Scarcely has the Cambridge version begun, however, when we find this sentence, referring to the supposed author in the third person: 'Et hoc fuit studium Mag. J. Mauduth, qui ad instanciam magistrorum quatuor breves tabulas adinvenit in civitate Oxoniensi, Anno Domini 1310, quibus faciliter possit hoc in omni regione operari.' The Melk version is more embroidered: John Maudith becomes the 'best man in the science of astronomy', but the year becomes 1410! The version is exceedingly poor in many other respects, and may generally be discounted.

One can scarcely prove that Maudith did not supply canons for his tables of 1310,[3] but we suspect that the tables were often to be found without them. The tables of MS. Laud misc. 674 are the only plain instance, but in Bodleian MS. Digby 168, at f. 76^{r}, there is a footnote to Maudith's table of ascensions for Oxford, a footnote which was subsequently incorporated into a copy of the text to which it was added, namely a *theorica planetarum* ascribed to Roger of Hereford. The copy, of the fifteenth century, is to be found in MS. Bodley 300, ff. 1^{r}–19^{v}; see f. 11^{v}. The common passage includes the words 'tabula (Magistri) Johannis Maudith facta Oxon. de ascensionibus signorum, per illam operaberis sicut dicam tibi . . .'. If Maudith left a canon for this table,[4] one writer at least did not intend to follow it. Then again, when Simon Bredon bequeathed a set of Maudith's tables to Richard Camsale in his will of 1368,[5] there was no mention of any associated work by Maudith. In British Museum, MS. Egerton 889, however, which is a rather disjointed manuscript, Maudith's third, first, fourth, and second tables are to be found (in that order, and separated) with fragmentary, but in part

[1] I am most grateful to Dr. R. P. Lorch for sending me his transcripts of this text, for other information relating to Maudith, and for the reference to the Dublin manuscript cited below.

[2] As, for example, John Leland, *Comm. de Script. Brit.*, p. 329, cap. 337; *Collectanea*, iv. 20; Pits, *Rel. Hist. de Rebus Angl.* i. 455; and so on, through Twyne, Tanner, Brodrick, and Thorndike, to name a few. Bale (*Index Brit. Script.*, pp. 231–2) follows Leland in crediting Maudith with a tractate *De corda recta et umbra*, formerly in the library of Clare College, Cambridge. This will be referred to again in due course.

[3] On this date see Appendix 25.

[4] It is discussed below at pp. 13–15.

[5] See F. M. Powicke, *The Medieval Books of Merton College*, 1931, p. 84.

at least derivative, canons.[1] Similar canons are to be found beneath and to the side of Maudith's tables in Trinity College, Dublin, MS. D. 2. 11 (392), ff. 23^v–28^v.[2] Arzachel and Ptolemy are there mentioned.

There is room for doubt, therefore, as to Maudith's authorship of the *Canones*, and no positive reason for thinking that he wrote the *Tractatus*. Not only did Richard say that he wrote a work fitting the description of both, but the contents of both have a marked similarity to sundry passages from his attested writings, notably to the opening paragraph of *Albion*, where arithmetical and geometrical instruments are contrasted. The corresponding passages in the *Canones* and *Tractatus* are of great interest, and will be considered subsequently. The rules now given for determining certain astronomical quantities are—as in *Exafrenon*—provided with no theoretical justification—but this is to be found in *Quadripartitum*, and would in any case be inappropriate in brief canons to tables. And here we may anticipate the opening sentence of *Quadripartitum*: 'Because the *Canones* [we take this to be a specific reference rather than one to canons in general] do not perfectly explain the evaluation of the sine and its use, I purpose to do so in this four-part work.'[3] This reference, which might equally be to what we are calling *Tractatus*, surely echoes (or is echoed by) that of *Exafrenon*.

We note that the opening words of the *Canones* ('Quia sciencia astronomie sine debitis instrumentis non completur . . .') bear a marked resemblance to those of a number of treatises of the time ('Cum sciencia . . .', 'Quoniam sciencia . . .'), ascribed to Profatius, John of Linières, and others, and there is little hope of deriving a probable date of composition from this consideration. We know that Profatius was writing at least a decade earlier, and that his work was known to Richard of Wallingford. There is no evidence that Richard knew the work of John of Linières, whom he never cited. On the date of composition of the *Canones*, and of the *Tractatus* if it be assumed to relate to the same tables, we observe that the table of ascension of signs is dated 1316, like the table of stars, and six years after a date often linked with the other tables. Since *Albion* seems to draw on the two texts rather than conversely, whereas neither of them mentions the albion as an example of a geometrical instrument (as does another derivative passage quoted elsewhere[4] from Bodleian MS. Digby 57), we may probably take the year of composition to lie within the ten years following 1316. Assuming

[1] See ff. 19^v, 23^r–24^v, 24^v–25^r, and 30^{r-v} respectively. At f. 19^r the collection of tables, which includes new material, begins: 'Canones super tabulam sequentem stellarum', and follows with an explanation closely resembling that of our *Canones* as regards the adjacent star table.

[2] This section of the codex appears to date from the early years of the sixteenth century.

[3] See p. 25 below.

[4] See vol. ii, p. 272.

Richard to have been the author, textual dependences suggest the following order of composition:

(i) *Canones* and/or *Tractatus*;
(ii) *Exafrenon* and *Quadripartitum*;
(iii) *Albion* and *Rectangulus*.

That both *Canones* and *Tractatus* are by the same author is suggested not only by their closely related contents, but by the fact that the former is—very appropriately on this hypothesis—referred to as 'parvus tractatus' in the rubric to the Cambridge version. Finally, we are reminded once more that the only copy of the *Tractatus* follows immediately a copy of *Exafrenon*, while the two versions of the *Canones* are in manuscripts containing a copy of *Albion*. The Melk manuscript of *Albion* also contains the tables excerpted from *Exafrenon* (at ff. 190–1), with a short canon ('Si vis investigare horam introitus Solis . . .') taken from towards the end of *Exafrenon*, cap. 1.[1]

John Maudith as a writer on trigonometry

Several writers have repeated a story to the effect that John Maudith was the 'first European writer on trigonometry', and that his writings were shortly followed and greatly improved upon by Richard of Wallingford, also as often as not said to be of the same Oxford college—or, worse still, of the 'Merton School'. In a reasonably typical quotation, we find George Sarton writing that 'Mauduith was apparently the real initiator of Western trigonometry preceding both Levi ben Gerson and Wallingford'. The 'Parvus tractatus',[2] he maintains, 'is especially important, as it was the source of Wallingford's Quadripartitum.'[3] From the point of view of trigonometry, however, the tables are alone of any real importance, the canons being mathematically trivial and containing, as pointed out above, nothing more than procedural instructions, devoid of explanations. There is nothing in these rules which an intelligent astronomer of the time could not have found reasonably easily, whether in Ptolemy or the canons of Arzachel. Even the tables are hardly sufficient to give Maudith a very secure place in the history of the subject, but this remains to be seen; and of course no significant treatise on the subject of trigonometry can be ascribed to him.

Part of the Maudith myth undoubtedly originates with Leland and Bale, who ascribed to him a treatise (formerly in Clare College library, but now lost) entitled *De corda recta et umbra*.[4] Moritz Cantor repeated

[1] See p. 188 below, where the text reads *invenire* for *investigare*.
[2] Called here *Canones*.
[3] G. Sarton, *Introduction to the History of Science*, vol. iii, 1947, pp. 660–1.
[4] See p. 4 above.

the reference,[1] which was subsequently reiterated by Duhem and others. Whether or not the Clare manuscript is irretrievably lost, there can be little doubt that Leland was looking either at the *Canones* or even at a rubric to the first table ('De corda et arcu recto et verso et umbris'). There is no reason for thinking Leland to have had significantly fuller information than is available today, and he would have been misled by the title to the *Canones* as all subsequent readers have been misled by it.

By the standards of his time, John Maudith was probably a good mathematical astronomer, and one perhaps not over-anxious to reveal his sources. As the *Catalogus vetus* of Merton College describes him, he was 'a good astronomer who composed astronomical tables, and a physician'.[2] Leland (d. 1552) has it that Maudith's tables were still in use in his own time, while, long before, John Walter (a fellow of New College from 1386 to 1393) had referred to them as subtle and useful.[3] A fellow of his college from 1309, Maudith vacated the fellowship to become Rector of Aston Cantlow in 1319. His only original treatise of any size to have survived is a *Tractatus de Doctrina theologica*, completed in 1342 whilst he was Rector of Hatfield in Yorkshire.[4] I have given reasons elsewhere, however, for thinking that John Maudith edited a version of Profatius on the new quadrant,[5] and this, together with his star tables, is some slight evidence for his astronomical rather than purely mathematical competence. However, whether or not he wrote the canons to his tables, they contain nothing essential which he could not have written, and in one respect at least—namely in the use of the tangent function after the manner of the Toledan tables, or those of al-Khwārizmī—the canons mark an interesting stage in the development of trigonometry in northern Europe. In addition, we know from his several references to Maudith that Richard was appreciably influenced by him, although the details of their relationship are pitifully few.

A comparison of the contents of the Canones *and the* Tractatus

The average length of the two versions of the *Canones* is approximately 3,300 words, while the surviving fragments of the *Tractatus*, accounting for four out of its six planned chapters, contain about 6,400 words. The over-all plan of the two is the same, and is adequately summarized in the table of contents prefixed to the *Tractatus*:

ISTE TRACTATUS in sex capitula dividitur. *Primum capitulum* est de diversis

[1] *Vorlesungen über die Geschichte der Mathematik*, 2nd ed., vol. ii, 1900, p. 111.

[2] See Emden, pp. 1243–4, for this and the following information.

[3] There are several copies of John Walter's tables, with the canons from which this quotation comes. See, for instance, Bodleian MS. Digby 97, f. 50^{v}.

[4] Sarum Cathedral MS. 167, art. 3.

[5] See p. 12 n. 2 below.

motibus in celo et de diversis instrumentis ad eos ordinatis, et de excellencia harum tabularum respectu omnium aliorum instrumentorum. *Capitulum secundum* de quibusdam suppositis ad cognicionem eorum que per has tabulas extrahuntur ut de vero gradu Solis inveniendo, et de altitudinibus sumendis, et de composicione cuiusdam instrumenti ad hec ordinati. *Capitulum tercium* de officio prime tabule, scilicet de invencione corde recte et verse ex arcubus suis, et econtra, et de invencione umbre ex altitudine, et econtra, et de mensuracione rerum ex suis umbris et ubi et quando potest umbra rei erecte a Sole et ubi et quando non, et de altitudine meridiana. *Capitulum quartum* de officio secunde tabule, scilicet de generali modo intrandi hanc secundam tabulam et ad extrahendum arcum equinoxialem elevatum, et de horis equalibus et inequalibus extrahendis, et de conversione horarum equalium in horas inequales, et econverso, et ad habendum arcum diurnum, et quot hore sunt in die longissima, et quot in die brevissima, et de altitudine meridiana et latitudine regionis et gradu Solis ignoto. *Capitulum quintum* de officio tercie tabule, scilicet de modo generali intrandi illam tabulam, et de altitudine stellarum fixarum que ponuntur in hac tabula cum fuerit maxime altitudinis meridiane, et quomodo latitudo regionis et altitudo equinoctialis in illo orizonte poterint reperiri, et etiam qualiter invenio arcum diurnum stelle et quantum de equinoctiali est elevatum ab ortu stelle usque ad tempus date altitudinis et arcum equinoctialem elevatum a principio noctis, et ascendens et horas equales et inequales. *Capitulum sextum* de officio quarte tabule, scilicet de modo intrandi hanc tabulam, et de modo inveniendi ascendens de die et similiter de nocte, et de equacione 12 domorum, et de verificacione stellarum fixarum et planetarum, et quomodo nomina stellarum ignota cognosci possunt.

The treatise proper opens with the words: 'In speris celestibus esse duos modos motuum a philosophis non solum naturalibus sed etiam astrologis est concessum . . .' It is explained in a very commonplace way that not only natural philosophers but also astrologers allow two motions to the celestial spheres, the first that of the first sphere (*prima spera*), from east to west, this sphere moving the lower spheres simultaneously by its (diurnal) motion. The second (planetary) motion, belonging to the other spheres, is on the contrary from west to east. The two are taken separately by the astronomer (*astrologus*), one motion relating to problems of the hours of day and night, the distribution of the houses, and so on, the other motion relating to mean and true places of the planets, their latitudes, and conjunctions, eclipses, and the like. It is said that Ptolemy gave over his first two books of *Almagest* to the first motion, and the rest to the second sort. The text then runs as follows: 'Sed quia hec sciencia non completur sine debitis instrumentis, necesse fuit astrologo habere quedam instrumenta per que ad ista poterit pertingere. Instrumenta autem eius duplicia sunt, geometrica scilicet et arithmetica . . .' Here at last we find the words resembling those with which the shorter treatise opens, namely:[1] 'Quia sciencia astronomie sine debitis instrumentis non completur, plures philosophi instrumenta ad hanc ordinaverunt; quedam geometrica . . . et quedam alia instrumenta arsmetrica . . .' Both treatises are alike in naming as geometrical instruments the astrolabe ('of Ptolemy', add the *Canones*), the saphea, the quadrant, new and old, the armillae, and the torquetum.[2] Both later add the chilindrum to their lists. The *Canones* mention as arithmetical instruments the tables of Arzachel, Profatius ('such as the almanac'), and others. For the moment, the *Tractatus* simply explains the character of an arithmetical table, and mentions Boethius as providing the means of calculating it.

The *Canones* pass at once to a new point, reached at length in the *Tractatus* after a suitable number of rather trite flourishes: the number of minutes of arc in a quadrant being 5,400, it is not possible to divide the limb absolutely uniformly and well. In addition, there is the problem of interpolation when the crucial point of intersection (*nodulus*)[3] falls between two hour lines or almucantars on the quadrant, astrolabe, or chilindrum. The resulting difficulties make it desirable to have tables —hence John Maudith's four tables of 1310, say the *Canones*, whilst the *Tractatus* mentions instead that on the authority of the Philosopher, in the first book of his *Mechanica* (an early pseudo-Aristotelian work, in fact), such arithmetical instruments are more certain than geometrical ones. The *Tractatus* goes on to say that the arithmetical tables, used with the help of the following canons, are straightforward, while the geometrical instruments are exceedingly costly if made of copper: 'You can do the same thing with four sheets of parchment as with a quadrant, and do it more accurately.'

In the later parts of the treatises, the altitude of the Sun will be needed, and its position in the zodiac. The latter will be found from an almanac, 'together with the motion of the eighth sphere' (*Tractatus*, cap. 2). The *Canones* hint at two methods for altitude: the use of an astrolabe, and a shadow method. The *Tractatus* is more precise: a pair of vanes (pinnules) with sighting holes and plumb-line, or alternatively a gnomon (divided into twelfths), or even a third method may be used, involving hinged rules. The last method is explained at great length, with letters as though corresponding to a diagram, although there is no such diagram in the one known manuscript. The instrument is, in fact, that often

[1] Neither copy is very good, but since the two differ in the precise form of a large number of sentences, I have followed the earlier and better (Cambridge) MS. throughout, except where indicated.

[2] *Turketum, torketum*. See our commentary on *Rectangulus*. The saphea will be mentioned frequently in connection with *Albion*.

[3] This word, unrecorded in the present sense, cannot mean 'little knot', much less 'jackdaw'. It is later found in places where it is distinguished from 'almuri', and thus cannot refer to a sliding bead.

called the 'parallactic instrument', 'Ptolemy's rules', or 'triquetrum' (but is not to be confused with the torquetum). It was to be made of copper, and the evidence suggests an instrument less than eighteen inches high.[1] The graduation may, it was pointed out, be effected more accurately on the straight rule (the third and lowest) than on an arc.[2] (Copernicus used a large triquetrum for the same reason, but Tycho, who made two of even larger size, favoured it less than his much finer instruments with graduated arcs.)[3] There are obvious references to Euclid I, 4, 5, 12, and later to Dulcarnon (I. 47), and then again to a statement 'in primo libro Gebri in commento suo super *Almagesti*', to *Almagest* I. 6, to Alfraganus, cap. 4, and finally to the commentator[4] on *De celo*, towards the end of Book II. We are referred to this last for further details: here 'enough has been said'—in fact far more than was said of the other instruments for measuring altitude. None of this appears in the *Canones*. The references are not to Ptolemy's description of the instrument, but are all apparently to passages in which is emphasized the smallness of the Earth, 'which is as a point to the sphere of the fixed stars'. As Ptolemy said, sundials and the centre of an armillary sphere may be regarded as the Earth's centre for sighting purposes, and similar remarks apply to gnomons, whose shadows are used for calculating altitudes.

With chapter 3 of the *Tractatus*, and the corresponding section of the *Canones*, we come to the use of the tables. At the outset, instructions are given for interpolation, where a chord (right or versed) of some argument lying between the integral arguments tabulated is required. Simple linear interpolation is used, of course, and the same applies when an arc is being found, corresponding to a given chord, right or versed. Needless to say, the explanation of this procedure is long and tedious, but one must remember that the tedium is usually avoided in elementary books today by our leaving such explanations to the spoken word. The first table is described as 'de sciencia cordandi arcum et arcuandi cordam et de umbris' (*Canones*), and we are reminded that the verbs *cordare* and *arcuare* are those used in the early chapters of *Albion*. *Procul dubio*, to use one of Richard's favourite phrases which occurs in the *Canones*, there is

[1] Ptolemy, who described his ὄργανον παραλλακτικόν in *Almagest* V. 12, suggested rods not less than four cubits in length, that is to say, somewhat greater than the height of most men. Ptolemy's name for the instrument is occasionally badly translated as 'instrument for determining parallaxes'.

[2] The first arm of the triquetrum was vertical, and could be turned about an axis in its line (i.e. through the zenith). The second (of length L) was affixed to the top by a hinge. The third was hinged in the same plane at distance L below the second. The line of sight was along the second, and its lower end was held against the graduated edge of the third rule in accordance with the zenith distance of the object; that is, the angle between the first and second rules is found from the known base of the isosceles triangle, using a table of chords.

[3] See J. L. E. Dreyer, *Tycho Brahe*, 1890, p. 322.

[4] Most likely a reference to Averröes; or possibly to the translation by William of Moerbeke (1271) of the commentary by Simplicius.

a marked stylistic similarity to his work in many more respects, but it is difficult to know whether or not this is due to the highly stylized conventions of Oxford mathematical writing at the time.[1]

To the first four rules of interpolation a fifth and sixth are added for determining *umbra recta* for a given angle,[2] and conversely. Shadows were to be measured in feet, palms, or digits, 'or other suitable measure', not exactly suggestive of accuracy. If the altitude of the Sun be independently known, shadow length may be used as a measure of the height 'of any thing' (*Canones*), and 'of the tower of Babel' (*Tractatus*). Over the problem of accuracy, the writer is insistent that more precise altitudes may be found with an astrolabe or quadrant, but ends a lengthy explanatory chapter of the *Tractatus* with the words 'when you do not have the use of these instruments you will be able to do as I have told you'. This, surely, is the answer to the general medieval dislike of the tangent function, which is not once mentioned, for example, in *Quadripartitum*. Conceived in real terms, it was rarely appreciated in its role as a convenient expression for the quotient of a sine and a cosine. As for the reasons why *umbrae* first found an important role in Hindu astronomy, this seems to have something to do with the Hindu preference for working with plane triangles *within* the celestial sphere, rather than with arcs on the sphere and chords of arcs. Expertise in Ptolemaic methods seems thus to have inhibited the development of trigonometrical concepts.

We are given due warning at this point in our two treatises that the table of *umbrae* will later prove invaluable for compiling tables of ascensional differences.[3]

An indication of the contents of John Maudith's first table will be given subsequently. The last surviving chapter of the manuscript with the *Tractatus*, chapter 4, concerns the use of the second table, which is for the evaluation of the *corda recta* of the meridian altitude of the Sun, the *corda versa* of the semidiurnal arc, and the arc itself. The theoretical justification for the procedures recommended will be found in *Quadripartitum* IV. 11, whilst in IV. 12 will be found a justification for the remainder of the chapter, which gives a rule to find the equinoctial arc from sunrise to the moment of observation of the Sun's altitude, and also rules for deducing the hour, whether equal or canonical. Some fairly obvious corollaries are to be found here in the *Tractatus*, such as a rule for meridian altitude, and another for the latitude of the region, in each case using a knowledge of the complementary information.

[1] It would be a rash linguist indeed who tried to determine authorship of a modern mathematical text on the strength of style, word frequency, or the like.

[2] In the *Canones* there is an authorial slip here, where the fraction of the interpolation interval is given wrongly inverted.

[3] See p. 13 below.

There are also some obvious and simple interpolation instructions 'where you have fractions of a degree in the Sun's position [in the zodiac], on the assumption that you wish to work precisely'. Interpolation instructions in the *Canones* are on yet another occasion slightly misleading as a result of over-abbreviation, but this is a trifling error, since the sense is clear enough.

The surviving part of the *Tractatus* now ends,[1] with chapter 4 not quite completed. We may be reasonably certain of the character of the remainder, since the table of contents corresponds precisely with the remainder of the *Canones*, which continue, one must presume, in the form of an abbreviated version of the longer work. Chapter 4 ended by indicating the possibility of finding the degree of the Sun in the zodiac from a comparison of meridian altitude, the meridian altitude of the vernal point, and the geographical latitude.

Chapter 5 may be summarized very briefly by saying that it rephrases all the rules of chapter 4, so that they can be used with a select number of stars rather than with the Sun. Maudith's third table was now to be used, and in general this appears to have contained eleven or twelve stars, although there is a longer star table in that Cambridge manuscript which contains the *Canones*.[2] It should go without saying that the problem is essentially simpler with stars (trepidation excluded), each of which has a constant semidiurnal arc, meridian altitude, rising-point, mediation, and latitude. These are the quantities tabulated.

Chapter 6, the last chapter, is related to the fourth table, and permitted the determination of the ascendent, given the Sun's position in the zodiac and its altitude; it allowed the equation of the twelve astrological houses; and finally the determination of stellar and planetary coordinates, and hence the name of any unrecognized star which happens to be in the catalogue, and whose altitude is known simultaneously with that of a known star. This last group of rules is not to be found in either version of the *Canones*, but it was clearly related to the problems treated in *Quadripartitum* IV. 9 and 10. The earlier parts of what is equivalent to chapter 6, however, are perhaps the most interesting section of the whole work, for they introduce distinctly non-Ptolemaic calculational techniques, as will be explained.

Ascensional differences by the use of the tangent

By reference to *Quadripartitum* IV. 8, to our comments on the same,

[1] Since the missing part might well survive, note that the catchwords are recorded at the end of f. 25v of B.M., Royal MS. 12.C.18: *modus sic intra* (?).

[2] MS. 1572 (Gg. VI. 3). I have discussed this in my 'Medieval star catalogues', *Archives internationales d'histoire des sciences*, 78–9, Janvier–Juin, 1967, 71–83, where I have also given reasons for thinking Maudith the editor of a version of Profatius on the new quadrant (inc.: 'Accipe ergo tabulam . . .'). The essential points of the argument will be found in Appendix 25 below, where some crucial printing errors in my original review are corrected.

and to Appendix 22 on the terminology of ascensional differences $(\alpha - \alpha_0)$, we see that Richard chose what was essentially Ptolemy's method of *Almagest* II. 7. It is easily seen, however, that the formula for $\mathrm{Sin}(\alpha - \alpha_0)$ may be immediately rewritten in terms of tangents:

$$\mathrm{Sin}(\alpha - \alpha_0) = R \tan \delta . \tan \phi,$$

where ϕ is the geographical latitude and δ is the declination of the ascendent point of the zodiac, of longitude λ, right ascension α, and oblique ascension α_0. Now ϕ is the complement of the meridian altitude of the Sun when this is at the equinoxes,[1] and the shadow (*umbra*) of a gnomon of conventional length G (usually taken to be 12^p) is then $G \tan \phi$. John Maudith's table IV (i) tabulated *umbrae* (with $G = 12^p$); taking meridian altitude as argument, that is, it gave the cotangent of the altitude. His table IV (ii) gave $(R/G)\tan \delta$, as a function of λ, this time with $G = 12^p$ and $R = 150^p$, as in Arzachel's writings, rather than the 60^p of *Almagest*. The product of the special value (constant for a given latitude) taken from the table IV (i) with the appropriate figure from the other table therefore gives immediately $\mathrm{Sin}(\alpha - \alpha_0)$, as the *Canones* explain.[2] Again we find Maudith referred to in the third person: 'Per tabulam differencie ascensionum universe Terre, que ponitur inter tabulas Mauduth, poteris cognoscere ascensiones cuiuslibet signi et cuiuslibet gradus eius in qualibet regione universe Terre, sic operando . . .' An example is given for the town of Oxford, suggesting that the author was at least an Oxford man.

There is a curious reference to the town of Cremona which is ostensibly easy to explain, since the latitude of Cremona, taken as 45°, has a tangent of 12^p, and with a very simple adjustment of the values in the table of ascensional differences we have a table of ascensions for Cremona. The text reads, however, 'Hec tabula facta est ad regionem cuius latitudo est 45 gra., cuius umbra est 12 puncta, cuius est Cremonia [*sic*].' This might be thought to hint at the origin of the table. We know that there were Cremona tables, added to the Toledan tables (and that Richard, for example, probably owned a set,[3] although this has no direct bearing on Maudith's possible use of them). Turning to one particularly fine set of such tables, those of Bodleian MS. Laud misc. 644, we find the very tables we are seeking, together with a canon for their use which is shorter than that of the *Canones*, but very similar to it. More specifically, at f. 45^r is *tabula umbre*, identical with Maudith's table IV (i); on

[1] The *Canones* refer to the 'first degree of Aries'.

[2] Actually the Cambridge MS. has what is apparently (judging by the Melk MS.) a small lacuna of no importance, but the Melk version ends in the middle of the explanation with the words: 'Et nota quod hec non est alia operacio a predicta. Et sic est finis.' For reference, the Cambridge ending is: 'Et non oportuit dividere quia in composicione tabule facta fuit divisio illa.'

[3] See Appendix 13.

f. 45^v is *tabula differencie ascensionum universe Terre*, identical with Maudith's table IV (ii); while at ff. 21^v–25^r is *tabula sinus declinacionis*, giving the declination of each degree of the zodiac, and the sine of the declination. Each of these three, moreover, seems to have been drawn or modified from the tables of al-Khwārizmī, although the second and most interesting is missing from the edition by H. Suter, A. Bjørnbo, and R. Besthorn.[1] It would be foolish to speculate on the transmission of these tables, for the subject has as yet been insufficiently studied. We observe, however, that Arzachel's canons refer to a standard radius of 150^p, that such a radius was accepted in the calculation of the Toledan table of universal 'ascensional differences', but that the derived value for the obliquity does not accord with Arzachel's chosen value of 23° 33′ 30″.[2] This might only mean that Arzachel adopted an earlier table or modified it by introducing a constant multiplier (5/2). Elsewhere, in an almanac ascribed to Arzachel on some occasions and to one 'Humenuz' on others, there is a table of universal ascensional differences of exactly the form considered here, but calculated with $\epsilon = 23° 33'$ and $R = 60^p$.[3] We further note that the Toledan table of sines (at least in the manuscript considered earlier) uses a reading of 60^p, whereas Maudith follows the supposed Arzachel convention of 150^p, perhaps himself having introduced a multiplier into what otherwise agrees with the Toledan table and (Arzachel's?) table. Maudith's table of *umbrae* (cotangents) is again to be found both in the Toledan tables[4] and in those of al-Khwārizmī,[5] if we overlook a considerable number of differences plainly due to miscopying. Needless to say, the appropriate canons (how to find solar altitudes from shadow lengths, and conversely) are in both earlier sources, although much shorter than in our *Tractatus*.

The medieval habit of compilation without regard to real internal consistency is pathetically obvious throughout all this, and Maudith's reputation does not suffer alone. Before leaving the subject of his tables and the associated canons, it is worth remarking on his Parisian near-contemporary, John of Linières, who had tables closely resembling his, but only because both clearly used the Toledan tables as a source.

[1] *Die astronomischen Tafeln des Muḥammed ibn Mūsā al-Khwārizmī*, Copenhagen, 1914. On the need for such a table see chapter 26^a, pp. 21 and 76.

[2] The last entry, for $\lambda = 90°$, is (for what we know to be equal to $(R/G)\tan\epsilon$, where ϵ is the obliquity) 5^p 31′ 34″. Since G is 12^p, and ϵ is approximately 23° 30′, R must be 150^p. Then $\tan\epsilon$ is precisely found as 0·44208. There is no reason to suppose a value of ϵ to have been accepted to seconds of arc, but the value corresponding to this tangent is 23° 50′ 58″·9. Astronomers who accepted a figure of 23° 51′ include Theon on occasion, and al-Khwārizmī.

[3] J. M. Millás Vallicrosa, *Estudios sobre Azarquiel*, Madrid, 1943–50, p. 225. $(R/G)\tan\epsilon = 2^p$ 10′ 46″. A strict evaluation of ϵ from this value leads to 23° 33′ 06″·5, but we know from the source that 23° 33′ 0″ was the figure accepted.

[4] For example, at f. 45^r of MS. Laud misc. 644 (late 13th cent.).

[5] Suter *et al.*, Tab. 60, p. 174.

Indeed, John of Linières's borrowings were so extensive, in his *Canones tabularum primi mobilis* (to take Maximilian Curtze's title),[1] and in part at least have been manifest for so long,[2] that the removal of extravagant epithets from his name in mathematical histories is long overdue. John of Linières's compilation of 1322 is much longer than our *Tractatus*, but resembles it in several respects, as in the attention paid to different instruments for taking altitudes. It is uncertain whether either work influenced the other directly, but it is odd that if this did not happen, two writers should have opened similar works with identical words.

Finally, John of Linières seems to have been one principal source for the writer of a short text, composed for the year 1428, and reprinted with a translation by Lynn Thorndike from Bodleian MS. Ashmole 191.[3] This drew the attention of Otto Neugebauer and Olaf Schmidt, who pointed out several traces of Hindu astronomy which it contains (such as the sine function, the notion of ascensional difference, the conventional division of the radius into 150 parts, and several chronological points).[4] They were especially at pains to draw attention to the *tabula differencie ascensionum universe Terre* in this short text, since this was known to be missing from al-Khwārizmī. They could have found it in at least one version of the Toledan tables, as we have explained. They printed and commented on the table, however, which Thorndike had failed to do, and also a *tabula sinus declinacionis*.[5] Both of the former they thereby made easily accessible for the first time.[6] They were not aware of Maudith's tables, nor of an apparent connection, whether immediate or not, between the Newminster tables and his. The reasons for claiming such a connection are, first, that the three tables are exactly as described in the canons to Maudith's tables, and second, that there is Maudith material earlier in MS. Ashmole 191, and this Thorndike overlooked. Admittedly the two or three preceding texts are capable of standing alone,[7] but they have the appearance of work from the same hand; and what is unmistakably a star catalogue deriving from Maudith's, incomplete in an absolutely characteristic manner, is to be found at f. 136^{r}.

[1] 'Urkunden zur Geschichte der Trigonometrie im christlichen Mittelalter', *Bibliotheca Mathematica* (3rd series), i (1900), 391–404.

[2] Namely in Curtze's partial publication of the work, op. cit., which may be compared with pp. 338 sqq. of the same article (containing parts of Arzachel's canons).

[3] 'Astronomical and chronological calculations at Newminster in 1428', *Annals of Science*, vii (1951), 275–83. Newminster was a house of Cistercian monks, 15 miles north of Newcastle upon Tyne.

[4] 'Hindu astronomy at Newminster in 1428', *Annals of Science*, viii (1952), 221–8.

[5] Both are in op. cit., p. 226.

[6] For the *tabula altitudinis et umbre* they referred the reader to al-Khwārizmī and John of Linières. The three tables are at ff. 137^{v}–138^{r} of MS. Ashmole 191. The text follows at ff. 138^{v}–140^{r}.

[7] At f. 123^{r} comes 'Scias quod ascendens est significator . . .', and at f. 128^{v}, 'Cognita res est quod opus . . .'

(More or less the same catalogue is to be found in MS. Egerton 889, at f. 19^{v}, and it will be recalled that there we find in addition other Maudith tables and parts of the associated *Canones*.) Here is yet another stage in a long process of transmission covering half the known world.

A brief account of the tables known as John Maudith's

The tables are to be found in similar form in the MSS. numbered (i) and (iii) at the outset of this chapter. The *Canones* are sufficient warrant for their completeness. It will here be convenient to distinguish between the Oxford and Cambridge manuscripts by the use of the letters 'O' (= (iii)) and 'C' (= (i)). C is the earlier copy, apparently dating from the latter half of the fourteenth century, while the relevant section of O is of the following century (perhaps a little before 1438, a date mentioned in glosses).

Both O (f. 72$^{r, v}$) and C (ff. 37^{r}–38^{r}) contain a *tabula ascensionis signorum in circulo obliquo Oxonii cuius latitudo est 51 gra. et 50 min., verificata Oxon. Anno Domini* 1310 *proximo*. (The title is from O, that in C being virtually illegible. Added to the information on latitude we find 'cuius hore sunt 16, 29 min. 36 sec.', referring to the length of the longest day, the maximum diurnal arc also being quoted as 247° 24'.)[1] This table gives the oblique ascension of each degree of the zodiac, together with the corresponding size of the astrological house. It is apparently calculated with the help of the last section of the *Canones* and tables IV below. This first table of Oxford ascensions[2] does not, it seems, fall within the terms of reference of the *Canones* and *Tractatus* proper, however, and those tables which do will be assigned suitable roman numerals (I to IV), with subdivision where necessary:

I. The first table to be mentioned in the *Canones* is *Tabula sinus et arcus* (a longer illegible title is found in C), this being found at ff. 68^{v}–70^{v} in O, but not included in C. The argument runs from 0° 30' (or 179° 30') at 30' intervals to 180° 0' (or 0° 0'). The table gives (i) half-chords, that is to say, sinus (or sinus versus when taking the complementary angle) to three sexagesimal places,[3] with a standard radius of 60^{p}. Interpolation of the half-degree values has been naïvely done by simply averaging adjacent values for integral degrees. This table also gives (ii) *umbrae* in points and minutes, and this part of the table may also have been

[1] In C, the *explicit* to the table adds to the date the information that the motus of the eighth sphere was 9° 31' 42" according to the almanac of Profatius. See Appendix 25.

[2] It is this which is mentioned as an example of a table drawn up for a special latitude in a *Theorica planetarum* of MS. Bodley 300, at f. 11^{v} (15th cent.), and MS. Digby 168, at f. 76^{r} (14th cent.), where the reference is in a footnote.

[3] The equivalent title in MS. Egerton 889 leaves out the seconds and converts the parts and minutes to minutes.

regarded as demanded by the last part of the *Canones*, that is, as the table referred to below as IV (i).

A poor textual link between al-Khwārizmī and Maudith means that we cannot be absolutely sure, but it seems almost certain that in tables I (i) and (ii) we have the tables 58, 58a, and 60 of Suter's edition (op. cit.).

II. The second table to be mentioned in the *Canones* is to be found in C at ff. 39^r–40^r, and in O at f. $71^{r, v}$. The titles in both C and O are fragmentary or illegible, but in MS. Egerton 889 the title is *Tabula extractionis arcus equatoris elevati supra orizontem et gradu Solis et eius altitudinem.*[1] The argument of the table is the degree of zodiac, and there are three other columns, giving (i) the versed chord (in minutes of a part, $R = 60^p$) of the semidiurnal arc when the Sun is at that degree; (ii) the right chord (also in minutes) of the meridian altitude of the Sun; and (iii) the semidiurnal arc itself (in degrees and minutes), all tables being appropriate to the supposed latitude of Oxford (51° 50′).

It seems highly probable that these tables were calculated by Maudith. They are reasonably accurately made—sufficiently so, in fact, for us to be reasonably confident that he always made sufficiently good interpolations for sinus and *umbra*. The values of the table for semidiurnal arc show that Maudith accepted the Toledan 23° 33′ 30″ for the obliquity,[2] notwithstanding table IV (ii) below, for example.

III. The third table to be mentioned in the *Canones* is the list of stars to be found in C at f. 48^v, but missing from O.[3] This was meant to provide for each star (i) the versed chord (in minutes, as in the second table) of its semidiurnal arc at Oxford; (ii) the right chord (in minutes) of its meridian altitude at Oxford; (iii) the semidiurnal arc itself (in degrees and minutes), again at Oxford; (iv) the size of an unequal hour (i.e. one sixth part) of that arc (in degrees and minutes); (v) the degree (sign, degrees, and minutes) of the zodiac with which the star rises over the Oxford horizon; (vi) the mediation[4] (sign, degrees, minutes); (vii) the star's declination[5] (degrees and minutes), and whether north or south; (viii) the ecliptic longitude[6] (sign, degrees, minutes); (ix) the ecliptic

[1] The title in C is in the margin of the preceding folio, where the words are added 'et est tercia tabula de tabulis Magistri Johannis Mauduth'.

[2] Thus when $\lambda = 90°$, and then $\epsilon = \delta$, the table gives as the semidiurnal arc 123° 42′. Then tan ϵ, which is the product of cot 51° 50′ (latitude) and sin(123° 42′ − 90°), is 0·43609, and ϵ is 23° 33′ 41″. There are many sources of potential inaccuracy in John Maudith's working.

[3] As stated earlier, the star table occurs in MS. Egerton 889, at f. 19^v. It includes twelve stars, but without the information numbered (viii) to (xi) below. A rather similar version is at f. 136^r of MS. Ashmole 191. A list of twenty-four stars ascribed to Maudith is in Hertford College (Oxford), MS. E.4, f. 157, and this is found unascribed in MSS. Erfurt, Amplon. 4° 351, and Munich, Nationalbibl., CLM 588, f. 21^r.

[4] *Gradus cum quo mediat celum.* See Appendix 21.

[5] *Latitudo stelle ab equinoxiali.*

[6] *Verus gradus stelle in zodiaco.*

latitude[1] (in degrees and minutes), and whether north or south; (x) the meridian altitude (degrees and minutes); (xi) the star's magnitude.

In C the table has this rubric and heading:

> *Tabula Johannis Mauduth has pone in quadranto.*
> Notandum quod hec tabula stellarum fixarum verificata est pro anno Christi 1316 pro novo quadrante ad latitudinem Oxonie, que est 51 gra. 50 min.[2]

I have shown elsewhere[3] that this table (for 11 stars in C) is based on a much longer table (86 stars) by Maudith, also for 1316; that, in compiling both, the adjustment made to the longitudes of the stars in the lists of *Almagest* VII and VIII is 16° 40′; and that this must have been calculated using a theory of trepidation.[4]

Some further remarks on John Maudith's list of stars are made in the commentary on the *Albion* tables (see IV. 12), where it is clear that Richard took over Maudith's list.

IV. What the *Canones* and the *Tractatus* regard as the fourth and last table is in fact three, as we have already seen, and these are written out separately rather than combined as in the earlier tables. They are:

(i) The table of *umbrae* of the meridian altitude (placed here in neither O nor C; but see I);
(ii) *Tabula differencia ascensionum universe Terre* (not in O, but in C at f. 40^r^);
(iii) *Tabula declinacionum Solis in qualibet die anni* (not in O, but in C at f. 40^v^).

In C the table is ascribed to Arzachel,[5] and shows a maximum solar declination of 23° 33′ 30″, in agreement with the equivalent table of MSS. Ashmole 191 and Egerton 889.[6] We may suppose that this was the version Maudith intended, although, as already explained,[7] it is inconsistent with table IV (ii), which is calculated for an obliquity of 23° 51′. It provides not only the declination (degrees, minutes, seconds) of each degree of the ecliptic (rather than for each day of the year, as the title suggests), but is also a table of sinus (parts, minutes, seconds; $R = 150^p$). We considered al-Khwārizmī and the Toledan tables as the most probable source of IV (i)–(iii), in the last section. The fact that IV (iii) uses $R = 150^p$ suggests that the Toledan tables are the

[1] *Latitudo stelle a zodiaco.*

[2] An illegible phrase probably gives the longest day at Oxford once more.

[3] In the article cited earlier (p. 12 n. 2), relations with lists in treatises on the new quadrant are considered.

[4] See Appendix 25.

[5] In fact immediately following IV (ii) and preceding IV (iii) in C we have 'Expliciunt tabule Magistri J. Maudith'.

[6] The latter MS. has another table of solar declinations at f. 18^v^, ascribed to Simon Bredon (max. 23° 29′ 0″).

[7] See p. 17.

immediate source, although they were in turn probably formed from al-Khwārizmī's by introducing the multiplier 150:60.

In conclusion, the tables linked with John Maudith's name may be reassembled with reasonable certainty as to their precise contents, largely with the help of the *Canones* and *Tractatus*, neither of which seems likely to be by Maudith himself, and both of which were probably by Richard of Wallingford. No complete collection of the tables survives, and the extant fragments are invariably out of correct sequence. The table of Oxford ascensions was in all probability calculated by Maudith or under his direction, using Toledan trigonometrical tables, and the same applies to table II. Like them, the star table would have required perseverance and some slight astronomical, not to say textual, ability, but is not remarkable except perhaps in the context of the much longer star table from which it seems to have been drawn. Finally, tables II and IV were borrowed from the Toledan tables (T) or those of al-Khwārizmī (K). The borrowings and over-all inconsistencies in this compilation may be seen here at a glance:

Table	R	ϵ	Potential source
I (i)	60	—	K
I (ii) = IV (i)	—	—	K, T
II (i) = (ii)	60	—	
II (iii)	—	23° 33′ 30″	
IV (ii)	150	23° 51′	T, K[1]
IV (iii)	150	23° 33′ 30″	T, K[2]

[1] But table lacking in Latin versions of K.
[2] K unlikely.

The fact that such a mixture of tables was still in use in Leland's day speaks more for the poor mathematical discrimination of succeeding generations of university men than it does for John Maudith himself.

II

QUADRIPARTITUM
SIVE
QUATUOR TRACTATUS DE CORDA VERSA ET RECTA

and

TRACTATUS DE SECTORE

SIGLA

C Cambridge University Library, MS. 1705 (Ii.1.13)
M Oxford, Bodleian Library, MS. Digby 190.

For full discussion of the manuscripts, Introduction to the text, and Commentary see vol. ii, pp. 23–82. The diagrams are printed in vol. iii, pp. 3–14. See also Plates II–IV in vol. iii.

QUADRIPARTITUM

QUATUOR TRACTATUS DE CORDA VERSA ET RECTA ET DE SINIBUS DEMONSTRATIS

I

Incipit[a] *tractatus de sinibus demonstratis.*

QUIA canones non perfecte tradunt noticiam sinus, nec utilitatis eius, intendo eam tradere in hoc opere quadripartito, in quorum primo dicam plane de relacione circuli ad eius diametrum et cuiuslibet porcionis circuli ad eius cordam. In secundo tractatu ostendam proporcionem cuiuslibet sinus ad alterum sinum, et hoc per manifestacionem eius in numeris datis, quia proporcio in magnitudine geometrica non potest esse nota nisi per ea que sunt in 5° libro Euclidis ab arsmetrica mutuata. In tercio tractatu ostendam illa esse in lineis que demonstrata sunt in numeris commensurabilibus; ita quod cuilibet capitulo secundi tractatus correspondebit capitulum tercii tractatus, quia unum est principium demonstracionum alterius. In quarto tractatu applicabo ad proporcionem et ostendam quod in lineis arcualibus est proporcio secundum proporcionem suarum cordarum, ita quod nota cordarum proporcione non poterit ignorari proporcio arcuum. Et hec est tradicio perfecta.

Est igitur sinus rectus et sinus versus et sinus duplatus. Sinus duplatus est linea recta cuius duo termini sunt simul tantum[b] cum duobus terminis porcionis circuli; et se habent sicut corda et arcus. Sinus rectus est medietas illius corde duplate respectu medietatis illius arcus. Sinus

[a] C 5λ [b] C λ

QUADRIPARTITUM

FOUR TREATISES ON RIGHT AND VERSED CHORDS AND ON THE PROVING OF SINES[1]

I

Here begins the treatise on the proving of sines.

BECAUSE the *Canones*[2] do not perfectly explain the evaluation of the sine and its use, I purpose to do so in this work of four parts. In the first treatise I shall discuss clearly the relation of a circle to its diameter, and of any portion of a circle to its chord. In the second treatise I shall show the ratio[3] of any one sine to another, and I shall do so by a proof for the case where we are dealing with numbers, since ratio as between geometrical magnitudes cannot be known except through those theorems of the fifth book of Euclid borrowed from arithmetic. In the third treatise I shall show that what is proved of commensurable numbers is true also of lines; so that to any chapter of the second treatise there will correspond a chapter of the third, there being one principle of proof common to both. In the fourth treatise I shall turn to proportion, and I shall show that the ratio of arcs is related to the ratio of their chords, so that the ratio of the chords being known, the ratio of the arcs cannot be unknown. And these are the complete contents.

There is, then, a sinus rectus, a sinus versus, and a sinus duplatus.[4] The sinus duplatus is a straight line whose two ends coincide with the ends of a part of the circle; and [the line and part of the circle] are related as a chord to its arc. The sinus rectus is a half of the double chord with respect to half the arc. The sinus versus is always that part

[1] This is a composite title, based on MSS. C, M, and D, but amply justified from the treatise itself. It seems that the word *demonstrare* is here as often elsewhere a vehicle for two ideas, that of demonstration in a logical sense, and that of evaluation, in an arithmetical sense. The English word 'prove' has both meanings. 'On the sines which are proved', on the other hand, while literal, is too vague.

[2] Since it is not unlikely that this is a reference to a particular set of canons, or rules for the use of tables, the word is left as though referring to a specific work. There are reasons for supposing that Richard is here referring to his own canons to John Maudith's tables. Either of the works discussed (under the titles *Canones* and *Tractatus*) might have been intended by the opening reference here. For further details see Section I.

[3] *proporcio*.

[4] These words and phrases will be taken henceforth as though they were English (but their plurals will be formed as in Latin), whenever there is risk of confusion with their modern counterparts.

versus est semper illa pars diametri que abscindit tam sinum duplatum quam arcum eius, et est sicut sagitta[a] respectu corde et arcus eius.

Verbi gracia, fiat circulus *ABD*[b] super centrum *O* et protrahatur diameter *AC* super quam cadat perpendiculariter *BD* super punctum *E*. Dico ergo quod linea *BED* est sinus duplatus arcus ipsius *BAD*; et linea *BE* est sinus rectus porcionis[c] *BA*; et linea *EA* est sinus versus tam arcus *BA* quam porcionis[d] *AD*, ut patet in primo circulo.

PRIMA PROPOSICIO

Cognito sinu recto alicuius porcionis sinum versum eiusdem contingit cognoscere.

Sit ergo circulus *CGPE*[e] et sit *FE* sinus rectus porcionis *CE* mihi notus. Dico quod possum ex hoc cognoscere sinum *CF* qui est sinus versus eiusdem porcionis.

Probacio.[f] Quia quod fit ex *EF* in seipsam per 34^am tercii Euclidis valet id quod fit ex *CF* in *FP*, ergo per primam secundi valet quod fit ex *CF* in lineam[g] *FD* et *CF* in *DP*, et per quos valet id quod fit ex *CF* in *FD* et *CF* in *CD*, cum sint equalia *CD* et *DP*. Set quod fit ex *CF* in *CD* per eandem primam secundi (vel[h] per terciam secundi) valet quod fit ex *CF* in seipsam et *CF* in *FD*. Ergo quod fit ex *EF* in seipsam valet id quod fit ex *CF* in seipsam et duplum eius quod fit ex *CF* in *FD*.[i] Set quod fit ex *CD* in seipsam per quartam secundi valet id quod fit ex *CF* in seipsam et *FD* in seipsam et duplum eius quod fit ex *CF* in *FD*. Ergo quod fit ex *FE* in seipsam deficit ab eo quod fit ex *CD* in seipsam per id quod fit ex *FD* in seipsam.

Ergo cum *CD* sit mihi nota (est enim totus sinus 150 minutorum cum sit medietas diametri), accipiam ergo quadratum *CD* ex quo quadrato subtraham quadratum *FE* mihi note, et remanebit quadratum *FD*. Inveniam igitur radicem illius quadrati que erit quantitas linee *FD*. Subtraham ergo lineam *FD* que nunc est mihi nota a linea *CD*, et remanebit linea *FC* mihi nota. Hoc fuit propositum, per noticiam sinus recti qui est *FE* devenire in noticiam *CF* qui est sinus versus eiusdem porcionis.

2^a PROPOSICIO

Cognito sinu verso alicuius porcionis sinum eius rectum ignorare non debes.

Hec est conversa precedentis. Maneat disposicio figure que prius. Dico ergo quod si fuerit mihi notus sinus[j] *FC* erit necessario notus sinus *FE*, quod sic patet:

[a] C Sagittarius [b] C *ABC* [c] C λ [d] C arcus [e] C *CPAG*
[f] C λ [g] M 19λ+*PD* et *OF* [h] C 4λ [i] M *PD* [j] C λ

of the diameter which cuts both the sinus duplatus and its arc, and is like an arrow in relation to the chord and its arc.

For example, let *ABD* be a circle with centre *O*, and let *AC* be drawn as a diameter on which *BD* falls perpendicularly at the point *E*. I say, then, that the line *BED* is the sinus duplatus of the arc of *BAD*; and the line *BE* is the sinus rectus of the arc *BA*; and the line *EA* is the sinus versus both of the arc *BA* and the part *AD*, as is clear in the first circle [fig. I. Prol.].[1]

I. 1

The sinus rectus of any arc being known, to discover its sinus versus.

Let there be a circle *CGPE*, with *FE** (= Sin *CE*).[2] To prove *CF** (= VSin *CE*).

Proof.[3]

(1)	$EF^2 = CF.FP$	Euclid III. 34[4]
(2)	$EF^2 = CF.FD + CF.DP$	(1), Euclid II. 1
(3)	$EF^2 = CF.FD + CF.CD$	(2), $DP = CD$
(4)	$CF.CD = CF^2 + CF.FD$	Euclid II. 1, or II. 3
(5)	$EF^2 = CF^2 + 2CF.FD$	(3), (4)
(6)	$CD^2 = CF^2 + FD^2 + 2CF.FD$	Euclid II. 4
(7)	$EF^2 = CD^2 - FD^2$	(5), (6)
(8)	$FD^2 = CD^2 - FE^2$	(7)
(9)	$FD = \sqrt{(CD^2 - FE^2)}$	(8)
(10)	$CF = CD - FD$	
(11)	CF^*	(9), (10), $CD = 150'$, FE^*.

This is what was proposed, that is to say, from a knowledge of a sinus rectus, *FE*, to arrive at *CF*, the sinus versus of the same arc.

I. 2

The sinus versus of any arc being known, you should not be ignorant of its sinus rectus.

This is the converse of the preceding [proposition]. The arrangement of the figure is to remain as before. I say that if *CF** [= VSin *CE*], then *EF** [= Sin *CE*].

[1] We may write: $BED = 2BE = 2 \text{ Sin } BA$,
and $EA = \text{VSin } BA$.
This notation is used subsequently to abbreviate the translation. See Appendix 20.

[2] The notation of the translation follows that explained in Appendix 20. The asterisk has the force of the phrase 'is known to me' (cf. the opening sentence's 'sit . . . mihi notus').

[3] Steps in the proof are numbered in translation so that in the final column the justification of a particular step may be given somewhat more fully than by the mere 'ergo' of the Latin version.

[4] Campanus's edition is presumably being followed. (As in some Arabic versions, this conflates III. 11 and 12 of the Greek.) In Heiberg's edition (and Heath's translation) the same proposition is numbered III. 35.

Subtraham *CF* de toto *CD* et remanebit *FD*. Cum ergo demonstratum sit quod quadratum *FE* deficit a quadrato *CD* per quadratum *FD*, subtraham quadratum *FD* de quadrato *CD* et residui queram radicem, que necessario erit quantitas linee *EF*, que[a] est sinus rectus porcionis *CE*. Et hoc est quod ostendere curavi.

3ª PROPOSICIO

Dato alicuius porcionis sinu recto, medietatis eiusdem porcionis sinum rectum facile erit invenire.

Sit ergo circulus *KCDE* et sit mihi notus *PB* sinus rectus porcionis *KB*, et posito quod *KA* sit medietas porcionis *KB*. Ego inveniam sinum rectum eius, scilicet *QA*. Nam per 34ᵃᵐ tercii libri Euclidis quod fit ex *PB* sinu porcionis *KB* in[b] seipsam valet quod fit ex *KP* in lineam *PD*. Et per penultimam primi libri, quod fit ex linea *KB* in seipsam valet quod fit ex *KP* in seipsam et *PB* in seipsam. Ergo *KB* in seipsam valet quod fit ex *KP* in seipsam et *KP* in *PD*. Et per terciam secundi, modo verso, quod fit ex *KB* in seipsam valet quod fit ex *KP* in *KD*. Sed secundum scienciam communem, si totum est equale toti et quartum unius erit equale quarto alterius. Nunc autem quarta pars eius quod fit ex *KP* in *KD* est illud quod fit ex *KP* in *KR*, cum *KR* sit quarta pars totius *KD*, ut bene patet[c] ex 3ª tercii libri; et quarta pars eius quod fit ex *KB* in[d] seipsam est id quod fit ex eius medietate in seipsam, scilicet ex *KT* in seipsam. Multiplicabo igitur *KP*, que potest esse mihi nota per primam huius operis, cum sit mihi nota *PB*, multiplicabo, inquam, *KP* in *KR*, que erit quarta pars diametri et illius summe queram radicem que necessario erit medietas *KB*, scilicet *KT*. Set *KT* necessario est equalis *QA*, cum utraque sit sinus rectus porcionis *KA*, que *KA* est medietas porcionis *KB* per diffinicionem sinus, et hoc erat propositum. Quod autem *KT* et *QA* linee sint sinus recti porcionis *KA* apparet si protrahatur linea a centro totius circuli ad *A* punctum per medium *T*.

Corelarium ex hoc sequitur notabile quod illud quod fit ex aliquo sinu recto in seipsum valet id quod fit ex quarta parte diametri in sinum versum duplicis porcionis.

Et nota quod quamdiu invenero sinum cuiuslibet gradus usque ad quartam circuli non intendo loqui de sinu porcionis circuli excedentis quartam. Et semper quando nomino sinum per se amodo semper intelligendum est de sinu nisi aliquid addidero ut 'versum' vel 'duplicis porcionis'.

[a] CM *qui*   [b] C 2λ   [c] M λ   [d] M 2λ

Proof.

(1)	$FD = CD - CF$	
(2)	$EF^2 = CD^2 - FD^2$	(7) of I. 1
(3)	$EF = \sqrt{(CD^2 - FD^2)}$	(2)
(4)	EF^* (= Sin CE)	CD^*, FD^* (by (1), and CF^*).

And this is what I have been concerned to show.

I. 3

The sinus rectus of any arc being given, it will be a simple matter to find the sinus rectus of half the same arc.

Let there be a circle *KCDE*, and PB^* (= Sin *KB*), and let *KA* be half of arc *KB*. I shall find its sinus rectus, that is, *QA*.

(1)	$PB^2 = KP.PD$	Euclid III. 34
(2)	$KB^2 = KP^2 + PB^2$	Euclid I. 47 (penult. prop.)
(3)	$KB^2 = KP^2 + KP.PD$	(1), (2)
(4)	$KB^2 = KP.KD$	(3), Euclid II. 3 converse
(5)	$KR = \frac{1}{4}.KD$	Def.
(6)	$\frac{1}{4}.KP.KD = KP.KR$	(5), common notion (if the wholes are equal the quarters are equal)
(7)	$\frac{1}{4}.KB^2 = KT^2$	Euclid III. 3
(8)	$KT = \frac{1}{2}.KB = \sqrt{(KP.KR)}$	(4), (6), (7)
(9)	$QA = KT$	each is Sin *KA*, and $KA = \frac{1}{2}KB$
(10)	KP^*	PB^*, I. 1 above
(11)	QA^*	(9), (8), KR^* (= 75′), (10).

A notable corollary follows from this, namely, that the square of any sinus rectus is equal to the product of a quarter of the diameter by the sinus versus of twice the arc.

And note that when I have found the sinus of any degree up to a quadrant I do not intend to speak of the sinus of any arc exceeding a quadrant. Also, when I name the sinus as such, it must henceforth be understood as sinus rectus, unless I add such words as ‘versus’ or ‘of twice the arc’.

4ª PROPOSICIO

Manifestato alicuius porcionis sinu recto sinum duplicis porcionis poteris invenire.

Sit datus circulus *ADEF*, et sinus porcionis *AB* sit *HB* sitque mihi notus. Dico quod deveniam in cognicionem linee *KC*, qui est sinus *AC* porcionis duplicis respectu *AB*, quod sic patet:

Multiplicabo *HB* in seipsam, et summam inde provenientem dividam per *AR*,[a] qui est medietas semidiametri; et exibit inde *AK*, scilicet sinus versus duplicis porcionis ad *AB*, ut patet per corelarium 3ᵉ conclusionis huius operis. Set habito sinu verso alicuius porcionis, scilicet *AK*, per secundam huius habere[b] potero sinum rectum eiusdem, scilicet *KC*, qui est sinus rectus duplicis porcionis ad porcionem *AB*. Et hoc est quod promisi.

5ª PROPOSICIO

Noto sinu recto alicuius porcionis, notus erit sinus rectus porcionis alterius[c] que[d] tantum deficit a tota quarta quantum est porcio sinus noti.

Sit circulus *CEGH*, et porcio *CB*, cui porcioni est dupla *AB*, sit equalis[e] porcioni *DE*, cui est dupla porcio *DF*. Dico quod si scio sinum porcionis *CB* possum scire sinum porcionis *CD*, que porcio minor est tota quarta *CE*[f] per tantum quantum est *CB*, cum *CB* ponitur esse equalis *DE* porcioni. Hec conclusio sic patet:

Multiplicabo sinum rectum *CB*, scilicet *OB*, in seipsam, quam subtraham de quadrato *CP* semidiametri, et remanebit quadratum *OP*, ut patet per ea que dicta sunt in prima conclusione illius. Igitur quadrati queram radicem et erit linea *OP*. Set *OP* est equalis *KD* qui est sinus *CD* porcionis quam scilicet querebamus. Igitur per talem operacionem patet propositum.

Incidens. Quod autem *OP* sit equalis *KD* probo. Nam linea *AB* est equalis linee *DF*, per 28ᵃᵐ tercii,[g] cum porcio *AB* posita sit equalis porcioni *DF*. Ergo per 3ᵃᵐ tercii latus *OB* trianguli *OBC* est equale lateri *DN* trianguli *DEN*; et[h] latus *DE* est equale lateri *CB* per 28ᵃᵐ tercii; et angulus *OBC* est equalis angulo *EDN* per 26ᵃᵐ tercii. Ergo per quartam primi libri Euclidis *EN* est equalis *OC*. Per consequens *OP* est equalis *PN*, quod patet per illud principium 'si ab equalibus equalia demas . . . et cetera'. Set *NP* est equalis *KD* per 34ᵃᵐ[i] primi. Igitur *KD* est equalis *OP*, quod est propositum.

Corelarium ex hoc evidenter patet quod cum equaliter ex fine et principio quarte accipitur porcio quod equaliter utrobique secant diametrum quia equalium porcionum equales sunt corde verse sicut et sinus recti.

[a] M *AT* [b] M habens [c] C alicuius [d] M qui [e] M circulus [f] C *C* [g] C secundi [h] C 17λ [i] CM 24ᵃᵐ

I. 4

The sinus rectus of any arc being known, you may find the sinus of twice the arc.

Let there be a circle *ADEF*, and suppose HB^* (= Sin *AB*). I say that I shall derive *KC*, namely Sin *AC*, where $AC = 2AB$:

(1) $AK = \text{VSin } 2AB$	by the construction
(2) $AK = HB^2/AR$	I. 3 above (*AR* is a quarter-diameter)
(3) $(\text{Sin } 2AB)^*$	HB^*, AR^*, (2), (1), I. 2 above.

And this is what I promised.

I. 5

The sinus rectus of any arc being known, I shall know the sinus rectus of the complementary arc.[1]

Let there be a circle *CEGH*, with arc *AB* twice the arc *CB* and equal to the arc *DF*, which is twice *DE*. I say then that if I know the sinus of the arc *CB* I may determine the sinus of *CD* (which arc is less than the quadrant *CE* by the amount *CB*, since $DE = CB$). This *conclusio* is proved thus:

(1) $CP^2 - OB^2 = OP^2$	I. 1 (7) above
(2) $OP = \sqrt{(CP^2 - OB^2)}$	(1)
(3) $KD = OP$	each is the sinus of the same arc (and see below)
(4) KD^*	(2), (3), CP^*, OB^*.

By such an operation the proposition is made clear.

Lemma. I prove, moreover, that $KD = OP$:

(1) $AB = DF$	Euclid III. 28, arc AB = arc DF
(2) $OB = DN$	Euclid III. 3[2]
(3) $DE = CB$	Euclid III. 28
(4) angle OBC = angle DEN	Euclid III. 26
(5) $EN = OC$	(2), (3), (4), Euclid I. 4
(6) $PN = OP$	(5), 'If equals are taken from equals . . . etc.'
(7) $PN = KD$	Euclid I. 34
(8) $KD = OP$	(6), (7).

This was proposed.

A corollary follows evidently from this: When an arc is cut equally from the beginning and end of a quadrant, [corresponding] projections on each of the two [bounding] diameters are equal, being cordae versae or sinus recti of equal arcs.

[1] Literally, 'that other arc which with the given arc makes a quadrant'; that is to say, the cosine of an angle is found from its sine.

[2] q.v. It is misleading to name also the triangles *OBC*, *DEN*, at this juncture.

6ª PROPOSICIO

Omnium kardagarum coniunctim et divisim sinum rectum[a] investigare facile est.

Incidens. Set antequam agam quod propono dicam unum notabile, scilicet quod licet omnis porcio circuli quantumcumque minima habeat sinum, et econverso. Tamen quia nullus sinus est nobis notus nisi quantum communicat cum diametro, quam dividimus in 300 minuta, non potest nobis precise esse notus quicumque sinus nisi tamen[b] ille de quo demonstrare possumus[c] quot minuta et secunda diametri continet. Quod facere non possumus de sinu qui est incommensurabilis[d] diametro vel aliter[e] pars aliquota linee incommensurabilis[f] diametro, cuius est sinus rectus 45 graduum[g] secundum quod totum circulum dividimus in 360 partes equales, quia ille sinus se habet ut costa quadrati cuius diameter est semidiameter circuli. Quod patet si inscribatur quadratum in circulo, secundum 6am quarti, et ducatur diameter ab angulo ad angulum, que diameter erit communis circulo[h] et quadrato; et patet[i] impossibilitas per 2am decimi libri. Verumptamen ita precise operari poteris per radicum extractionem in talibus numeris surdis per addicionem multarum ciphrarum, ut docet canon ad hoc factus; quia per addicionem multarum ciphrarum in tantum subtiliare poteris radicem quod 1000ma pars minuti physici non deficiet.

Redeo ad propositam conclusionem ostendendam. Scias igitur primo quod kardaga est porcio circuli continens 15 graduum. Fiat ergo circulus *AGHI*, cuius quarta pars *AG* dividatur in 6 partes equales ad 6 notas *A*, *B*, *C*, *D*, *E*, *F*, et a puncto *B* protrahatur linea *BQ* equidistans semidiametro *APH*.[j] Similiter protrahatur linea *CR* et linea *DS* et linea *ET* et linea *FU* perpendiculariter cadentes super *PG*. Erit ergo *GF* kardaga prima, *FE* secunda, *ED* tercia, *DC*[k] quarta, *CB* quinta, *BA* sexta. Cum igitur vis invenire sinum prime kardage, scilicet *GF* que est 24ª pars circuli, scilicet[l] 15 graduum, sumas porcionem *EX* quadruplam ad *GF*, duplari scilicet ex utraque parte que erit sexta pars circuli. Igitur ex corelario penultimo 15e[m] quarti Euclidis linea *EX* est medietas diametri. Ergo per diffinicionem sinus, sinus porcionis *EG* que[n] est duarum kardagarum est nobis notus que est medietas semidiametri, scilicet 75 minutorum, quod est *ET*. Igitur per operacionem 3e huius inveniam sinum *FU*, qui est sinus medie porcionis, et erit sinus *FG*[o] porcionis

[a] C λ [b] Cλ [c] M + est [d] M incommensurabili [e] C aliqua
[f] M incommensurabili [g] C 7λ [h] M circuli [i] M inpatet
[j] C *AP* [k] M *DE* [l] C 19λ [m] M 3 λ [n] M quod [o] C λ

I. 6

It is a simple matter to investigate the sinus rectus of all kardagas,[1] *compound or simple.*

Remark. Before I do what I propose, I shall mention a notable fact, namely, that every arc, no matter how small, may have a sinus, and conversely. However, because no sinus is known to us except in terms of a diameter (which we divide into 300 minutes), no sinus can be accurately known to us except one where we can show how many minutes and seconds of the diameter it contains. We cannot do this for a sinus which is incommensurable with the diameter, or otherwise, when an aliquot part of the line is incommensurable with the diameter. [Into this last category comes] the sinus of 45°, since that sinus is as the side of a square whose diagonal[2] is a radius of the circle. This is clear if a square be inscribed in a circle in accordance with Euclid IV. 6. We draw a diagonal from angle to angle, which diagonal will be common to the circle and the square. The impossibility [of commensurability] is evident from Euclid X. 2. However, you will be able to work so precisely in the extraction of the root of such surds, by the addition of many zeros[3] (as the canon for doing this teaches), that there will be no appreciable error; for by the addition of many figures[4] you will be able to extract the root with such precision that you will not be in error by the thousandth part of a minute.[5]

I return to the proposition which is to be proved. Know then first that a kardaga is a part of a circle containing 15°. Let there be a circle *AGHI*, of which the quadrant *AG* is to be divided into six equal parts at the six points *A*, *B*, *C*, *D*, *E*, and *F*. From the point *B*, a line *BQ* is drawn parallel to the semidiameter *APH*. In the same way, lines *CR*, *DS*, *ET*, *FU* are drawn, falling perpendicularly on *PG*. Then *GF* will be the first kardaga, *FE* the second, *ED* the third, *DC* the fourth, *CB* the fifth, and *BA* the sixth. When, therefore, you wish to find the sinus of the first kardaga (that is to say, the sinus of *GF*, the 24th part of a circle, or 15°), it is convenient to take *EX* four times as great as *GF*, that is to say, twice *GF* on each side [of *G*]. Then by the penultimate corollary to Euclid IV. 15, line *EX* is half the diameter. By the definition of sinus, therefore,

$$ET = \text{Sin}\, EG = R/2 = 75', \text{ where } EG = 2 \text{ kardagas } [= 2^k].$$ [6]

[1] See the second paragraph, and also the commentary.

[2] Literally 'diameter'.

[3] *ciphra.*

[4] 'The addition of many figures' was presumably a standard phrase descriptive of the arithmetical technique of root extraction.

[5] *minutum physicum.*

[6] This abbreviation for kardaga is used henceforth in the translation.

scilicet prime[a] kardage. Et habito sinu *FG* porcionis, per 5^{am} huius inveniam sinum *GB* porcionis, qui est *BQ* sinus scilicet 5 kardagarum, eo quod *GB* deficit a tota quarta[b] *GA* per quantitatem *FG* cui est equalis *AB*. Et tunc per 4^{am} huius duplabo illum sinum, scilicet *GF*, et habebo sinum *GE* notum, scilicet sinum duarum kardagarum, qui erit *ET*,[c] et per eandem 4^{am} et per 5^{am} huius habebo sinum porcionis *GC*,[d] scilicet 4 kardagarum, qui erit *RC*. Postea accipiam sinum rectum maximum, scilicet *PG* qui est sinus 6 kardagarum; et secundum[e] 3^{am} huius dimidiabo ipsum et habebo sinum *GD* quod est 3 kardagarum, qui[f] est linea *DS*.

Et si volueris cuiuslibet kardage per se sinum, accipe ut secunda pars huius conclusionis[g] proponit:

Subtrahas *FU* sinum prime kardage de sinu duarum kardagarum, scilicet de sinu *ET*, et remanet linea *OE* qui dicitur sinus secunde kardage vel excessus sinus duarum kardagarum super sinum unius kardage, quod idem est postea. Subtrahe sinum *TE* de sinu 3 kardagarum qui[h] est *SD* et remanet *ND*, scilicet sinus 3^{e} kardage; et sic usque 6^{am} kardagam, cuius sinus esset *KA*. Et sic plane patet propositum quod intendimus et cetera.

7^{a} PROPOSICIO

Diversarum porcionum cordis cognitis, unam cordam communiter omnibus illis subtensam porcionibus invenire non erit difficile.

Sit circulus *ADEG* et sit mihi nota corda *EF* et corda *FG* et corda *GB*. Dico ergo quod erit mihi nota corda *BE* subtensa tribus porcionibus *EF*, *FG*, *GB*, quod sic patet:

Protraham lineam *FKC* diametrum circuli et lineam *GKD* que[i] etiam erit diameter circuli cum *K* fuerit[j] centrum eius. Protraham[k] lineam *GA* et lineam *AC* et lineam *FA*[l] et lineam *GC*. Item, cum ex ypotesi nota fuerit mihi linea *EF*, necessario nota mihi erit linea *FA*, per penultimam primi. Est enim angulus *EFA* rectus per 30^{am} tercii. Et si scio lineam *FA*, per eandem penultimam primi sciam lineam *AC*, cum angulus *CAF* sit similiter rectus per 30^{am} tercii. Et etiam cum positum sit quod scio lineam *FG*,[m] per eandem racionem sciam lineam *GC*.[n]

Et tunc imaginor unum quadrangulum inscriptum in dicto circulo quod est *ACFG*, et in isto quadrilatero protracte sunt duo diametri, scilicet *GC* et *FA*. Et quia, ut probabo convenienter, cuiuslibet quadranguli inscripti in circulo qualitercumque fuerit quadrilaterum si protrahantur in eo diametri, quod fit ex una diametro in aliud valet quod fit ex ductu oppositorum laterum in opposita latera eiusdem quadrilateri. Unde quod fit ex *FA* in *GC* valet illud quod fit ex *GA* in *FC*, et

[a] M post [b] C λ [c] M *CE* [d] M *GT* [e] C +scienciam
[f] CM que [g] M quovis [h] C quod [i] CM qui
[j] M +protraham [k] M extraham [l] C *PA* [m] C *PG* [n] CM *GT*

Since ET^*, then FU^*, by I. 3 above; and $FU = \text{Sin}\, FG = \text{Sin}\, 1^k$. Wherefore BQ^* ($= \text{Sin}\, BG = \text{Sin}\, 5^k$), by I. 5 above. And then ET^* ($= \text{Sin}\, EG = \text{Sin}\, 2FG = \text{Sin}\, 2^k$), and CR^* ($= \text{Sin}\, CG = \text{Sin}\, 2EG = \text{Sin}\, 4^k$), by I. 4 and I. 5 above. Subsequently I shall take the greatest sinus PG ($= \text{Sin}\, 6^k$), and in accordance with I. 3 above I shall halve [the arc], wherefore DS^* ($= \text{Sin}\, DG = \text{Sin}\, 3^k$).

Should you require the sinus of any kardaga *per se*,[1] follow the second part of this proposition:

EO (sinus of the second kardaga *per se*) $= ET - FU = \text{Sin}\, 2^k - \text{Sin}\, 1^k$;
DN (sinus of the third kardaga *per se*) $= DS - ET = \text{Sin}\, 3^k - \text{Sin}\, 2^k$;

and so on, up to the sixth kardaga, whose sinus [*per se*] is AK.

And this is what we intended, clearly propounded.

I. 7

The chords of a number of arcs being known, it will not be difficult to find the single chord subtended by all those chords taken together.

Let there be a circle $ADEG$, with chords EF^*, FG, GB^*. I say, then, that BE^*, this being the chord subtended by the three arcs EF, FG, GB. This is explained as follows.

I shall draw lines FKC and GKD as diameters of the circle, with centre K, and shall join GA, AC, FA, and GC. Then

(1) FA^*	Euclid I. 47, EFA a right angle by Euclid III. 30, EF^* (hyp.), EA^* (diam.)
(2) AC^*	Euclid I. 47, CAF right by III. 30, FA^* by (1), FC^* (diam.)
(3) GC^*	cf. (1) and (2), FG^* (hyp.)
(4) $FA . GC = GA . FC = GF . AC$	$ACFG$ a cyclic quadrilateral, theorem to be proved below
(5) $GA = (FA . GC - GF . AC)/FC$	(4)

[1] This is left untranslated since there is no modern equivalent terminology in use, unless we rephrase the sentence completely in terms of orthogonal projections.

illud quod fit ex *GF* in *AC*. Subtrahatur ergo ex eo quod fit ex *FA* in *GC* (cum utraque supra[a] sit mihi nota) id[b] quod fit ex *GF* in *AC* (que etiam utraque est mihi nota, ut patet per ea que dicta sunt) et tunc remanebit summa que fit ex *FC* in lineam *GA*. Set quia *FC* est mihi nota, est enim tota[c] diameter, scilicet 300 minutorum, dividam[d] igitur summam dictam illam per *FC*; et deveniam in cognicionem linee *GA*.

Tunc protraham lineam *GE* que per penultimam primi est mihi nota postquam fuerit mihi nota *GA*, angulus enim *EGA* est rectus. Et cum scivero[e] *GA* per eandem penultimam, sciam lineam *AD*, et cum per ypotesim scio lineam *GB*, per eandem penultimam sciam lineam *BD*. Cum igitur istius quadrilateri, *ABGD*, sunt mihi nota tria latera *GD* et *GB* et *AD* et duo diametri *GA* et *BD*, erit igitur necessario mihi notum quartum[f] latus[g] *BA*. Cum igitur angulus *ABE* fuerit rectus, noto *BA* erit mihi nota *BE*,[h] quod proposui ostendere.

Notandum quod inveni in margine contra istam conclusionem istam figuram et istam lineam sequentem.

Per idem, cognitis duabus cordis in circulo que ex altera parte conterminales existunt, terciam cordam cognovero que arcui inter duo reliqua extrema intercepta subtenditur. Necesse est ut si cognoscas cordam *AB* et cordam *AC* necessario cognosces cordam arcus *BC*, si vis[i] discurrere[j] per premissa. Quod patet si ab *A* protrahas diametrum in *D* et aliam lineam a *B* in *D* et alteram a *D* in *C*. Cum sit tibi nota *AD* et *AB*, erit tibi nota *BD*. Item cum *AD* sit nota et *AC* igitur similiter *DC*. Igitur si cognoscas[k] *AB* et *DC* et *AC* et *BD* et diametrum *AD*, quomodo poteris ignorare lineam *BC*? Quod est propositum.

Igitur, cognitis duobus lateribus alicuius trianguli in circulo, necesse est cognoscere tercium latus. Et nota hoc, non plus inveni in margine. Sequitur continuacio precedencium.

Incidens. Nunc autem probemus illud suppositum in quo dependet maxima vis huius demonstracionis, scilicet quod *illud quod fit ex ductu unius diametri quadrilateri descripti in circulo in alterum diametrum eiusdem quadrilateri valet illud quod fit ex oppositis lateribus in opposita latera*; ex quo sequitur quod illud quod fit ex *AC* in *BD* valet illud quod fit ex *AD* in *BC* et ex *AB* in *CD*. Similiter, quod fit econverso ex *AD* in *BC* . . . et cetera. Suppositum, hec conclusio sic probatur:

Per 23^{am} primi libri faciam angulum *EAB*[l] equalem angulo *CAD*, et tunc imaginor duos triangulos equiangulos *ABE* et *ADC*.[m] Nam angulus *A* unius et angulus *A* alterius ex ypotesi sunt equales, set per 26^{am} tercii libri angulus *EBA* est equalis angulo *DCA*, quia sunt constituti super eandem circuli porcionem, scilicet *DA*. Ergo per 32^{am}[n] primi, tercius

[a] C λ [b] C item [c] M totus [d] M λ [e] M sonerio
[f] C 2λ+quadrilaterum [g] M latum [h] C *BC* [i] C vel [j] C ascendere
[k] M cognoscit [l] M *CAB* [m] C *ABC* [n] CM 3^{am}

(6) GA^*	FC^* (300′), GF^* (hyp.), (1), (2), (3), (5)
(7) GE^*	Euclid I. 47, *AGE* a right angle, (6), AE^* (diam.)
(8) AD^*	(6), cf. (7) for method
(9) BD^*	BG^* (hyp.), method of (1) and (2)
(10) BA^*	*ABGD* a cyclic quadrilateral, diagonals GA^*, BD^* (by (6) and (9)), sides GD^*, GB^*, AD^* (diam., hyp., (8)), then method of (4) and (5)
(11) BE^*	Euclid I. 47, *ABE* a right angle, (10), AE^* (diam.).

And this it was proposed to show.

It is to be noted that I found in the margin against this proposition that figure and that line following:[1]

By the same method, two adjacent chords being known (AB^*, AC^*), I shall know the third chord (*BC*) subtended by the two non-adjacent ends of the chords. This is evident if you draw a diameter from *A* to *D* and another from *D* to *C*. Then

(1) BD^*	AD^* (diam.), AB^* (hyp.), Euclid I. 47
(2) DC^*	AD^* (diam.), AC^* (hyp.), Euclid I. 47
(3) BC^*	(1), (2), AB^*, AC^*, [theorem on cyclic quadrilateral (cf. (4), (5) above)].

Thus two sides of any triangle inscribed in a circle being known, the third side must be known. And note this: I found no more in the margin. A continuation of the preceding truths follows.

Lemma. But now let us prove that supposition on which the greatest force of this proof depends, namely that ***the product of diagonals of a quadrilateral inscribed in a circle is equal to the sum of the two products of opposite sides***;[2] that is to say, [where *A*, *B*, *C*, and *D* are concyclic],

$$AC.BD = AD.BC+AB.CD,$$

reading the equation in either direction. The problem has been set, and the proposition is proved thus:

(1) angle *CAD* = angle *EAB* [*E* on *DB*]	construction of Euclid I. 23
(2) angle *ABD* = angle *DCA*	Euclid III. 26

[1] The writer is presumably Richard of Wallingford, although this has the appearance of an incorporated gloss. It concerns a special case of the foregoing proposition. A similar passage is certainly to be found incorporated as a corollary in *De sectore*. A second corollary to the same chapter appeared subsequently as a gloss to Lewis of Caerleon's copy (see Comm. I. 7).

[2] This is generally known as Ptolemy's theorem (*Almagest* I. 10, where it is referred to as a lemma).

angulus trianguli *AEB* est equalis tercio angulo trianguli *ADC*, scilicet totalis *D* et parcialis *E*. Ergo per[a] 4[am][b] sexti libri, sicut se habet *AB* ad *AC*[c] sic se habet *BE*[d] ad *CD*.[e] Ergo per 16[am] sexti, quod fit ex *AB* in *DC* valet id quod fit ex *AC* in *BE*. Proporcio linearum patet quia respiciunt angulos propositos equales.

Tunc imaginor duos alios triangulos *ABC* et *AED* et sunt equianguli. Nam per 26[am] tercii, angulus *ACB* est equalis angulo *ADE*, quia angulus *C* et[f] angulus *D*[g] sunt super eandem porcionem circuli, scilicet *AB*. Et cum ex ypotesi positum sit quod angulus *A* unus est equalis angulo *A* alterius, addam utrique angulo angulum *CAE*. Erit igitur angulus *CAB* totalis equalis angulo *EAD* totali. Ergo cum duo anguli unius trianguli sint equales duobus angulis alterius per 32[am] primi erit tercius angulus unius equalis tercio angulo alterius. Ergo per 4[am] sexti, sicut se habet *DA* ad *AC*[h] sic se habet *DE*[i] ad *CB*.[j] Per igitur 16[am][k] sexti ut prius, quod fit ex *DA* in *BC* valet quod fit ex *CA* in *DE*. Sed per primam secundi, quod fit ex *AC* in *DE* et ex *CA* in *EB* valet illud quod fit ex *AC* in *DB*. Et ita patet propositum, scilicet quod fit ex diametro quadrilateri in diametrum valet quod fit ex oppositis lateribus in opposita latera. Et ex hoc manifeste patet quod cum sciuntur tria latera alicuius quadrilateri collocati infra circulum, et sciuntur etiam duo eius diametri, quod necessario scietur quartum latus eiusdem quadrilateri.

8[a] PROPOSICIO

Duplicis porcionis sinu recto cognito eius porcionis in qua una excedit aliam sinum ostendere potes.

Sit circulus *ABC*, et sit mihi notus sinus porcionis *BO* qui est *GO*, et similiter sit mihi notus[l] sinus porcionis *BD*, scilicet *FD*. Dico quod ego deveniam in cognicionem sinus porcionis *OD* in qua porcione *BD* excedit *BO*. Nam si subtrahatur *BO* de porcione *BD* solum remanet porcio *OD*, cuius sinus sic ostenditur postquam ego cognoscam sinum versum *BG*. Et per eandem cognoscam sinum *BF*, postquam cognosco sinum rectum *FD*. Subtraham igitur sinum versum *BO* porcionis, scilicet *BG*, de sinu verso maioris porcionis, scilicet de linea *BF*, et remanebit *GF* notum. Et per consequens erit mihi nota *OE* sibi equalis de racione parallelogrammi, ex 34[a] primi. Tunc subtraham sinum rectum porcionis minoris, scilicet porcionis *BO*, qui[m] est *GO*, de sinu recto porcionis maioris, scilicet de *FD*, et remanebit *ED* mihi nota; nam *FE* est equalis *GO* per eandem 34[am] primi. Et ideo nota erit mihi linea residua *ED*, et linea *EO* est mihi nota. Ergo necessario erit mihi nota linea *OD* per Dulcarnon, que est penultima primi, quia si quadravero *ED* et *EO* et coniungam utrumque numerum illius radix erit quantitas linee *OD*.

[a] C 17λ [b] M 16[am]; C λ [c] M *DC*; C λ [d] M *AC*; C λ
[e] M *BE*; C λ [f] M 2λ+triangulus [g] C λ [h] CM *BC*
[i] CM *CA* [j] CM *DE* [k] CM 15[am] [l] M λ [m] CM quod

(3) angle ADC = angle AEB	(1), (2), Euclid I. 32
(4) triangles EAB, DAC are similar	(1), (2), (3)
(5) $AB:AC = BE:CD$	(4), Euclid VI. 4
(6) $AB.DC = AC.BE$	(5), Euclid VI. 16.

(The ratio of the lines is evident, for they face angles taken to be equal.)

(7) angle ACB = angle ADE	Euclid III. 26
(8) angle BAC = angle EAD	add angle CAE to, [or subtract it from], the equal angles at A (cf. (1))
(9) angle CBA = angle DEA	(7), (8), Euclid I. 32
(10) triangles ABC, AED are similar	(7), (8), (9)
(11) $DA:CA = DE:BC$	(10), Euclid VI. 4
(12) $DA.BC = CA.DE$	(11), Euclid VI. 16
(13) $AC.DE+AC.EB = CA.DB$	Euclid II. 1
(14) $AC.BD = AD.BC+AB.CD$	(6), (12), (13).

The proposition is therefore evident. And from this it is manifestly clear that when three sides of any cyclic quadrilateral are known, and also its two diagonals, the fourth side of the same quadrilateral is necessarily known.

I. 8

The sinus of two arcs being known, you may determine the sinus of their difference.

Let there be a circle ABC with GO* (= Sin BO) and FD* (= Sin BD). I say that I shall determine Sin OD, that is, Sin($BD-BO$).

(1) BF*	FD* (hyp.), [I. 5]
(2) BG*	GO* (hyp.), [I. 5]
(3) GF*	$GF = BF-BG$, (1), (2)
(4) OE*	$OE = GF$ by Euclid I. 34
(5) DE*	$DE = FD-GO$ by Euclid I. 34; FD* (hyp.), GO* (hyp.)
(6) $OD = \sqrt{(DE^2+OE^2)}$	Euclid I. 47[1]
(7) OD*	(4), (5), (6)

[1] 'Dulcarnon'—see p. 251 n. 4 below.

Habeo igitur sinum duplum porcionis *OD*; ergo, per diffinicionem sinus, medietas linee *OD* erit sinus rectus medietatis porcionis *OD* quod erit *OR*. Igitur per quartam huius inveniam sinum duplicis porcionis, et ita habebo sinum rectum tocius *OD*. Et hoc est quod promisi.

9ª PROPOSICIO

Si linea medietati diametri addita et in totam iam compositam lineam ipsa pars addita ducta fuerit, et sic fecerit quadrangulum equale quadrato semidiametri, ipsa linea illi linee necessario quoequatur que 36 gradibus eiusdem circuli constat esse subtensam.

Dividatur linea *KD* per 11ᵃᵐ secundi ita quod una pars ducta in totam *KD* faciat quadrangulum equale quadrato alterius partis. Verbi gracia, sit circulus *KPQS* et dividatur eius diameter *KQ* in *D* ita quod linea *OD* ducta in *KD* faciat quadrangulum equalem quadrato semidiametri, scilicet *KO*, per 11ᵃᵐ secundi. Dico ergo quod si linea *OD* applicetur ad circulum erit linea subtensa 36 gradibus circuli, que est 10ª pars circuli.

Applicetur ergo ad circulum linea equalis linee *OD*, que sit *KL*, secundum doctrinam prime quarti libri, et protrahatur linea *KL* in continuum et directum donec fiat equalis linee *KD* secundum doctrinam secunde primi libri, et sit linea *KC*. Igitur erit linea *KC*[a] equalis linee *KD*; ergo ex ypotesi nostra quod fit ex *KL* in *KC* erit equale ei quod fit ex *LC* in seipsam, et per consequens ex *KO* in seipsam; ergo per 16ᵃᵐ sexti, sicut se habet *KC* ad *KO* sic se habet *KO* ad *KL*; ergo duo latera trianguli parcialis, scilicet trianguli *LOK*, sunt proporcionalia duobus lateribus alterius trianguli totalis, scilicet *KOC*, et angulus proporcionalibus lateribus contentus unius, scilicet angulus *K*, est equalis angulo alterius trianguli contenti in suis proporcionalibus lateribus, scilicet angulo *K*; nec mirum quia idem est et communis utrique triangulo in quo communicat. Igitur per 6ᵃᵐ sexti illi duo trianguli sunt equianguli. Sed cum triangulus parcialis *KOL* sit equilaterus, quia *OK* et *OL* exeunt ab eodem centro ad eandem circumferenciam, ergo oportet quod triangulus totalis *KCO* sit equilaterus, aliter enim non essent equianguli. Cum ergo *KO* sit minor quam *KC* vel *OC*, erit *KC* equalis *OC*; ergo per 5ᵃᵐ primi angulus *CKO* est equalis angulo *COK*. Sed cum angulus *OKL* sit equalis angulo *KLO* per eandem 5ᵃᵐ primi, ergo angulus *OLK* trianguli parcialis est equalis angulo *KOC* trianguli totalis; ergo tercius, scilicet angulus *KOL*, est equalis angulo *OCK*.

[a] CM *KL*

Definition: R is the mid point of arc OD.

(8) (Sin OR)*	(7), Sin $OR = OD/2$
(9) (Sin OD)*	(8), I. 4 above.

And this is what I promised.

I. 9

If a line be added to a semidiameter, and then multiplied by the sum, and if when this is done the rectangle[1] *formed be equal to the square of the semidiameter, that line will necessarily be equal to a chord of the circle subtended by 36 degrees.*[2]

Let the line KD be divided into two parts, by Euclid II. 11,[3] so that the rectangle contained by one part and the whole is equal to the square of the other part. For example, let there be a circle $KPQS$, and let its diameter KQ be divided at D by Euclid II. 11 so that

(1) $OD.KD = KO^2$.

I say, then, that OD as a chord of the circle is subtended by 36°, or one-tenth part of a circle.

A chord KL is to be drawn equal to OD (by Euclid IV. 1), and let KD be produced to the point C such that $KC = KD$. Then

(2) $KL.KC = LC^2$	
$= KO^2$	(1), constructions
(3) $KC{:}KO = KO{:}KL$	(2), Euclid VI. 16
(4) triangles LOK, KOC are equiangular	(3), Euclid VI. 6
(5) triangle KLO is isosceles[4]	equal radii KO, LO
(6) triangle KCO is isosceles	(4), (5)
(7) $KC = OC$	(6), KO is less than both
(8) angle CKO = angle COK	(6), (7), Euclid I. 5
(9) angle OKL = angle OLK	(5), Euclid I. 5
(10) angle OLK = angle KOC	(8), (9)
(11) angle KOL = angle OCK	(10)[5]

[1] Literally 'quadrangle'.

[2] Algebraically expressed:

$$\text{If } x(x+R) = R^2, \text{ then } x = \text{Crd } 36°.$$

[3] The solution offered there is required in Euclid IV. 10, 11, where a construction is found for a regular cyclic pentagon. The method appears again in Euclid VI. 30 (on dividing a line in so-called 'extreme and mean ratio').

[4] *equilaterus*. (This does not have the same meaning as our word 'equilateral'.)

[5] Of the angles KOL, LCO, LOC, and KLO, each of the first two is the difference between the third and fourth.

Sed tunc imaginor alium triangulum, scilicet triangulum *OLC*, qui est equilaterus eo quod ex ypotesi habetur quod linea *LC* est equalis linee *OL*, quia *LC* equatur semidiametro. Igitur per 5^{am} [a] primi angulus *LOC* est equalis angulo *LCO*. Sed per 32^{am} primi angulus *OLK* valet ambos illos, cum sit angulus extrinsecus trianguli. Igitur angulus *OLK*[b] est duplus ad angulum *OCL*, et per consequens est duplus ad angulum sibi equalem, scilicet *LOC*; ergo similiter est duplus ad angulum *KOL*, cum probatum sit quod angulus *KLO* est equalis totali angulo *KOC*. Sed per 32^{am} [c] primi angulus *QOL* est duplus ad angulum *KLO*, eo quod valet tam angulum *K* quam *L*, ergo angulus *LOQ* erit quadruplus ad angulum *KOL*. Dividam ergo ipsum angulum *LOQ* in quatuor partes equales, et hoc secundum doctrinam 9^{e} primi libri, dividendo scilicet ipsum primo per medium et postea utramque medietatem per medium. Et protraham lineas *OM*, *ON*, et *OP*, ergo ad centrum *O* concurrunt quinque anguli equales; ergo per 25^{am} tercii arcus super quos fundantur erunt equales; ergo porcio *KL* est equalis porcioni proxime *LM* et illa porcioni *MN* [. . . et cetera]. Igitur totus semicirculus *KQ* dividitur in 5 partes equales, ergo linea *KL* est subtensa 10ma parti circumferencie, quod est 36 graduum secundum quod totus circulus dividitur in 360 partes, cuius pars 10ma est 36[d] graduum. Et hoc est quod propositum erat.

10^{a} PROPOSICIO

Si fuerint duo quadrata quorum unum est quintuplum ad reliquum, erit radix minoris medietas semidiametri, et radix maioris continebit eandem medietatem semidiametri eiusdem circuli et insuper cordam subtensam decime parti[e] *circumferencie eiusdem circuli; et econverso.*

Secundum 11^{am} [f] secundi libri protrahatur linea *AB* que ducta in se facit quadratum *AE*, et dividatur linea *AB* in duo equalia super *C*. Deinde protrahatur *CD* faciens triangulum rectangulum *DAC*. Deinde protrahatur linea *AC* versus *G* ad equalitatem *CD*, per secundam[g] primi, et quadretur *GA* et fiat quadratum *GF*.[h] Probatum est ergo per 11^{am} [i] secundi et per 10^{am} quarti quod linea *FD* est corda porcionis 36 graduum, id est latus[j] decagoni circuli cuius *AF* est semidiametrum. Fiat igitur circulus secundum exigenciam *AF* que[k] dividatur in duas partes equales super punctum *I*.

Deinde linee *IF* addatur linea equalis linee *FD*. Deinde quadretur tam *ID* quam *AD* et sit quadratum linee *AD*, scilicet *ADCB*, et quadratum *ID* linee est *IDEH*. Dico ergo quod quadratum linee *ID*[l] est quintuplum ad quadratum linee *IF*; et[m] econverso, si quadratum *ID* sit quintuplum ad quadratum *IF*, tunc ideo *ID* constabit ex medietate semidiametri et ex latere decagoni eiusdem circuli; quod sic probatur:

[a] C 4^{am}; M 10^{am} [b] M *OLC* [c] M 33^{am} [d] M 30 [e] C λ
[f] C 13^{am} [g] M terciam [h] M *GP* [i] C 13^{am} [j] M latum
[k] M quod [l] C λ [m] M 10λ

(12) triangle *OLC* is isosceles		*LC* = *OL* by the construction
(13) angle *LOC*	= angle *LCO*	(12), Euclid I. 5
(14) angle *OLK*	= angle *LOC* + angle *LCO*	Euclid I. 32
(15) angle *OLK*	= 2 angle *LCO*	
	= 2 angle *LOC*	(13), (14)
(16) angle *OLK*	= 2 angle *KOL*	(10),[1] (15)
(17) angle *QOL*	= 2 angle *KLO*	(9), Euclid I. 32
(18) angle *QOL*	= 4 angle *KOL*.	

Divide now angle *QOL* into four equal parts by Euclid I. 9 (i.e. by repeated bisection), and draw the [radii] *OM*, *ON*, *OP*; then

(19) arcs *KL*, *LM*, *MN*, *NP*, *PQ* are equal Euclid III. 25.

The whole semicircle is divided into five equal parts, and the chord *KL* is subtended by one-tenth part of the circumference, or 36°. And this is what had been proposed.

I. 10

If there are two squares, one of which is five times the other, the square root of the lesser being half a semidiameter, the root of the greater will contain half a semidiameter in excess of a chord subtended by one-tenth of the circumference of the same circle; and conversely.[2]

In accordance with Euclid II. 11, let *AB* be drawn (the square formed from it being *AE*), and let it be divided into two equal parts at *C*. Let *CD* be drawn to form the right-angled triangle *DAC*. The line *AC* is to be produced in the direction of *G*, so that *GC* = *CD* (Euclid II. 2), and *GA* is to be squared (resulting in the square *GF*). Now it was proved in Euclid II. 11 and IV. 10[3] that the line *FD* is Crd 36°, that is, the side of a regular decagon inscribed in a circle of radius *AF*. Let there be a circle with radius *AF* (*I* being its mid-point). Square both *ID* (*IDEH*) and *AD* (*ADCB*).[4] I maintain then that the square of *ID* is five times the square *IF*. And conversely, if the square of *ID* is five times the square of *IF*, then *ID* will contain half the radius and the side of an inscribed decagon. This is proved as follows (fig. I. 10 (b)):

[1] Or, more immediately, from (11).

[2] In algebraic terms,

$$\frac{5R^2}{4} = \left(\frac{R}{2} + \text{Crd } 36^\circ\right)^2.$$

[3] This is the first reference to the main source of the proof given for the previous proposition.

[4] Notice the change of notation at this point of the chapter, the change being incorporated in the second of the two figures.

AD est quadratum, ut supponitur, et *ID* est quadratum, et habent angulum communem *D*. Ergo diameter[a] transiens *AD* in *B* necessario transibit per *H*, per secundum principium primi et per 24^{am}[b] sexti. Tunc protrahatur linea *FGK* equidistans *AB*; et linea *IH* compleatur in *L* ab *I*, que sit *IHL*; et lineam *EM*[c] complebo in *M*; et quadrabo *FD* per 45^{am} primi, et sit quadratum *FDNO*; et complebo lineam *NO* et sit linea *NOPQ*. Et necessario sic linea *AD* est equalis linee *DC*, ergo et linee *FK*; et *FO*[d] est equalis *FD*, ergo *OK* est equalis *FA*. Ergo de racione quadrati *HB*, linea *GK* est equalis linee *FI* (mediante *H*, *L*) et *IA*; ergo *OG* et *GK* sunt equales; ergo per 36^{am}[e] primi libri sive per primam sexti *GC* et *OE* parallelogramma sunt equalia. Sed per 43^{am} primi libri *IO* et *OE*[f] sunt equalia, ergo per communem scienciam[g] *IO* et *GC* sunt equalia; ergo quadratum[h] linee *ID*, quod est *IE*, valet parallelogrammum[i] *FC* et quadratum *OH*. Sed superficies *OH* valet quadratum *FI*, per corelarium[j] 4^{e} secundi, quia linea *OP* est equalis linee *FI*, per 34^{am} primi, ergo valebit *OH* quadratum [medietatis] semidiametri.

Ergo habeo quod quadratum *IE* valebit superficiem *FC* et quadratum *FI*. Sed ut probatum est per 4^{am} secundi quadratum *FA* valebit quatuor quadrata *FI*. Cum igitur ex ypotesi nostra quadratum *IE* valet quinque quadrata *FI* ergo quadratum *FA* et quadratum *FI* equantur quadrato *IE*. Sed cum probatum sit quod superficies *FC* cum quadrato *OP* equantur eidem quadrato *IE*, per illud principium 'que uni et eidem sunt equalia . . . et cetera' erit quadratum *FA* equale superficie *FC*. Sed superficies *FC* constituitur ex ductu *FD* in *DA*, mediante *DC*; ergo . . . et cetera.

Sic autem invenies faciliter sinum duplatum decime partis circuli et 5^{e} partis. Similiter fiat semicirculus *KSQ*, cuius diameter circuli sit *QOK*,[k] et a puncto *S* ducatur *SO* perpendicularis super centrum *O* dividens *KQ* in duo equalia. Deinde dividamus *OK* in duo equalia ad punctum *T* et protrahatur linea *ST*. Deinde, posito centro in *T*, fac lineam in diametro, scilicet lineam *TD* ad equalitatem *ST*. Deinde ab *S* in *D* ducatur linea. Dico ergo quod linea *DS* est corda 5^{e} partis circuli et *DO*[l] est corda decime partis eiusdem, sicut patet per 6^{am} secundi libri Euclidis et per penultimam primi et per 9^{am} terciidecimi libri et per 3^{am}[m] tercii et per precedencia.

[a] M diametrum [b] M 226^{ti} [c] M *EH* et lineam [d] C *FG*
[e] CM 35^{am} [f] M *KE* [g] M lineam [h] C λ
[i] M parallella [j] M 10λ+7 [k] C *QCK* [l] M de *O*
[m] M 30^{am}

(1) *D*, *H*, *B* collinear	Euclid, Postulate 2 and VI. 24.

Produce *FGK* parallel to *AB*; take *L* on [*BC* and on] *IH* produced; produce [*EGH*] to *M* [on *AB*]; square *FD* (*FDNO*), by Euclid I. 45;[1] produce *NO* to [meet *IL* in] *P* and [*AB* in] *Q*. Then

(2) $AD = DC = FK$	constr. [Euclid I. 34]
(3) $FO = FD$	constr.
(4) $OK = FA$	(2), (3)
(5) $OG = GK$	(4), *OKBQ* square, *IH* bisecting opposite sides
(6) ▱OE = ▱GC	(5), Euclid I. 36
(7) ▱IO = ▱OE	Euclid I. 43
(8) ▱IO = ▱GC	(6), (7), axiom [1][2]
(9) ▱IE = ▱FC + ▱OH	(6), (7), (8)
(10) area ▱$OH = FI^2 = R^2/[4]$	$OP = FI$ by Euclid I. 34, and ▱*OH* is a square
(11) ▱IE = ▱$FC + FI^2$	(9), (10)
(12) $FA^2 = 4.FI^2$	Euclid II. 4
(13) ▱$IE = 5.FI^2$	hyp. (initial constr.).

[This proves the theorem. We now turn to the converse, assuming (13) as hypothesis, and first proving (11) and (12):]

(14) ▱$FA + FI^2$ = ▱IE	(11), (12), (13)
(15) ▱FA = ▱FC	(11), (14), the principle 'things equal to one and the same . . . etc.'
(16) ▱$FC = FD.DA$	$DC = AD$
[(17) $FD.DA$ = ▱$FA = FA^2$	(15), (16).

And this proves the converse, if taken together with the former proposition, I. 9 above.]

So, moreover, you will easily find the sinus duplatus of a tenth part of a circle and of a fifth part. In like manner, let *KSQ* be a semicircle of which the diameter is to be *QOK*, and from the point *S* drop a perpendicular *SO*, dividing *KQ* at *O* into two equal parts. Then let us divide *OK* into two equal parts at the point *T*, and draw the line *ST*. Then, with *T* as centre and radius *TS*, describe an arc cutting *KQ* [internally] at *D*. I say, then, that *DS* is the chord of a fifth part of a circle, and that *DO* is the chord of a tenth part of the same circle (Euclid II. 6, I. 47, XIII. 9, II. 3, and the preceding discussions).

[1] This reference seems quite redundant. The proposition concerns the construction of a parallelogram equal in area to any given rectilineal figure (by subdivision of the latter into its component triangles).

[2] *sciencia communis*, common notion.

11ª PROPOSICIO

Suppositis decem conclusionibus loco suo habitis de demonstracione sinus recti ad arcum suum et ad totam diametrum, sic componere poteris tabulas de corda recta et versa tam secundum modum Ptholomei, qui utitur sinu duplato ad arcum suum,[a] et similiter[b] secundum modum Arzachelis, qui utebatur sinu medio respectu arcus[c] medii; ita quod illud Ptholomeus posuit in *Almagesti* in tabulis suis contra aliquem arcum Arzachel[d] posuit[e] eius medietatem contra arcum mediatum.

Disponas ergo tibi lineam numeri ab uno gradu usque ad 90 gradus, vel usque ad 180, quod est medietas circuli. Et postea docebo componere tabulas secundum modum Arzachelis. Quo habito faciliter habebis modum Ptholomei, sicut[f] postea docebo te certissime. Sit igitur hec conclusio demonstranda, que est 11ª in ordine:

Ab uno gradu usque ad complecionem quarte circuli sinum rectum ostendere

Supposito in principio te nosse ex 6ª conclusione incidenter quod non omnis arcus habet sinum notum noticia demonstrativa[g] (quia aliquis sinus arcus est assimetrus et incommensurabilis diametro) procedam tamen tali arte in demonstracione huius conclusionis quod non oportet deficere a veritate per unam 9000ªᵐ partem unius tercii, si volueris. Et tradam tibi tres regulas quibus procedere poteris circa huius investigaciones, quarum secunda est subtilior prima, et tercia subtilior secunda. Et hoc exempli gracia, circa invencionem sinus unius gradus de gradibus circuli habentis 360 partes equales, et eius diameter 300 minuta, vel 120 partes:

Supposita igitur quantitate diametri 300 min., accipe medietatem semidiametri tibi notam, que est 75 min., quam quadrabis ducendo in se et erit numerus 5625; quod quadratum quintuplabo, et quadratum quintuplum erit 28125. Huius ergo quadrati quintupli extrahe radicem, que[h] radix per 10ªᵐ eiusdem tractatus continebit precise[i] medietatem semidiametri et latus decagoni, id[j] est cordam duplatam arcus 36 graduum circuli. Cum ergo hec radix sit mihi nota, ab ista subtraham medietatem semidiametri, scilicet 75 min., et residuum est mihi notum, videlicet corda duplata 36 graduum, quod est corda posita in tabulis Ptholomei contra tot gradus.

Nota igitur quod semper continue quando inveneris in isto processu cordam, id est cordam rectam mediatam alicuius arcus, illam statim scribe contra talem arcum in linea numeri tabule quam preparasti pro opere Arzachelis. Et quando habueris cordam[k] duplatam alicuius arcus, scribe illam[l] statim contra arcum in tabula quam preparasti pro sinibus

[a] C sinum [b] M sit [c] M λ [d] M +id [e] CM + contra duplicem arcum [f] C 5λ [g] C demonstratiam [h] M quod [i] C λ [j] C 2λ+in [k] C λ [l] M illum

I. 11

Having accepted the ten propositions, proved in the appropriate place, on the relation of the sinus rectus to its arc and to the whole diameter, you will thus be able to draw up tables of corda recta and versa according to the method of Ptolemy, who used the sinus duplatus of the arc, and according to the method of Arzachel, who used an arc corresponding to half the sinus. Half of what Ptolemy put in *Almagest* against any arc was therefore put by Arzachel against half the arc.[1]

Arrange, then, your line of number[2] from 1° to 90°, or to 180°, a semicircle. Afterwards I shall show how to draw up tables according to the method of Arzachel, and this having been understood, you will easily grasp Ptolemy's method, as I shall subsequently most convincingly teach you. This proposition is then to be proved, the eleventh in order:

To derive the sinus rectus, from 1° to 90°.

It being initially presumed that you know from Proposition 6 that not every arc has a sinus known by demonstrative knowledge (for some arcs are *assimetri*,[3] that is, incommensurable with the diameter), I shall nevertheless proceed in such a manner with the proof of this proposition that there is no reason to err from the truth by as much as one 9000th part of a third,[4] should you wish it so. I will give you three rules by which you may proceed with the investigation, the second rule being more accurate than the first, and the third more accurate than the second. Here is an example of how to find the sinus of one degree (taking the circle as having 360 degrees, and a diameter of 300 minutes, or [alternatively] 120 parts):

The diameter being assumed to contain 300′ ($= 2R$), take half the semidiameter (75′) and square it, giving the number 5625. I multiply by 5, giving 28125. I take the root of this, namely of $5R^2/4$, which root (by I. 10 above) will contain exactly $R/2+\text{Crd}\ 36°$. When I have worked out the root, therefore, I shall subtract $R/2$ (that is, 75′), leaving Crd 36°, which in value is what Ptolemy put opposite 36° in his tables.

Note, then, that whenever you find a chord by this method, that is, the half-chord of any arc, write that chord against the appropriate arc in the line of number of the sheet which you prepared for an Arzachel-type table. And when you have found the duplate chord of an arc, write it at once against the arc in the table you prepared for sinus duplati

[1] Since $\text{Sin}\ a/2 = \frac{1}{2}\ \text{Crd}\ a$.

[2] The left-hand line containing the tabular argument.

[3] The meaning of the adjective *assimetrus* is given clearly enough, but the derivation is not. (Note that the MSS. generally make nonsense of this and the following word(s). Its form is deduced from a later passage in this chapter for which see p. 49 n. 2.)

[4] A standard radius here contains 150 minutes (Arzachel), and the error quoted is consequently one part in 486×10^7. The figure 9000 is quoted again later in the chapter without any clear reason. No doubt a *possible* accuracy is understood. The maximum error anticipated for Crd 1° below is 40‴, which is one part in 5655.

duplatis secundum modum operandi Ptholomei. Et nota hic ulterius: mediabis illam lineam iam notam, que est sinus duplatus 36 grad., et eius medietas est sinus rectus 18 grad. Arzachelis, per diffinicionem sinus recti. Pono igitur hanc differenciam, quod sinus rectus vocetur sinus mediatus Arzachelis et sinus duplatus sit sinus Ptholomei.

Ultra accipies sinum rectum unius kardage per 6^{am} conclusionem huius, et statim per 3^{am} conclusionem huius scire potes[a] sinum rectum medie porcionis, scilicet sinum rectum 7 grad. et dimidii; [quam] scribe. Ultra, per 6^{am} huius, invenias sinum rectum 30 grad., hoc est 2 kardagarum; a quo arcu subtrahe 7 grad. et dimidium, et per 8^{am} huius invenias sinum rectum residui, scilicet 22 grad. et dimidium. Scribe. Ulterius statim per 3^{am} huius invenire poteris sinum rectum medietatis eius, scilicet 11 grad. et 15 min. Scribe. Ultra minue arcum istum, scilicet 11 grad. et 15 min., de arcu 18 grad., et per 8^{am} huius invenias sinum rectum residui, scilicet 6 grad. et 45 min., quod [45 min.] est tres 4te unius[b] gradus. Ultra accipe arcum 7 grad. et dimidii, et per 3^{am} huius invenies sinum rectum medietatis, scilicet 3 grad.[c] et 3 quartarum, id est 45 min. Ultra extrahe istum arcum 3 grad. et 45 min. de arcu 6 grad. et 45 min., et per 8^{am} huius invenias sinum rectum residui, scilicet 3 grad. Precise hoc tamen levius haberi potest isto modo, dimidiando arcum 18 grad. et per 8^{am} huius inveniendo sinum residui, scilicet 9[d] grad., quod similiter haberi potest per 3^{am} huius; item per 3^{am} huius invenire poteris sinum medietatis eius, scilicet 4 grad. et dimidii; et si hoc subtraxeris de arcu 7 grad. et dimidii remanent 3 grad., cuius sinum rectum invenies per 8^{am} huius.

Quo habito, per 3^{am} huius invenire poteris sinum medietatis, scilicet 1 grad. et dimidii. Hoc igitur habito, patet nobis via ad ostendendum sinum unius gradus absque errore sensibili, ut predixi. Quod ostendam tibi, ut promisi, primo per modum quo ostendit commentator hoc super ultimam proposicionem primi *Almagesti*, capitulo[e] 6°, que etiam erit tibi exemplum ad procedendum consimiliter in omnibus cordis assimetris.

Accipe cordam rectam unius gradus et dimidii, que potest vere determinari ut patet per premissa, et sit illa porcio, gracia exempli, *AG* mihi nota. Sit eius corda dupla vel simpla, non curo, que secundum demonstraciones Ptholomei[f] est 1 grad. 34 min. et 15 sec., et sit *AB* porcio unius gradus cuius sinus est mihi ignotus. Cum igitur ut probatur illo 9°[g] capitulo primi libri *Almagesti* quod proporcio arcus ad arcum maior est quam proporcio corde ad cordam, non dubites si arcus et linee sint inequales. Sed ex ypotesi nostra, proporcio arcus *AG* ad arcum *AB* est sesquialtera, ergo proporcio *AG* corde ad cordam *AB* necessario erit minor quam[h] sesquialtera. Cum ergo constet ex demonstracione nota quod corda porcionis *AG* est 1 grad. 34 min. et 15 sec., secundum quod tota diameter est 120 grad., unde cum 1 grad. 34 min. 15 sec. sesquial-

[a] CM possum [b] C septimi [c] C kardagarum [d] CM trium
[e] M 2λ+66 M potest 5 omni [g] M 2λ+66 [h] C λ

according to Ptolemy's method of working. And furthermore note that you will halve the known line, the sinus duplatus of 36°, and half of it is the Sin 18° of Arzachel, by the definition of sinus rectus. I make this distinction therefore, in that 'sinus rectus' is the name given to Arzachel's half-sine, and 'sinus duplatus' is the name given to Ptolemy's sine.

You will further take the sinus rectus of one kardaga, by I. 6 above, and immediately, by I. 3, you may know the sinus rectus of half the arc, that is Sin $7\frac{1}{2}°$. Write this down. In addition, by I. 6, find Sin 30° (Sin 2^{k}). Subtract $7\frac{1}{2}°$ from the arc, and by I. 8 find the sinus rectus of the remainder, that is, Sin $22\frac{1}{2}°$. Write it down. Furthermore, by I. 3 you will at once be able to find the sinus rectus of half of it, that is, Sin 11° 15′. Write it down. Further, subtract this arc (11° 15′) from 18°, and by I. 8 you will find Sin 6° 45′, or Sin $6\frac{3}{4}°$. Take the arc of $7\frac{1}{2}°$, and by I. 3 find the sinus rectus of half, Sin $3\frac{3}{4}°$ or Sin 3° 45′. Subtract this arc of 3° 45′ from 6° 45′, and by I. 8 find the sinus rectus of the remainder, Sin 3°. This result is more easily found otherwise, however, by halving 18°, and finding (by I. 8) Sin 9°, which may similarly be found by I. 3. By I. 3 you may find the sinus of half, Sin $4\frac{1}{2}°$; by I. 8 you will find (since if you subtract $4\frac{1}{2}°$ from $7\frac{1}{2}°$ you get 3°) Sin 3°.

When you have Sin 3°, by I. 3 you will be able to find half, namely Sin $1\frac{1}{2}°$. And when this is secured, the way is open to us to find Sin 1°, without sensible error, as was said before. And as I promised, I will show you this first by the method of the Commentator on the last proposition of *Almagest*, [which is to be found in *Parvum Almagestum*] I. 6.[1] This will also be an example for you for proceeding similarly in the case of all incommensurable chords.[2]

Take Crd $1\frac{1}{2}°$, which may be exactly determined, as is clear from what has gone before. Let this arc (*AG*) be known, for example. Let its chord be double or single, I do not mind which. According to Ptolemy's proof,[3] it is 1° 34′ 15″. Let *AB* be an arc of 1°, whose sinus is unknown to me. Since, then, as is proved in *Almagest* I. 9, the ratio of one arc to another is greater than the ratio of their chords, do not have any qualms should the ratio of the arcs and the lines be unequal. But from our hypothesis, the ratio *AG*:*AB* is 3:2,[4] and thus the ratio Crd *AG*: Crd *AB* will necessarily be less than 3:2. Since to a diameter of 120 it

[1] That this is a reference to *Parvum Almagestum* has been verified by reference to British Museum MS. Harley 625.

[2] *in omnibus cordis assimetris*—see p. 47 n. 3.

[3] *Almagest* I. 10.

[4] That is, sesquialterate ratio.

tera ad 1 grad. 2 min. 50 sec. erit necessario corda *AB* maior quam 1 grad. 2 min. et 50 sec.

Rursus ponamus *AC* arcum dimidii gradus et 15 min., cuius sinus potest mihi esse notus per 3^{am} huius, cum arcus *AC* sit medietas *AG*; ipse[a] vero arcus *AB*[b] est unius gradus, igitur proporcio *AB* ad *AC* est proporcio sesquitercia. Sed faciliter potest ostendi quod arcus *AC* corda est 47 min.[c] 8 sec. ad quem numerum erit sesquitercius 1 grad. 2 min. 50 sec. 40[d] ter. Igitur[e] corda arcus *AB* minor erit quam 1 grad. 2 min. 50 sec. 40[f] ter. et maior quam 1 grad. 2 min. 50 sec.[g] Igitur quid erroris esset ponere cordam subtensam uni gradui esse 1 grad. 2 min. 50 sec. 20 ter., quod minus est 2 terciis unius secundi[h] in errore, quare multo minus quam in uno secundo? Sed in inquisicione cordarum quod minus est quam unum secundum abicitur, ideoque ponit Ptholomeus cordam dimidii gradus esse 31 min. 25[i] sec.

Aliter et subtilius poteris invenire cordam rectam unius gradus. Sic habita corda recta 1 grad. et dimidii per premissa, statim invenias per 3^{am} huius sinum rectum eius medietatis, scilicet trium quartarum, id est 45 minutorum. Item statim per eandem 3^{am} invenias sinum rectum eius medietatis, scilicet sinum trium[j] octavarum [id est] 22[k] min. 30 sec. et ceterum; eodem modo sinum istius medietatis, scilicet sinum sexdecimarum quod est medietas trium octavarum, scilicet 11 min. 15 sec. (que sunt 3 sexdecime, quarum tercia pars, scilicet 3 min. 45 sec., sunt sexdecima pars unius gradus). Deinde adde arcum trium sexdecimarum ad arcum trium quartarum, et per 7^{am} huius invenias sinum unum communiter subtensum arcui composito ex utrisque, qui arcus est unius gradus excepta una sexdecima, videlicet 3[l] min. 45 sec. (que sunt una sexdecima gradus). Sunt 56 min. 15 sec.[m] arcus unius gradus, sicut satis manifestum est. Item huic arcui addam arcum trium sexdecimarum et habebo sinum 1 gradus et 2 sexdecimarum, per 7^{am}. Extraham igitur terciam partem differencie que est inter istos duos sinus, videlicet inter sinum arcus 1 gradus una[n] 16^{a} minus et sinum unius gradus et 2 sexdecimarum,[o] et habebo sinum unius gradus; qualiter autem devenies ad noticiam sinus 1 gradus et 2 sexdecimarum, non est difficile per 7^{am}, ex quo notus est tibi sinus 1 gradus minus una sexdecima et sinus trium sexdecimarum.[p]

Tercio[q] et subtilissime et absque errore sensibili[r] ab oculo mentis, sic operabis in omnibus huius. Habito sinu 3 16^{arum} secundum premissa per 3^{am} huius, inveneris[s] sinum rectum medietatis eius, scilicet 3 32^{arum}.[t] Quo habito eius medietatis sinum rectum queras per eandem 3^{am}, scilicet 3 64^{arum}; qui sinus inventus, scilicet sinus 3 64^{arum}, addatur ad sinum 15 16^{arum}, qui est arcus unius gradus excepta una 16ma, et

[a] CM ipsum [b] CM 4λ; C +*AB* unum gradus; M ad unum gradum [c] M 2λ +nota tam [d] C 4 [e] M 15λ [f] C 4 [g] C +40 ter.
[h] CM tercie [i] M 15 [j] C λ [k] CM 42λ; M +bene novi
[l] CM 12λ; M +15 16^{arum} [m] C 4λ+ 16^{arum} [n] C 3λ
[o] C decimarum [p] M tergintarum (?) secundarum sinum rectum invenias
[q] M 32λ [r] C λ [s] C eius [t] C λ

is established that Crd *AG* is $1^\circ\,34'\,15''$ ($= 3/2 \times 1^\circ\,2'\,50''$), therefore Crd *AB* is greater than $1^\circ\,2'\,50''$.

Again, let *AC* be an arc of $45'$, the sinus of which may be found by I. 3, since arc *AC* is half *AG*. The arc *AB* is 1°, and thus the ratio *AB*:*AC* is 4:3.[1] But it is easily shown that Crd *AC* is $47'\,8''$, to which $1^\circ\,2'\,50''\,40'''$ bears the ratio 4:3. Therefore Crd *AB* is less than $1^\circ\,2'\,50''\,40'''$, and greater than $1^\circ\,2'\,50''$. What error could there be in fixing the chord subtended by 1° at $1^\circ\,2'\,50''\,20'''$, which is less than $\frac{2}{3}''$ in error, and for this reason much less than $1''$? But in the search for chords, anything less than $1''$ is rejected, for which reason Ptolemy fixes Crd $\frac{1}{2}^\circ$ at $31'\,25''$.

You may find Crd 1° in another and more accurate way. Having found Crd $1\frac{1}{2}^\circ$ by the foregoing arguments, you will immediately find Sin $\frac{3}{4}^\circ$, that is, Sin $45'$ (by I. 3); and then Sin $\frac{3}{8}^\circ$, or Sin $22'\,30''$ (again by I. 3); and in the same way Sin $\frac{3}{16}^\circ$, or Sin $11'\,15''$ (and $\frac{1}{16}^\circ = 3'\,45''$). Then add arc $\frac{3}{16}^\circ$ to arc $\frac{3}{4}^\circ$, and you will find (by I. 7) the one sinus of the two arcs combined, making 1° less $\frac{1}{16}^\circ$, 1° less $3'\,45''$, or $56'\,15''$, as is clear enough. Next I[2] shall add $\frac{3}{16}^\circ$ to the arc, and shall have Sin $1\frac{2}{16}^\circ$ (by I. 7). I shall then find one-third part of (Sin $1\frac{2}{16}^\circ$ − Sin $\frac{15}{16}^\circ$), and [by adding it to Sin $\frac{15}{16}^\circ$] I shall have Sin 1°. You proceed as when you found Sin $1\frac{2}{16}^\circ$, which is easily done by I. 7, from which you found Sin $\frac{15}{16}^\circ$ and Sin $\frac{3}{16}^\circ$.[3]

Thirdly, and most accurately, and without error such as would be detected by the mind's eye, you will proceed as follows for all [the calculations instanced in] this [treatise]. Having earlier found Sin $\frac{3}{16}^\circ$ by I. 3, you will find Sin $\frac{3}{32}^\circ$, the sinus of half the former arc. Having found this, again by I. 3, you will seek Sin $\frac{3}{64}^\circ$, the sinus of half the arc.

[1] That is, sesquitertiate ratio.

[2] The change from 'you' to 'I' is made with complete abandon. Although usually preserved here—although not within a single sentence—the changes so surprised the copyist of MS. C that he occasionally left the verbs without endings.

[3] The Latin corresponding to this last sentence is properly omitted in some of the later MSS., for it is obscure and wrong in a trivial but careless way.

per 7^{am} huius invenias sinum huius aggregati quod erit sinus 63 64^{arum}, ut bene novi. Ulterius mediabis arcum 3 64^{arum}, quod est arcus 3 128^{arum},[a] cuius sinum invenias per 3^{am} huius. Item mediabis[b] huius arcum.[c] Per eandem 3^{am} quere sinum, qui erit sinus 3 256^{arum}. Addas igitur hunc arcum arcui 63 64^{arum}, et per 7^{am} huius[d] quere sinum rectum[e] aggregati, et invenies sinum id est unius gradus excepta una 256ma parte unius gradus. Iterum resume opus et accipe sinum arcus 3 256^{arum}, et per 3^{am} huius quere[f] sinum rectum[g] medietatis eius, scilicet 3 512^{arum}. Item medietatis huius arcus per[h] eandem 3^{am} quere sinum, qui[i] erit sinus scilicet 3 1024^{arum}. Quem arcum adde ad priorem arcum, scilicet ad arcum 255 256^{arum}, et per 7^{am} invenias sinum aggregati et habebis sinum 1 gradus excepta una 1024^{a} parte unius gradus. Et sic procede usque ad partem 9000 unius gradus, vel in[j] infinitum si volueris subtiliando. Hoc est opus subtilissimum, sed primum est commendabilius.

Habito ergo sinu unius gradus, sic facies tabulas tuas expedite, quia per 5^{am} huius habere poteris sinum 90 grad. excepto uno gradu.[k] Statim duplabo arcum unius gradus, et per 4^{am} huius inveniam sinum 2 grad., et per 5^{am} habebo sinum 88 grad. Iterum accipiam arcum unius gradus et per eius sinum deveniam in noticiam sinus dimidii gradus, per 3^{am} huius. Hunc arcum auferam de arcu duorum graduum, et per 8^{am} inveniam sinum residui, scilicet unius gradus et dimidii; et isto habito, per 4^{am} huius inveniam sinum sui dupli, scilicet 3 grad., et per 5^{am} statim sciam sinum 87 grad. Et per sinum 2 grad., ex 4^{a} huius sciam sinum 4 grad., et per 5^{am} huius sciam sinum 86 grad. Statim ab arcu 3 grad. abstraham arcum dimidii et per 8^{am} huius statim sciam sinum 2 grad. et dimidii; et per 4^{am} huius sciam[l] sinum dupli arcus, scilicet 5 grad., et per 5^{am} huius sciam sinum arcus 85 grad. Et sic procedam operando donec complevero tabulas Arzachelis de corda recta.

Deinde, secundum corelarium[m] 13^{e} huius, abstraham cordam rectam 1 gradus de semidiametro et residuum est corda versa 89 grad.; vel si illam cordam unius gradus addidero ad semidiametrum, collectum erit corda versa 91 grad. Et sic de duobus gradibus, et tribus, usque habuero cordam versam 180 graduum.

Quo habito, sic perficiam tabulas Ptholomei. Accipiam cordam rectam unius gradus, et cordam versam eiusdem, utramque cordam ducam in seipsam; et illa duo quadrata coniungam, et collecti[n] queram radicem que erit corda duplata 1 gradus, quam ponit Ptholomeus contra 1 gradum in tabulis suis. Et sic deinceps faciam de 2 gradibus, et 3, usque ad complecionem semicirculi 180 graduum. Et hoc est quod promisi. Si[o] igitur merui gracias, redde illas anime studenti naturale.[p]

[a] M 12 graduum huius [b] CM medietatis [c] C arcus [d] C + radicem
[e] C λ [f] CM λ [g] CM λ [h] M 3λ [i] M 3λ [j] M λ [k] M λ
[l] C λ [m] M corelare [n] M intellecti [o] C 9λ [p] M vale

Now find $\text{Sin}(\frac{3}{64}+\frac{15}{16})°$ by I. 7, which—as I am quite sure—is to say find $\text{Sin}\,\frac{63}{64}°$. Lastly, halve the arc of $\frac{3}{64}°$ giving $\frac{3}{128}°$, whose sinus you will find by I. 3. Next you will halve the arc of this. By I. 3, seek $\text{Sin}\,\frac{3}{256}°$. Find therefore $\text{Sin}(\frac{63}{64}+\frac{3}{256})°$, by I. 7, that is to say, you will find $\text{Sin}(1-\frac{1}{256})°$. Again resume the work, take $\text{Sin}\,\frac{3}{256}°$, and by I. 3 seek $\text{Sin}\,\frac{3}{512}°$. Likewise by I. 3 seek the sinus of half this arc, that is, of $\frac{3}{1024}°$. By I. 7 you will find $\text{Sin}(\frac{255}{256}+\frac{3}{1024})°$, or $\text{Sin}(1-\frac{1}{1024})°$. And proceed in this way even to the 9000th[1] part of a degree, or even to the infinitely small, if by working minutely you wish to do so. This method of working is the most precise, but the first is the more to be recommended.

Having found Sin 1°, you will be able to compile your tables quickly, since by I. 5 you will be able to find Sin(90—1)°. I shall at once double the arc of 1°, by I. 4 find Sin 2°, and by I. 5 I shall have Sin 88°. Next I shall find $\text{Sin}\,\frac{1}{2}°$ from Sin 1°, by I. 3. By I. 8 I find $\text{Sin}(2-\frac{1}{2})°$, that is, $\text{Sin}\,1\frac{1}{2}°$, and thence Sin 3°, by I. 4. By I. 5, I at once find Sin 87°. From Sin 2° I shall find Sin 4°, by I. 4; and thence Sin 86°, by I. 5. I shall learn $\text{Sin}(3-\frac{1}{2})°$, that is $\text{Sin}\,2\frac{1}{2}°$, by I. 8; and Sin 5° by I. 4; and Sin 85° by I. 5. And so I shall go on working until I have completed the tables of Arzachel's corda recta.

In accordance, then, with a corollary to I. 13[2] of this treatise, I shall find VSin 89° as the difference (R—Sin 1°), or VSin 91° as the sum (R+Sin 1°). And so for 2°, 3° . . . and so on until I have VSin 180°.[3]

Having done this, I shall draw up the Ptolemy tables as follows. I shall express the corda duplata of 1° as

$$\sqrt{(\text{Sin}^2 1°+\text{VSin}^2 1°)},$$

and this is what Ptolemy places opposite 1° [in the line of number] in his tables. I shall continue doing this for 2°, for 3°, and so on, until I have completed a semicircle of 180°. And this is what I promised. If, therefore, I have merited thanks, render them to a mind naturally studious.

[1] Three more operations would bring us within an 8192nd part of a degree, and four within a 16384th part. The source of the 9000 is obscure. Cf. p. 47 n. 4 above.

[2] This relates Sin a to VSin(90—a), and so on.

[3] Note that the '*corda*' of the text is here rendered by 'sinus', since the argument of the function quoted is clearly meant to be the *half-angle*, rather than the full angle subtended by the chord.

12ª PROPOSICIO

Cognito perfecte unius quarte sinu recto, trium aliarum quartarum sinum cognoscere non est grave.

Sinus rectus cum sit medietas corde porcionis circuli, et[a] non sit maior corda in circulo quam diameter, sicut nec maior porcio quam semicirculus, erit maximus sinus rectus sinus 90 graduum, scilicet unius quarte circuli, cum sibi respondet semidiameter. Ergo cum per precedencia haberi possit noticia sinus recti unius quarte, patebit noticia sinus recti trium aliarum quartarum, quod sic patet:

Sit[b] circulus *ACOF* et sciam sinum tocius *AC* quarte, et velim[c] scire aliquem sinum porcionis maioris quarta. Tunc si ista porcio sit minor semicirculo debebo subtrahere istam porcionem de[d] porcione semicirculo, id est de 180 graduum, et residui queram sinum rectum secundum premissam. Et iste erit sinus porcionis quesite, quia regula est generalis sic: *idem est sinus cuiuslibet porcionis minoris semicirculo et tocius residui de semicirculo, et econverso.* Unde si velim[e] invenire sinum porcionis *AD*, subtraham *AD* de porcione *AO* et remanet porcio *DO*. Istius queram sinum rectum, qui erit *MD*. Dico ergo quod porcionis *AD* est *MD* sinus. Sed *MD* potest faciliter sciri, cum sit equalis sinui *HB*. Quia postquam subtraxero *AC* de *AD* remanebit *CD*; si igitur iterum subtraxero *CD* de porcione *AC* remanebit *AB*. Sed *OB*[f] est equalis *AD*, ergo dempto communi *BD* remanebunt porciones equales *AB* et *DO*. Ergo eorum sinus tam rectus quam versus sunt equales. (Et idem patet per 34^am^ primi[g] libri.)

Iterum si voluero invenire sinum porcionis *ACE* subtraham de *AE*[h] *AO* et remanebit *OE*, cuius sinus est *EM* qui est equalis *MD*; ergo per premissam equalis *HB*. Cognito igitur sinu *HB*, cognoscitur sinus porcionis *ADE*.

Iterum, si voluero cognoscere[i] sinum *ADG*, subtraham illud de toto circulo et remanet *AG*[j] cuius sinus est *GH*, qui est equalis *HB*. Et patet propositum. Et possunt hec omnia probari de equalitate[k] si protrahas lineas *AB*, *AG*, *OD*, *OE*: patet demonstracio per triangulos rectangulos per operacionem 5^am^ huius.

[a] M λ [b] C 2λ [c] M voleri [d] M 3λ+semicirculi [e] M volo
[f] CM *CB* [g] C tercii [h] M *EC* [i] M +*EC* [j] C alia
[k] M equaliteraret

I. 12

The sinus rectus in one quadrant being known, it is not hard to find the sinus in the other three quadrants.

Since the sinus rectus is one-half the chord of an arc of the circle, and there is no greater chord in a circle than the diameter, just as there is no greater arc than a semicircle, the greatest sinus rectus is Sin 90°, that is, the sinus of one quadrant, since the semidiameter corresponds to it. Since it is possible to know the sinus rectus in one quadrant, therefore a knowledge of those in the other quadrants will easily be had, as follows:

Let *ACOF* be a circle, and suppose that I know the sinus within the entire quadrant *AC*. Suppose I want to know the sinus of any arc greater than the quadrant. Then if that arc is less than a semicircle, I shall have to subtract it from the arc of the semicircle, that is, from 180°, and find the sinus of the remainder in accordance with what has been said previously. This will be the sinus of the arc which was required, since it is a general rule that *the sinus of any arc is the same as that of its supplement, and conversely*. Whence if I wish to find Sin *AD*, I shall express it [see fig. I. 12] as

$$\mathrm{Sin}(AO-AD) = \mathrm{Sin}\, DO = MD.$$

But *MD* may be easily known, for it is equal to *HB*. Furthermore,

$$CD = AD-AC,$$

and therefore

$$AB = AC-CD.$$

Also, since *OB* and *AD* are equal, therefore, by removing *DB* from each, equal arcs *AB* and *DO* will remain. The sinus [of an angle and that of its supplement] are therefore equal, both rectus and versus.[1] (The same follows from Euclid I. 34.)[2]

Next, if I wish to find the sinus of arc *ACE*, I find *OE*, by subtracting *AO* [i.e. 180°] from [the whole arc] *AE*; and Sin *OE* is equal to *EM*, which is equal to *MD*, and therefore to *HB* (by the foregoing discussion). The sinus *HB* being known, the sinus of the arc *ADE* is known.

Next, if I wish to know the sinus of *ADG*, I subtract the arc from the whole circle, leaving *AG*, whose sinus is *GH*, which is equal to *HB*. And what was proposed is clear. And all these things concerning the equality [of sinus] may be proved if you draw lines *AB*, *AG*, *OD*, *OE*: the proof by right-angled triangles is evident from the method of I. 5 above.

[1] It was a mistake to include sinus versus, for $\mathrm{VSin}(180°-a) = 2R-\mathrm{VSin}\, a$.

[2] The 'III. 34' of most MSS. is almost certainly a mistake, for this theorem (in Campanus's edition) equates the products of segments of intersecting chords in a circle (= III. 35 in the original).

13ª PROPOSICIO[a]

Porciones equales in fine et in principio unius quarte in sinu recto et verso modo retrogrado coequentur.

Hoc patet quia si *AR* sit equalis *CD* in circulo *ADFG*, erit *SR*, que est corda recta *AR* porcionis, equalis linee *QC*, que est corda recta[b] *DC* porcionis, sicut satis prius patuit. Ex hoc nota quod cum sinus versus prime kardage equatur sinui recto ultime kardage tunc sinus versus secunde kardage equabitur sinui recto quinte kardage, et sic deinceps. Nota etiam quod si volo habere sinum versum maioris porcionis quam est una quarta circuli debeo pro quarta accipere totum sinum, scilicet medietatem diametri; et pro eo quod excedit quartam debeo accipere tantumdem porcionis a principio quarte, et iungere ambo simul. Et satis patet operacio inspicienti.

Ex hoc sequi potest tanquam corelarium quod *sinus rectus porcionis minoris quarta et sinus versus residui eiusdem quarte semidiametro coequantur; et sinus rectus quarte cum sinu recto porcionis semicirculi excedentis quartam coequantur sinui verso aggregati ex utraque porcione*. Hoc patet quia in circulo *ADFG* linea *OC*, que est sinus rectus *AC* porcionis, cum sinu verso *CD*, quod est *QD*,[c] coequatur toti *MD*. Et equaliter linea *KA*, qui est sinus versus *AB*, cum linea *BR'*, qui est sinus rectus residui arcus[d] *BD*, coequatur toti *AM*; vel *MD* cum linea *QT*, que est corda porcionis *DT*, coequatur toti *AF'*, que est corda versa tocius *AT*.

[a] Proposition entirely omitted from M [b] C versa [c] C λ [d] C λ

I. 13

If equal arcs be taken at the beginning and end of a quadrant, they are to be equated in sinus rectus and sinus versus modo retrogrado.[1]

This is evident, since (fig. I. 13) if *AR* be equal to *CD* on the circle *ADFG*, then *SR* (= Sin *AR*)[2] is equal to *QC* (= Sin *DC*), as has been sufficiently proved before. And note that since the sinus versus of the first kardaga is equated to the sinus rectus of the last, then the sinus versus of the second will be equal to the sinus rectus of the fifth, and so on.[3] Note also that if I want the sinus versus of a greater arc than a quadrant, I should take the whole sinus for the quadrant (that is the semidiameter). I should take the excess of the arc over a quadrant and set it at the beginning of the quadrant, [find its sinus rectus, and] add this to the semidiameter. The operation is sufficiently clear to anyone inspecting [the figure].[4]

From this there follows the corollary: *the sinus rectus of an arc less than a quadrant and the sinus versus of its complement together make a semidiameter; and the sinus rectus of a quadrant with that of the excess of an arc over a quadrant equals the sinus versus of the arc.*[5]

This is clear since, in the circle *ADFG*, the lines *OC* (= Sin *AC*) and *QD* (= VSin *CD*) make together *MD*. And equally, the lines *KA* (= VSin *AB*) and *BR'*[6] (= Sin *BD*) together make *AM*; or *MD* and *QT* (= Sin *DT*) together make *AF'* (= VSin *AT*).[7]

[1] Does this merely mean 'in opposite ways'? One might imagine that the phrase is taken from arithmetic, denoting some sort of relationship between terms. On the other hand, it might simply refer to the symmetry in the equations relating the two sinus, a sinus rectus being the sinus versus of the complementary angle, and conversely.

[2] The words *corda* and *sinus* are used here, but as is often the case elsewhere, both refer to *sinus*.

[3] This sentence, with its (for us) idiosyncratic terminology, is best understood with the help of I. 6 above. It is not essential to the understanding of the remainder of this chapter.

[4] We assume here that more than pure contemplation was needed for the understanding of the problem, which is explained in one of the most obscure sentences of the treatise. The verb *inspicere* is taken to have most probably carried the idea of visual inspection.

[5] By 'the sum of both arcs' the author means merely the arc which exceeds 90°, conceived of as 90° plus the excess.

[6] The MSS. have *BR*, changed here to *BR'* since the letter *R* was inadvertently used twice in the diagram, which is presumably the author's. Cf. the following note.

[7] The author has used the letter *F* for two different points, one on the circle and one (here *F'*) on the diameter.

II

Hic incipit secunda pars huius tractatus.

In secunda parte huius tractatus oportet investigare proporcionem[a] sinus ad sinum, ad quod faciendum permitto quasdam diffiniciones proporcionum ex 5 Euclidis, que valent ad propositum. Prima est:

Proporcio est duarum quantitatum eiusdem generis ad invicem habitudo.

Secunda est:

Cum duarum quantitatum eiusdem generis una dividit aliam, quod exiit per divisionem dicitur denominacio proporcionis divise ad dividentem.

Tercia est hec:

Proporcionem componi ex proporcionibus est denominacionem[b] componi ex denominacionibus proporcionem ex quibus illa componitur cum una ducitur in aliam.

Et quarta est hec:

Proporcionem dividi per proporcionem aut dividentem abici ex dividenda est denominacionem dividende dividi per denominacionem dividentis.

II. 1

Si denominacio proporcionis quorumlibet duorum extremorum producatur in unum extremum, producitur alterum extremum.

Quia enim per secundam diffinicionem, diviso primo per secundum exit denominacio ergo ducta denominacione in secundum producitur primum, quod patet tam in numeris quam in continuis; quia *B* est denominacio proporcionis *C*[c] ad *A*, ducta *B* in *A* producitur *C*, sicut diviso *C* per *B* exibit *A*.

[a] C proporcionum [b] C denominacionum; M denominacionibus [c] CM 3λ+*A* ad *C*

II

Here begins the second part of the treatise.

In the second part of this treatise it is necessary to investigate the ratio of sinus to sinus, to which end I give certain useful definitions on ratio, from Euclid, Book V:

1. A *ratio* is a mutual relation between two quantities of the same kind.
2. When one of two quantities of the same kind divides the other, what results from the division is called the *denomination*[1] of the ratio of the dividend[2] to the divisor.[3]
3. A ratio [is said to be] *compounded*[4] of ratios when the product of the denominations gives rise to some denomination.
4. A ratio [is said to be] *divided*[5] by a ratio when the quotient of the denominations gives rise to some denomination.

II. 1

If the denomination of a ratio of any two extrema[6] *be multiplied by* [*the second*][7] *extremum, the other extremum results.*

By Definition 2, the first being divided by the second a denomination results, and therefore multiplying by the second produces the first. This is evident both with numbers and continuous[8] [magnitudes], since when B is the denomination of $C{:}A$,[9] then $BA = C$, just as $C{:}B = A$.[10]

[1] *denominacio.* This is not to be translated 'denominator' (as of a fraction). The denomination is clearly an integral or rational number expressing the ratio in its lowest terms. Cf. the Greek πυθμήν (T. L. Heath, *Greek Mathematics*, vol. i (1921), pp. 55–7, etc.). For another Greek word with the same function, see the note below on the interpolation in Euclid VI. Def. 5, corresponding to our Def. 3.

[2] *divisa.*

[3] *dividens.*

[4] The verb is *componere.* Although generally indicating the formation of a (multiplication) product, it may—like our word 'product'—have a wider meaning. See our commentary.

[5] *dividere.*

[6] The *extrema* are the terms of the *proporcio.*

[7] Not justified by the text, but required mathematically. See the note at the end of the proof.

[8] A frequent addition. The distinction is that between the fifth and seventh books of Euclid's *Elements.*

[9] The MSS. have $A{:}C$, which would imply that the 'denomination' function was one of reciprocation. This cannot be, for in II. 2 below, and subsequently, we find both 'sit E proporcio B ad C' and 'E est denominacio proporcionis B ad C'. Whatever historical reasons there may be for distinguishing denominations from ratios elsewhere, they are not sufficiently strong here for us to express these two ideas by anything but $E = B{:}C$.

[10] This ending suggests that the '*producatur*' of the statement of the proposition is best translated 'multiplied or divided by', but if this is what the writer meant then the ambiguity is intolerable.

II. 2

Duobus quibuslibet extremis interposito medio cuius ad utrumque duorum sit aliqua proporcio, componetur proporcio[a] *primi ad tercium ex primi ad secundum et secundi ad tercium proporcionibus.*

Id etiam patet tam in continuis quam in discretis: habeant se *A*, *B*, *C* consequenter sitque ipsius *B* medii[b] ad utrumque aliqua proporcio. Erunt ergo ex prima diffinicione *A*, *B*, *C* eiusdem generis quare per eandem erit *A*[c] ad *C* aliqua proporcio. Dico ergo eam componi ex duabus proporcionibus, ex ea que est *A* ad *B* et ex ea que est *B* ad *C*; quod sic patet:

Sit *D* proporcio *A* ad *B*, et sit *E* proporcio *B* ad *C*, et sit *F* proporcio *A* ad *C*. Cum ergo ex *E* in *C* fit *B*, per primam huius, eo quod *E* est denominacio proporcionis *B* ad *C*. Et eadem racione ex *D* in *B* fit *A* et similiter ex *F* in *C* fit *A*; igitur per primam huius erit *F* ad *E* sed *A* ad *B*; quare *D* cum sit denominacio proporcionis *A* ad *B* erit etiam denominacio proporcionis *F* ad *E*; quare per eandem primam ex *D* in *E* fit *F*, ergo denominacio *A* ad *C* fit denominaciones[d] *A* ad *B* et *B* ad *C*, ergo, per terciam diffinicionem, *A* ad *C* proporcio composita est ex *A* ad *B*, *B* ad *C* proporcionibus.

II. 3

Quotlibet mediis inter duo extrema positis, erit proporcio primi ad ultimum ex omnibus intermediis composita.

Sint inter *A* et *D* duo media, scilicet *B* et *C*. Dico tunc quod proporcio *A* ad *D* componitur ex *A* ad *B*, et *B* ad *C*, et *C* ad *D* proporcionibus. Per precedentem enim constat quod *A* ad *D* componitur ex *A* ad *B* et *B* ad *D* proporcionibus, sed *B* ad *D* constat ex *B* ad *C* et *C* ad *D*, igitur a primo erit *A* ad *D* ex *A* ad *B* et *B* ad *C* et *C* ad *D*. Et hoc est verum generaliter, tam in continuis quam in discretis, et hoc sive media se habeant consequenter in proporcione sive non; ut si *A* sit 4 et *B* 12 quicumque numerus assignetur *C* et *D* interposita inter *A* et *B*, erit proporcio *A* ad *B* composita ex proporcione *A* ad *C* et *C* ad *D* et *D* ad *B*. Sic enim potest hoc declarari: sit *C* 16, *D* 2, patet quod proporcio 4 ad 12 componitur ex proporcione 4 ad 16 et 16 ad 2 et 2 ad 12. Est enim proporcio 4 ad 16 subquadrupla, ponatur igitur pro ea 4. Item proporcio

[a] C λ [b] C medium [c] M 16λ [d] C ex denominacione

II. 2

Suppose there to be two extrema (to each of which a mean is in some ratio). Then the ratio of the first to the third will be compounded of [the ratios of] the first to the second and the second to the third.

This is evident both in continuous and discrete magnitudes. Let A, B, and C be in order of magnitude, and let B be the mean, in some ratio to each of A and C. By Definition 1, A, B, and C [must have been] of the same kind, wherefore (by the same definition) there will be a ratio $A:C$. I say then that it is compounded from the two ratios $A:B$ and $B:C$, which is proved as follows:

(1) $D = A:B, E = B:C, F = A:C$	Defs. of D, E, F
(2) $E.C = B$	II. 1, (1)
(3) $D.B = A$	II. 1, (1)
(4) $F.C = A$	II. 1, (1)
(5) $F:E = A:B$	(2), (4), II. 1
(6) $D = F:E$	(5), $D = A:B$ by (1)
(7) $D.E = F$	II. 1
(8) $A:C = (A:B).(B:C)$	(7), substitution from defs. (1).

II. 3

If two means are interposed between two extrema, the ratio of the first to the last will be compounded of the intermediates.

B and C are to be two means between A and D. I maintain then that

$$A:D = (A:B).(B:C).(C:D).$$

For

(1) $A:D = (A:B).(B:D)$	II. 2
(2) $B:D = (B:C).(C:D)$	II. 2
(3) $A:D = (A:B).(B:C).(C:D)$	(1), (2).

This is true in general both of continuous and discrete magnitudes, and it is true whether or not the means are sequential in ratio.[1] Thus if $A = 4$ and $B = 12$, $A:B$ will be $(A:C).(C:D).(D:B)$, which may be expounded as follows. Let C be 16, and D be 2, then it is clear that

$$4:12 = (4:16).(16:2).(2:12).$$

[1] *se habeant consequenter in proporcione.* This awkward expression is presumed to signify the extrema of the ratios (A, B, D, C above) in order of increasing (or decreasing) magnitude; that is, in medieval terms, that they are all 'ratios of lesser (or greater) inequality'. Cf. the similar expression in II. 2 above.

16 ad 2 est octupla, ponatur ergo ibi 8. Item proporcio 2[a] ad 12 est subsextupla. Cum ergo proporcio *A* ad *C* se habet secundum sub ad[b] supra, et proporcio *C* ad *D* se habet econverso secundum superius ad inferius, docet proporcio *A* ad *D* componi[c] ex divisione prime proporcionis ad secundam. Dividatur ergo 8 per 4, et surgunt 2,[d] que est proporcio *A* ad *D*. Item cum proporcio *A* ad *D* sit subdupla et proporcio *D* ad *B* erit sextupla. Dividatur 6 per 2 et surgunt 3, dico igitur quod *A* ad *B* se habet in tercia proporcione. Quia quociens media inter alia extrema continue crescunt[e] vel continue decrescunt[f] cum extremis, componetur proporcio primi ad ultimum ex ductu omnium denominacionum in alternas,[g] ut si proporcio 2 ad 4 ducatur in proporcione 4 ad 8, et illud collectum[h] in proporcione 8 ad 16, habebitur proporcio 2 ad 16 sive crescendo sive decrescendo. Sed si una proporcio fuerit inter numeros crescentes et sequens inter numeros decrescentes, vel econtra, erit proporcio primi ad ultimum investigata per divisionem unius proporcionis per aliam. Et hec est plana declaracio.

II. 4

Si aliqua proporcio componitur ex duabus, eius conversa componetur ex conversis.

Quod sic probatur: sint *A*, *B* composita proporcio ex proporcionibus *C* ad *D* et *E* ad *F*. Dico quod *B* ad *A* componetur ex *D* ad *C* et ex *F* ad *E*. Continuentur enim *C* ad *D* et *E* ad *F* ut *G* ad *H* et *H* ad *K*, ut sit proporcio *C* ad *D*, sicut proporcio *G* ad *H*, et proporcio *H* ad *K* sicut erit proporcio *E* ad *F*. Erit itaque *A* ad *B* proporcio composita ex *G* ad *H* et *H* ad *K*, quare[i] per secundam[j] huius erit *A* ad *B* ut *G* ad *K*; ergo econverso *B* ad *A* erit ut *K* ad *G*. Sed *K* ad *G* constat ex *K* et *H* et *H* ad *G*, per eandem secundam huius. Sed *F* ad *E* se habet ut *K* ad *H*, et *D* ad *C* ut *H* ad *G*; igitur *B* ad *A* constabit ex *D* ad *C* et *F* ad *E*, quod est propositum.

II. 5

Cum sex quantitatum proporcio prime ad secundam composita fuerit ex proporcione tercie ad quartam et quinte ad sextam accident 360 species composicionis, quarum 36 sunt utiles, cetere autem inutiles,[k] *id est 324.*

Hoc autem manifestum fiet si ponamus proporcionem *A* ad *B* componi ex ea que est *C* ad *D* et ea que est *E* ad *F*. Cum enim termini sunt sex

[a] M 212 [b] C et [c] C λ [d] C λ [e] M querunt [f] C λ
[g] C alternites [h] M erit collectionem [i] CM quere [j] M terciam
[k] C λ

For 4:16 is subquadruple, 4 is to be put for it. Next, 16:2 is octuple, therefore 8 is to be put for it. And 2:12 is sextuple.[1] When, therefore, $A{:}C$ is taken as denominator to numerator,[2] and $C{:}D$ conversely as numerator to denominator,[3] then $A{:}D$ is formed from the division of the first ratio into the second. Let 8 be divided by 4, giving 2. This is the ratio $A{:}D$. Next, if $A{:}D$ be subduple, then $D{:}B$ will be sextuple. Let 6 be divided by 2. The result is 3, and therefore I say that $A{:}B$ is in tertiate ratio. If the mean quotient between the other extrema increases or decreases continuously with the extrema, the ratio of the first to the last will be compounded out of the product of all denominations by the alternates, as in the case when 2:4 is multiplied by 4:8, and the result multiplied by 8:16, giving 2:16 whether [the ratios are all] increasing or decreasing. But if one ratio is between increasing numbers and the following between decreasing numbers, or the other way round, then the ratio of the first to the last is investigated by the division of one ratio by the other. And the exposition is now clear.

II. 4

If a ratio is compounded of two [others], its inverse[4] will be compounded of [their] inverses.

Proof. Suppose that $A{:}B = (C{:}D).(E{:}F)$. Then I am maintaining that $B{:}A = (D{:}C).(F{:}E)$.

(1) $C{:}D = G{:}H,\ E{:}F = H{:}K$	Defs.
(2) $A{:}B = (G{:}H).(H{:}K)$	(1), hyp.
(3) $A{:}B = G{:}K$	(2), II. 2
(4) $B{:}A = K{:}G$	(3), inversion[5]
(5) $K{:}G = (K{:}H).(H{:}G)$	II. 2
(6) $F{:}E = K{:}H,\ D{:}C = H{:}G$	(1), inversion
(7) $B{:}A = (D{:}C).(F{:}E)$	(4), (5), (6).

And (7) is what was proposed.

II. 5

When, of six quantities, the ratio of first to second is compounded of the ratios of third to fourth and of fifth to sixth, there are 360 species of compound, of which 36 are of use, but the remaining 324 are of no use.

This will become clear if we suppose $A{:}B$ to be compounded of $C{:}D$ and $E{:}F$. Since there are six terms, the ratio of any two may be regarded

[1] This is consistently preferred in the text to the (more correct) 'subsextuple'.

[2] *se habet secundum sub ad supra.*

[3] *secundum superius ad inferius.*

[4] *conversa.* This has been translated throughout as 'inverse' since 'converse' and 'conversion' are terms of the medieval theory of proportions with totally different meanings.

[5] *ergo econverso* . . . The principle is discussed in our commentary.

termini, proporcio quorumlibet duorum potest intelligi componi ex duplici proporcione inter quatuor residuos existente;[a] quod etiam posse fieri manifeste declarabitur.

Ex his enim sex terminis provenient 30 diversa intervalla; nam ex *A*, si comparem[b] ad residuos, proveniunt 10 intervalla ex *B* 8, ex *C* 6, ex *D* 4, ex *E* 2, ex *F* nullum, quoniam prius sumpte sunt omnes, sicut patet in hac presenti tabula. Sunt enim 5 tabule in quarum prima *A* comparatur ad reliquos terminos, et ipsi ad *A*. In secunda, *B* comparatur ad reliquos, et ipsi ad *B*. In tercia tabula, *C* comparatur ad residuos, et econverso. In quarta, *D* comparatur ad alios et ipsi ad *D*. In quinta, *E* comparatur ad alios et econverso.

Fiunt ex singulis terminis duobus duo intervalla, ut ex *A* et *B* fit unum intervallum *A* ad *B* et aliud quod est *B* ad *A*; et[c] sic de ceteris. Quia ergo sex erant termini, amotis duobus primis, scilicet *A* et *B*, quorum proporcio componitur ex reliquis, erunt residui 4 termini, quorum erunt 24 ordines diversi, secundum quos potest intelligi proporcio *A* ad *B* componi ex eis; et isti ordines facient tantum 12 intervalla, quod ut plane pateat. Amoveantur *A* et *B* facientes proporciones sive intervallum, quod est *A* ad *B* et *B* ad *A* conversam eius. Remanebunt[d] ergo quatuor termini *C*, *D*, *E*, *F*, quorum erunt 24 ordines componentes tantum 12 intervalla, que in hac presenti patent tabula, in cuius prima linea ponitur numerus ordinum qui concordant et ideo non componunt diversa intervalla sicut patet; quia nonus ordo[e] et quintus sunt idem et 16us et sextus sunt idem. Nam eadem proporcio componitur ex ea que est *D* ad *E* et *C* ad *F* quam denotat nonus ordo ex ea que est *C* ad *F* et *D* ad *E* que precedit. Per numeros igitur illius linee patet quot[f] componuntur[g] ordines qui[h] ad composicionem proporcionum geminati[i] sunt, qui sit primus, secundus, tercius, quartus, quintus sextus, 7us, 8us,[j]

[a] C existentes; M resistente [b] C comparetur [c] C 4λ [d] M remanebit
[e] C λ [f] CM quod [g] M habentur [h] M quantum [i] M seminati
[j] C +9us

as compounded of the double ratio existing between the four remaining ratios. It will be clearly explained how this is so.

From these six terms come 30 different intervals;[1] for from *A*, if I compare it[2] with the others, there arise 10 intervals,[3] from *B* 8, from *C* 6, from *D* 4, from *E* 2, and from *F* none.[4] The reason is that at first all are assumed, as is clear from the table.[5] There are five tables, in the first of which *A* is compared with the remaining terms,[6] and they with it. In the second, *B* is compared with the others, and they with *B*. In the third table, *C* is compared with the remainder, and conversely. In the fourth, *D* is compared with the others and they with *D*. In the fifth, *E* is compared with the others, and conversely.

From two individual terms two intervals result. Thus from *A* and *B* come the intervals $A:B$ and $B:A$; and so for the others. Since there were six terms, therefore, having removed the first two, namely *A* and *B*, whose ratio may be compounded of the remainder, there are four terms remaining, and these can be arranged in 24 different ways,[7] in accordance with which the ratio $A:B$ may possibly be compounded. These arrangements make 12 intervals,[8] which is quite as obvious.[9] Let *A* and *B* be removed, these making ratios or an interval, which is to say $A:B$ and its inverse $B:A$. Four terms (*C*, *D*, *E*, and *F*) will remain, of which there will be 24 component arrangements, but 12 intervals, as is clear from the table. In the first column of this table is placed the number of any arrangements which agree, and therefore do not form different intervals, as is clear.[10] The ninth and fifth arrangements are the same, as are the sixteenth and sixth, for the same ratio is compounded from $D:E$ and $C:F$ (ninth arrangement) as from $C:F$ and $D:E$ which came earlier [in the tables]. Through the numbers therefore on that line it is clear how many arrangements are compounded which are twinned, namely the first, second, third, fourth, fifth, sixth, seventh, eighth, tenth,

[1] *intervallum*, here sometimes meaning 'simple ratio', but at other times apparently meaning 'distinct ratio'. $A:B$ and $B:A$, for instance, count as two ratios but one interval, and the product of two ratios makes, with its commutant, only one distinct (compound) ratio, that is to say, interval.

[2] That is, make it the other term ('extremum') of a simple ratio.

[3] There are five other terms, and *A* may be the antecedent or consequent (numerator or denominator in modern parlance) of the ratio.

[4] The antecedent may be chosen in six different ways, with each of which there are five ways of choosing the consequent, making thirty ways in all. But the text argues otherwise: *A* may be the antecedent or consequent in 10 different ways; *B* only in 8, since it has already occurred in two ratios involving *A*; and so on.

[5] The tables are illustrated in Plate II.

[6] The one line '*ABA*' means that there are *two* ways of conjoining *A* with *B*, namely as $A:B$ and $B:A$.

[7] Which we may easily see, since $4.3.2.1 = 24$. But the method followed is to write down the twenty-four possibilities (sixth table).

[8] Since for every compound ratio in the twenty-four, there is also its inverse.

[9] But the verb '*pateo*' almost always heralds a proof, as is the case here!

[10] However, in the MSS, many entries carry a zero, which is a mistake. The fifth row reads '9 | 5 | *c* | *f* | *d* | *e*' and the ninth *should* read '5 | 9 | *d* | *e* | *c* | *f*', which should serve to explain the author's point. See the following sentence.

10us, 12us, 14us, 16us; quare ex omnibus 24 ordinibus qui ponuntur in secunda linea tabule tantum 12 sunt diversi, qui[a] extracti patent in sequenti tabula.

Proporcio ergo *A* ad *B* et eius conversa que est *B* ad *A* potest intelligi componi ex duabus proporcionibus inter quatuor terminos residuos *C*, *D*, *E*, *F* 12 modis diversis tantum, et, eodem modo, unaquaque predictarum aliarum; quare cum sint in toto 30, ut patet per primam tabulam, erunt omnes combinaciones duodecies 30, quod est 360. Sed harum omnium posito quod *A* ad *B* componatur ex *C* ad *D* et *E* ad *F*, demonstrabo tibi tantum 36 esse utiles combinaciones quantum ad proporcionum[b] extrahendam, ceteras autem omnes[c] inutiles, et non tenere, que sunt in numero 324; et hoc sit:

Compositarum quas prius ostendimus esse 30 singularum binarum erit idem indicium cum 15 illarum sint aliarum 15 converse. Nam quecumque[d] conversa proporcio ex conversis omnium earum componitur, ut patet per 4 huius partis; quare, solum 15 earum cognitis cetere patebunt que sunt earum converse. Prime ergo 15 *AB*, *AC*, *AD*,[e] *AE*, *AF*,[f] *BC*, *BD*, *BE*, *BF*,[g] *CD*, *CE*, *CF*,[h] *DE*, *DG*, *EF*,[i] sicut[j] patet per primas omnes combinaciones prime tabule, quarum 15 combinacionum erunt 9 ex aliquibus duabus inter 4 reliquos terminos existentibus composite, sex autem ex nullis, ut probabo. Et composite quidem sunt *AB*, *AC*, *AE*, *BD*, *BF*, *CD*, *CF*, *DE*, *EF*; relique autem sex non componuntur. Quecumque autem componuntur duobus modis componuntur tantum videlicet ex proporcione tercii ad quartum et quinti ad sextum, item que ex proporcione tercii ad sextum et quinti ad quartum, quare cum 9 earum sunt composite 18 fient composiciones. Sed et conversarum erunt totidem ut superius patet: erunt igitur in toto 36 modi utiles. Que autem non componuntur sunt sex, ergo 12 combinaciones inutiles et conversarum totidem igitur in toto sunt 24.

[a] M *OS* [b] C proporcionem [c] M λ [d] M quem [e] C λ
[f] C λ [g] C λ [h] M λ [i] C λ [j] C 42λ+Reliquarum

twelfth, fourteenth, and sixteenth.[1] Thus out of all 24 of the arrangements which are placed on the second line of the table, as many as 12 are different, and these may be taken into the second table.

The ratio $A:B$, therefore, and its inverse $B:A$ may be known to be compounded out of two ratios between the four remaining terms C, D, E, and F in as many as 12 different ways; and in the same way each one of the others earlier [in the table corresponds to each of these]. From this it follows that if there be 30 in all—as is obviously the case from the first table—the sum total of all combinations[2] is twelve times thirty, namely 360. But I will prove to you that of these, for every occasion on which $A:B$ is compounded of $C:D$ and $E:F$, 36 are useful combinations of ratios to be extracted, while the others are inapplicable,[3] and are not to be retained, and these are 324 in number; and this [is shown] as follows.

I have first shown that there are 30 ratios between two terms, 15 of them the inverses[4] of the other fifteen. For any inverse ratio is compounded from the inverses of all [the ratios encompassing the other fourteen], as is evident from I. 4; wherefore, only 15 of these being known, the others (their inverses) will be manifest.[5] The first 15 [are] therefore *AB, AC, AD, AE, AF, BC, BD, BE, BF, CD, CE, CF, DE, DG, EF*, which are clearly all the first combinations of the first table. Nine of the 15 combinations were compounds of two [ratios] involving the four remaining terms; six, however, may not be [compounds of any such], as I shall prove.[6] The compounds are *AB, AC, AE, BD, BF, CD, CF, DE, EF*; but the other six may not be compounded. However, those which are compounded are compounded in two ways, namely from the ratio of the third to the fourth and the fifth to the sixth, or from the ratio of the third to the sixth and of the fifth to the fourth; wherefore if 9 of them are compounded, 18 compounds result. But there were just as many inverses as was evident above: there are therefore altogether 36 applicable modes. Six, however, are not compounded, and there are therefore 12 inapplicable combinations and 12 inverses, making together 24.

[1] We may go down to the eighth row without duplication; but the fifth was the 'twin' of the ninth, the third of the eleventh, the twelfth of the thirteenth, and the sixth of the fifteenth. It is necessary, therefore, to go down to row sixteen before we have found twelve compounds none of which is a mere rearrangement of its constituent ratios.

[2] *combinacio.*

[3] *inutilis*—they are 'useless' in different ways which will emerge at the end of Part II. Essentially they are unwanted because they are not derivable by a valid transformation from the standard proportionality, $A:B = (C:D).(E:F)$.

[4] *conversa.*

[5] The argument is that, if we have kept the ratio $A:B$, we do not need $B:A$ (which will be among the discarded 15), since with $A:B$ alone we are already prepared to cover the cases (for example) $A:B = (C:D).(E:F)$ and $A:B = (D:C).(F:E)$. Taking the case where $B:A$ appears on the left-hand side, we should (by I. 4) be merely repeating these proportionalities.

[6] See II. 8, below.

II. 6

Supposita igitur prima quod *A* ad *B* proporcio constat ex *C* ad *D* et *E* ad *F*, ostendam tibi secundam composicionem quod scilicet proporcio *A* ad *B* componitur ex *C* ad *F* et *E* ad *D*. Ponam inter *C* et *F*, *D* et *E*: erit ergo proporcio *C* ad *F* composita ex proporcione *C* ad *D*, *D* ad *E*, et *E* ad *F*, per terciam huius; quare *C* ad *F* et *E* ad *D* proporciones constabunt ex *C* ad *D* et *D* ad *E* et *E* ad *F* et super *E* ad *D*. Sed *C* ad *D* et *D* ad *E* et *E* ad *D* componunt eam que est *C* ad *D*, quod patet si ponamus *D* et *E* inter *C* et *D*, necessario erit proporcio *C* ad *D* composita ex *C* ad *D* et *D* ad *E* et *E* ad *D*; quare patet quod *C* ad *F* et *E* ad *D* sunt sicut *C* ad *D* et *E* ad *F*. Sed *A* ad *B* constat ex *C* ad *D* et *E* ad *F*, igitur constabit ex *C* ad *F* et *E* ad *D*, que sunt illis eedem coniuncte. Et sic patet *primus* modus et *secundus*.

Tercius autem modus est[a] ut *A* ad *C* componatur ex *B* ad *D* et *E* ad *F*. Quod cito patet quia, posito *B* inter *A* et *C*, erit *A* ad *C* composita ex *A* ad *B* et *B* ad *C* per secundam huius secundi. Sed proporcio *A* ad *B* componitur ex proporcione *C* ad *D* et *E* ad *F*, ut patet superius; igitur *A* ad *C* constabit ex *C* ad *D* et *E* ad *F* et *B* ad *C*, ut patet ad oculum. Sed *B* ad *C* et *C* ad *D* componunt eam que est *B* ad *D*; igitur *A* ad *C* constabit ex *B* ad *D* et *E* ad *F*, quod intendimus tibi ostendere.

Quartus modus procedit ex consimiliter partibus combinatis ad modum secundi, ita quod proporcio *A* ad *C* componitur ex *B* ad *F* et *E* ad *D* positis inter *B*, *F*, *D*, *E*; et eodem modo procede arguendo sicut superius fecisti in secunda demonstracione secunde composicionis, ut patet.

Quintus autem modus est ut *A* ad *E* componatur ex *B* ad *F* et *C* ad *D*. Posito enim *B* inter *A* et *E*, fiat argumentum tercii, quia constabit *A* ad *E* ex *A* ad *B* et *B* ad *E*. Sed *A* ad *B* constat ex *E* ad *F* et *C* ad *D* sicut primo positum est in principio capituli. Constat igitur *A* ad *E* ex *B* ad *E* et ex *C* ad *D* et *E* ad *F*. Sed *B* ad *E*[b] et *E* ad *F* componunt *B* ad *F*; igitur *A* ad *E* constabit ex *B* ad *F* et *C* ad *D*.

Sextus autem modus, scilicet quod *A* ad *E* componitur ex *B* ad *D* et *C* ad *F*, patet per comparacionem secunde partis, *F* et *C* mediis inter *B* et *D*.

Septimus modus est ut *B* ad *D* componatur ex *A* ad *C* et ex *F* ad *E*. Cum enim *A* ad *B* componatur ex *C* ad *D* et *E* ad *F*, componetur *B* ad *A* ex *D* ad *C* et *F* ad *E*, per 4 huius. Posito igitur *A* inter *B*, *D*, constabit *B* ad *D* ex *B* ad *A* et *A* ad *D*. Sed *B* ad *A* est ut *D* ad *C* et *F* ad *E*, igitur *B* ad *D* constabit ex *A* ad *D* et ex *D* ad *C* et *F* ad *E*. Sed *A* ad *D* et *D* ad *C* componunt *A* ad *C*; igitur *B* ad *D* componetur ex *A* ad *C* et *F* ad *E*, quod intendebamus.

[a] C λ [b] CM +componens

II. 6

The first supposition being that the ratio $A:B$ is compounded of $C:D$ and $E:F$,[1] I will show you a second arrangement,[2] namely

$$A:B = (C:F).(E:D).$$

I place D and E between C and F. Then

(1)	$C:F = (C:D).(D:E).(E:F)$	II. 3
(2)	$(C:F).(E:D) = (C:D).(D:E).(E:F).(E:D)$	(1)
(3)	$(C:D).(D:E).(E:D) = C:D$	II. 3
(4)	$(C:F).(E:D) = (C:D).(E:F)$	(2), (3)
(5)	$A:B = (C:D).(E:F)$	first mode
(6)	$A:B = (C:F).(E:D)$	(4), (5).

The *first* and *second* modes, [(5) and (6)], are thus evident.

The *third* mode is $A:C = (B:D).(E:F)$. What I write down is evident since, putting B between[3] A and C,

(7)	$A:C = (A:B).(B:C)$	II. 2
(8)	$A:B = (C:D).(E:F)$	first mode
(9)	$A:C = (C:D).(E:F).(B:C)$	(7), (8)
(10)	$(B:C).(C:D) = B:D$	II. 2
(11)	$A:C = (B:D).(E:F)$	(9), (10).

And this is what I intended to show you.

The *fourth* mode is derived in the same way as the second, so that $A:C = (B:F).(E:D)$. The argument is [from (11)], by the process leading from [(5) to (6)].

The *fifth* mode is $A:E = (B:F).(C:D)$. Taking B [as a mean] between A and E, follow the argument [establishing] the third [mode]:

(12)	$A:E = (A:B).(B:E)$	II. 2
(13)	$A:B = (E:F).(C:D)$	first mode
(14)	$A:E = (B:E).(C:D).(E:F)$	(12), (13)
(15)	$(B:E).(E:F) = B:F$	II. 2
(16)	$A:E = (B:F).(C:D)$	(14), (15).

The *sixth* mode, namely $A:E = (B:D).(C:F)$, is clear [from (16), by the sort of argument which led from (5) to (6)], putting F and C as means between B and D.

The *seventh* mode is $B:D = (A:C).(F:E)$. For

(17)	$B:A = (D:C).(F:E)$	first mode, II. 4
(18)	$B:D = (B:A).(A:D)$	II. 2
(19)	$B:D = (A:D).(D:C).(F:E)$	(17), (18)
(20)	$(A:D).(D:C) = A:C$	II. 2
(21)	$B:D = (A:C).(F:E)$	(19), (20).

which we meant [to show].

[1] The first mode is $A:B = (C:D).(E:F)$. [2] *composicio.* [3] That is, as a mean.

Octavus modus, scilicet quod *B* ad *D* componitur ex *A* ad *E* et *F* ad *C*, patet per modum arguendi secundi. Si ponantur *C* et *F* inter *A* et *E*, patet liquide.

Nonus autem modus est ut *B* ad *F* constet ex *A* ad *E* et *D* ad *C*, quia enim constat tibi quod *B* ad *A* constat ex *D* ad *C* et *F* ad *E*. Posito *A* inter *B* et *F*, erit *B* ad *F* composita ex *A* ad *F* et ex *F* ad *E* et *D* ad *C*, sicut satis patet ex prioribus, eo quod ipsa componatur ex *B* ad *A* et *A* ad *F*. Sed *B* ad *A* fit ex *F* ad *E* et *D* ad *C*; quare *B* ad *F* necessario erit ex *A* ad *F* et *F* ad *E* et *D* ad *C*. Sed *A* ad *F* et *F* ad *E* componit *A* ad *E*, quod patet si ponatur inter *A* et *E*, *F*; quare *B* ad *F* componetur ex *A* ad *E* et *D* ad *C* quod fuit propositum.

Decimus autem modus est scilicet quod proporcio *B* ad *F* componitur ex *A* ad *C* et *D* ad *E*. Patet per modum arguendi secundi. Positis *D* et *E* inter *A* et *C* et patet argumentum deducenti.

Undecimus modus est ut *C* ad *D* componatur ex *A* et *B* et *F* ad *E*. Quia enim per 3^{am} huius capituli[a] *A* ad *C* componetur ex *B* ad *D* et *E* ad *F*, componetur *C* ad *A* ex *D* ad *B* et *F* ad *E*. Posito ergo *A* inter *C* et *D* erit per premissa in 2^{a}[b] conclusione *C* ad *D* ex *C* ad *A* et *A* ad *D*. Ergo erit *C* ad *D* ex *A* ad *D* et *D* ad *B* et *F* ad *E*. Sed *A* ad *D* et *D* ad *B* sunt ut *A* ad *B*; quare *C* ad *D* erit ex *F* ad *E* et *A* ad *B*.

Duodecimus modus, scilicet quod *C* ad *D* fit ex *A* ad *E* et *F* ad *B*, per argumentum patet secundi.

Terciusdecimus modus est ut *C* ad *F* constat ex *A* ad *B* et *D* ad *E*, quod patet positis *D* et *E* inter *C* et *F*. Constabit *C* ad *F* ex *C* ad *D* et *D* ad *E* et *E* ad *F*. Sed *C* ad *D* et *E* ad *F* componunt eam que est *A* ad *B*; quare palam est quod *C* ad *F* constabit ex *A* ad *B* et *D* ad *E*.

Quartusdecimus modus, scilicet quod *C* ad *F* constat ex *A* ad *E* et *D* ad *B*,[c] patet per modum arguendi in secunda conclusione huius capituli.

Quintusdecimus modus est ut *D* ad *E* componatur ex *B* ad *A* et *C* ad *F*. Positis enim *C* et *F* inter *D* et *E*, erit *D* ad *E* composita ex *D* ad *C* et *C* ad *F* et *F* ad *E*. Sed *D* ad *C* et *F* ad *E* componunt *B* ad *A*, eo quod earum converse componunt *A* ad *B* per primam ypotesim; quare *D* ad *E* constabit ex *B* ad *A* et *C* ad *F*.

Sextusdecimus modus, scilicet *D* ad *E* fit ex *B*, *F*, et *C*, *A*, patet per[d] modum arguendi secundi.

Septimusdecimus modus est scilicet quod *E* ad *F* constabit ex *A* ad *B* et *D* ad *C*, quod ex prima ypotesi et argumento 11° patet. Cum enim constat *A* ad *B* ex *E* ad *F* et *C* ad *D* per ypotesim, fiet per similitudinem argumenti undecimi *E* ad *F* ex *A* ad *B* et *D* ad *C*.

Modus autem *octavusdecimus* extrahitur ex modo arguendi secundi, scilicet quod *E* ad *F* fit ex *A* ad *C* et *D* ad *B*, ut patet satis clare.

[a] M λ [b] CM 3ª [c] C +quod [d] M 2λ

The *eighth* mode, namely $B:D = (A:E).(F:C)$, is found as is [(6) from (5)]. If C and F be placed between A and E [in (21)], it follows plainly.

The *ninth* mode is $B:F = (A:E).(D:C)$, which follows from (17). (We first show that $B:F = (A:F).(F:E).(D:C)$.)

(22)	$B:F = (B:A).(A:F)$	II. 2
(23)	$B:A = (F:E).(D:C)$	(17)
(24)	$B:F = (A:F).(F:E).(D:C)$	(22), (23)
(25)	$A:E = (A:F).(F:E)$	II. 2
(26)	$B:F = (A:E).(D:C)$	(24), (25).

This was the proposition.

The *tenth* mode is $B:F = (A:C).(D:E)$. It follows [from (26)] as, in the argument above, [(6) follows from (5)].

The *eleventh* mode is $C:D = (A:B).(F:E)$.

(27)	$A:C = (B:D).(E:F)$	(11) (= third mode)
(28)	$C:A = (D:B).(F:E)$	II. 4
(29)	$C:D = (C:A).(A:D)$	II. 2
(30)	$C:D = (A:D).(D:B).(F:E)$	(28), (29)
(31)	$(A:D).(D:B) = A:B$	II. 2
(32)	$C:D = (F:E).(A:B)$	(30), (31).

The *twelfth* mode, namely $C:D = (A:E).(F:B)$, follows [from (32)] as [(6) from (5)].

The *thirteenth* mode is $C:F = (A:B).(D:E)$, and is clear when D and E are placed between C and F.

(33)	$C:F = (C:D).(D:E).(E:F)$	II. 3
(34)	$A:B = (C:D).(E:F)$	first mode
(35)	$C:F = (A:B).(D:E)$	(33), (34).

The *fourteenth* mode, namely $C:F = (A:E).(D:B)$, follows [from (35)] as [(6) from (5)].

The *fifteenth* mode is $D:E = (B:A).(C:F)$. Putting C and F between D and E,

(36)	$D:E = (D:C).(C:F).(F:E)$	II. 3
(37)	$(D:C).(F:E) = B:A$	(17)
(38)	$D:E = (B:A).(C:F)$	(36), (37).

The *sixteenth* mode, namely $D:E = (B:F).(C:A)$, is evident from the argument [leading from (5) to (6), and applied to (38)].

The *seventeenth* mode, namely $E:F = (A:B).(D:C)$, follows from the first mode as in the argument concerning the eleventh mode.[1] For since by hypothesis $A:B = (E:F).(C:D)$, therefore by the same argument $E:F = (A:B).(D:C)$.

The *eighteenth* mode, namely $E:F = (A:C).(D:B)$ is derived [from the seventeenth] by an argument of the type [leading from (4) to (5)].

[1] Now merely interchange C with E and D with F, throughout the proof.

II. 7

Quoniam autem demonstratum est unamquamque predictarum 9 compositarum proporcionum quas utiles esse prediximus duobus modis ex reliquis componi, cum ille ex quibus poterat unaquaque illarum intelligi componi essent 12, nunc restat ut ostendamus nullam istarum ex ceteris 10 posse componi. Et nobis de una earum pro omnibus sufficit demonstracio, nam in omnibus erit idem modus. Sumamus ergo eam que est *A* ad *B*, quam per secundam precedentis capituli monstravimus componi ex tercia *C* ad sexta *F* et ex quinta *E* ad quarta *D*, sicut componitur ex tercia ad quartam et quinta ad sextam.

Dico ergo eam, scilicet *A* ad *B*, non posse componi ex aliqua illarum decem; quia enim ipsa componitur ex *C* ad *D* et ex *E* ad *F*, non poterit componi ex *C* ad *D* et *F* ad *E*. Et insuper, quia componitur ex *C* ad *F* et *E* ad *D*, sicut patet in secundo modo, non potest componi ex *C* ad *F* et *D* ad *E*; quare neque componitur ex ea que est *C* ad *D* et *F* ad *E*. Et rursum quia non componitur ex ea que est *C* ad *D* et *F* ad *E*, non componitur ex ea que est *C* ad *E* et *F* ad *D*; alioquin, argumento secundi oportet ipsum componi ex *C* ad *E* et *F* ad *D* si componatur ex *C* ad *D* et *F* ad *E*.

Amplius ex quo componitur ex *C* ad *D* et *E* ad *F*, non componetur ex *D* ad *C* et *E*[a] ad *F*. Et racio quare ex *D* ad *C* et *F* ad *E* est eo quod eius conversa componitur ex eisdem, scilicet *B* ad *A*, ex quo sequitur quod nec[b] similiter componetur ex *D* ad *E* et ex *F* ad *C*; quia tunc per argumentum secundi componetur ex *D* ad *C* et ex *F* ad *E*, quod non potest, eo quod *B*, *A*, componitur ex hoc.

Item non potest *A* ad *B* componi ex *D* ad *F* et *E* ad *C*; nam tunc per argumenta secundi modi componetur ex *D* ad *C* et *E* ad *F*, quod fieri non potest, ut patet; at vero neque ex *E* ad *C* et *F* ad *D* componitur, quia tunc per argumentum secundi componetur ex *E* ad *D* et *F* ad *C*. Hoc autem probatum est non posse fieri, igitur neque ex *E* ad *D* et *F* ad *C*, quia tunc componetur ex *E* ad *C* et *F* ad *D*, quod non potest.

Et secundum hanc similitudinem probantur omnes 10 residue reliquarum non posse tenere.

[a] M *A* [b] M 28λ

II. 7

Since it has been shown that any one of the aforesaid nine ratios which we have said are useful is compounded in two ways out of the remainder, it now remains for us to show that if there were twelve ways in which any one of them might be known to be compounded, none may in fact be compounded in the other ten ways. If we show this for one, then this is sufficient proof for all, for the method will always be the same. Let us take, therefore, $A:B$, which by the second mode of the last chapter we showed to be compounded from $C:F$ and $E:D$ (the third to the sixth and the fifth to the fourth), just as it is compounded of the third to the fourth and the fifth to the sixth.[1]

I say then that $A:B$ may not be compounded in any other of those ten ways, for since it is compounded of $C:D$ and $E:F$, it may not be compounded of $C:D$ and $F:E$. And in addition, since it is compounded from $C:F$ and $E:D$ (second mode), it may not be compounded of $C:F$ and $D:E$.[2] And since it is not compounded of $C:D$ and $F:E$, neither is it of $C:E$ and $F:D$; for otherwise we should be able to pass from $(C:D).(F:E)$ to $(C:E).(F:D)$ by an argument such as was used to obtain the second mode.

Moreover, from the fact that $A:B = (C:D).(E:F)$, it follows that $A:B \neq (D:C).(E:F)$. And from $B:A = (D:C).(F:E)$, it follows that $A:B \neq (D:E).(F:C)$. For otherwise, by the argument leading to the second mode, we should have $A:B = (D:C).(F:E)$, which is not possible, since $B:A$ is compounded in this way.

Next, $A:B = (D:F).(E:C)$ is impossible; for otherwise, by the argument of the second mode, we should have $A:B = (D:C).(E:F)$, which cannot be admitted, as is clear. And neither can we have

$$A:B = (E:C).(F:D),$$

or by the argument of the second mode we should have

$$A:B = (E:D).(F:C);$$

but it has been proved that this is impossible, therefore neither is the [premiss] possible.

And in like manner it is proved impossible for any of the ten [not considered thus far] to hold.

[1] That is, in the basic first mode.

[2] The words 'wherefore neither is it compounded from $C:D$ and $F:E$' are omitted here, as redundant.

II. 8

Determinatis igitur 9 modis utilibus, restat ostendere reliquorum[a] 6 inutilitates quos[b] ex nullis aliorum monstrabimus esse compositos.[c] Hec[d] autem sunt proporcio *A* ad *D*, et *A* ad *F*, et *B* ad *C*, et *B* ad *E*, et *C* ad *E*, et *D* ad *F*. Ostendemus primo proporcionem *A* ad *D* ex nullis aliarum posse componi.

Ducatur enim *A* in *D* et fiat *G*, qui dividatur per aliquod maius vel minus ipso *A*, et[e] ipsum dividens[f] sit *H*; et quod exit per talem divisionem sit *K*. Erit ergo *K* maius vel minus ipso *D*. Quoniam enim ex *H*[g] in *K* fiat *G*, ex prima[h] proporcione huius, et ex *A* in *D* fiat idem, scilicet *G*, per ypotesim, erit *H* ad *A* sicut *D* ad *K*; quare[i] cum *H* sit maior vel minor ipso *A* erit *D* maior vel minor ipso *K*.

Posito igitur *A* inter *H* et *B*, erit *H* ad *B* composita ex *H* ad *A* et *A* ad *B*. Sed *A* ad *B* est ut *C* ad *D* et *E* ad *F*; et *H* ad *A* est ut *D* ad *K*. Ergo[j] *H* ad *B* composita erit ex *D* ad *K* et *C* ad *D* et *E* ad *F*. Sed *C* ad *D* et *D* ad *K* sunt ut *C* ad *K*; ergo necessario *H* ad *B* constat ex *C* ad *K* et *E* ad *F*. Si ergo prime ad quartam proporcio componeretur ex aliquibus duabus proporcionibus existentibus inter reliqua extrema, sit *H* ad *K* tanquam *A* ad *D*; cum enim cetera quatuor extrema utrobique simpliciter maneant et idem[k] ordo eorum, ex quo componitur proporcio *A* ad *D* secundum petitionem hanc, ex eisdem componitur proporcio *H* ad *K*. Erit ergo *H* ad *K* ut *A* ad *D*; permutatim *H* ad *A* ut *K* ad *D*. Sed *H* ad *A* erat ut *D* ad *K*, ut patet, ergo sequitur quod proporcio *K* ad *D* erit sicut proporcio *D* ad *K*, et per consequens non erit *K* maior *D* nec minor, quare[l] neque ipse *A* erit [*H*] maior aut minor, cuius tamen oppositum suppositum est.

[a] CM reliquarum [b] CM quas [c] C compositas [d] CM hee
[e] M in [f] C 3λ; M λ [g] C +primo [h] C λ
[i] M 72λ [j] C 14λ [k] M racionem [l] M 3λ+quia in ipsa

II. 8

The nine useful modes having been determined, it remains to show the remaining six inapplicable. These six we shall prove to be [derivable from] none of the others.[1] They are, however, the ratios $A:D$, $A:F$, $B:C$, $B:E$, $C:E$, and $D:F$.[2] We will first show that $A:D$ may not be compounded out of any other [ratios between the other four terms].

Let $A.D = G$ and let G be divided by anything greater than or less than A, and let this divisor be H, and the result of the division K. Then K will be greater than or less than D. Since $H.K = G$, from the first ratio,[3] and $A.D = G$, by hypothesis, then $H:A = D:K$; wherefore since H is greater than or less than A, D will be greater than or less than K.

Having placed A [as a mean] between H and B, we shall have

(1)	$H:B = (H:A).(A:B)$	II. 2
(2)	$A:B = (C:D).(E:F)$	first mode
(3)	$H:A = D:K$	hyp.
(4)	$H:B = (D:K).(C:D).(E:F)$	(1), (2), (3)
(5)	$(C:D).(D:K) = C:K$	II. 2
(6)	$H:B = (C:K).(E:F)$	(4), (5).

If, therefore, the ratio of the first to the fourth[4] were compounded from a certain two ratios existing between any of the other extrema, $H:K$ would be as $A:D$; for the other four extrema on both sides[5] remain simply the same,[6] and likewise in their order, the order being that from which $A:D$ is compounded according to our premiss,[7] so $H:K$ will be compounded from the same [elements in the same way]. Thus

(7)	$H:K = A:D$	
(8)	$H:A = K:D$	(7), 'permutatim'.

But $H:A = D:K$, as is clearly the case,[8] and so it follows that

$$K:D = D:K,$$

and in consequence K will be neither greater than nor less than D,

[1] The sentence is written in this way for the sake of consistency in interpreting the sentence of the Latin text, in which the scribes have floundered with a variety of noun endings. '*Aliarum*' and '*reliquarum*' are no doubt correct, since the author is thinking of writing six (and nine) *ratios* in terms of other ratios; but he begins the sentence by speaking of modes, and this is how we have continued it.

[2] It is a matter of definition that the inverses of the ratios are inapplicable (*inutilis*). Briefly, the standard (first) mode may be written $\frac{B}{A}\cdot\frac{C}{D}\cdot\frac{E}{F} = 1$, from which it is impossible to find ratios between any two out of B, C, and E, or any two out of A, D, and F.

[3] Note that he does not write '*ex prima proporcionalitate*'.

[4] This is an oblique way of referring to $A:D$ in the standard first mode, or $H:K$, in (6), for example.

[5] *utrobique*—a reference, perhaps, to the two 'sides' of the verbal equivalent of our equation (6).

[6] As between the first mode and (6).

[7] *peticio*—apparently a reference to the first mode, which would have been seen as a fundamental hypothesis (cf. (2)).

[8] This was shown at the outset, as a consequence of the definition of H and K.

Quaslibet[a] enim duas proporciones quatuor terminorum tantum 12 modis posse fieri monstramus, *H* autem esse maius aut minus *A* infinitis potest modis contingere, quare *H* plusquam 12 modis assumpto. Cum semper fiat *H* ad *K* ex aliquibus duarum proporcionum ipsorum 12 existentium inter quatuor residuos terminos, scilicet *B*, *C*, *E*, *F*, necesse erit duas earum inter *H* et *K* plures assumi; ex quo sequitur quod erit *H* dividentis, ergo[b] *K* exiens, sicut econverso *K* dividentis et[c] *H* exiens; quod est impossibile, quia quanto maius est dividens, tanto minus est exiens, quare oportet minora maioribus adequari cum regula sit generalis quod 4 numerorum proporcionalium productum ex primo in quartum sit equale producto ex secundo in tercium (ut si sit *C* ad *D* sicut *E* ad *F* erit productum ex *C* in *F* equale producto ex *D* in *E*). Non igitur componetur[d] proporcio primi ad quartum ex aliquibus duabus existentibus inter[e] quatuor terminos residuos proporcionibus.

Per hoc insuper manifestum erit quod etiam proporcio *A* ad *F* ex nullis reliquorum terminorum componi potest. Constat enim per secundum modum precedentis capituli quod *A* ad *B* componitur ex *C* ad *F* et *E* ad *D*. Sed iam monstravimus quod primi *A* ad quartum *D* proporcio ex nullis terminis quatuor reliquorum posset componi, quare neque *A* ad *F* proporcio potest ex aliquibus eorum[f] componi.

Similiter quoque, ac eadem racione, palam est quod *B* ad *C* ex nullis aliorum quatuor terminorum proporcionibus componi potest. Et hoc quia, cum *A* ad *B* constet ex *C* ad *D* et *E* ad *F*, constabit *B* ad *A* ex *D* ad *C* et *F* ad *E*; quare sequitur quod *B* primi ad *C* quartum non componetur ex proporcionibus aliquibus inter 4 terminos residuos; nec, per consequens, *B* ad *E*. Constat enim quod *B* ad *A* fit ex *F* ad *E* et *D* ad *C*, quare primi *B* ad quartum *E* non constabit proporcio ex aliquibus residuorum quatuor. Item *C* ad *E* proporcio ex nullis residuorum componi potest duabus[g] proporcionibus, cum enim, ex argumento duodecimi modi, *C* ad *D* componitur ex *A* ad *E* et *F* ad *B*, palam est ex his que hic tociens dicta sunt ⟨quod⟩ proporcio *C* primi ad *E* quartum non componetur ex residuis. Item, nec eciam ultimum, quod

[a] CM 3λ+Quelibet enim due [b] M +*G* ad [c] CM ad [d] C proponetur
[e] M λ [f] C earum [g] M 2λ+duorum proporcionalium

wherefore also A will be neither greater nor less [than H]—and yet the opposite of this was supposed to be the case.

We proved that any two ratios of four terms may be in 12 different modes, but that H is greater or less than A may come about in an infinite number of ways; wherefore, by the assumption, H [may enter into][1] more than 12 modes. If $H:K$ is always compounded out of certain others of the two ratios from among these same 12, existing between the four remaining terms, namely B, C, E, and F, it will be necessary that *two* more such ratios be assumed. It follows that [in one of the relations] H will be the divisor and K the numerator,[2] [while in the other], conversely, K will be divisor and H the numerator. This is impossible, since inasmuch as the divisor is greater, by a corresponding amount is the result less; and therefore lesser quantities must be equated to greater, if it be a general rule that, of four proportional numbers, the product of the first and fourth is equal to the product of the second and third (as when $C:D = E:F$ then $C.F = D.E$). The ratio, therefore, of the first to the fourth may be compounded from no two others existing between the four remaining terms.

From this it will be manifest in addition that $A:F$ may not be compounded from the remaining terms,[3] for by the second mode of the last chapter $A:B = (C:F).(E:D)$. But we have already shown that the ratio of the first (A) to the fourth (D) may not be compounded from the remaining four terms, and therefore neither may $A:F$ be compounded from any of them.

Similarly too, and by the same argument, it is clear that $B:C$ may be compounded out of no ratios between the remaining four terms. This is because if $A:B = (C:D).(E:F)$, then $B:A = (D:C).(F:E)$. From this it follows that the ratio of the first (B) to the fourth (C) is not compounded from ratios between the four remaining terms; nor, in consequence, is $B:E$. For $B:A = (F:E).(D:C)$, and thus $B:E$ will not stand compounded out of the remaining four. Next, $C:E$ may be compounded out of no two proportions between the remaining [terms]. For if from the argument on the twelfth mode [we take] $C:D = (A:E).(F:B)$, it is clear from what has so often been said here that the ratio of C (the first) to E (the fourth) is not compounded from the rest. Next, and[4] also last,

[1] This passage I find obscure.

[2] *exiens*. Literally 'result'; but there are what appear to be lapses on the part of the author within this sentence, reflected in variants to the MS. readings.

[3] Literally (here and elsewhere) 'may be compounded from none of the remaining terms'; but this is not what was intended.

[4] *nec*—an ungrammatical but understandable introduction to the negation which follows, and of a sort often encountered in medieval Latin.

est *D* ad *F* proporcio non potest componi ex alia proporcione inter residuos; nam ex 16° modo palam est quod *D* ad *E* componitur ex *B* ad *F* et *C* ad *A*, quare *D* ad *F* non componetur ex residuorum quatuor proporcionibus.

Et hec[a] de propositis conclusionibus sufficiunt.

Explicit[b] secundus tractatus

III

Et hic incipit tercius tractatus.

SI FUERINT sex quantitates continue, sive linee, vel superficies, erit proporcio prime ad secundam composita ex proporcione tercie ad quartam et ex proporcione quinte ad sextam, si alie quatuor quantitates ponantur inter primam et secundam, quia eadem est racio in continuis et discretis. Et fiunt 36 proporciones, quarum 18[c] sunt coniuncte et 18 disiuncte sive distincte, secundum quod monstravimus in numeris. Suppono ergo proposicionem unam, scilicet ex 4ª libri[d] sexti Euclidis, quod si alique due linee parallele exierint a duobus punctis alicuius linee, et aliqua tercia linea exierit a tercio puncto eiusdem linee secans parallele, quod talis erit proporcio unius linee parallele ad aliam, qualis est proporcio linee[e] totius secantis ad alteram eius partem priorem.

III. C. I

Ponam igitur duas lineas exeuntes ab *A* puncto, que[f] sunt *AC*, *AD*, et alias duas lineas exeuntes a terminis *C* et *D* et intersecantes se in *G* puncto, que erunt[g] *CE* et *DB*. Dico ergo primo quod proporcio linee *AC* prime ad lineam *AB* secundam componitur ex proporcione *CE* tercie ad *GE* quartam et ex proporcione *DG* quintam ad *DB* sextam, quod sic probatur:

Protraham enim a puncto *B* unam lineam *BF*[h] equidistantem linee *CG*,[i] et ponam lineam *GE* mediam in proporcione inter *CE* et *BF*.

[a] M *H* [b] C 3λ [c] CM 8 [d] C 2λ [e] C λ
[f] CM qui [g] M erit [h] M *DF* [i] C *EG*

$D:F$ may not be compounded as a ratio between the remainder; for from the sixteenth mode it is clear that $D:E = (B:F).(C:A)$, and therefore that $D:F$ may not be compounded from ratios between the remaining four [terms].

And enough has now been said concerning the propositions.

Here ends the second part

III

Here begins the third part.

IF there are six continuous quantities, or lines, or areas, the ratio of the first to the second will be compounded from the ratio of the third to the fourth and from the ratio of the fifth to the sixth, when four other quantities are placed between the first and second, the reasoning[1] being the same for continuous as for discrete quantities. There are 36 relations, of which 18 are conjoint and 18 disjoint (or distinct), as we showed for numbers.[2] I presuppose now Euclid VI. 4: If any two parallel lines leave two points of any line, and a third line leaves a third point of the same line and cuts the parallels, whatever be the ratio of one parallel line to the other, the ratio of the corresponding intercepts on the third line will be the same.[3]

III. C. 1[4]

I take two lines *AC* and *AD*, leaving the point *A*, and two other lines *CE* and *DB*, leaving the termini *C* and *D* and meeting at the point *G*. I say then, first, that the ratio of the first line *AC* to the second line *AB* is compounded of the ratios *CE* (third) to *GE* (fourth) and *DG* (fifth) to *DB* (sixth), which is proved thus:

I shall draw a line *BF* from the point *B* parallel to line *CG*, and I shall set line *GE* as a mean in the ratio between *CE* and *BF*.

[1] *racio.*

[2] In Part II.

[3] Some MSS. quote here Euclid VI. 2 rather than VI. 4. The latter is more appropriate to our column of justifications below, as being somewhat more general (similar triangles have corresponding sides proportional).

[4] This third part of the work contains thirty-six chapters, for clarity numbered here C. 1–C. 18 and D. 1–D. 18. The first eighteen concern what we call the theorem of Menelaus, where the transversal (*AED*) cuts all three sides of the triangle externally (*cata coniuncta*), while D. 1–D. 18 deal with cases where only one side is cut externally (*cata disiuncta*).

Argumento igitur ex secunda secunde huius tractatus, proporcio *CE* ad *BF* fit ex proporcione *CE* ad *GE* et ex *GE* ad *BF*. Sed per quartam sexti libri [Euclidis] proporcio *GE* ad *BF* est sicut proporcio *DG*[a] ad *DB*;[b] ergo proporcio *CE* ad *BF* componitur ex proporcione *CE* ad *GE* et ex proporcione *DG* ad *DB*. Sed notum est ex 4ª sexti quod proporcio *CE* ad *BF* est sicut proporcio *AC* ad *AB*; ergo proporcio *AC* ad *AB* est composita ex proporcione *CE* ad *GE* et ex proporcione *DG* ad *DB*. Et hoc est quod probare volebamus.

Et est hec prima proposicio in continuis, que eadem racione probatur sicut prima probatur in numeris. Et eadem est regula utrobique scientibus applicare.

III. c. 2

Proposicio secunda est in qua maneat eadem disposicio que prius. Dico ergo quod proporcio *AC* ad *AB* componitur ex proporcione *CE* ad *DB* et ex proporcione *DG* ad *GE*. Et hec est[c] similis determinacio secunde demonstracioni capitulo sexto[d] in numeris, que sic probatur:

Ponamus *BD* inter *CE* et *BF*, igitur proporcio *CE* ad *BF* componitur ex proporcione *CE* ad *DB*[e] et ex proporcione *DB* ad *BF*. Sed proporcio *CE* ad *BF* est sicut proporcio *AC* ad *AB*; ergo proporcio *AC* ad *AB* componitur ex proporcione *CE* ad *DB* et proporcione *DB* ad *BF*. Sed proporcio *BD* ad *BF* est sicut proporcio *GD* ad *GE*, ergo proporcio *AC* ad *AB* componitur ex proporcione *CE* ad *DB* et ex proporcione *DG* ad *GE*. Et hoc probare volebamus.[f]

III. c. 3

Proposicio tercia est quod proporcio *AC* ad *CE* componitur ex proporcione *AB* ad *GE* et ex proporcione *DG* ad *DB*, quod sic probatur:

Ponam enim lineam *GE* mediam in proporcione inter *AB* et *BF*; igitur proporcio *AB* ad *BF* componitur ex proporcione *AB* ad *GE* et *GE* ad *BF*. Sed proporcio *GE* ad *BF* est sicut proporcio *DG* ad *DB*; ergo proporcio *AB* ad *BF* componitur ex proporcione *AB* ad *GE* et proporcione *DG* ad *DB*. Sed *AB* ad *BF* est sicut proporcio *AC* ad *CE*; ergo proporcio *AC* ad *CE* composita est ex proporcione *AB* ad *GE* et proporcione *DG* ad *DB*. Et hoc est propositum huius tercie conclusionis.

III. c. 4

Proporcio quarta est quod proporcio *AC* ad *CE* composita est ex *AB* ad *DB* et ex *DG* ad *GE*, quod sic probatur:

[a] CM *DB* [b] CM *GD* [c] M λ [d] C λ; M suo
[e] C +*BF* componitur ex proporcione *CE* ad *DB* (dittogr.)
[f] M here interpolates a very confused explanatory passage: Nota circa 6 . . . *CE* ad *DB*

(1) $CE:BF = (CE:GE).(GE:BF)$ II. 2
(2) $GE:BF = DG:DB$ Euclid VI. 4
(3) $CE:BF = (CE:GE).(DG:DB)$ (1), (2)
(4) $CE:BF = AC:AB$ Euclid VI. 4
(5) $AC:AB = (CE:GE).(DG:DB)$ (3), (4).

And this is what we wished to prove.

This is the first proposition on continuous magnitudes, proved by the same reasoning as the first on numbers. And the rule is the same for both, for those who know how to apply it.

III. C. 2

The second proposition concerns the same arrangement as the first. I say then that $AC:AB = (CE:DB).(DG:GE)$. The proof is similar to the second demonstration in II. 6, and is proved as follows. We put BD [as a mean in the proportion] between CE and BF.

(1) $CE:BF = (CE:DB).(DB:BF)$ II. 2
(2) $CE:BF = AC:AB$ Euclid VI. 4
(3) $AC:AB = (CE:DB).(DB:BF)$ (1), (2)
(4) $BD:BF = GD:GE$ Euclid VI. 4[1]
(5) $AC:AB = (CE:DB).(DG:GE)$ (3), (4).

And this is what we wished to prove.

III. C. 3

The third proposition is that $AC:CE = (AB:GE).(DG.DB)$, and is proved thus. I shall put line GE as a mean in the proportion between AB and BF.

(1) $AB:BF = (AB:GE).(GE:BF)$ II. 2
(2) $GE:BF = DG:DB$ Euclid VI. 4
(3) $AB:BF = (AB:GE).(DG:DB)$ (1), (2)
(4) $AB:BF = AC:CE$ Euclid VI. 4
(5) $AC:CE = (AB:GE).(DG:DB)$ (3), (4).

And this is what was proposed as the third *conclusio*.

III. C. 4

The fourth proposition is that $AC:CE = (AB:DB).(DG:GE)$, and is proved thus. I shall put line GF [in a new figure] parallel to the line AC,

[1] No explicit allusion is made to similar triangles, although a spurious interpolation at III. D.3 in some MSS. uses the idea.

Ponam enim lineam *GF* equidistantem linee *AC*, et ponam lineam *DG* mediam in proporcione inter *GF* et *GE*. Proporcio igitur *GF* ad *GE* componitur ex proporcione *GF* ad *DG* et *DG* ad *GE*. Sed proporcio *GF* ad *GD* est sicut proporcio *AB* ad *BD*; igitur proporcio *GF* ad *GE* componitur ex proporcione *AB* ad *BD* et ex proporcione *GD* ad *GE*. Sed proporcio *GF* ad *GE* est sicut proporcio *CA* ad *CE*; ergo proporcio *AC* ad *CE* componitur ex proporcione *BA* ad *BD* et ex proporcione *GD* ad *GE*. Et id est quod proposuimus.

Et nota quod proporcio *AC* prime ad *GE* quartam non componitur ex aliquibus duabus proporcionibus reliquorum quatuor terminorum, sicut etiam ostendimus in numeris loco suo: et ideo non oportet hic ostendere nisi utiles modos 9 et eorum dupplicem modum, sicut ibi ostendebamus suppositiones sufficienter esse detractas de 6 modis inutilibus, quia in omni ordine posito quod primi ad 4 ostendatur inutilis coniugato patet quod primi ad 6 erit inutilis.

III. c. 5

Proposicio quinta. Et dico quod *AC* ad *GD*[a] componitur ex *AB* ad *BD* et ex *CE* ad *GE*, cuius probacio est hec:

Protraham lineam *GF* equidistantem *AC* ut prius, et ponam *GF* mediam in proporcione inter *AC* et *GD*, igitur proporcio *AC* ad *GD* est composita ex proporcione *AC* ad *GF* et *GF* ad *GD*. Sed proporcio *AC* ad *GF*[b] est sicut proporcio *CE* ad *GE*, et proporcio *GF* ad *GD* est sicut proporcio *AB* ad *BD*, ergo proporcio *AC* ad *GD* componitur ex proporcione *AB* ad *BD* in proporcione *CE* ad *GE*.

III. c. 6

Proposicio sexta, pro qua dico quod proporcio *AC* ad *GD*[c] componitur ex *AB* ad *GE* et ex *CE* ad *BD*, quod sic probatur:

Ponam enim *BF* equidistantem *CE*, et ponam duas lineas *CE* et *BD* medias inter *AC* et *GD*: igitur proporcio *AC* ad *GD* est composita ex proporcione *AC* ad *CE* et *CE* ad *BD* et *BD* ad *GD*. Sed proporcio *AC* ad *CE* est sicut proporcio *AB* ad *BF*,[d] et proporcio *BD* ad *GD* est sicut proporcio *BF* ad *GE*. Sed proporcio *BA* ad *GE* componitur ex proporcione *AB* ad *BF* et ex proporcione *BF* ad *GE*; ergo proporcio *AB* ad *GE* componetur ex proporcione *AC* ad *CE* et ex proporcione *BD* ad *GD*. Ergo proporcio *AC* ad *GD* est composita ex proporcione *BA* ad *GE* et proporcione *CE* ad *BD*. Et hoc est quod proposuimus. Et nota quod

[a] M *AD* [b] CM ɪɪλ [c] C *ED* [d] CM *BD*

and the line DG as a mean in the proportion between GF and GE,

(1)	$GF:GE = (GF:DG).(DG:GE)$	II. 2
(2)	$GF:GD = AB:BD$	Euclid VI. 4
(3)	$GF:GE = (AB:BD).(GD:GE)$	(1), (2)
(4)	$GF:GE = CA:CE$	Euclid VI. 4
(5)	$AC:CE = (BA:BD).(GD:GE)$	(3), (4).

And this is what we proposed.

And note that the ratio of the first[1] (AC) to the fourth (GE) is not compounded from any two ratios involving the other four terms, as we also showed for numbers in the appropriate place: and therefore it is unnecessary to show here any but the nine useful modes and the double of those modes, just as we showed there sufficiently well the invalidity of the six useless modes. Whatever their order [in a valid proportionality], the ratio of the first to the fourth being assumed inapplicable,[2] that of the first to the sixth will also be so.

III. c. 5

The fifth proposition. And I say that $AC:GD = (AB:BD).(CE:GE)$, the proof being as follows. I shall draw the line GF parallel to AC, as before, and I shall put the mean GF in the proportion between AC and GD, and therefore

(1)	$AC:GD = (AC:GF).(GF:GD)$	II. 2
(2)	$AC:GF = CE:GE$	Euclid VI. 4
(3)	$GF:GD = AB:BD$	Euclid VI. 4
(4)	$AC:GD = (AB:BD).(CE:GE)$	(1), (2), (3).

III. c. 6

The sixth proposition, which I express $AC:GD = (AB:GE).(CE:BD)$, and which is proved thus: I shall draw BF parallel to CE, and shall put the two lines CE and BD as means between AC and GD.

(1)	$AC:GD = (AC:CE).(CE:BD).(BD:GD)$	II. 3
(2)	$AC:CE = AB:BF$	Euclid VI. 4
(3)	$BD:GD = BF:GE$	Euclid VI. 4
(4)	$BA:GE = (AB:BF).(BF:GE)$	II. 2
(5)	$AB:GE = (AC:CE).(BD:GD)$	(2), (3), (4)
(6)	$AC:GD = (BA:GE).(CE:BD)$	(1), (5),

which is what we proposed.

[1] The numbering here and throughout is that of the basic relation (III. c. 1), namely $AC:AB = (CE:GE).(DG:DB)$.

[2] Or 'not derivable' (*inutilis*).

AC ad *BD* non componitur ex aliquibus duabus proporcionibus reliquarum quantitatum pro eo quod est unus de modis inutilibus.

III. c. 7

Proposicio septima, in qua dico quod proporcio *BA* ad *GE* componitur ex proporcione *AC* ad *CE* et proporcione *BD* ad *GD*, quod sic probatur:

Supposita[a] figura precedenti, in qua protrahitur linea *BF* equidistans linee *EC*; et ponam eam in proporcione mediam inter *AB* et *GE*. Proporcio ergo *AB* ad *GE* componitur ex *AB* ad *BF* et *BF* ad *GE*. Sed proporcio *AB* ad *BF* est sicut proporcio *AC* ad *CE*, et proporcio *BF* ad *GE* est sicut *BD* ad *GD*. Ergo proporcio *AB* ad *GE* componitur ex *AC* ad *CE* et ex *BD* ad *GD*. Et hoc est quod volui ostendere.

III. c. 8

Proposicio octava est hec, quod proporcio *AB* ad *GE* componitur ex *AC* ad *GD* et ex *BD* ad *CE*, quod sic probatur:

Protrahatur iterum linea *BF* equidistans linee *CE*, et ponam illam mediam in proporcione inter *BD* et *CE*: igitur proporcio *BD* ad *CE* fit ex *BD* ad *BF* et *BF* ad *CE*. Sunt autem linee due *AC*, *GD* medie inter *BA* et *GE*: patet ergo quod proporcio *BA* ad *GE* componitur ex *BA* ad *AC* et *AC* ad *GD* et *GD* ad *GE*. Sed proporcio *BA* ad *CA* est sicut proporcio *BF* ad *CE*; et proporcio *GD* ad *CE*[b] est sicut proporcio *BD* ad *BF*. Ergo proporcio *BD* ad *CE* est composita ex proporcione *BA* ad *AC* et ex *GD* ad *GE*; ergo proporcio *BA* ad *GE* fit ex *AC* ad *GD* et ex *BD* ad *CE*. Et hoc voluimus probare.

Nota etiam hic quod proporcio *BA* ad *DG* non componitur ex aliquibus quatuor reliquarum proporcionibus, sicut patet in priori tractatu.

III. c. 9

Proposicio nona, in qua ostendo quod proporcio *BA* ad *BD* componitur ex *AC* ad *CE* et ex *GE* ad *GD* quod sic patet:

Protraho lineam *BF* equidistantem *CE*, omnino ut prius, quam pono mediam inter *BA* et *BD*: igitur proporcio *BA* ad *BD* componitur ex *BA* ad *BF* et *BF* ad *BD*. Sed proporcio *BA* ad *BF* est sicut proporcio

[a] M supponam [b] C *GE*

And note that $AC{:}BD$ is not compounded from any two ratios among the remaining quantities, since that is one of the inapplicable modes.

III. c. 7

The seventh proposition, in which I say that

$$BA:GE = (AC:CE).(BD:GD),$$

is proved as follows. I shall go back to the preceding figure, in which the line BF is parallel to EC, and shall put [BF] as a mean in the proportion between AB and GE.

(1) $AB:GE = (AB:BF).(BF:GE)$	II. 2
(2) $AB:BF = AC:CE$	Euclid VI. 4
(3) $BF:GE = BD:GD$	Euclid VI. 4
(4) $AB:GE = (AC:CE).(BD:GD)$	(1), (2), (3).

And this is what I wanted to show.

III. c. 8

The eighth proposition is this, that $AB:GE = (AC:GD).(BD:CE)$, which is proved as follows. Again let the line BF be drawn parallel to CE, and I shall put [BF] as a mean in the proportion between BD and CE.

(1) $BD:CE = (BD:BF).(BF:CE)$	II. 2
(2) $BA:GE = (BA:AC).(AC:GD).(GD:GE)$	II. 3
(3) $BA:CA = BF:CE$	Euclid VI. 4
(4) $GD:GE = BD:BF$	Euclid VI. 4
(5) $BD:CE = (BA:AC).(GD:GE)$	(1), (3), (4)
(6) $BA:GE = (AC:GD).(BD:CE)$	(2), (5).

And this we wished to prove.

Note also here that $BA:DG$ is not compounded from the ratios of the four others, as is evident from the earlier treatise.

III. c. 9

The ninth proposition, in which I show that

$$BA:BD = (AC:CE).(GE:GD),$$

which is evident thus: I draw the line BF parallel to CE, as at first, and put [BF] as a mean between BA and BD. Then

(1) $BA:BD = (BA:BF).(BF:BD)$	II. 2
(2) $BA:BF = AC:CE$	Euclid VI. 4

AC ad *CE*; et proporcio *BF* ad *BD* est sicut proporcio *GE* ad *GD*. Igitur proporcio[a] *BA* ad *BD* componitur ex *AC* ad *CE* et *GE* ad *GD*. Et hoc voluimus ostendere.

III. C. 10

Proposicio decima est in qua ostendam quod proporcio *AB* ad *BD* componitur ex *AC* ad *GD* et ex *GE* ad *CE*, quod sic patet:

Protraho lineam *GF* equidistantem linee *AC*, et ponam *AC* mediam in proporcione inter *GF* et *GD*; proporcio ergo *GF* ad *GD* fit ex proporcione *GF* ad *AC* et *AC* ad *GD*. Sed proporcio *GF* ad *AC* est sicut proporcio *GE* ad *CE*; igitur proporcio *GF* ad *GD* fit ex *GE* ad *CE* et *AC* ad *GD*. Sed *GF* ad *GD* est sicut proporcio *BA* ad *BD*, ergo proporcio *BA* ad *BD* componitur ex proporcione *GE* ad *CE* et ex proporcione *AC* ad *GD*, quod proposuimus.

III. C. 11

Proposicio undecima in qua patebit quod proporcio *CE* ad *GE* fit ex proporcione *AC* ad *BA* et ex *BD* ad *GD*, quod sic probatur:

Protraham lineam *BF* equidistantem *CE*, quam pono mediam inter *CE* et *GE*; igitur *CE* ad *GE* fit ex *CE* ad *BF* et ex *BF* ad *GE* proporcionibus. Sed proporcio *CE* ad *BF* est sicut *AC* ad *BA*; et proporcio *BF* ad *GE* est sicut *BD* ad *GD*. Igitur *CE* proporcio ad *GE* componitur ex *AC* ad *BA* et ex *BD* ad *GD*. Et hoc est quod querebamus.

III. C. 12

Proposicio duodecima, in qua patebit quod *CE* ad *GE* fit ex *AC* ad *GD* et ex *BD* ad *BA*, quod sic probatur:

Protrahatur linea *GF* equidistans *AC*, et ponam *GD* mediam in proporcione inter *AC* et *GF*: igitur proporcio *AC* ad *GF* fit ex *AC* ad *GD* et *GD* ad *GF*. Sed proporcio *GD* ad *GF* est sicut *BD* ad *BA* et proporcio *CA* ad *GF* est sicut proporcio *CE* ad *GE*. Igitur proporcio *CE* ad *GE* componitur ex *AC* ad *GD* et ex *BD* ad *BA*, et hoc voluimus.

[a] CM + *GE* ad *GD* est proporcio

(3) $BF:BD = GE:GD$ Euclid VI. 4
(4) $BA:BD = (AC:CE).(GE:GD)$ (1), (2), (3).

And this we wished to show.

III. c. 10

The tenth proposition is that in which I shall show that

$$AB:BD = (AC:GD).(GE:CE),$$

which is evident thus. I draw the line *GF* parallel to *AC*, and put *AC* as a mean in the proportion between *GF* and *GD*. Therefore

(1) $GF:GD = (GF:AC).(AC:GD)$ II. 2
(2) $GF:AC = GE:CE$ Euclid VI. 4
(3) $GF:GD = (GE:CE).(AC:GD)$ (1), (2)
(4) $GF:GD = BA:BD$ Euclid VI. 4
(5) $BA:BD = (GE:CE).(AC:GD)$ (3), (4),

which we proposed.

III. c. 11

The eleventh proposition, in which it will be evident that

$$CE:GE = (AC:BA).(BD:GD),$$

which is proved as follows. I shall draw the line *BF* parallel to *CE*, and I put [*BF*] as a mean between *CE* and *GE*. Therefore

(1) $CE:GE = (CE:BF).(BF:GE)$ II. 2
(2) $CE:BF = AC:BA$ Euclid VI. 4
(3) $BF:GE = BD:GD$ Euclid VI. 4
(4) $CE:GE = (AC:BA).(BD:GD)$ (1), (2), (3).

And this is what we sought.

III. c. 12

The twelfth proposition, where it will be evident that

$$CE:GE = (AC:GD).(BD:BA),$$

which is proved as follows. Let the line *GF* be drawn parallel to *AC*, and I shall make *GD* a mean in the proportion between *AC* and *GF*. Then

(1) $AC:GF = (AC:GD).(GD:GF)$ II. 2
(2) $GD:GF = BD:BA$ Euclid VI. 4
(3) $CA:GF = CE:GE$ Euclid VI. 4
(4) $CE:GE = (AC:GD).(BD:BA)$ (1), (2), (3).

And this we wanted.

Et nota quod proporcio *CE* ad *GD* non componitur ex proporcionibus aliquarum quantitatum reliquarum.

III. c. 13

Proposicio terdecima est hec quod *CE* ad *BD* componitur ex *AC* ad *BA* et ex *GE* ad *GD* quod sic probatur:

Protrahatur linea *BF*[a] equidistans *CE*, ut in penultima figura omnino quam pono mediam in proporcione inter *CE* et *DB*: igitur proporcio *CE* ad *BD* fit ex *CE* ad *BF* et ex *BF* ad *BD*. Sed *AC* ad *AB* est sicut *CE* ad *BF*,[b] et *GE* ad *GD* est sicut *BF* ad *BD*. Igitur *CE* ad *BD* fit ex *AC* ad *AB*, et *GE* ad *GD*, quod voluimus ostendere.

III. c. 14

Proposicio quartadecima, quod *CE* ad *BD* componitur ex *AC* ad *GD* et *GE* ad *BA* quod sic patet:

Protrahatur linea *GF* equidistans linee *AC*, et ponam duas lineas *GE* et *BA* in proporcione medias inter duas lineas *CE* et *BD*: igitur proporcio *CE* ad *BD* fit ex *CE* ad *GE* et *GE* ad *BA* et ex *BA* ad *BD*. Sed proporcio *CE* ad *GE* est sicut proporcio *AC* ad *GF*; et proporcio *BA* ad *BD* est sicut proporcio *GF* ad *GD*. Sed proporcio *AC* ad *GD* componitur ex *AC* ad *GF* et *GF* ad *GD*, igitur proporcio *CE* ad *BD* fit ex *AC* ad *GD* et ex *GE* ad *BA* quod probare[c] promisimus.

III. c. 15

Proposicio quindecima est hec quod *GE* ad *GD* proporcio fit ex *AB* ad *AC* et ex *CE* ad *BD*, quod sic probatur:

Protraham enim *BF* equidistantem *CE*, et ponam *CE* in proporcione mediam inter *BF* et *BD*. Fit igitur proporcio *BF* ad *BD* ex proporcione *BF* ad *CE* et *CE* ad *BD*. Sed proporcio *BF* ad *CE* est sicut *BA* ad *AC*, igitur *BF* ad *BD* componitur ex proporcione *BA* ad *AC* et ex *CE* ad *BD*. Sed *BF* ad *BD* est sicut *GE* ad *GD*; igitur proporcio *GE* ad *GD* componitur ex *AB* ad *AC* et ex *CE* ad *BD*, quod monstrare spopondimus.

[a] M λ   [b] M *BD*   [c] 3 2λ+volemus

Note that $CE:GD$ is not compounded of ratios between any of the remaining quantities.

III. c. 13

The thirteenth proposition is this, that $CE:BD = (AC:BA).(GE:GD)$, which is proved as follows. Let the line BF be drawn parallel to CE, as in the penultimate figure, and [BF] I put as a mean in the proportion between CE and DB. Thus

(1)	$CE:BD = (CE:BF).(BF:BD)$	II. 2
(2)	$AC:AB = CE:BF$	Euclid VI. 4
(3)	$GE:GD = BF:BD$	Euclid VI. 4
(4)	$CE:BD = (AC:AB).(GE:GD)$	(1), (2), (3),

which we wanted to show.

III. c. 14

The fourteenth proposition, that $CE:BD = (AC:GD).(GE:BA)$, is shown thus. Let the line GF be drawn parallel to the line AC, and I shall make the two lines GE and BA means in the proportion between CE and BD. Therefore

(1)	$CE:BD = (CE:GE).(GE:BA).(BA:BD)$	II. 3
(2)	$CE:GE = AC:GF$	Euclid VI. 4
(3)	$BA:BD = GF:GD$	Euclid VI. 4
(4)	$AC:GD = (AC:GF).(GF:GD)$	II. 2
(5)	$CE:BD = (AC:GD).(GE:BA)$	(1), (2), (3), (4),

which we promised to prove.

III. c. 15

The fifteenth proposition is this, that $GE:GD = (AB:AC).(CE:BD)$, which is proved as follows. I shall draw BF parallel to CE, and I shall make CE a mean in the proportion between BF and BD. Therefore

(1)	$BF:BD = (BF:CE).(CE:BD)$	II. 2
(2)	$BF:CE = BA:AC$	Euclid VI. 4
(3)	$BF:BD = (BA:AC).(CE:BD)$	(1), (2)
(4)	$BF:BD = GE:GD$	Euclid VI. 4
(5)	$GE:GD = (AB:AC).(CE:BD)$	(3), (4),

which we promised to show.

III. c. 16

Proposicio sextadecima dicit quod *GE* ad *GD* fit ex *AB* ad *BD* et ex *CE* ad *AC*, quod sic probatur:

Protraham lineam *GF* equidistantem *AC*, ut sepe dixi, et ponam *GF* mediam in proporcione inter *GE* et *GD*. Patet igitur quod proporcio *GE* ad *GD* fit ex *GE* ad *GF* et *GF* ad *GD*. Sed proporcio *GF*[a] ad *GD* est sicut proporcio *BA* ad *BD*, et proporcio *GE* ad *GF* est sicut *CE* ad *CA*. Igitur proporcio *GE* ad *GD* componitur ex proporcione *AB* ad *BD* et ex *CE* ad *AC*. Et hoc intendimus ostendere.

Et nota ex hoc quod proporcio *GE* ad *BD* ex nullis quatuor reliquarum componi potest, quia primi ad quartum nulla est ex aliis proporcio.

III. c. 17

Proposicio septimadecima probat quod *GD* ad *BD* proporcio componitur ex *AC* ad *AB* et ex proporcione *GE* ad *EC*, quod sicut probatur:

Protrahatur linea *BF* equidistans *CE*, ut prius, et ponam *CE* mediam in proporcione inter *GE* et *BF*; igitur *GE* ad *BF* componitur ex *GE* ad *CE* et *CE* ad *BF* proporcionibus. Sed proporcio *CE* ad *BF* est sicut proporcio *AC* ad *AB*, igitur proporcio *GE* ad *BF* componitur ex proporcione *AC* ad *AB* et ex proporcione *GE* ad *CE*. Sed proporcio *GE* ad *BF* est sicut *GD* ad *BD*, igitur proporcio *GD* ad *BD* componitur ex *AC* ad *AB* et *GE* ad *CE*, et hec promisimus demonstrare.

III. c. 18

Proposicio octavadecima et ultima coniunctarum probat quod proporcio *GD* ad *BD* sit ex proporcione *AC* ad *CE* et ex *GE* ad *AB*, cuius racio est hec:

Protraham lineam *BF*, ut prius, equidistantem linee *CE*, et ponam lineam *AB* mediam in proporcione inter *GE* et *BF*. Constat igitur quod proporcio *GE* ad *BF* componitur ex *GE* ad *AB* et ex *AB* ad *BF*. Proporcio vero *AB* ad *BF* est sicut proporcio *AC*[b] ad *CE*, igitur proporcio *GE* ad *BF* componitur ex proporcione *GE* ad *AB* et ex *AC* ad *CE*. Sed *GE* ad *BF* est sicut *GD* ad *DB*; igitur proporcio *GD* ad *DB* componitur ex *AC* ad *CE* et ex *GE* ad *AB*. Et hoc est quod ostendere curavimus.

[a] CM *CG* [b] M 3λ+*CE* ad *AC*

III. c. 16

The sixteenth proposition says that $GE:GD = (AB:BD).(CE:AC)$, and is proved as follows. I shall draw the line GF parallel to AC, as I have so often said, and I shall make GF a mean in the proportion between GE and GD. It is evident, therefore, that

(1)	$GE:GD = (GE:GF).(GF:GD)$	II. 2
(2)	$GF:GD = BA:BD$	Euclid VI. 4
(3)	$GE:GF = CE:CA$	Euclid VI. 4
(4)	$GE:GD = (AB:BD).(CE:AC)$	(1), (2), (3).

And this we intended to show.

Note from this that $GE:BD$ may not be compounded from any of the other four terms, since there is no proportion of the first to the fourth.[1]

III. c. 17

The seventeenth proposition proves that

$$GD:BD = (AC:AB).(GE:EC),$$

as follows. The line BF is drawn parallel to CE, as at first, and I shall make CE a mean in the proportion between GE and BF. Therefore

(1)	$GE:BF = (GE:CE).(CE:BF)$	II. 2
(2)	$CE:BF = AC:AB$	Euclid VI. 4
(3)	$GE:BF = (AC:AB).(GE:CE)$	(1), (2)
(4)	$GE:BF = GD:BD$	Euclid VI. 4
(5)	$GD:BD = (AC:AB).(GE:CE)$	(3), (4),

and this I promised to prove.

III. c. 18

The eighteenth and last proposition of conjuncts[2] proves that $GD:BD = (AC:CE).(GE:AB)$, the argument for which is as follows. I shall draw the line BF, as at first, parallel to the line CE, and I shall make the line AB a mean in the proportion between GE and BF. Then

(1)	$GE:BF = (GE:AB).(AB:BF)$	II. 2
(2)	$AB:BF = AC:CE$	Euclid VI. 4
(3)	$GE:BF = (GE:AB).(AC:CE)$	(1), (2)
(4)	$GE:BF = GD:DB$	Euclid VI. 4
(5)	$GD:DB = (AC:CE).(GE:AB)$	(3), (4).

And this is what we were at pains to show.

[1] GE and BD are the 'first' and 'fourth' in (4) above.

[2] *coniuncta*. A gloss to MS. D runs 'scilicet catharum'. For the word '*cata*' see our commentary.

III. D. 1

Nunc postquam ostendimus cathas coniuncte proporcionis, restat dicere de disiunctis proporcionibus; et recitabo figuram.

Pono igitur duas lineas *AG*, *CG* concurrentes in puncto *G*; et ab *A* et *C* terminis protraham alias duas lineas intersecantes se super punctum *D* et determinantes altrinsecus in lineis *AG* et *CG* super puncta duo *B* et *E*, que linee sunt *ADE* et *CDB*.

Dico ergo quod proporcio *AB* prime ad *BG* secundam componitur ex proporcione *AD* tercie ad *DE* quartam et ex proporcione *CE* quinte ad *CG* sextam, quod sic probatur:

Protraham lineam *EH* super *CB* equidistantem linee *GA*, et ponam *EH* mediam inter *AB* et *GB*. Dico ergo quod proporcio *AB* ad *GB* fit ex *AB* ad *EH* et ex *EH* ad *BG*. Sed proporcio *AB* ad *EH* est sicut proporcio *AD* ad *DE*; et proporcio *EH* ad *BG* est sicut proporcio *CE* ad *CG*. Igitur proporcio *AB* ad *BG* fit ex *AD* ad *DE* et ex *CE* ad *CG*, quod querebamus.

III. D. 2

Secunda proposicio est hec, quod proporcio *AB* ad *BG* componitur ex *AD* ad *CG* et ex *CE* ad *DE*, quod sic patet:

Protraham lineam *GH* equidistantem linee *AE*; et producam lineam *CB* usque concurrat in *H* cum lineam *GH* super punctum *H*; et ponam *GC* mediam in proporcione inter *AD* et *GH*: igitur proporcio *AD* ad *GH* fit ex *AD* ad *GC* et ex *GC* ad *GH*. Sed proporcio *GC* ad *GH* est sicut proporcio *EC* ad *DE*. Igitur proporcio *AD* ad *GH* componitur ex *AD* ad *GC* et ex *EC* ad *DE*. Sed proporcio *AD* ad *GH* est sicut proporcio *AB* ad *BG*; igitur *AB* ad *BG* componitur ex *AD* ad *CG* et ex *CE* ad *DE*, quod ostendere proposuimus. Et declaracio eius est notabilis.

* * * * *

III. D. 1

Now after we have shown the catas of conjunct proportion, it remains to tell of disjunct proportions. I shall rehearse the figure.

I take two lines, AG and CG, meeting at G; and I shall draw two other lines from the points A and C, intersecting in D, and serving to determine points B (on AG) and E (on CG). These are the lines ADE and CDB.

I say then (AB being now the first, BG the second, etc.) that

$$AB:BG = (AD:DE).(CE:CG),$$

which is proved as follows. I shall draw the line EH (H on CB) parallel to the line GA, and I shall take EH as a mean between AB and GB. I maintain then that

(1) $AB:GB = (AB:EH).(EH:BG)$	II. 2
(2) $AB:EH = AD:DE$	Euclid VI. 4
(3) $EH:BG = CE:CG$	Euclid VI. 4
(4) $AB:BG = (AD:DE).(CE:CG)$	(1), (2), (3),

which is what we asked.

III. D. 2

The second proposition is that $AB:BG = (AD:CG).(CE:DE)$, which is evident from what follows. I shall draw the line GH parallel to the line AE, and produce the line CB until it meets GH (in H). And I make GC a mean in the proportion between AD and GH. Then

(1) $AD:GH = (AD:GC).(GC:GH)$	II. 2
(2) $GC:GH = EC:DE$	Euclid VI. 4
(3) $AD:GH = (AD:GC).(EC:DE)$	(1), (2)
(4) $AD:GH = AB:BG$	Euclid VI. 4
(5) $AB:BG = (AD:CG).(CE:DE)$	(3), (4),

which we proposed to show. And the theorem is a noteworthy one.

[III. D. 18] Sed *ED* ad *GH* est sicut *EC* ad *GC*; . . . Ergo proporcio *EC* ad *GC* fit ex *AB* ad *AD* et ex *ED* ad *BG*; quod promisimus ostendere.

Explicit[a] *tercius tractatus*

[a] C 3λ

III. D. 3–D. 18

[The treatise continues with geometrical proofs of the same character, closely following those of C. 3–C. 18, with the difference that the propositions are now a rehearsal of the different cases of Menelaus' theorem for a transversal (*BDC*, in fact), intersecting two sides of the triangle (*AGE*) internally, and one externally. Previously all three intersections were external. (The triangle was then *BCG*, with transversal *AED*.) The language of the proofs in III. D is very close to that of III. C, and precious little purpose would be served by giving both *in extenso*. The following table summarizes the propositions, to five of which (4, 6, 8, 12, 16) a caveat is added, drawing attention to impossible combinations of terms, *modi inutiles*. In the column headed 'Pattern of proposition', we represent *AB* by 1, *BG* by 2, *AD* by 3, *DE* by 4, *EC* by 5, and *CG* by 6. The fundamental proposition (the first), that is,

$$AB:BG = (AD:DE).(EC:CG),$$

is then written simply as 123456. Note that the patterns of propositions under III. C. 1–18 are exactly as here (although the basic proposition is written as $AC:AB = (CE:GE).(DG:DB)$), but that they deviate in both places from the modes in Part II to the extent that the patterns of modes 9 and 10 are interchanged (that is, III. C. 9 and III. D. 9 are in mode 10, while III. C. 10 and III. D. 10 are in mode 9). The last chapter of the third part of the treatise (III. D. 18) now begins and ends thus:]

But *ED* to *GH* is as *EC* to *GC* . . . therefore *EC* to *GC* is compounded of *AB* to *AD* and *ED* to *BG*, [that is, $EC:GC = (AB:AD).(ED:BG)$], which we promised to show.

Here ends the third part

[Table]

Chapter	*Figure*	*Pattern of proportion*
D. 1	i	1 2 3 4 5 6
2	ii	1 2 3 6 5 4
3	i	1 3 2 4 5 6
4	ii	1 3 2 6 5 4
5	i	1 5 2 6 3 4
6	ii	1 5 2 4 3 6
7	i	2 4 1 3 6 5
8	ii	2 4 1 5 6 3
9	ii	2 6 1 3 4 5
10	i	2 6 1 5 4 3
11	ii	3 4 1 2 6 5
12	i	3 4 1 5 6 2
13	ii	3 6 1 2 4 5
14	i	3 6 1 5 4 2
15	ii	4 5 2 1 3 6
16	i	4 5 2 6 3 1
17	i	5 6 1 2 4 3
18	ii	5 6 1 3 4 2

IV

IV. 1

Cum aliquis semicirculus dividitur in duos arcus, quolibet modo contingat, erit corda dupli arcus unius equalis corde dupli alterius.

HEC conclusio patet prima facie. Sit circulus *ABCD* et dividatur in[a] duos semicirculos per diametrum *AC*. Tunc dividatur semicirculus *ABC* super punctum *B* ubicumque volueris quod cadat *B*, et protrahatur linea [ab] *B* usque ad diametrum perpendiculariter in *E*. Erit ergo *BE* medietas corde dupli arcus utriusque porcionis quod patet si *BE* protrahatur ultra usque in *D*:[b] tunc est arcus *BAD* duplus ad arcum *BA*, et insuper arcus *BCD* est duplus ad *BC*. Sed *BD* est corda utriusque arcus duplati, ergo propositum patet.

IV. 2

Si diameter circuli secaverit cordam alicuius porcionis circuli, erit proporcio unius partis corde ad aliam qualis est proporcio corde dupli arcus unius partis[c] porcionis secate ad cordam duple porcionis alterius partis.

Hec conclusio probatur in primo libro *Almagesti* in fine, tamen[d] pono hic et recito eam quod indigebimus ea. Sit ergo circulus *ABCD* et sit porcio *DAB* secata per diametrum *AC*. Dico ergo quod proporcio *DF* partis corde ad partem *FB* est sicut proporcio corde dupli arcus *DA* ad cordam dupli arcus *BA*, quod patet sic:

Protrahatur[e] ad diametrum *AC* due linee perpendiculariter, scilicet *DG* et *BH*, et cum linee *BH* et *DG* sint equidistantes, inter quas cadit linea *DFB*, patet quod[f] trianguli *DGF* et *BHF* sunt eiusdem racionis; et per hec que dicta sunt in tercio tractatu patet conclusio, quia *DG* est medietas corde dupli arcus *DA*, et similiter *BH* est medietas corde dupli arcus *BA*.[g]

IV. 3

Si ab aliquo puncto semicirculi protrahatur linea que[h] extra circulum concurrat cum diametro, erit proporcio linee infra circulum ad totam illam lineam qualis est proporcio corde dupli arcus inter ipsam et diametrum intercepti ad cordam dupli arcus totius porcionis a principio linee.

Hec conclusio probatur in primo libro *Almagesti*, quam solum hic

[a] C 2λ [b] C *DE* [c] M parte [d] M item [e] M +linea
[f] M λ [g] M λ [h] C λ

IV

IV. 1

When a semicircle is divided into two arcs, in whatsoever way it be done, the chord of double the one arc is equal to the chord of double the other.

This proposition is apparent at a glance. Let *ABCD* be a circle, divided into two semicircles by the diameter *AC*. Let a semicircle then be divided at any point *B*, taken wheresoever you wish, and let a perpendicular from *B* upon the diameter meet it at *E*. Then *BE* will be half the chord of double either arc, which is plain to see if *BE* be produced to [meet the circle] in *D*: then arc *BAD* is double the arc *BA*, while, moreover, the arc *BCD* is double *BC*. But *BD* is the chord of either double arc, so the proposition is clear.

IV. 2

If the diameter of a circle cut the chord of any arc of the circle, the ratio of one part of the chord to the other will be the ratio of the chord of double the arc of the one part cut off the arc to the chord of double the arc of the other part.

This proposition is proved in the first book of the *Almagest* at the end, but I put it here and review it because we shall need it. Let *ABCD*, then, be a circle, and let the arc *DAB* be cut by the diameter *AC*. I say then, that

$$DF{:}FB = \text{Crd } 2DA{:}\text{Crd } 2BA,$$

which is clear from what follows.

Let two lines, *DG* and *BH*, be drawn at right angles to the diameter *AC*. The lines *BH* and *DG* being parallel, and between them lying the line *DFB*, it is clear that the triangles *DGF* and *BHF* are similar.[1] By this, what was said in Part III makes the proposition obvious, for *DG* is half the chord of the double arc *DA*, and similarly *BH* is half the chord of the double arc *BA*.

IV. 3

If a line be drawn from any point of a semicircle so as to intersect [its limiting] diameter outside the circle, then the ratio of the line outside the circle to the whole of the line will be as the ratio of the chord of the double arc intercepted between it and the diameter to the chord of the whole double arc from the end of the line.

This proposition is proved in the first book of the *Almagest*, and is

[1] Literally 'of the same reckoning' (*racio*).

recito propter evidenciam. Sit igitur circulus *ABCD*, et protrahatur diameter *AD* in continuum et directum versus *E*. Similiter[a] a *B* puncto protrahatur linea *BCE*, que necessario dividet porcionem *BD* in *C* puncto. Dico ergo proporcio *CE* ad *EB* est sicut proporcio corde dupli arcus *CD* ad cordam dupli arcus *BD*, quod patet; quia protrahantur ad *AD* diametrum[b] perpendiculares *CG* et *BF*, que[c] necessario erunt equidistantes, ergo sicut patet ibi et similiter ex predictis in proximo tractatu, proporcio *CE* ad *EB* est sicut proporcio *CG* ad *BF*, cum sint equidistantes—ergo et ad eorum duplum. Et patet conclusio.

IV. 4

Cum alicui arcui minori semicirculo subtendatur corda cui a centro circuli protrahatur equidistans, erit corda dupli arcus inter illas intercepti equalis corde dupli arcus compositi ex eo cui corda prima subtenditur et ex intercepta.

Sit circulus *ABCE* et sit arcus *AB* minor semicirculo cui subtendatur corda *BHA*. Sitque centrum circuli *D*, a quo protrahatur *DC*, *DE* equidistantes *BHA*. Dico ergo quod corda dupli arcus *BC*[d] est equalis corde dupli arcus *AC*, quod sic probatur:

Protrahatur ab *A* linea per *D* centrum usque in *F*, que necessario erit diameter.[e] Erit igitur necessario arcus *CF* equalis arcui *EA*, quia totus arcus *EAC* est equalis arcui *ABF* (dempto communi *ABC*, que relinquuntur erunt equalia, scilicet, *CF* et *AE*). Sed *BC* est equalis *AE*, propter equidistanciam linee *BA* ad *CE* lineam; cum igitur corda dupli arcus *CA* est equalis corde dupli arcus *AE*, per primam istarum quatuor proposicionum; igitur et *CF* erit equalis. Et[f] sic patet proposicio, nichilominus Ptholomeus.

IV. 5

Ex hiis quatuor proposicionibus[g] sufficienter patebit demonstracio figure sectoris qua usus est Abrachis,[h] Ptholomeus, Gebir, et alii in invencione declinacionis cuiuslibet gradus ecliptice ab equinoctiali circulo, et[i] cuiuslibet stelle sive puncti, tam in longitudine quam in latitudine. Ex supposita sciencia tradita in tribus tractatibus precedentibus satis patebit racio et modus operandi et[j] figurandi; enim illa ante invencione predictorum fuit difficilis valde ad intelligendum, necnon insufficienter fuit ab eis exposita.[k]

[a] M 3λ+sub *AB* [b] M diametri [c] C qui [d] C *AE* [e] C diametrum
[f] C 6λ [g] M proporcionibus [h] C Nicholas; M 2λ
[i] M in [j] C 2λ+Figura [k] C composita

proved at this point here only for the sake of clarity. Let there be a circle *ABCD*, and let the diameter be produced to *E*. Similarly, from the point *B* let the line *BCE* be drawn, which divides the arc *BD* at the point *C*. I say then that

$$CE{:}EB = \text{Crd } 2CD{:}\text{Crd } 2BD.$$

This is clear, since (*CG* and *BF* being perpendiculars to the diameter *AD*, and therefore being of necessity parallels) it is evident both there[1] and from what has been said in the adjacent treatise that

$$CE{:}EB = CG{:}BF,$$

and therefore $CE{:}EB = 2CG{:}2BF$—whence the proposition.

IV. 4

If a chord be subtended by any arc less than a semicircle, and a parallel be drawn from the centre of the circle, the chord of double the arc intercepted between them will be equal to the chord of double the arc comprising that arc subtended by the first chord and that intercepted.

Let there be a circle *ABCE*, with *AB* an arc (less than a semicircle) at which the chord *BHA* is subtended. Let *D* be the centre of the circle, from which point are to be drawn *DC* and *DE* parallel to *BHA*. I say then that

$$\text{Crd } 2BC = \text{Crd } 2AC,$$

which is proved as follows.

A line is to be drawn from *A* through the centre *D* as far as *F*. This will necessarily be a diameter. Necessarily, therefore, since arcs *EAC* and *ABF* are equal, arcs *CF* and *EA* are equal (taking away the common [arc] *ABC*, equal arc lengths *CF* and *AE* are left). But $BC = AE$ (because *BA* is parallel to *CE*); wherefore Crd $2CA =$ Crd $2AE$ (by IV. 1). Also *CF* will be equal [to *BC* and *AE*]. Thus the proposition follows. It is not in Ptolemy.

IV. 5

From these four propositions the proof of the sector figure will be evident enough. Hipparchus, Ptolemy, Gebir, and others used it to determine the declination of any degree of the ecliptic from the equinoctial circle,[2] and of any star or point in latitude and in longitude. From the hypotheses given in the three preceding treatises, the reasoning and method of working and constructing the figure will be clear. Before meeting with the previous propositions it was exceedingly difficult to understand [this reasoning], which was insufficiently explained.

[1] Presumably in *Almagest*.

[2] That is, the celestial equator.

Cum igitur inter duos arcus minores semicirculo, scilicet *AB* et *BC*, qui sunt de maioribus circulis in spera, intersecaverint se alii duo arcus in puncto *F*, qui sunt *AFD*,[a] *CFE*, erit proporcio corde dupli arcus *AE* ad cordam dupli arcus *EB* composita ex duplici proporcione, scilicet ex proporcione corde dupli arcus *AF* ad cordam dupli arcus *FD*, et ex proporcione corde[b] dupli arcus *CD* ad cordam dupli arcus *CB*, quod statim probabitur.

Duo tamen nota, quod semper fit hic sermo de arcu minori[c] semicirculo, ita quod eius duplum nec compleat totum circulum nec maius eo; aliud nota quod omnes arcus de quibus loquimur hic sunt imaginati arcus de maioribus circulis[d] totius spere, quorum[e] quilibet[f] dividit speram in duo equalia. Ergo omnium circulorum debet intelligi[g] unum centrum communem *G*. Licet igitur hoc in plano. Describatur arcus super aliud centrum quam *G*; suppone tamen *G* esse centrum eius, quia in plano impossibile est describere plures[h] circulos equales super idem centrum.

Quia, igitur, figura huius demonstracionis tripliciter variatur et Ptholomeus non tetigit nisi unum modum[i] arguentis,[j] secundum illum nos (ut nichil relinquamus insufficiens) etiam in moribus omnes tres modos persequamur, per ordinem primo modum Ptholomei, deinde utrumque modum a predecessoribus nostris omissum.

Fiant igitur ut premissi duo arcus orbium[k] maiorum[l] minores semicirculo *AB*, *BC*, intelligendo eorum centrum *G*, inter quos fiant alii duo arcus, *CE*, *AD*, intersecantes se super punctum *F*. Intelligantur fieri super idem centrum *G*,[m] quod necesse est fieri in spera et impossibile in plano: supponamus tamen, propter demonstracionem. Protrahatur ergo linea *GC* et linea *BD*. Iste igitur due linee, cum sint in una superficie, videlicet circuli *CDB*, aut concurrent aut equidistabunt; necessario quod si concurrant[n] aut ergo hoc erit ex parte *A* aut ex parte *C*. Cum enim arcus *CB*, secundum ypotesim, sit minor semicirculo, aut ergo quod deest ad complendum semicirculum est equale *CD*, vel maius, vel minus. Si ipsum sit maius *CD*,[o] tunc *CG* et *DB* concurrent ex parte *C*. Concurrant ergo super punctum *H*, et tunc patet per argumentum Ptholomei, in fine prime dictionis, quia ipse arguit[p] solum ex illo supposito cuius est hec figura perfecta, et eius argumentum recitabo priusquam ad alios duos modos[q] me transferam.

Probat enim quod proporcio corde dupli arcus *AE* ad cordam dupli arcus *EB* componitur ex proporcione corde dupli arcus *AF* ad cordam dupli arcus *FD*[r] et ex proporcione corde dupli arcus *CD* ad cordam

[a] CM 2λ; C +*AD*, *FE*; M +*AD*, *CFE* [b] C λ [c] M minor [d] CM λ
[e] CM quarum [f] M quelibet [g] C λ [h] M pluros
[i] C medium [j] M argiens [k] C λ [l] M maiores [m] C *F*
[n] CM concurrant [o] M *AD* [p] C λ [q] M 2λ+metrans femina (!)
[r] M 12λ

If between two arcs (*AB* and *BC*) of less than a semicircle, these being parts of great circles on a sphere, two other arcs (*AFD* and *CFE*) meet at *F*, then

$$\text{Crd } 2AE : \text{Crd } 2EB = (\text{Crd } 2AF : \text{Crd } 2FD) . (\text{Crd } 2CD : \text{Crd } 2CB),$$

which will be proved at once.

Two notes: It is important always to lay down the condition that the arc be smaller than a semicircle, so that its double be neither equal to nor greater than a whole circle. And second, note that all arcs of which we speak here are pictured as arcs of great circles of the whole sphere, any of which divides the sphere into two equal parts. Thus all circles are known to share a common centre, *G*. [The figure] may be taken to be in a plane. An arc will then be described over some other centre than *G*; you must suppose, however, that *G* is its centre. Of course, in a plane it is impossible to describe several equal circles about the same centre.

The figure of this proof may be given three different forms, and since Ptolemy touched on only one mode of reasoning, we (so that we leave nothing incomplete), applying his method, shall investigate the character of all three, in this order: first Ptolemy's mode, and then each mode omitted by our predecessors.

Let there then be two arcs *AB*, *BC* of great circles as premised, less than a semicircle. Their centre is *G*, and between them there are to be two other arcs, *CE* and *AD*, intersecting each other in *F*. They are known to be described about the same centre *G*, which is necessarily the case with a sphere, but impossible in a plane: let us imagine it, however, for the sake of the proof.[1] Let the lines *GC* and *BD* be drawn. Since they are in one plane, namely that of the circle *CDB*, they either meet or are parallel. If they meet, this will either be to the side *A* or to the side *C*. If the arc *CB*, by hypothesis, be less than a semicircle, its supplement is equal to, greater than, or less than *CD*. If that is greater than *CD*, then *CG* and *DB* will meet to the side *C*. Let them meet at *H*. Then [the proposition?] is clear by Ptolemy's argument, at the end of the first declaration—although he only argues this case, which I shall rehearse initially, before turning my attention to the other two possibilities.

He proves that

$$\text{Crd } 2AE : \text{Crd } 2EB = (\text{Crd } 2AF : \text{Crd } 2FD) . (\text{Crd } 2CD : \text{Crd } 2CB),$$

and he does so as follows. From the point *B* two lines *BA* and *BH* are

[1] The style of the diagrams of the MSS. is reproduced here together with more easily appreciated perspective drawings for comparison. There are many obvious merits in the medieval examples—which, for instance, make it a simple matter to study the plane triangles involved in the proof. They are constructed from straight lines and circular arcs alone, and in principle they could be drawn in an entirely acceptable way, namely as stereographic projections from a point on the great circle *CA*. In this case, however, figures IV. 5b and IV. 5c would correspond only with mirror symmetry.

dupli arcus *CB*, et hoc sic: a puncto *B* procedunt due linee *BA* et *BH*, et ab eiusdem terminis *H* et *A* reflectuntur due linee *HN* et *AD*, intersecantes se in puncto *M*. Igitur sicut probatur in prima proposicione in disiunctis proporcionibus tercii tractatus, que est 19ª proposicio tocius tercii,[a] proporcio *AN* ad *NB* componitur ex proporcione *AM* ad *MD* et ex proporcione *HD* ad *HB*; hoc bene intellige. Item, ex quo arcubus *AB*, *AD*, subtenduntur due corde, protrahunturque linee *GC* et *BD* concurrentes in puncto *H*, et a centro *G* protrahantur linee *GF* et *GE*, intersecantes predictas cordas *DA* et *BA* in duobus punctis *M* et *N*. Necessario omnia tria puncta *H*, *M*, *N* sunt in superficie trianguli *ABH* et in superficie circuli *CEF*, que due superficies sese secant. Erunt igitur *H*, *M*, *N* in eadem linea recta necessario cum nulla puncta possunt poni communia duabus superficiebus planis intersecantibus se nisi illa puncta situantur in linea communis sectionis, et omnis talis linea est recta, igitur . . . et cetera. Quod autem superficies trianguli *ABH* et superficies circuli *CFE* intersecant[b] se secundum lineam *NM*. (Patet quia triangulus *ABH* secat[c] superficiem circuli *CFE*, que superficies[d] non est superficies circuli habentis eandem diametrum cum diametro circuli *AEB*:[e] oportet quod intersecent se alibi quam in diametris, ergo oportet quod sit punctis communibus[f] utrisque superficiebus, qui sunt *M* et *N*; ergo . . . et cetera.)

Nunc autem ad propositum. *GE* semidiameter intersecat cordam arcus *BEA* in puncto *N*, ergo proporcio *BN* ad *NA* est sicut proporcio corde dupli arcus *BE* ad cordam dupli arcus *AE*, per secundam demonstracionem huius quarti tractatus, que est conclusio Ptholomei. Item, a puncto *B* in arcu minori semicirculo *CDB*, protrahitur linea occurrens diametro *GC* extra circulum, scilicet in puncto *H*, ergo secundum terciam huius quarti tractatus erit proporcio *HD* ad *HB* totam qualis est proporcio corde dupli arcus *CD* ad cordam dupli arcus *CB*.[g]

Rursus, *GF* egreditur a centro ad circumferenciam circuli *AFD*, intersecans arcum *AD*, igitur per secundam huius quarti, ut prius, qualis est proporcio *DM* ad *MA*, talis erit proporcio corde dupli arcus *DF* ad cordam dupli arcus *FA*. Ergo sicut prius probatum est quod proporcio linee *AN* ad *NB* componitur ex proporcione linee *AM* ad lineam *MD* et ex proporcione corde *BD*[h] ad totam lineam *HB*, cum eodem modo iam sit probatum in arcubus, patet quod proporcio corde dupli arcus *AE* ad cordam dupli arcus[i] *EB* componitur ex proporcione corde dupli arcus *AF* ad cordam dupli arcus *FD*, et ex proporcione corde[j] dupli arcus *CD* ad cordam dupli arcus *CB*.

[a] CM λ [b] C 13λ [c] M 2λ+iacet in superficie [d] CM +ex quo
[e] C *ACB* [f] M quibus [g] C *EB* [h] C *HD* [i] C *CB* [j] M λ

to be drawn, and from the ends *H* and *A* two lines *HN* and *AD* come back, intersecting in the point *M*, and therefore exactly as is proved in III. D. 1 (the nineteenth proposition of the third treatise as a whole),

$$AN:NB = (AM:MD).(HD:HB).$$

Mark this well.

Next, two chords are subtended by *AB* and *AD*; and the lines *GC* and *BD*, meeting in *H*, are drawn, while from the centre *G* let the lines *GF* and *GE* be drawn, cutting the aforesaid chords *DA* and *BA* in the two points *M* and *N*. Of necessity, all three points *H*, *M*, and *N* are in the plane of the triangle *ABH*,[1] and in the plane of the circle *CEF*,[2] which the planes intersect. The points *H*, *M*, and *N* will necessarily be in the same straight line, since no points may be shared by two intersecting plane surfaces unless they are situated on the line of common section, and every such line is straight, therefore . . . etc.[3] The plane of the triangle *ABH* and that of the circle *CFE* intersect along the line *NM*. (It is obvious since the triangle *ABH* cuts the plane of the circle *CFE*. This plane is not the plane of the circle having the same diameter as the circle *AEB*. It is necessary that they intersect elsewhere than along diameters, and that therefore the intersection be in points common to both surfaces, namely *M* and *N*; therefore . . . etc.)

Now, however, to the proposition. The semidiameter *GE* cuts the chord of the arc *BEA* at the point *N*, and therefore

$$(1)\quad BN:NA = \text{Crd } 2BE:\text{Crd } 2AE \qquad \text{IV. 2.}$$

(IV. 2 is Ptolemy's proposition.) Next, from the point *B* on the arc *CDB* (which is less than a semicircle), a line is drawn meeting the diameter *GC* outside the circle, namely in *H*, and therefore

$$(2)\quad HD:HB = \text{Crd } 2CD:\text{Crd } 2CB \qquad \text{IV. 3.}$$

Again, *GF* is drawn from the centre towards the circumference of the circle *AFD*, cutting the arc *AD*, and therefore

$$(3)\quad DM:MA = \text{Crd } 2DF:\text{Crd } 2FA \qquad \text{IV. 2.}$$

Therefore, exactly as it was previously proved that

$$(4)\quad AN:NB = (AM:MD).(BD:HB),$$

if now this same proof be given in terms of arcs, it is clear that [by (1), (2), (3), and (4)]

$$(5)\quad \text{Crd } 2AE:\text{Crd } 2EB = (\text{Crd } 2AF:\text{Crd } 2FD).(\text{Crd } 2CD:\text{Crd } 2CB).$$

[1] They are on the sides of the triangle *ABD*.

[2] They are on radii (produced if necessary) of the circle.

[3] Some MS. marginalia refer to the theorem, Euclid XI. 3. Others vary the text at this point, incorporating some very diffuse explanatory remarks, promoted, no doubt, by some vagueness in the following sentences.

Et nota quod hec demonstracio posset 18 modis coniunctim et 18 modis disiunctim variari; sed quia isti 36 modi sufficienter determinantur in secundo per numeros, et in tercio huius per lineas, nichil restat nisi in ista eadem figura omnes illos modos designare[a] in circulis. Et hoc facillime sicut forte exemplariter discurram in aliis duobus casibus, in quibus hec figura variari potest, de quibus nullus adhuc dixerit[b] quicquam.

Secundus[c] *modus.* Sit ergo residuum semicirculi minus quam *CD*, et compleantur semicirculi *CB* et *CE*, intersecantes se in punctis *K* et *C*. Protrahanturque linee *GC* versus *K* et *DB*, qui concurrent ex parte *A*. Quia linea *CG* et linea *BD* magis distant apud *D* quam apud *B*, ergo concurrent si ultra *B* protrahatur linea *BD*; et sit hoc in puncto *H*. Et argumento ut prius, iuxta modum arguendi Ptholomei, eo quod probatum[d] est quod *M*, *N*, *H* erunt in eadem linea recta. Si protrahantur corde *AB* et *AD* et a centro *G* due linee exeuntes ad puncta *E* et *F*, intersecabunt cordas predictas in punctis *M* et *N*; que propter racionem predictam erunt in una linea recta, quia sunt puncta communia duabus superficiebus intersecantibus se: deinde sic ut prius.

Nam inter arcus duos *KD* et *AD* intersecant se alii duo arcus que sunt *AB* et *KF* in puncto *E*; ergo proporcio corde dupli arcus *AF*[e] ad cordam dupli arcus *FD* componitur ex proporcione dupli arcus corde *AE*[f] ad cordam dupli arcus *EB*[g] et ex proporcione corde dupli arcus *KB* ad cordam dupli arcus *KD*, quod patet eo quod a *D* puncto procedunt due[h] linee *DH*, *DA*, a quarum[i] terminis *H* et *A* reflectuntur due linee *AB*, *HM*, igitur proporcio *AM* ad *MD* fit ex proporcione *AN* ad *NB* et ex proporcione *HB* ad *HD* per priorem demonstracionem Ptholomei quam satis plane premisi; igitur, per undecimum[j] modum proporcionum, proporcio corde dupli arcus *AE*[k] tercii ad cordam dupli arcus *EB*[l] quarti componitur ex proporcione corde dupli arcus *AF* primi ad cordam dupli arcus *FD* secundi et ex proporcione corde dupli arcus *KD* sexti ad cordam dupli arcus *KB* quinti. Sed corda dupli arcus *KD* est equalis corde dupli arcus *CD* cum *CD* sit residuum semicirculi per primam huius quarti. Et corda dupli arcus *KB* est equalis corde dupli arcus *CB*[m] eadem racione; ergo proporcio corde dupli arcus *AE* ad cordam dupli arcus *EB* constat ex proporcione corde dupli arcus *AF* ad cordam dupli arcus *FD* et ex proporcione corde dupli arcus *CD* ad cordam dupli arcus *CB*.

Item, per terciam proposicionem de disiunctis potest probari quod proporcio corde dupli arcus *AF* ad cordam dupli arcus *AE* componitur ex proporcione corde dupli arcus *FD* ad cordam dupli arcus *EB* et ex

[a] C assignare [b] M duxerit [c] C 2λ [d] C protractum; M propposituro (?)
[e] C *AE* [f] C *AC* [g] CM *AB* [h] M et [i] C cuius
[j] CM secundum [k] C *AC* [l] M *ECB* [m] C *EB*

Note that this proof may be modified in 18 conjunct and 18 disjunct modes; but since these 36 modes are sufficiently determined in Part II for numbers and in Part III for lines, nothing remains except to cast into that same figure all those modes for circles. And this is most easily done, just as I shall discuss at length in two other cases, as it happens, by way of example. In these cases, the present figure may be modified; and concerning them, nothing whatever has thus far been said.

The second case. Let the remainder of the semicircle be then less than *CD*, and let there be semicircles *CB* and *CE*, meeting in the points *K* and *C*. Let the lines *GC* and *DB* be produced, the former towards *K*, to meet on the side on which *A* is found.[1] Since the lines *CG* and *BD* are farther apart at *D* than at *B*, therefore they meet if the line *BD* be produced beyond *B*. Let this happen at the point *H*. We follow the argument as before, in keeping with Ptolemy's reasoning, since it has been proved that *MNH* will be in the same straight line. If the chords *AB* and *AD* be drawn, and two lines from centre *G* to the points *E* and *F*, they will intersect the aforesaid chords in the points *M* and *N*. According to the previous reasoning they will be in one straight line since they are points common to two mutually intersecting planes. From this stage the argument proceeds as before.

Between the two arcs *KD* and *AD*, two other arcs, *AB* and *KF*, intersect at the point *E*; therefore,

(1) $\text{Crd } 2AF : \text{Crd } 2FD = (\text{Crd } 2AE : \text{Crd } 2EB) . (\text{Crd } 2KB : \text{Crd } 2KD)$,
which follows from the fact that two lines, *DH* and *DA*, proceed from the point *D*. From the ends *H* and *A* of these there return two lines *AB* and *HM*, and therefore

$$(2)\quad AM : MD = (AN : NB) . (HB : HD),$$

by Ptolemy's first proof. Then, by the eleventh mode of the ratios [with pattern 341265],[2]

(3) $\text{Crd } 2AE : \text{Crd } 2EB = (\text{Crd } 2AF : \text{Crd } 2FD) . (\text{Crd } 2KD : \text{Crd } 2KB)$.
But

(4) $\text{Crd } 2KD = \text{Crd } 2CD$ IV. I

(5) $\text{Crd } 2KB = \text{Crd } 2CB$ IV. I

(6) $\text{Crd } 2AE : \text{Crd } 2EB = (\text{Crd } 2AF : \text{Crd } 2FD) . (\text{Crd } 2CD : \text{Crd } 2CB)$ (3), (4), (5).

Next, it may be shown that

(7) $\text{Crd } 2AF : \text{Crd } 2AE = (\text{Crd } 2FD : \text{Crd } 2EB) . (\text{Crd } 2KB : \text{Crd } 2KD)$ II. D. 3;

[1] Looking at the sphere with *AC* on the nearer side.

[2] As applied to (1).

corda dupli arcus *KB* ad cordam dupli arcus *KD*; et sic de singulis, sicut satis ibi expresse traditur in omnibus simili modo determinando.

Tercius modus. Sit ergo tercius quod residuum semicirculi *CB*[a] sit equale arcui *CD*. Erunt ergo *BD* et *CG* linee equidistantes, cuius probacio tangitur in quarta proporcione huius quarti. Si ergo hoc, et *N* [et] *M* erunt equidistantes ab utrisque. Cum enim fuerint tres linee quarum quelibet due fuerint in superficie una et non tamen omnes tres in una fuerint, si due earum sint equidistantes erunt omnes equidistantes. Sunt enim tres linee *p*, *q*, *r*, quarum quelibet due in una superficie sunt. Omnes tamen tres non sunt in una alia superficie, et due illarum, scilicet *p* et *r*, sunt equidistantes per ypotesim. Eo quod arcus *CD* est equalis arcui *KB*, dico ergo quod *q* linea equidistabit ab utraque, quod sic probatur:

Quia aut equidistat *q* ipsi *p* aut concurret cum eo, cum sint in superficie una, scilicet semicirculi *CDBK*,[b] et si equidistet[c] ei equidistabit ipsi *r*, cum positum sit *p* et *r* esse equidistantes. Si autem concurrat cum eo,[d] hoc erit in aliquo puncto, et similiter concurret cum linea *r* in aliquo puncto, cum sit secum in superficie una, scilicet circuli *CFEK*. Pone ergo quod concurrat cum utraque linea tam *p* quam *r*, aut ergo in puncto uno aut in diversis. Si in puncto uno, igitur *p* et *r* concurrent in eodem puncto, quod est contra ypotesim. Si in diversis, igitur duo puncta[e] *q* linee sunt in superficie duarum linearum *p* et *r*; ergo tota linea *q* est in eadem superficie. Erunt igitur omnes tres linee in eadem superficie, quod est contra ypotesim. Relinquitur ergo quod istarum trium linearum *DB*, *MN*[f] et *CG* quecumque[g] due sunt in superficie una.

Nam *BD* et *CG* sunt in superficie arcus *CB*;[h] item *MN* et *DB* sunt in superficie trianguli *ADB*; et *CG* et *MN* sunt in superficie circuli *CFE*. Et non sunt iste superficies una, quod patet per diversitatem circulorum. Nunc autem due istarum linearum, scilicet *DB* et *CG*, sunt equidistantes; ergo *MN* erit equidistans utrique.

Ergo, per consequens, erit proporcio *AN* ad *NB* qualis est proporcio *AM* ad *MD*. Sed *AN* ad *NB* proporcio est sicut proporcio corde dupli arcus *AE* ad cordam dupli arcus *EB*, per secundam huius quarti (cum *GE* linea transeat a centro super porcionem *AB*); et proporcio linee *AM*[i] ad *MD* est sicut proporcio corde[j] dupli arcus *AF* ad cordam dupli arcus *FD*, per[k] eandem racionem. Ergo proporcio corde dupli arcus *AE*

[a] M *OB* [b] C *EDBK* [c] M equidistat [d] C ea [e] CM + *G*
[f] CM λ [g] M queque [h] CM *AB* [i] M 3λ+*AN* ad *AM* lineam in *D*
[j] M cordam [k] C 26λ

and so for the different individual cases, as is there [in Part III] sufficiently fully propounded for all cases, with a similar method of proof.

The third case. Third, let it be supposed that the supplement of *CB* is equal to the arc *CD*. Then the lines *BD* and *CG* will be parallel. The proof of this is touched on in IV. 4 above. If, therefore, this is true, both *N* and *M* will be equidistant from each. When there are three lines of which any two (but not all three) are in one plane, and two of them are parallel, then all will be parallel.[1] There are three lines, *p*, *q*, and *r*, of which any two are in one plane. Not all three, however, are in separate planes, and two of them, namely *p* and *r*, are by hypothesis parallel. Since the arc *CD* is equal to the arc *KB*, I hold that the line *q* is parallel to both, which is proved as follows.

Since *q* will either be parallel to *p*, or will meet it, if they be in the same plane (namely that of the semicircle *CDBK*), then if [*q*] be parallel to [*p*], it will be parallel to *r*, since *p* and *r* were taken to be parallel by hypothesis. If, however, [*q*] meets [*p*], this will be in some point, and [*q*] will likewise meet with *r* in some point, since they are together in the same plane, that is to say, the plane of the circle *CFEK*. Suppose, then, that [*q*] meets with *p* and *r*, whether at a single point or at separate points. If in a single point, then *p* and *r* will meet at the same point, which is contrary to hypothesis. If at separate points, then two points on the line *q* are in the plane of the two lines *p* and *r*; and therefore the whole line *q* is in the same plane. The three lines are therefore all in the same plane, which is contrary to hypothesis. It only remains, therefore, that any two of the three lines *DB*, *MN*, and *CG* are in a single plane.

BD and *CG* are in the plane of the arc *CB*; next, *MN* and *DB* are in the plane of the triangle *ADB*; and *CG* and *MN* are in the plane of the circle *CFE*. And these planes are not one, as is clear from the circles, which are different. But now two of the lines, namely *DB* and *CG*, are parallel; and therefore *MN* will be parallel to both.

In consequence,

(1) $AN:NB = AM:MD$ Euclid VI. 4.

But

(2) $AN:NB = \text{Crd } 2AE:\text{Crd } 2EB$ IV. 2

(for the line *GE* passes from the centre over the arc *AB*). And

(3) $AM:MD = \text{Crd } 2AF:\text{Crd } 2FD$ IV. 2

[1] Cf. Euclid XI. 9.

ad cordam dupli arcus *EB* est sicut proporcio corde dupli arcus *AF* ad cordam dupli arcus *FD*, et sicut proporcio corde dupli arcus *CD* ad cordam dupli arcus *CB*; sicut satis patet per antedicta, et tunc ultra per quartam huius quarti, cum *CG* et *DB* sint equidistantes erit corda dupli arcus *CD* equalis corde dupli arcus *CB*, ergo inter cordam dupli arcus *CD* et cordam dupli arcus *CB* nulla est proporcio, quia sunt idem. Relinquitur ergo quod proporcio corde dupli arcus *AE* ad cordam dupli arcus *EB* componitur et est omnino sicut corda dupli arcus *AF* ad cordam dupli arcus *FD* tantummodo.

Et sic poteris in hac figura discurrere per omnes 18 proporciones in disiunctis proporcionibus, sicut satis exemplificavimus.[a]

IV. 6

Et postquam declaravimus demonstracionem huius figure secundum omnes modos eius quibus posset variari et eius assignacionem in proporcione disiuncta, palam erit eius[b] sciencia in proporcione coniuncta. Sint ergo, ut prius, inter duos arcus *AB* et *CB* qui sint de maioribus circulis in spera, alii duo arcus similiter de maioribus, qui[c] sint *CE* et *AD*, sese intersecantes super punctum *F*; sintque omnes minores[d] semicirculo. Dico ergo, per modum prime proporcionis tercii tractatus quod proporcio corde dupli arcus *AB* ad cordam dupli arcus *BE* componitur ex proporcione corde dupli arcus *AD* ad cordam dupli arcus *DF* et ex proporcione corde dupli arcus *CF* ad cordam dupli arcus *CE*, quod sic probatur:

Compleantur hii duo semicirculi *AB* et *AD* concurrentes ad punctum *K*.[e] Quia ergo inter duos arcus *KE* et *EC* intersecant se duo alii arcus *KF* et *CB* super punctum *D* de maioribus circulis in spera et semicirculo minores, erit proporcio corde duplicis porcionis *KB* ad cordam duplicis porcionis *BE* composita ex proporcione corde duplicis porcionis *KD* ad cordam duplicis porcionis *DF* et ex proporcione corde duplicis porcionis *CF* ad cordam dupli arcus *CE*, ut patet ex prioribus. Quia igitur corda duplicis arcus *AB* est equalis corde duplicis arcus *KB*, per primam huius quarti,[f] est enim *KB* residuum semicirculi. Item, corda dupli arcus *AD* est equalis corde duplicis arcus *KD*, per eandem racionem. Patet manifeste quod corda dupli arcus *AB* ad cordam dupli arcus *BE* composita ex proporcione corde dupli arcus *AD* ad cordam dupli arcus *DF* et ex proporcione corde dupli arcus *CF* ad cordam dupli arcus *CE*. Et hec conclusio similis est in omnibus prime conclusioni tercii huius; que ostenditur in lineis rectis. Patet ex precedentibus qualiter per quatuor primas proposiciones huius quarti ostenditur equalis proporcio et similis in omnibus in lineis rectis et in his lineis arcualibus que sunt de maioribus circulis spere et de minoribus porcionis semicirculo.

[a] C explicavimus [b] C λ [c] C 2λ [d] M maiores [e] C λ [f] M sexti

(4) Crd $2AE$: Crd $2EB$ = Crd $2AF$: Crd $2FD$[1] (1), (2), (3)

(5) Crd $2AE$: Crd $2EB$ = (Crd $2AF$: Crd $2FD$).(Crd $2CD$: Crd $2CB$).

For when *CG* and *DB* are parallel, Crd $2CD$ = Crd $2CB$, by IV. 4, wherefore there is no ratio between them, for they are the same.[2] [Statement (4)] remains unmodified, therefore.

And thus you will be able to propound the 18 proportions in disjunct form for this figure, as we have sufficiently well exemplified [previously].

IV. 6

After we have stated the proof of this figure, for all its possible variations, and its treatment in terms of disjunct ratio, its theory in terms of conjunct ratio will be clear. As before, let there be two mutually intersecting arcs of great circles on the sphere, *CE* and *AD*, meeting at *F*, and lying between two arcs of great circles—*AB* and *CB*. As before, all are to be less than a semicircle. I say, then, that in the manner of III. c. 1,

Crd $2AB$: Crd $2BE$ = (Crd $2AD$: Crd $2DF$).(Crd $2CF$: Crd $2CE$),

which is proved as follows.

Let these two semicircles, *AB* and *AD*, be completed, meeting in *K*. Since between the two arcs *KE* and *EC* the other two arcs *KF* and *CB* intersect at the point *D* (the arcs being arcs of great circles on the sphere, and less than a semicircle),

(1) Crd $2KB$: Crd $2BE$ = (Crd $2KD$: Crd $2DF$).(Crd $2CF$: Crd $2CE$),

as is apparent from what has gone before.[3] Also

(2) Crd $2AB$ = Crd $2KB$ IV. 1

(3) Crd $2AD$ = Crd $2KD$ IV. 1

(4) Crd $2AB$: Crd $2BE$ = (Crd $2AD$: Crd $2DF$).(Crd $2CF$: Crd $2CE$) (1), (2), (3).

This result is entirely similar to II. c. 1, which proposition is shown for straight lines. By means of the four initial propositions of Part IV, the proportion in straight lines is shown alike in all respects to that in terms of arc-lines, these being portions of great circles of a sphere which are less than a semicircle in length.

[1] The text reads as though, in addition, the first ratio was being identified with
Crd $2CD$: Crd $2CB$.
This was certainly not intended, but it seems that the ratio is needed at stage (5) below. Perhaps the phrase 'et sicut proporcio corde dupli arcus *CD* ad cordam dupli arcus *CB*' has been displaced in the manuscripts from the end of the paragraph, where it rightly belongs.

[2] It is interesting to note that a ratio was apparently regarded as essentially a relation between *discrepant* quantities.

[3] That is, by applying IV. 5 to the spherical triangle *KEF*, with transversal *CDB*.

Nunc gracia exempli adhuc ostendamus similem conclusionem secunde proposicioni[a] huius tercii tractatus in coniunctis. Ostensum est hic quod proporcio corde dupli arcus *AB* ad cordam dupli arcus *BE*[b] fit ex *AD* ad *DF* et ex *CF* ad *CE*. Ex eo per modum arguandi secunde proposicionis,[c] erit proporcio dupli arcus *AB* ad cordam dupli arcus *BE* composita ex proporcione corde dupli arcus *AD* ad cordam dupli arcus *CE*, et ex proporcione corde dupli arcus *CF* ad cordam dupli arcus *DF*; quod sic probatur:

Converte te ad primam figuram, que fuit figura Ptholomei, ubi omnes linee protrahuntur, et ibi videbis figuram in lineis rectis, ex qua tu argues que est iste facta ex quatuor lineis *BH* et *BA* et aliis duabus reflexis *HN*, *AD* intersecantibus se super punctum *M*. Et notum est tibi, ex secunda tercii huius, quod proporcio *AB* ad *BN* componitur ex *AD* ad *HN* et ex *HM* ad *MD*. Sed constat tibi ex prima figura istarum trium, que est Ptholomei, quod sicut se[d] habet *AN* ad *BN* ita se habet corda dupli arcus *AE* ad cordam dupli arcus *BE*. Ergo similiter sicut se habet tota[e] *AB* linea ad *BN* parcialis ita se habet corda dupli arcus totius[f] *AEB* ad cordam dupli arcus *BE* parcialis. Et sic multipliciter vertere potes demonstraciones tuas.

IV. 7

Premissa sciencia quam promisi de cathis.

Nunc restat applicare[g] ad opus in omni calculacione tabularum que docent distanciam cuiuslibet puncti firmamenti a quocumque, sive fuerit stelle a stella, sive[h] declinacionis cuiuscumque gradus vel puncti a quocumque, vel porcionum circuli ascensionum, tam in circulo recto quam obliquo; et quodlibet istorum tradam tibi in exemplo.

Et primo de invencione ascensionum in circulo directo: ymaginaberis ergo catham factam ex 4 porcionibus[i] circulorum[j] sic: Prima porcio circuli sit quarta pars equinoxialis, illa que est a principio Arietis usque in fine Geminorum in equinoxiali; et sit ista linea *BD*. Secunda linea sit[k] quarta pars coluri transiens per maximam declinacionem; et sit linea *AD*. Ergo *A* est polus mundi et iste due linee concurrunt in *D*, et sint iste due principales linee dicte cathe, ab extremitatibus autem istarum linearum, que sunt *A* et *B*. *A* dico polus, et *B* caput Arietis. Imaginentur[l] alie due quarte quarum illa pars ex *B* terminetur in linea

[a] M proporcioni [b] M *DE* [c] M proporcionis [d] C 2λ [e] CM totum
[f] C λ [g] C applicari [h] C sit [i] C proporcionibus
[j] M +equalibet [k] M ex [l] CM 9λ+et

Now, by way of an example, let us here[1] show a proposition similar to II. c. 2. It has been shown here that (4) holds good, and from that, by the argument concerning ratios of the second mode,

(5) Crd $2AB$:Crd $2BE$ = (Crd $2AD$:Crd $2CE$).(Crd $2CF$:Crd $2DF$), which is proved as follows.

Turn to the first figure,[2] which was Ptolemy's, where all the lines are drawn, and there you will see the figure in straight lines comprising the four lines BH, BA and the other two return lines HN, and AD (the latter intersecting at M). Also

(6) $AB:BN = (AD:HN).(MH:DM)$ III. 2

(7) $AN:BN$ = Crd $2AE$:Crd $2BE$ IV. 2

(8) $AB:BN$ = Crd $2AB$:Crd $2BE$ IV. 2.

Doing this[3] several times over you may convert your proofs.[4]

IV. 7

The theory of catas which I promised.

It remains to add to the work something on the calculations of those tables which give the distance of any one point of the firmament from any other, whether this be [the distance] of star from star, or whether it be of the declination of any degree or point from any other, or whether of arcs of the circle of ascensions, both right and oblique;[5] and for each of these things I will give you an example.

First, on finding right ascensions. You will picture to yourself a cata formed from four circular arcs as follows. The first is to be an arc of a circle, a quarter of the equinoctial, namely that which goes from the beginning of Aries to the end of Gemini on the equinoctial; and let this be the line BD. The second line is to be the fourth part of the colure[6] crossing the second line at its maximum declination; and let this be the line AD. Then A is the pole of the world, and the two lines meet in D, and they make the two chief lines which go under the name *cata*. Of the ends of the lines A and B, I call A the *pole* and B the *head of Aries*.

Two other quadrants are to be pictured, of which an arc from B is

[1] *adhuc*—used idiosyncratically here and elsewhere by Richard of Wallingford.

[2] Fig. IV. 5b (or c).

[3] Continuing a process which involves, for instance, substituting in (6) from (8).

[4] Thus yielding theorems in terms of chords.

[5] The ascensions are 'at the right circle' or 'at the oblique circle'. '*Circulus rectus*' often refers to the equator (which is more generally here '*circulus equinoctialis*'), but here refers to the *horizon rectus* (see p. 113 n. 1), while '*circulus obliquus*' could mean either the ecliptic (for which it is the standard name) or the horizon (*horizon obliquus*). Our phrases 'right ascension' and 'oblique ascension' are explained below in notes—but see also Appendix 22. There is ambiguity in the Latin, but the difficulties of translation do not appear here until the end of the next chapter, q.v.

[6] *colurus*—a great circle through the poles.

coluri *AD*, que erit ecliptica, et terminabitur in puncto *E*, igitur *ED* est quantitas totius declinacionis. Quarta linea reflectatur ab *A* polo, et terminetur in linea *BD* que est quarta equinoxialis, et transeat per illum gradum ecliptice cuius volumus invenire ascensiones—ut verbi gracia per finem tocius Arietis, si tocius Arietis voluimus[a] habere ascensiones.

Et sit ista linea *AC* que transeat per eclipticam super punctum *F*. Ergo *FB* est totum signum Arietis[b] in ecliptica, et *CB* sunt eius ascensiones in circulo directo, quia *CB* pars equinoxialis simul elevatur super orizontem[c] rectum cum *FB*; et quantitatem[d] *BC* volumus invenire. Intellige hanc catham componi ex hiis quatuor lineis, sicut vides in hac figura. Sit ergo *A*[e] polus mundi, *B* vero principium Arietis, *D* autem finis[f] Geminorum in circulo equinoxiali, *E* vero finis[g] Geminorum in ecliptica; et sit *AD* quarta coluri distinguentis solstitia. Linea *BD*[h] sit quarta equinoxialis que est a principio Arietis usque in finem Geminorum in equinoxiali. Iste due linee terminantur in eodem puncto, scilicet in puncto *D*. Item, *BE* sit quarta zodiaci, a principio Arietis usque in finem Geminorum in ecliptica.

Sit ergo nobis propositum invenire elevaciones[i] tocius Arietis ad circulum directum; et ideo a puncto *A* ducatur colurus[j] per illum gradum cuius volumus invenire elevacionem. Et sit quarta coluri illius *AC*, transiens per *F*, id est per finem Arietis in ecliptica. Dico ergo quod porcio[k] circuli equinoxialis intercepta inter principium Arietis et istum colurum, scilicet *BC*, simul[l] elevatur in orizonte recto cum illa porcione ecliptice que intercipitur inter principium Arietis et illum colurum, scilicet *FB*; ita quod simul oriuntur *BF* (porcio zodiaci) et *BC* (porcio equinoxialis) ad illos qui habent orizontem rectum. Hanc ergo porcionem, scilicet *BC*, sic invenies.

Intellige ergo catham sic: proporcio *DE* ad *EA* est composita ex proporcione *CF* ad *FA* et ex proporcione *DB* ad *CB*, sicut patet per primam proposicionem de cathis disiunctis conversam. Sed que est proporcio *DE* ad *EA*, eadem est proporcio *CF* ad aliquid aliud. Illud igitur aliud sic invenies. Multiplica medium in medium, id est *EA* in *CF*, et quod surgit ex tali multiplicacione divide per primum, scilicet per *DE*; et exibit quiddam ad quod se habebit *CF* sicut *DE* se habet ad *EA*. Illud ergo sit *P*. Est ergo proporcio *CF* ad *P* sicut proporcio *DE* ad *EA*. Sed proporcio *DE* ad *EA* erat composita ex[m] proporcione *CF* ad *FA* ducta in proporcione *DB* in *CB*; ergo similiter proporcio *CF* ad *P* erit composita ex eisdem, scilicet ex *CF* in *FA* et ex *DB* in *CB*.

[a] C volueris [b] C λ [c] M orientem [d] C quarta [e] C λ
[f] M seminis [g] M seminis [h] M +Linea *BG* [i] C ascensiones
[j] M +quod [k] C proporcio [l] C super [m] C 20λ+in

to be terminated on the line of the colure *AD*; and this will be the ecliptic, and it will end at the point *E*, and therefore *ED* is the total declination. The fourth line returns from the pole *A*, and is to end on the line *BD*, which is a quadrant of the equinoctial, and crosses through that degree of the ecliptic whose ascensions we want to find—as in the following example for the end of Aries, assuming that we wanted to know the ascensions for the whole of Aries.

Let *AC* be the line which crosses the ecliptic at the point *F*. Then *FB* is the whole sign of Aries on the ecliptic, and *CB* is its right ascension, since the part *CB* of the equinoctial crosses the right horizon[1] at the same time as *FB*. We wish to find the quantity *BC*. You must realize that the cata is composed of the four lines such as you see in the figure.[2] *A* is to be the pole of the world, *B* the beginning of Aries, *D* the end of Gemini on the equinoctial, but *E* the end of Gemini on the ecliptic; and let *AD* be the quadrant of the colure marking out the solstice. The line *BD* is to be the quadrant of the equinoctial from the beginning of Aries to the end of Gemini on the equinoctial. Those two lines end at the same point, namely *D*. Next, let line *BE* be a quadrant of the zodiac, from the beginning of Aries to the end of Gemini on the ecliptic.

Let it then be proposed that we find the right ascension[3] of the whole of Aries. From point *A*, therefore, let the colure be drawn through the degree of which we wish to find the ascension. Let a quadrant of that colure be *AC*, crossing at *F*, that is, the end of Aries on the ecliptic. I say, then, that the arc *BC* of the equinoctial circle intercepted between the beginning of Aries and the colure rises at the right horizon at the same time as that part of the ecliptic *FB* intercepted between the beginning of Aries and the colure; so that *BF* (a part of the zodiac) and *BC* (a part of the equinoctial) rise at the same time to those who have [the same] right horizon. This is how you should find the portion *BC*.

Find the cata thus:

$$(1)\quad DE:EA = (CF:FA).(DB:CB) \qquad \text{converse, III. D. I.}$$

But the ratio *DE*:*EA* is the ratio of *CF* to some other thing, which other thing you then find as follows.

Multiply mean into mean, that is *EA* by *CF*, and divide the product by the first, that is, by *DE*. The result is something to which *CF* bears the ratio *DE*:*EA*. This will be *P*, wherefore *CF*:*P* = *DE*:*EA*. But

$$DE:EA = (CF:FA).(DB:CB),$$

and therefore similarly

$$(2)\quad CF:P = (CF:FA).(DB:CB).$$[4]

[1] *horizon rectus*, the great circle (a colure) running from the pole (*A*) to the eastern point of the (oblique) horizon.

[2] The figure is omitted from all MSS.

[3] *elevaciones . . . ad circulum directum*. Note the use of the plural, found also with the more common synonym, *ascensiones* (which some of the MSS. accept here).

[4] Note the unheralded occurrence of the symbol *P* to denote the unknown.

Istud modum[a] conclusum servabis[b] in memoria. Et intellige quod inter *CF* et *P* cadit quedam quantitas media, scilicet *FA*, ita quod proporcio *CF* ad *P* componitur ex proporcione *CF* ad *FA* et *FA* ad *P*. Si ergo proporcio *CF* ad *P* componitur ex *CF* ad *FA* ducta in proporcione *DB* ad *CB*, et similiter eadem proporcio *CF* ad *P* componitur ex *CF* ad *FA* et *FA* ad *P*, erit proporcio *FA* ad *P* eadem cum proporcione *DB* ad *CB*; quia *FA* ad *P* cum proporcione *CF* ad *FA* constituit idem quod constituit proporcio *DB* ad *CB* cum eadem, scilicet cum proporcione *CF* ad *FA*, quia utraque multiplicacio duarum multiplicancium constituit proporcionem *DE* ad *DA*.[c]

Dicamus ergo sic: que est proporcio *FA* ad *P* illa est proporcio *BD* ad *CB*.[d] Sed proporcio *FA* ad *P* est tibi nota per premissa, cum sit tibi nota proporcio *CF* ad *P*, quia *DE* ad *EA*; et nota est proporcio *CF* ad *FA* et proporcio *CF* ad *P* componitur ex *CF* ad *FA* nota et ex proporcione *FA* linee note ad *P*. Et *DB* est tibi nota, quia semidiameter est enim sinus quarti circuli. Ergo reliquum[e] erit tibi notum, sic multiplica medium in medium, id est per *P*[f] multiplica *DB*, hoc est multiplica numerum qui exit in prima operacione per 150 minuta (que sunt minuta universitatis semidiametri secundum tabulas de corda recta et versa), et quod sic[g] exit tali multiplicacione divide per *FA*, id est per numerum quantitatis *FA*, et exibit quiddam quod erit quantitas *BC*; quia illud quod exiit se habebit ad *DB* sicut se habet *FA* ad *P*. Et hoc est *BC*[h] sinus qui respondet arcui intercepto inter principium Arietis in equinoxiali et colurum transeuntem per gradum cuius elevacionem querebas. Huius ergo sinus quere[i] arcum, et habebis quod queris, scilicet porcionem equinoxialis que[j] elevatur ad circulum directum cum gradu cui numerasti.

Et sic operaberis ad singula.

IV. 8

Iam restat disponere catham ad elevacionem in circulo obliquo. Fiat ergo circulus meridianus qui diametetur duabus diametris *AL*, *MB*,

[a] C modo [b] M servabum [c] C *EA* [d] M *EB* [e] M reliquando
[f] M *F* [g] CM 2λ+ex [h] C +et est *BC* [i] M λ
[j] C qui

This intermediate conclusion you will keep in mind. And you should understand that between *CF* and *P* there falls a mean quantity, namely *FA*, such that

$$(3)\ CF{:}P = (CF{:}FA).(FA{:}P) \qquad \text{II. 2.}$$

Therefore

$$(4)\ FA{:}F = DB{:}CB \qquad (2), (3),$$

since *FA*:*P* constitutes with *CF*:*FA* the same as does *DB*:*CB* with the same thing (*CF*:*FA*), each giving as a product of two multiplicands the ratio *DE*:*DA*.

I say then that (4) is true. Then

(5) (*DE*:*EA*)* hyp.[1]
(6) (*CF*:*P*)* (5), equal by def.
(7) (*CF*:*FA*)* hyp.[2]
(8) (*FA*:*P*)* (3), (6), (7)
(9) *DB** 'a semidiameter is the sinus of a quadrant'.[3]

Therefore what remains to be found [namely *CB*] will be known to you [by (4), (8), and (9)]. So multiply mean into mean, that is *P* into *DB*, which is to say multiply the number which resulted from the first operation[4] by 150′ (the number of minutes in the whole semidiameter, according to the tables of corda recta and versa), and divide the product by *FA*, that is, by the number of the quantity of *FA*, and the outcome is the magnitude of *BC*; for the result has to *DB* the ratio *FA*:*P*. And this is Sin *BC*, corresponding to the arc along the equinoctial intercepted between the beginning of Aries and the colure passing through the degree whose elevation you were seeking. Find the arc corresponding to this sinus, and you will have what you seek, namely that part [*BC*] of the equinoctial which is crossing the right circle[5] with the degree which was your concern.

And you proceed in this way for each individual degree.

IV. 8

We now come to the arrangement of the cata which relates to the problem of oblique ascension.

Let there be a 'meridian' circle[6] divided[7] by the two diameters *AL*

[1] This is known since the obliquity of the ecliptic is known.
[2] It is never made absolutely clear that this is indeed presupposed known, but if not, the proof is insufficient.
[3] This statement is important, as showing that the author regarded these 'lines' (ostensibly straight) as the arcs of great circles, which of course the correct solution of the problem required.
[4] (*EA*.*CF*)/*DE*, or rather this quantity expressed in sinus (or chords).
[5] *elevatur ad circulum directum*, a reference to the right horizon (*orizon rectus*), rather than to the equator (*circulus rectus*, occasionally *directus*).
[6] It is perverse to call this *circulus meridianus*. It is the colure of the equinoxes, not a meridian.
[7] *diametetur*.

et sit *A* polus mundi septentrionalis, *H* punctum primum Cancri in ecliptica, ita quod linea *MHB* fit ecliptica. *AG* sit elevacio poli super orizontem et *GF*[a] residuum de 90 gradibus. Quia *MB* est linea equinoxialis, sit *C* locus ille in equinoxiali ubi ad ascensionem totius Arietis ecliptice, orizon obliquus intersecat equinoxialem; et sit *I* summus Arietis. *D* sit locus ille equinoxialis ubi colurus transiens per finem Arietis ecliptice intersecat equinoxialem. *E* sit principium Cancri in equinoxiali, et *F* sit punctus equinoxialis distans a gradu equinoxiali qui est in orizonte, scilicet a *C* per 90 graduum in illa hora in quo totus Aries est super orizontem. Et *G* est punctus ubi orizon intersecat meridianum in parte septentrionali, ex parte scilicet poli elevati super orizontem obliquum tuum. Est igitur linea *AE* colurus transiens per caput Cancri, scilicet *H*; linea *BE* est quarta equinoxialis ab Ariete in Cancrum;[b] linea *CF* est quarta equinoxialis, scilicet porcio 90 graduum ab orizonte versus angulum terre, id est usque ad punctum medie noctis. Linea vero *NGIC* est orizon obliquus; linea *DIA* est colurus transiens per gradum 30^m[c] zodiaci orientem in[d] circulo obliquo; linea *BH* est quarta zodiaci ab Ariete in Cancrum. *AB*[e] est colurus transiens per caput Arietis; linea *AG* est latitudo regionis, vel elevacio poli, quod idem est; linea *GF* est residuum latitudinis regionis; linea *BD* est illud quod elevatur de equinoxiali cum toto Arietis in circulo directo; linea *CB* est illud quod elevatur de toto circulo equinoxiali cum toto Arietis in circulo obliquo, et istam lineam querimus, linea enim *CD* est differencia inter ascensiones Arietis in circulo directo et orizonte obliquo.

Et istius linee quantitatem inveniemus per eius subtractionem[f] a linea *BD* quod est nobis notum, quia sunt ascensiones in circulo directo. Remanebitque quantitas *BC* quod est nobis ignotum; erit tamen notum sic:

Intellige lineam *AF* et lineam *FC* utramque 90 graduum. Iste due linee constituunt angulum in puncto *F*, et erunt principales in catha. Et a termino *A* in linea *AF* reflectitur[g] linea *AD* 90 graduum super lineam *CF*; et similiter a termino *C*[h] reflectitur linea *CG* super lineam *AF*, sese intersecantes in puncto *I*. Igitur que[i] est proporcio *FG* ad *GA* fit[j] ex proporcione *DI* ad *IA* ducta in proporcione *FC* ad *DC*. Sed que est proporcio *FG* ad *GA*, eadem est *DI* ad aliquid aliud. Illud igitur aliud[k] invenias operando sicut superius, multiplicando medium in medium et productum dividendo per primum; et exibit illud quartum;

[a] CM *GM* [b] C Cancro [c] M λ [d] M 2λ+talem
[e] arcus *AH*; M alicuius *AB* [f] CM 4λ+remocionem acuto *KD*
[g] M replectitur [h] CM λ [i] CM 2λ [j] M 19λ [k] M autem

and *MB*, and let *A* be the north pole of the world, and *H* the first point of Cancer on the ecliptic, so that *MHB* forms the ecliptic. *AG* is to be the altitude of the pole above the horizon, and *GF* its complement. Since *MB* is the line of equinoxes, let *C* be the place on the equinoctial where, when the whole of Aries (on the ecliptic) has risen, the horizon[1] cuts the equinoctial. Let *I* be the end of Aries. *D* is to be that place on the equinoctial where the colure crossing from the end of Aries (on the ecliptic) cuts the equinoctial.[2] Let *E* be the beginning of Cancer on the equinoctial, and let *F* be point 90° removed from *C* at the moment when the whole of Aries is above the horizon. *G* is the point where the horizon meets the northward meridian—that is, the part where the pole is raised over your horizon. The line *AE* is therefore the colure through *H*, the head of Cancer; *BE* is a quadrant of the equinoctial from Aries to Cancer; *CF* is a quadrant of the equinoctial, namely those 90 degrees from the horizon towards the angle of the Earth,[3] that is, towards the midnight point.[4] The line *NGIC* is the (oblique) horizon; the line *DIA* is the colure through the 30th degree of the eastern zodiac (on the ecliptic); the line *BH* is a quadrant of the zodiac from Aries to Cancer, of which *AB* is a colure through the head of Aries; *AG* is the latitude of the region, or the elevation of the pole, which is the same thing; *GF* is its complement; *BD* is that part of the equinoctial which rises above the right horizon when the whole of Aries does so; *CB* is that part of the equinoctial above the [oblique] horizon at the same time as the whole of Aries—and this line *CB* is what we seek, for *CD* is the difference between the ascensions of Aries on the right horizon and the oblique horizon.

We shall find the size of the line [*BC*] by subtracting [*CD*] from *BD*, which latter is known; *BC* will remain, which is unknown to us. It is found thus:

You should know that *AF* and *FC* are both 90°. The two lines form the angle at the point *F*, and are the chief lines in the cata. From *A*, the end of *AF*, the line *AD* (of 90°) returns to *CF*; and similarly from *C*, the line *CG* returns to *AF*. The two, [*AD* and *GC*], intersect at the point *I*. Then

$$FG:GA = (DI:IA).(FC:DC).^{5}$$

But *FG*:*GA* is the same ratio as *DI* to some other thing, which you should find as you did above, multiplying mean into mean and dividing the product by the first [term]. The fourth [term] results. That is to say,

[1] *horizon obliquus.*

[2] This colure, explained so carefully, is the *horizon rectus*, which in IV. 7 above merited no explanation whatsoever.

[3] *angulus Terre.*

[4] *punctum medie noctis*—in this case, the lower point at which the equator crosses the meridian (this part of which was often called the 'midnight line').

[5] Applying the theorem of Menelaus (Δ*FAD*/*ICG*).

id est, multiplica *GA*, quod est latitudo regionis, in *DI*,[a] quod est declinacio gradus cuius queris ascensiones, et productum divide per *FG*, quod est residuum latitudinis regionis, et exibit quoddam aliud quod se habet ad *DI* sicut se habet *GA*[b] ad *FG*. Illud serva et sit illud *P*.

Sed primo nota quod in canonibus Arzachelis et aliorum est doctrina huius operacionis sub talibus verbis: Accipe sinum latitudinis regionis, et sit illud primum, et sinum residui, et sit illud secundum, sinum declinacionis gradus, et sit illud tercium, sinum residui declinacionis, et sit illud quartum. Unde illud quod est primum, scilicet residuum latitudinis regionis, in ordine proporcionis,[c] ponitur ibi secundum in ordine invencionis; et econverso illud quod est tercium[d] in ordine proporcionis, scilicet latitudo regionis, ponitur esse primum, quia est primum in ordine invencionis. Latitudo enim invenietur per se et eius residuum per eam. Unde non docet ibi multiplicare medium per medium et dividere per primum, sed docet multiplicare primum per tercium et dividere per secundum; quia quod est primum invencionis est secundum in proporcione, et econverso. Et eodem modo in canone de circulo directo.

Dico ergo quod que est proporcio *FG* ad *GA* eadem erit *DI* ad *P*. Sed proporcio *FG* and *GA* est producta ex proporcione *DI* and *IA* ducta in proporcione *FC* ad *DC*. Igitur proporcio *DI* ad *P* est composita ex similitere isdem, scilicet ex proporcione *DI* ad *IA* et ex *FC* ad *DC*. Istam conclusionem commenda memorie, et intellige quod inter *DI* et *P* cadit quedam quantitas media, scilicet *IA*; igitur proporcio inter *DI* et[e] *P* composita ex proporcione *DI* ad *IA* et ex *IA* ad *P*. Dico ergo quod que est proporcio *IA* ad *P* eadem est proporcio *FD* ad *DC*.

Probacio. Proporcio *DI* ad *IA*[f] multiplicata in proporcione *FC*[g] ad *DC* constituit proporcionem *DI* ad *P*, ita eadem proporcio *DI* ad *IA* ducta in proporcione *IA* ad *P* constituit[h] proporcionem *DI* ad *P*, ut prius probatur. Ergo eadem est proporcio *FC*[i] ad *DC* et proporcio *IA* ad *P*, quia utrumque ductum in eodem, scilicet in proporcione *DI* ad *IA*, producit idem, scilicet proporcionem *DI* ad *P*. Habeo igitur quod que est proporcio *FC*[j] ad *DC* illa est proporcio *IA* ad *P*, et econverso. Sed proporcio *IA* ad *P* est nota, quod utrumque est notum cum proporcio *FG* ad *GA* sit nota; et nota est etiam linea *DI*, quia declinacio gradus dati, igitur notum erit *P*. Similiter nota est ex hoc proporcio *IA* ad *P*, quia *P* iam notum est et *IA* nota est, cum sit residuum declinacionis, et iterum *FC* est notum, quia semidiameter est enim sinus 90 graduum, scilicet quarte circuli. Igitur reliquum,[k] scilicet *DC*, erit mihi notum sic:

[a] CM *DA* [b] CM 3λ+*FG* ad *GA* [c] M porcionis [d] C secundum [e] M ad
[f] M *EA* [g] C *FD* [h] M +in [i] CM *FD* [j] C *FD* [k] M +quando

multiply *GA* (the latitude of the region) by *DI* (the declination of the degree whose ascension you want), and divide the product by *FG* (the complement of the latitude of the region); and the result will be something which bears to *DI* the ratio *GA*:*FG*. Retain this, and let it be *P*.

But first note that in Arzachel's canons,[1] and in those of other writers, the doctrine of this chapter is given as follows: take the sinus of the latitude of the region (the first term), the sinus of the complement (the second), the sinus of the declination (third), and the sinus of its complement (fourth). It follows that what is first in the proportionality, namely the complement of the latitude of the region, is second in the order of discovery; and conversely, what is third in [our] proportionality (namely the latitude) is there given first, since it is first in the order of discovery. For the latitude is found by itself, and its complement is found from it. Thus it is not there taught to multiply mean by mean and divide by the first, but to multiply the first by the third and divide by the second; for what is first to be found is second in the proportionality, and conversely. And the same holds good in the canon on right ascensions.[2]

I say then that

(1)	$FG:GA = DI:P$	def.
(2)	$FG:GA = (DI:IA).(FC:DC)$	III. D[3]
(3)	$DI:P = (DI:IA).(FC:DC)$	(1), (2).

Commit this conclusion to memory. And you must realize that *IA* can be put as a mean between *DI* and *P*:

(4)	$DI:P = (DI:IA).(IA:P)$	II. 2.

I hold then that $FD:DC = IA:P$.

Proof. Consider (3) and (4). The separate products of two distinct ratios ($FC:DC$ and $IA:P$) and a third ($DI:IA$) give the same result ($DI:P$), and therefore the first two ratios are equal, that is,

(5)	$FC:DC = IA:P$	
(6)	FG^*, GA^*, $(FG:GA)^*$	obliquity*
(7)	DI^*	hyp.
(8)	P^*	(1), (6), (7)
(9)	IA^*	$(= 90° - DI^*)$, (7)
(10)	FC^*	Sin 90°[4]
(11)	DC^*	(5), (8), (9), (10).

[1] The following passage is merely concerned with the technique of solving a proportionality in six terms for a term on the right-hand side.

[2] Literally 'in the canon on the direct circle'.

[3] So much for the usefulness of Part III; but the pattern of this proportionality (653421) is not there represented—although it is immediately obtained by inverting III. D. 17 (561243).

[4] *FC* is known 'because the sinus of 90° is a semidiameter'. The arc *FC* is itself 90°, but in the entire chapter, as in the last, sinus, rather than arcs, are understood. This fact is formally announced for the first time at the end of the next paragraph!

multiplica medium in medium et productionem[a] divide per primum, scilicet multiplica *P* in *FC*[b], quod est multiplicare numerum servatum (qui[c] habitus fuit ex multiplicacione et divisione prima). Multiplica[d] per 150, que sunt minuta universitatis semidiametri secundum tabulas sinus, vel in 60 gradus secundum tabulas de corda et arcu, et productum divide per *IA*[e] (id est per residuum declinacionis gradus cuius querimus ascensiones) et exibit quantitas *DC*,[f] que est differencia recte ascensionis gradus quesiti inter orizontem rectum et obliquum illius regionis cuius supposuisti latitudinem. Huius autem sinus quere arcum, quia hoc quod invenisti est sinus. *Nam hoc intelligas* quod in omni operacione circuli[g] operamur cum sinibus et non cum arcubus. Licet enim dicamus '*BE* est quarta equinoxialis' vel '*AG* est latitudo regionis' intelligas quod non est nisi sinus quarte equinoxialis vel *AG* sinus latitudinis regionis.

Quere igitur arcum huius sinus in tabula equacionis sinus, si operatus es cum semidiametro 150 minutorum, vel in tabula de corda et arcu si multiplicavisti per 60 gradus quod ibi supponitur esse quantitas semidiametri, et redit in idem.

Et si illum arcum dempseris de arcu *DB*, scilicet de elevacione in circulo directo, remanebit arcus *CB*, quod est elevacio totius Arietis in circulo tuo obliquo.

Et similiter Piscium ad eandem regionem; et si illud idem addideris ad elevaciones Arietis in circulo directo habebis elevaciones suorum oppositorum ad eandem regionem, scilicet Libre et Virginis; et sic intellige de aliis signis, secundum quod docent canones, primo invenire differencias ascensionum omnium graduum primorum[h] trium signorum.[i] Et duo quilibet arcus equidistantes a duobus punctis[j] Arietis et Libre habent equales ascensiones,[k] duo autem quelibet signa et quilibet arcus equidistantes a duobus reliquis punctis, scilicet Cancri, Capricorni, arcus vel signa que sunt ex parte Arietis habent minores elevaciones quam que sunt ex parte Libre. Illi vero[l] arcus qui sunt ex parte Arietis habent minores elevaciones in circulis obliquis quam in rectis, et illi qui sunt ex parte Libre habent maiores in obliquis quam in rectis, in tantum in quantum alii habent maiores. Propter hoc igitur ad circulum obliquum[m] non est necessarium invenire elevaciones nisi trium signorum,

[a] C productionum [b] CM *FD* [c] M quia [d] M Illudendo (?) multiplicare
[e] M *A* [f] M *DO* [g] C tali [h] C λ [i] M +quelibet signa
[j] M 17λ [k] C elevaciones [l] M nomen [m] M obliquando

[In the last step] we multiply the number saved (obtained from a multiplication and a division earlier) by 150, the minutes of a whole semidiameter according to tables of sinus, or by 60 for tables of chord and arc. Divide the product by *IA* (the complement of the declination of the degree whose ascension we are seeking), and the result is *DC*, the difference in right ascension required, being that between the right and oblique horizons of that region, whose latitude you were supposed to know. Find the arc of this sinus, however, for what you have found is a sinus. *For note* that in every operation concerning circles we are working with sinus and not with arcs. If we say, for instance, that '*BE* is a quadrant of the equinoctial', or '*AG* is the latitude of the region', you are to understand this only as [implying] that [*BE* is the] sinus of a quadrant of the equinoctial, or *AG* is the sinus of the latitude of the region.

Look, therefore, for the arc corresponding to the sinus in the table of equation of sinus, if you worked with a semidiameter of 150 minutes, or in a table of chord and arc, if you multiplied by 60 degrees, which is there taken as the size of the semidiameter; and it amounts to the same thing.

If you now subtract that arc [*DC*] from arc *DB*, that is, from the right ascension, the arc *CB* will remain, this being the ascension[1] of the whole of Aries over your horizon.

And similarly for Pisces at the same region. And if you take them together with the right ascensions of points of Aries, you will have the right ascensions of their opposites (for the same region), namely of Libra and Virgo.[2] And so for the other signs, as the canons teach. They first show how to find the differences of the ascensions of all degrees of the first three signs. Any two equal arcs measured from the two points of Aries and Libra have equal ascensions. However, in the case of any two signs and arcs measured to equal distances from the two other points, of Cancer and Capricorn, arcs and signs towards Aries have smaller ascensions than those towards Libra.[3] But those arcs which are towards Aries correspond to smaller differences in oblique ascension than in right ascension, and those which are towards Libra correspond to greater differences in oblique ascension than in right ascension, [the relative magnitude being the same whether we reckon from Cancer or Capricorn].[4] On this account, it is not necessary to find oblique

[1] *elevacio*.

[2] As an example (for Oxford) for $\lambda =$ Aries 3°, $\alpha = 0^s\, 2^\circ\, 45'$, while for $\lambda =$ Libra 3°, $\alpha = 6^s\, 2^\circ\, 45'$ (following the table of *Albion* IV. 16). Ascensional *differences* (see n. 4) are therefore the same.

[3] In modern terms, the rate of change of ascension with respect to arc is less. See n. 4 and the graph of the commentary, vol. ii, p. 70.

[4] The meaning of the Latin sentence is obscure. The rates of change of what we are regarding as two functions with the names 'right ascension' (α) and 'oblique ascension' (α_0) are correctly described qualitatively. We have been led (in this and the previous chapter) to treat α and α_0 as quantities measured from the first point of Aries. The discussion now turns to the *differences* $\Delta\alpha$ and $\Delta\alpha_0$. When speaking in general of an arc *not commencing at the*

scilicet Arietis, Tauri, et Geminorum, quorum quodlibet potest inveniri per talem catham qualem tibi descripsi. Inventis autem elevacionibus horum trium signorum, habebitur facile omnis aliorum. Nam, cognitis elevacionibus istorum trium in circulo directo et obliquo per hanc secundam demonstracionem, ex differencia sue elevacionis in circulo directo et obliquo, demam illam differenciam de elevacione eius ad circulum directum (quia[a] minorem habet elevacionem ad circulum obliquum[b] quam directum, ut predixi); quod residuum[c] est elevacio duorum arcuum equidistantium a principio Arietis ex utraque parte, scilicet totius Arietis et Piscium. Item illam eandem differenciam addam super eandem elevacionem ad circulum directum, et quod provenit erit elevacio alicuius arcus equalis arcui tantum distans a Cancro quantum Aries distat ab eo, scilicet Virginis et Libre quod tantum distat[d] a puncto Capricorni quantum Piscium.[e] Similiter accipe differenciam elevacionis Tauri in circulo directo et obliquo quam si minueris de eius elevacione ad circulum directum, remanebit eius elevacio et Aquarii ad circulum obliquum.[f] Et si addideris eandem differenciam super ascensiones illius in circulo directo, habebis elevaciones Scorpionis et Leonis in circulo obliquo. Eodem modo si differenciam Geminorum subtraxeris ab eadem habebis elevaciones eius et Capricorni in circulo obliquo, et si eandem addideris habebis elevaciones Cancri et Sagittarii. Et propter hoc non docetur in canonibus invenire elevaciones nisi trium signorum ad circulum obliquum.[g] Nota etiam si istam catham volueris assignare in lineis arcualibus secundum quod iacent in superficie spere, tunc oportet declarare[h] per cordas duplicis arcus sic: proporcio corde duplicis arcus *FG* ad cordam duplicis arcus *GA* componitur ex proporcione corde duplicis arcus *DI* ad cordam duplicis arcus *IA*, ducta in proporcionem corde duplicis arcus *FC* ad cordam porcionis duplicis arcus *DC*. Et utraque operacio redit in idem, sicut predictum est superius in isto quarto tractatu.

Nunc restat declarare declinacionem gradus Solis ab equinoxiali; et concludam istum tractatum tanquam perfecte compilatum.

[a] M quem [b] C λ; M obliquando [c] C +fuerit; M +sinum [d] M λ
[e] CM Pisces [f] M obliquando [g] M obliquando [h] M +te

ascensions apart from those for three signs, namely Aries, Taurus, and Gemini, for which any [ascension] may be found by means of a cata of the sort I have described for you. Having found the ascensions for these three signs, however, weshall easily find them for the rest. For, the right and oblique ascensions for these three signs being knownby our second proof, but only as a difference between right and oblique ascension, I shall, [in the first three signs], subtract that difference from the right ascension (since the oblique is less than the right ascension, as I said before). The remainder is the ascension of two arcs equidistant from the beginning of Aries on either hand, that is, [following our earlier example], of the whole of Aries and Pisces.

[The argument from this point may be expressed in terms of the following notation. Let $\Delta\alpha$ be the difference between the right ascensions of the beginning and end of an ecliptic arc, and $\Delta\alpha_0$ be the corresponding difference in oblique ascensions. It is tacitly assumed that the ecliptic is divided into an integral number of equal arcs—as in Ptolemy's 10-degree division. Assume that λ is the mean longitude of one such arc. Then the difference derived from two differences like the *DC* of the earlier part of the chapter may (for $0° \leqslant \lambda \leqslant 90°$) be written

$$D(\lambda) = \Delta\alpha - \Delta\alpha_0.$$

Thus far it has been said that from Aries 0° to Cancer 0°, for longitude λ,

$$\Delta\alpha_0 = \Delta\alpha - D(\lambda).]$$

Then

from Cancer 0° to Libra 0°, for longitude $90°+\lambda$,

$$\Delta\alpha_0 = \Delta\alpha + D(90°-\lambda);$$

from Libra 0° to Capricorn 0°, for longitude $180°+\lambda$,

$$\Delta\alpha_0 = \Delta\alpha + D(\lambda);$$

from Capricorn 0° to Aries 0°, for longitude $270°+\lambda$,

$$\Delta\alpha_0 = \Delta\alpha - D(90°-\lambda).$$

By reason of all this, finding oblique ascensions is not taught in the canons except for three signs. Note also that if you want to express the cata in terms of arcs[1] such as lie on the surface of a sphere, then it is necessary to do so using cords of the double arc, as follows:

(2′) $\text{Crd } 2FG : \text{Crd } 2GA = (\text{Crd } 2DI : \text{Crd } 2IA) . (\text{Crd } 2FC : \text{Crd } 2DC)$.

Each method of working reduces to the same thing, as was said earlier in this fourth treatise.

It is now required that we speak of the declination of the Sun from the equinoctial; and I will conclude the treatise, giving it a semblance of completeness.

vernal point, we shall translate such phrases as '*elevacio ad circulum* (*directum*)' as 'difference in (right) ascension' (in symbols $\Delta\alpha$). We could alternatively have referred to the 'ascension of an arc'.

[1] *lineae arcuales*.

IV. 9

Ad demonstrandum autem declinacionem cuiuslibet gradus zodiaci ab equinoxiali, disponatur catha sic:[a] sit *A* polus, *C* principium Cancri in equinoxiali, *D* principium Arietis, *B* principium Cancri in ecliptica. Igitur linea *AC* est colurus transiens per caput Cancri, et *DC* est quarta equinoxialis a principio Arietis usque in finem Geminorum, et[b] *DB* est quarta ecliptice a principio Arietis usque in finem Geminorum vel principium Cancri, quod idem est. Sit etiam *E* gradus zodiaci cuius volumus inquirere declinacionem, et ab[c] *A* puncto per *E* transeat una linea usque ad *F* in equinoxiali. Habemus ergo catham ex *AC* et *DC* lineis principalibus concurrentibus in *C* puncto; et ab *A* polo reflectitur unus colurus super *DC* equinoxialem qui est *AF*, et ad[d] *AF* est a *D* puncto, quod est caput Arietis, in linea *DC* reflectitur altera super lineam *AC*, et iste due linee intersecant se super punctum *E*. Et est *CB* tota declinacio, *BA* est residuum declinacionis, *DE* est arcus cuius querimus declinacionem, *EB*[e] est residuum. Item *EF* est declinacio gradus *E* quod querimus, quod sic invenitur:

Per 15^{am} proposicionem de cathis coniunctis, patet quod proporcio *EF* ad *DE* componitur ex proporcione *CB* ad *CA* et ex proporcione *AF*[f] ad *DB*. Sed cum tam *DB*[g] quam *AF* utrumque sit sinus 90 graduum, et per consequens equalia, illorum nulla[h] est proporcio. Ergo qualis est proporcio *CB* (quod est sinus tocius declinacionis) ad *AC* (quod est sinus 150 minutorum vel 60 gradus, ut predixi; est enim semidiametri), talis est proporcio *EF* (quod est declinacio gradus assignati) ad *DE* (quod est arcus assignatus). Ergo econtra, qualis est proporcio *AC* ad[i] *BC* talis est proporcio *DE* ad *EF*; permutatim, qualis est proporcio[j] *DE* ad *AC*, talis est proporcio *EF* ad *BC*. Sic ergo invenies quantitatem *EF*, quod est declinacio tocius *DE*. Probatum est quod qualis est proporcio *AC* ad *BC*, talis est proporcio *DE* ad *EF*. Multiplica ergo medium per medium et divide per primum, et quod ex divisione resultat erit quantitas linee *EF*. Hoc est, pro sinu arcus cuius queris declinacionem multiplica [per sinum assignati] sinum tocius declinacionis et quod surgit divide per sinum totum secundum 150 minuta vel 60 gradibus; et quod surgit erit sinus illius declinacionis quam queris, cuius sinus quere circuli porcionem, et habes quod queris.

[a] M λ [b] M 12λ [c] C λ; M *AB* [d] C 3λ; M a *D* [e] M *CB*
[f] CM 3λ+*DB* ad *AF* [g] C *DF* [h] M nullam
[i] M +*AB* talis est proporcio *AC* ad
[j] C +*EC* talis est *EF*, quod est declinacionis gradus assignati ad *DE*, quod est arcus assignatus; ergo econtra, qualis est proporcio *AC* ad *BC*, talis est proporcio *DE* ad *EF*, et permutatim qualis est proporcio

IV. 9

To demonstrate the declination from the equinoctial[1] of any degree of the zodiac, arrange the catas as follows. Let *A* be the pole, *C* the beginning of Cancer on the equinoctial, *D* the beginning of Aries, and *B* the beginning of Cancer on the ecliptic. Then the line *AC* is the colure through the head of Cancer, and *DC* is a quadrant of the equinoctial from the beginning of Aries to the end of Gemini, and *DB* is a quadrant of the ecliptic from the beginning of Aries to the end of Gemini or the beginning of Cancer, which is the same thing. Let *E*, furthermore, be that degree of the zodiac whose declination we require, and let a line cross from *A* through *E* and as far as *F* on the equinoctial. We then have a cata formed out of the principal lines *AC* and *DC*, meeting at the point *C*. A colure *AF* returns from *A* (the pole) to *DC*, the equinoctial. The other line [*DB*] from *D* (the head of Aries) to *AC* and falling across *DC* crosses *AF* in *E*. Then *BC* is the total declination, [ϵ],[2] *BA* its complement, *DE* the arc [λ] whose declination we seek, and *EB* its complement. *EF* is the declination [δ] of the degree *E*, which we require, and which is found as follows.

$$(1)\ EF:DE = (CB:CA).(FA:DB) \qquad \text{II. c. 15.}$$[3]

Since, however, both *DB* and *AF* are Sin 90°,[4] there is no ratio between them, since they are equals. Thus

$$(2)\ CB:AC = EF:DE,$$

$BC = \text{Sin}\ \epsilon$, $EF = \text{Sin}\ \delta$, and $DE = \text{Sin}\ \lambda$. Therefore,

$$(3)\ AC:BC = DE:EF \qquad (2)$$
$$(4)\ DE:AC = EF:BC \qquad (2).$$

This, then, is how you will find *EF*, the declination of *DE*. It has been proved that $AC:BC = DE:EF$. Multiply therefore mean by mean and divide by the first, and *EF* will result. That is, since $AC = \text{Sin}\ 90°$,

$$(5)\ \text{Sin}\ \delta = (\text{Sin}\ \epsilon.\text{Sin}\ \lambda)/150',$$

or $\text{Sin}\ \delta = (\text{Sin}\ \epsilon.\text{Sin}\ \lambda)/60^p$.[5]

Look for the arc corresponding to the sinus, and you will have what you require.

[1] Henceforth simply 'declination'.

[2] That is, the obliquity of the ecliptic.

[3] In fact the arrangement is $\Delta ABE/DFC$. The MSS. all give the last ratio as *DB*:*FA*, and since the mistake is repeated below with chords, it must have been an author's mistake—but an unimportant one, as regards the result, since *DB* and *FA* are equal.

[4] We are reminded again here that when the context makes it clear, in these chapters concerning the sphere, *AB* (for example) is written as a shorthand for Sin *AB*.

[5] This result could have been obtained at once, had the sine rule been applied to the triangle *DEF*.

Si hoc idem volueris operari in catha facta ex arcubus sicut iacent in superficie spere, tunc operaberis per cordas duplicis porcionis sic: proporcio corde duplicis arcus *EF* ad cordam duplicis arcus *ED* componitur ex proporcione corde duplicis arcus *CB* ad cordam duplicis[a] arcus *CA*, et ex proporcione corde duplicis arcus *AF*[b] ad cordam duplicis arcus *DB*.[c] Et patet demonstracio superius in ultimo tractatu de cathis arcualibus, eo quod omnis conclusio probata de cathis rectilineis in tercio libro, que intelligi debent de cordis mediatis, illa eadem conclusio et omnes et singule patent in cathis arcualibus, si intelliguntur de cordis duplicis porcionis. Et hoc est tibi notabile.

Nunc sequitur ultimum capitulum de stellis, ad probandum eius latitudinem ab equinoxiali, et gradum[d] cum quo mediat celum, per verum gradum eius in ecliptica et per latitudinem eius ab ecliptica, si Deus voluerit.

IV. 10

Si habueris longitudinem stelle in ecliptica, et latitudinem eius ab ecliptica verificatas per antiquas tabulas et per motum octave spere, vel certius[e] per bona instrumenta ut armilla, facile erit tibi invenire latitudinem ipsius stelle ab equinoxiali et gradum illum cum quo provenerit ad medium celi hoc modo:

Describatur unus[f] circulus meridianus in quo sunt *A*, *B* duo poli mundi et *CD* duo poli zodiaci, ita[g] quod *C* distat ab *A*, polo septentrionali, per totam declinacionem; et similiter *D* a puncto *B*[h] distet per tantumdem. Deinde, super polos *C*,[i] *D*, describatur alius circulus *GH* qui erit vice zodiaci; et[j] super polos *A*, *B* describatur circulus *EF* pro equinoxiali. Ergo *E*, *G* puncta distabunt adinvicem similiter per totam declinacionem, et *F*, *H* similiter. Et ubi isti duo arcus exeuntes a duobus polis, scilicet equinoxialis et zodiaci, concurrunt[k] ponatur *T* stella, cuius querimus longitudinem et latitudinem secundum equinoxialem.

Transit arcus[l] enim a polo zodiaci per centrum stelle *T*,[m] ergo ille est verus gradus stelle ubi ille arcus abscindit eclipticam ubi ponatur *I*. Similiter a polo mundi *A*[n] transit arcus per centrum *T* super equinoxialem ad punctum *K*[o] secans[p] eclipticam super[q] *L*. Et ille erit gradus cum quo stella pervenit ad medium celi, et[r] hoc erit unum quod querimus. Item, *TI* dicit mihi latitudinem stelle ab[s] ecliptica, eo quod linea *CTIND* exit a polo zodiaci. Sed et linea *BT*[t] dicit mihi latitudinem *T* equinoxiali, eo quod linea *ATLKB* exit a polo mundi, et est unus colurus. Omnes isti circuli sunt de maximis circulis in spera, ut[u] supponitur.

[a] C λ [b] CM *DB* [c] CM *AF* [d] M gradu cum et gradum [e] C certe; M certa [f] M λ [g] M λ [h] M *D* [i] CM *G* [j] CM 11λ [k] CM λ [l] CM λ [m] M *C* [n] C 2λ [o] C ergo; M λ [p] C secat [q] C ad punctum [r] M in [s] M in *AB* [t] C *TK* [u] C 2λ

If you want the same thing set out as a cata comprising arcs of a sphere, you will work with chords of the double arc as follows:

$$\text{Crd } 2EF : \text{Crd } 2ED = (\text{Crd } 2CB : \text{Crd } 2CA) . (\text{Crd } 2AF : \text{Crd } 2DB).$$

The proof is given above, in the last treatise, which deals with catas involving arcs. Thus for any conclusion proved in Part III concerning rectilinear catas applies to half chords, that same conclusion, like every other [to be found there], is evidently true for catas understood in terms of chords of double arcs. And this is something of which you should take note.

There now follows, if God wills it, the last chapter concerning stars and relating to the declination[1] from the equinoctial, and the mediation,[2] [both being found from] ecliptic longitude[3] and latitude.

IV. 10

If you have the longitude and latitude of a star, verified from old tables and from the motion of the eighth sphere, or with greater certainty from good instruments such as the armilla, you will easily find its declination and mediation in the following manner.

Let a meridian circle be described on which *A* and *B* are the two poles of the world, and *C* and *D* the two poles of the zodiac, so that *C* is separated from *A*, the north pole, by the total declination;[4] and similarly *D* is at this same distance from *B*. Then another circle, *GH*, is to be described about the poles *C* and *D*, and this will stand for the zodiac; and about the poles *A* and *B* let a circle *EF* be drawn, standing for the equinoctial. The points *E* and *G* will thus be separated by the total declination, as will *F* and *H*. Let the star *T* be where the two arcs which leave the two poles (of the ecliptic and of the zodiac) meet. We seek the star's declination and right ascension.[5]

An arc crosses from the pole of the zodiac through *T*, the centre of the star, and this therefore is the true degree of the star; and let *I* be that point where the arc crosses the ecliptic. Similarly, an arc from the pole *A* through the centre *T* crosses the equinoctial at *K*, and the ecliptic at *L*. This will be the degree of the star's mediation, and will be what we are seeking. Next, I call *TI* the latitude of the star from the ecliptic, since the line *CTIND* goes trom the pole of the zodiac. I call *BT* the 'equinoctial latitude' [or declination], since the line *ATLKB* goes from the pole of the world, and is a colure. All these circles are taken to be great circles on a sphere.

1 *latitudo.*

2 *gradus cum quo mediat celum,* a cumbrous but expressive and conventional phrase. When the star crosses the meridian, so also does this particular point ('degree') of the ecliptic. See Appendix 21 for a survey of this and other coordinates of the sphere.

3 *verus gradus in ecliptica.*

4 The obliquity, ϵ, as before.

5 Literally, 'equinoctial latitude and longitude'.

Et ideo volumus declarare quantitatem *TK*, quod est declinacio *T* ab equinoxiali. Intellige unam catham factam ex duabus lineis principalibus *EC*, *EK*,[a] concurrentibus in *E*, et ex aliis duabus lineis, reflexis a punctis *C* et *K* intersecantibus sese super punctum *T*, scilicet linea *CN* et linea *KA*. Ergo sicut[b] probatum est antecedenter in prima proposicione de cathis coniunctis in lineis rectis, et consequenter in quinta proposicione libri quarti de cathis arcualibus,[c] proporcio *CE* ad *EA* componitur ex proporcione *CN* ad *NT* ducta in proporcione *KT* ad *KA*.[d] Sed cum *CG* et *AE* et *CI* et *AK* omnes sunt equales, quia quarte circuli, et cum *IN* sit declinacio gradus *I* nobis nota, et *TI* erit declinacio stelle ab[e] ecliptica nobis similiter nota, erit nobis notus totus arcus *NT*. Istius ergo cathe sunt nobis noti: arcus *CE*, et *EA*, similiter *CN*, et *NT*, et nota est nobis tota *AK*.[f] Ergo non possumus ignorare quantitatem *TK*, quod querere volebamus.

Exempli gracia. Sit ita quod *T* sit una stella que distat in latitudine septentrionali ab ecliptica per 22 graduum (ut Alaioch fere, quia distat per 22 grad. et 30 min., sed omitto[g] minuta), et sit eius longitudo ab Ariete 2 sig. et 6 grad. quod est linea *OI*. Videamus primo que sit proporcio corde duplicis porcionis *CE* ad *EA*: est enim *CE* porcio 114 grad. (supponendo ad presens quod declinacio sit 24 gradus precise), ergo dupla porcio erit 228,[h] cuius corda mediata secundum artem cordarum, est 54 grad. 48 min. 46 sec., secundum quod tota[i] diameter est 120 grad. ut sepe dixi; sed sinus totus graduum 228, secundum equaciones[j] sinus, est 222 min. 56 sec. 40 ter., secundum quod tota diameter supponitur esse 300[k] minutorum; nec est cura[l] cum quo operemur—redit enim ad[m] idem, excepto quod per sinus operamur cum duplici porcione, in tabulis de cordis[n] mediatis operamur per cordam rectam arcus dati. Accipio igitur ad presens sinus tabularum de sinu duplato cuius tercie sunt 802,600.[o] Corda autem dupli arcus *EA*[p] est sinus tot graduum[q] 180, cuius sinus est 300 minutorum, quia tota[r] diameter; et istos duos sinus divide, maiorem per minorem, et surgit proporcio unus ad alteram. Sunt enim tercie 300 minutorum tot 1,080,000, quod si diviseris per 802,600 exibit proporcio 13[pla] et remanent 31,200[s] quod quia minus est[t] medietate abice. Est ergo

[a] M *EH* [b] M sic [c] C (et passim) arcubus [d] C *KL* [e] CM ad
[f] C *NK* [g] C omitte [h] M 28 [i] M totus [j] C tabulam equacionis
[k] M 100 [l] C corda [m] M cum [n] M corde [o] M 402,600
[p] C cuius [q] C gradum [r] M totus [s] C 32,200 [t] C λ

We wish therefore to determine the magnitude of *TK*, that is, the declination[1] of *T* from the equinoctial. Take a cata formed from the two principal lines *EC* and *EK* meeting in *E*, and two further lines *CN* and *KA* returning from the points *C* and *K*, and intersecting in the point *T*. Then

(1) $CE{:}EA = (CN{:}NT).(KT{:}KA)$	III. c. 1, IV. 5
(2) $CG = AE = CI = AK = 90°$	constr.
(3) [$NT = IN + TI$	constr.]
(4) TI^*	[$= \beta$]
(5) IN^*	[from λ][2]
(6) NT^*	(3), (4), (5)
(7) CE^*	[$= 90° + \epsilon$].

Therefore from (2), (6), and (7) we may know[3] *TK*, which is what we require.

Example. Let *T* be a star with ecliptic latitude 22° (as is approximately the case with Alaioch,[4] of latitude 22° 30′, but I omit the minutes), and let its longitude from Aries (the line *OI*) be 2^s 6°. Let us first see what is the ratio Crd 2*CE*:Crd 2*EA*.[5] The arc *CE* is 114°, supposing for the moment that the obliquity is precisely 24°, and therefore the double arc will be 228°, the half-chord of which—according to the art of chords—is 54° 48′ 46″,[6] to a diameter 120^p, as I have often said; but according to the tables of sinus, 2 Sin 228° is 222′ 56″ 40‴,[7] taking a diameter of 300′. (Which [diameter] I choose is of no consequence, for both lead to the same result, except that with sinus we work with the double arc, whereas in the tables of half-chords we work with the right chord of the given arc.) I now take, then, the sinus from the tables of sinus duplati, which in thirds is 802,600.[8] But Crd 2*EA* = 300′. Divide the two sinus, the greater by the less, and the ratio of one to the other results. In 300′ there are 1,080,000 thirds, and divided by 802,600 this gives 13, remainder 31,200, which discard, since it is less than half.[9]

[1] The word '*declinacio*' is used at last.

[2] No reference is provided, nor is one implied by the statement that *IN* is the known 'declination' of *I*, for it is not the declination proper, in the modern sense, *IN* being orthogonal to the ecliptic, and not to the equator. We are therefore not entitled to suppose that the intention was to use IV. 9 for *IN*, a step we could take only if we knew *RN*.

[3] From (1).

[4] The star is that chosen as an example in what we designate as Tract II, in the *Rectangulus* section. See the further comments made there on the matter.

[5] The text should strictly be translated (Crd *CE*):*EA*; but the conventions for avoiding the tedium of writing proportions out in full should by now be fully appreciated.

[6] Crd 228° = Crd(360°−228°) = Crd 132°, and according to the table of *Almagest* I. 11, ½ Crd 132° = 54^p 48′ 46″, as here. This is accurate to five places of decimals.

[7] The Toledan tables give 111′ 28″ 25‴ (MS. Laud misc. 644) for Sin 228° (= Sin 48°), but this is not the quantity he should be seeking. See our commentary.

[8] This number of thirds comes from 222′ 56″ 40‴, or 2 Sin 228°.

[9] Note this rare medieval affirmation of the common modern rule of rounding off an approximate numerical value. We know that it was often ignored. Second, note that the division, like almost every step in this example, is miscalculated. See our commentary.

proporcio corde duplicis porcionis *EC* subterciadecima ad cordam duplate porcionis *EA*, et hoc bene nota.

Item *I*, quod supponitur esse 6 graduum Geminorum declinat ab equinoxiali per 21 grad. 24 min. 56 sec., et hec est quantitas linee *IN*; abicio ergo fractiones et suppono *IN* esse 21 grad. Similiter *IT* est latitudo stelle ab ecliptica, quod supponitur esse 22 grad.; quia *CI* est 90 grad. Nunc ergo similiter inquiramus proporcionem corde duplate dupli[a] arcus *CN* ad cordam duple porcionis *NT*;[b] est enim corda duplata duple porcionis *NC*, quod est 222 grad., [id est] 280 min. 4 sec. 30 ter., quod est in terciis 799,440, corda vero duple porcionis *NT*, quod est 86 grad. est 259 min. 18 sec. 10 ter., quod est in terciis tamen 433,490. Quod si primum dividatur per secundum exibit proporcio *CN* ad *NT* cordarum 2pla fere, que proporcio debet dividere priorem[c] proporcionem que est inter *EC* ad *EA*, videlicet subterciadecima et exibit 6pla et dimidium. Illa igitur est[d] proporcio corde duplicis porcionis *KT* ad cordam duplicis porcionis *KA*.

Sed scimus quod corda duplicis porcionis *KA* est 300 min., ergo corda duplicis porcionis *KT*[e] erit subsextupla,[f] et hoc est 50 min.; huius quere arcum duplatum vel medietatis arcum mediatum (et illud erit quantitas *KT* quod volebamus). Et hoc est 20 grad.

Hoc tibi posui gratiam exempli, non quod precise operatus sum, sed ut per hoc sencias quomodo poteris operari.

Iam restat disponere catham ad ostendendum cum quo gradu zodiaci *T* stella pervenit ad medium celi, et hoc est ostendere quis erit punctus *L*[g] in zodiaco. Fiat igitur catha que prius, et quia probatum est superius proporcione 17[h] de cathis disiunctis[i] quod proporcio corde dupli arcus *CA*[j] ad cordam dupli arcus *EC* componitur ex proporcione corde dupli arcus *AT* ad cordam dupli arcus *TK*[k] et ex proporcione corde dupli arcus *KN* ad cordam dupli arcus *NE*. Similiter sunt[l] in cathis de lineis rectis sed notum est et ex prioribus, quia proporcio *CA* ad *CE* fit ex proporcione *AT* ad *TK* et ex proporcione *KN* ad *NE*. Sed que est proporcio *CA*[m] ad *CE* eadem est proporcio *AT*[n] ad aliquid aliud. Illud aliud sic invenias. Multiplica medium in medium, et productum divide per primum, et exibit illud aliud; ad quod[o] sic se habet *AT* sicut se

[a] M duplum [b] C *AC* [c] C priori [d] C λ [e] M *BT*
[f] C sextupla [g] CM *I* [h] CM secunda [i] CM coniunctis
[j] M 12λ [k] CM *CK* [l] C λ [m] CM *CK*
[n] CM *AC* [o] M +scilicet *AT*

Thus

$$\text{Crd } 2EC : \text{Crd } 2EA = 1:13.$$

Next I, which is taken to be at 6° of Gemini, has declination[1] 21° 24′ 56″, and this is the size of line IN. I reject then the fractions, and suppose IN to be 21°. Similarly, IT is the ecliptic latitude of the star, taken to be 22°; and CI is 90°. We shall now therefore likewise ask for the ratio Crd $2CN$: Crd $2NT$, where $2CN$ is 222°, with chord 280′ 4″ 30‴ (or 799,440‴),[2] and where $2NT$ is 86°,[3] with chord 259′ 18″ 10‴ (or 433,490‴).[4] If the first be divided by the second, the ratio of chords with CN and NT is roughly 2 : 1, which ratio should divide the first ratio EC:EA, namely 1 : 13, giving $6\frac{1}{2}$: 1. This, then, is the ratio

$$\text{Crd } 2KT : \text{Crd } 2KA.^{5}$$

We know, however, that Crd $2KA$ is 300′, and therefore Crd $2KT$ will be $\frac{1}{6}$, or 50′. We look for the double arc corresponding to this, or the half-arc which corresponds to half of it (which will be the magnitude of KT which we wanted). This [KT?] is 20°.[6]

This I gave you as an example, not that I worked precisely,[7] but that you might be able to work out examples by means of it.

It is now left to us to set out a cata to show what degree comes to the meridian with the star T, which is to say, to show at what point of the zodiac is L.[8] Let there be the cata as before. It is shown above that

$$(8)\quad \text{Crd } 2CA : \text{Crd } 2EC = (\text{Crd } 2AT : \text{Crd } 2TK) . (\text{Crd } 2KN : \text{Crd } 2NE)$$

III. D. 17.[9]

The procedure is similar to that with catas involving straight lines, as where we have

$$(9)\quad CA:CE = (AT:TK).(KN:NE).$$

The ratio CA:CE is the same as the ratio of AT to some other thing, which you are to find as follows. Multiply mean by mean, divide by the first, and the other (to be known as P) results:

[1] The word is used in the modern sense, and the statement is strictly incorrect, as explained in our commentary.

[2] The chord is correctly given, but in thirds it reduces to 1,008,270‴, rather than to 799,440‴. The latter was obviously obtained, even so.

[3] Namely $2(IT+IN)$, or $2(22°+21°)$, if we accept 21° for IN.

[4] In fact Crd 86° = 204′ 35″ 58‴·2. We note, however, that the quoted chord is equal to 933,490‴, making the text's 433,490‴ wrong in the first digit—but it was clearly the figure calculated.

[5] Cf. (1) above.

[6] This answer is irredeemably wrong. See our commentary.

[7] This is an understatement, as we have now seen that there was an arithmetical error at almost every stage.

[8] This is a curious statement of intent, for the remainder of the chapter is devoted to finding the right ascension of T, that is, the point of the equator which comes to the meridian with the star. But see *gradus cum quo stella mediat celum* in the Latin glossary.

[9] The pattern of II. D. 17 is actually 561243, permuted here to 564312.

habet *CA* ad *CE*. Illud aliud sit *P*. Et intellige hoc bene, hoc est, multiplica sinum 90 et 24 grad., scilicet 114 grad., in sinum residui declinacionis stelle ab equinoxiali, et quod surgit divide per sinum tocius declinacionis 24 grad., et surget unus sinus *P* ad[a] quod se habet sinus residui declinacionis stelle ab equinoxiali sicut se habet sinus tocius declinacionis 24 grad. ad sinum arcus aggregati ex[b] declinacione et quarta circuli. Ergo que est proporcio *CA* ad *CE* illa est *AT*[c] ad *P*.

Sicut ergo proporcio *CA* ad *CE* componitur ex proporcione *AT*[d] ad *TK* ducta in proporcione *KN* ad *NE*, ita eodem modo proporcio *AT*[e] ad *P* componitur ex eisdem,[f] scilicet ex proporcione *AT*[g] ad *TK* ducta in proporcione *KN* ad *NE*. Sed scias quod proporcio *AT* ad *P* habet medium, scilicet *KT*,[h] igitur proporcio *AT* ad *P* componitur ex *AT* ad *KT* et *KT* ad *P*; igitur proporcio *KT*[i] ad *P* erit sicut proporcio *KN* ad *NE*, quia idem resultat ex *AT*[j] ad *TK* cum proporcione *TK* ad *P* sicut cum proporcione *KN* ad *NE*. Cum ergo proporcio *AT* ad *P* sit nota, quia nota est proporcio *CA* ad *CE*, et *AT* similiter est notum, ergo necessario *P* erit notum. Est enim *AT* residuum declinacionis stelle ab equinoxiali. Ex quo ergo similiter nota est linea *TK*, quia est declinacio stelle ab equinoxiali, quod iam in precedenti operacione invenimus, erit proporcio *KT*[k] ad *P* nota. Ergo similiter nota erit proporcio *KN* ad *NE*.

Sed linea *NE* potest esse nobis nota, cum fuerit nobis nota linea *IG*. Vides enim in hanc figura duas cathas, quarum una componitur ex[l] *EC* et *EK* principalibus et duabus lineis reflexis, *KA* et *CN*, intersecantibus se in puncto *T*, ita quod se *C* est vice poli zodiaci, *A* poli mundi, *G*[m] caput Cancri in zodiaco, *E* caput in equinoxiali, et *CI*[n] est colurus transiens a polo zodiaci per centrum *T*, docens[o] quis est verus gradus *T* stelle. Item aliam catham habebis factam ex *EA*, *ER*[p] principalibus et duabus reflexis una *RG* et altera *AK*, ita quod *R*[q] est caput Arietis, et *A* polus mundi, *RE*[r] est quarta equinoxialis a puncto Arietis usque in finem Geminorum et *RG*[s] est quarta zodiaci a principio Arietis in finem Geminorum.

Cum igitur cognoscis *IG*, per tabulam ascensionum in circulo directo poteris cognoscere *EN*, sicut dictum est prius in demonstracione

[a] CM 5λ+qui se habet ad sinum [b] C 2λ [c] C *AC* [d] C *AC*
[e] M *AC* [f] M eadem [g] C *AC*; M *AD* [h] M *KC* [i] M *KC*
[j] C *AE* [k] M *KC* [l] CM 2λ+*ET* [m] CM ergo [n] CM *EI*
[o] C dices [p] C *EI* [q] M *I* [r] M ergo *E* [s] M *IG*

$$(10)\ CA{:}CE = AT{:}P,\ P = (AT.CE)/CA.$$

Note this well. Form the product

$$\mathrm{Sin}(90^\circ-\delta).\mathrm{Sin}(90^\circ+24^\circ),$$

and divide the result by Sin ϵ, or Sin 24°, giving Sin P, which occurs in the proportionality (cf. (10))

$$(11)\ \mathrm{Sin}(90^\circ-\delta){:}\mathrm{Sin}\,P = \mathrm{Sin}\,24^\circ{:}\mathrm{Sin}\,114^\circ.$$

Just as we have (9), so in the same way

$$(12)\ AT{:}P = (AT{:}TK).(KN{:}NE) \qquad (2), (3).$$

But you should know that if KT be a mean between AT and P,

$$(13)\ AT{:}P = (AT{:}KT).(KT{:}P) \qquad \text{II. } 2,$$

and therefore

$$(14)\ KT{:}P = KN{:}NE \qquad (12), (13),$$

since both $KT{:}P$ and $KN{:}NE$ combine with $AT{:}KT$ to give the same result. Now

$$(15)\ (CA{:}CE)^* \qquad \epsilon^*$$
$$(16)\ AT^* \qquad \delta^*$$
$$(17)\ P^* \qquad (10), (15), (16).$$

AT is $(90^\circ-\delta)$. From the fact that

$$(18)\ KT^*,$$

KT being the declination, found in the first operation above, therefore

$$(19)\ (KT{:}P)^* \qquad (17), (18)$$
$$(20)\ (KN{:}NE)^* \qquad (14), (19).$$

Now the line NE may be determined, for we shall find the line IG. Consider in the figure two catas,[1] one having principal lines EC and EK, and the two return lines KA and CN, meeting in T, so that C stands for the pole of the zodiac and A for the pole of the world, G is the head of Cancer on the zodiac, E the same on the equinoctial, and CI the colure crossing from the pole of the zodiac through T, showing what is the true degree[2] of the star T. Next, you will have another cata formed by the principal lines AE and ER, and the two reflex lines, one RG and the other AK, so that R is the head of Aries and A the pole of the world, RE is a quadrant of the equinoctial, from the head of Aries to the end of Gemini, and RG is a quadrant of the zodiac from the beginning of Aries to the end of Gemini.

Since you know IG,[3] you may determine EN from the table of right ascensions, as was explained formerly in the proof dealing with right

[1] These seem to be merely described for our perusal, and are not made the basis of any further proportionalities. See the penultimate footnote to this chapter.

[2] The ecliptic longitude.

[3] From the longitude which it complements.

ascensionum in circulo directo; quia si *C* esset polus mundi, *EN* essent ascensiones *GI* in circulo directo. Nunc autem est vice versa: *C* est polus zodiaci, igitur *GI*[a] est ascensio porcionis *NE*. Vide ergo cui porcioni *IG* correspondet, et illa erit quantitas linee *EN*.

Ergo ad proporcionem que est proporcio *TK* ad *P*, illa erit *KN* ad *NE*. Sed nota est proporcio *TK* ad *P* et iam nota est linea *EN*. Ergo nota erit linea *KN*. Sed[b] ut invenias quod se habet ad *EN* sicut se habet *P* ad *TK*, multiplica medium per medium et quod surgit divide per primum; id est, multiplica *EN* per *TK*[c] et productum divide per *P*,[d] et surget numerus *NK*. Ergo nota erit tota linea *EK*, que sunt ascensiones gradus stelle in circulo directo. Sed cum eodem gradu provenit stella ad medium celi, et oritur in circulo directo; igitur habetur propositum bene intuenti, et cetera.

IV. 11

Nunc autem disponenda est catha ad horas cognoscendas per altitudinem Solis vel stelle. Sit igitur *AB* quarta meridiana, a cenith capitis usque ad punctum orizontis meridianum, ita quod *A* sit cenith et *B* punctus in orizonte. Et sit *BC* quarta orizontis a puncto meridiei usque ad *C*, qui est punctus orientis.[e] Erunt igitur *AB* et *BC* due linee principales a quarum terminis reflectuntur due linee alie intersecantes se in puncto *D*, scilicet linea *CE*, que erit vice linee equinoxialis et alia linea *AF*, que est linea coluri transiens a cenith capitis, per punctum *D*[f] qui erit corpus Solis vel stelle exeuntis ab[g] ortu equinoxiali. Est igitur linea *AB* linea 90 graduum a cenith usque ad orizontem, que vocatur 'linea meridiana'; et linea *BE* elevacio capitis Arietis super orizontem tuum; *EA* residuum, scilicet, latitudo regionis. Linea[h] *AF* similiter est linea transiens a cenith per corpus Solis usque in orizontem, igitur *AF* similiter est 90 graduum; et linea *FD* est altitudo accepta, et *DA* residuum. Item *CE* est medietas arcus diei, igitur *CD* est arcus pertransitus in hora quando *D* habet altitudinem *DF*, et similiter *CF*[i] est azimuth eiusdem in eadem hora; quorum utrumque querimus scilicet quis est arcus *CD* et[j] quis[k] est azimuth *CF* quando Sol elevatur per *DF*, quorum utrumque sic patet.

[a] M 12λ+in *ET* [b] C sic [c] CM *P* [d] CM *TK* [e] M orizontis
[f] C λ [g] C in [h] C λ [i] C *CE* [j] M 3λ [k] C 2λ

ascensions; for if *C* were the pole of the world, *EN* would be the ascension. Now, however, it is the other way round: *C* is the pole of the zodiac, and therefore *GI* is the ascension of arc *NE*. See, therefore, to what arc *GI* corresponds,[1] and that will be the size of the line *EN*:

(21)	*EN*	tables of ascension
(22)	*KN**	(20), (21).

To find *KN*, multiply mean by mean and divide by the first (cf. (14)):

$$KN = (NE.KT)/P.$$ [2]

The complete line *EK* will then be known [from (21), (22)], this being the [complement of][3] the right ascension of the degree of the star. But with this same degree the star comes to the middle of the sky, and rises on the right horizon.[4] The proposition is therefore plain to see.

IV. 11

[It is initially assumed that the Sun or star is on the equinoctial.][5] Now to set out a cata to discover the hour by the altitude of the Sun or of a star.[6] Let *AB* be a quadrant of the meridian, from the zenith[7] to the meridian point of the horizon,[8] with *A* the zenith and *B* the point on the horizon. Let *BC* be a quadrant of the horizon from the meridian point up to *C*, a point in the east. *AB* and *BC* will then be two principal lines, from the ends of which run two other lines, intersecting in *D*, namely *CE* (which represents the equinoctial), and *AF* (the line of the colure from the zenith through *D*, the body of the Sun or of the star as it leaves the east). The line *AB*, known as the 'meridian line', is then of 90°, from the zenith to the horizon; and the line *BE* is the [maximum] elevation[9] of the head of Aries above your horizon; and *EA* is its complement, that is to say, is the latitude of the region. The line *AF* is likewise a line crossing from the zenith. It passes through the body of the Sun and as far as the horizon, and is therefore also 90°. The line *FD* is the measured altitude,[10] and *DA* its complement. Next, *CE* is half the diurnal arc, and *CD* is the arc traversed at the time when *D* has the altitude *DF*. Similarly, *CF* is *D*'s azimuth at the same moment. We seek both of these, namely the arc *CD* and the azimuth *CF* when the Sun's altitude is *DF*. Each is evident from what follows.

[1] That is, use the tables of ascensions in reverse.

[2] The MSS. agree—suggesting a lapse on the part of the author—that $P = (EN.P)/KT$.

[3] Translating 'ascensions on the direct circle' as 'right ascensions' has disadvantages in a case of this sort. An ascension in the sense of the text could of course be any stretch of the equator. See Appendix 22.

[4] *circulus directus*.

[5] The configuration of the diagram leaves no alternative, but the assumption is not made explicit until the chapter is half completed.

[6] The promise of this opening is only partially fulfilled in this chapter, but completely so in the next.

[7] *cenith capitis*.

[8] *orizon* (without a qualifying '*obliquus*').

[9] *elevacio*—not an ascension in the sense of IV. 9.

[10] *altitudo accepta*.

Primo quantitas *DC*.[a] Scimus ex precedentibus quod proporcio *AB* noti ad *BE* notum componitur ex proporcione *AF* noti ad *FD* notum ex proporcione *DC* ignoti ad *CE* notum. Hec omnia nota bene. Ulterius, qualis est proporcio *AB* ad *BE* talis est proporcio *AF* ad aliquid aliud, quod[b] potest investigari per multiplicacionem *BE* in *AF* et divisionem producti per *AB*. Sed cum *AB* et *AF* sunt equales, utrumque 90 graduum, qualis est proporcio *AB* ad *BE* talis erit proporcio *AF* ad *BE*. Sed proporcio *AF* ad *BE* fit ex proporcione *AF* ad *DF* et *DF* ad *BE*, ergo proporcio *DF* ad *BE* similiter est proporcioni *CD* ad *CE*.

Similiter econverso, qualis est proporcio *EB* ad *DF*, talis est proporcio *CE* ad *CD*. Multiplica ergo medium in medium et divide per primum et surget quartum, scilicet *CD* ignotum. Hoc est multiplica cordam altitudinis accepte in corda recta medietatis arcus diei, scilicet *CE*, et productum divide per cordam rectam altitudinis meridiane, et exibit *CD*, quod est corda recta arcus[c] pertransiti ab ortu Solis usquequo *D* elevatur[d] per *DF* super orizontem. Et hec est una conclusio.

Et sequitur alia conclusio, scilicet que sit quantitas *CF*, qui est azimuth Solis in hora tante[e] altitudinis accepte. Est autem[f] azimuth distancia linee exeuntis a cenith capitis per centrum Solis usque in orizontem, secundum distanciam termini[g] eius a termino *C* orientis, versus[h] *B* meridiei si sit ante meridiem, vel a *B* in *G* si sit post meridiem, vel a *G*[i] occidente si sit post occidentem,[j] et cetera.

Linea ergo *CF* erit sic nota: per primam proporcionem de cathis disiunctis notum est tibi quod proporcio *AE* ad *EB* componitur ex *AD* ad *DF* et ex *CF* ad *BC*.[k] Sed que est proporcio *AE* ad *EB* talis est proporcio *AD* ad aliquid aliud. Sit illud *P*. Igitur proporcio *AD* ad *P* fit ex *AD* ducta in *DF* et ex *CF* ducta in *CB*. Nunc autem *AD* ad *P* fit ex *AD* ad *DF* et ex *DF* ad *P*, ergo eadem est proporcio *DF* ad *P* qualis est *CF* ad *CB*. Igitur econtra qualis est *P* ad *DF* talis erit *CB* ad *CF*. Sed nota est tibi proporcio *P* ad *DF* ex prioribus cum utrumque extremum est notum; igitur multiplica medium in medium et divide per primum *P* et surget quantitas *CF* quesitum. Hoc est, multiplica sinum altitudinis accepte in sinum totum quod est sinus *CB*, et productum divide per *P*, quod se habet ad *AD* (residuum altitudinis accepte) sicut se habet latitudo regionis ad altitudinem Arietis; et exibit azimuth altitudinis accepte.

[a] C +sic [b] C +sic [c] C alicuius [d] CM elevetur [e] M tandem
[f] M λ [g] C λ [h] C usque [i] CM *GE* [j] CM occidens
[k] CM *B*

First the quantity DC. We know from the preceding that

$$(1)\ AB:BE = (AF:FD).(DC:CE),$$

where all but DC are known. Note all this well.

Finally, the ratio $AB:BE$ is the ratio of AF to some other thing, which may be investigated by multiplying BE by AF and dividing the product by AB. Since AB and AF are equal (each being 90°),

(2)	$AB:BE = AF:BE$	
(3)	$AF:BE = (AF:DF).(DF:BE)$	II. 2
(4)	$DF:BE = CD:CE$	(2), cf. (1) with (3).

Conversely,

$$(5)\ EB:DF = CE:CD \qquad (4).$$

Multiply mean by mean, divide by the first, and the fourth results, namely the unknown CD. In other words, multiply the chord[1] of the measured altitude [DF] by the chord of half the diurnal arc, namely CE; divide the product by the chord of the meridian altitude, and we find CD, the chord of the arc traversed from the rising of the Sun to where D has an altitude of DF above the horizon. And this is one conclusion.

The other conclusion follows: to find CF, the azimuth of the Sun at the moment its altitude is that which was measured. The azimuth is a distance measured to the end-point [on the horizon] of the line from the zenith through the centre of the Sun to the horizon. The distance is measured from C in the east towards B on the meridian, if it be before noon, or away from B towards G if it be afternoon, or from G in the west if it be past the western point,[2] and so on.

The line CF will be known as follows.

(1)	$AE:EB = (AD:DF).(CF:BC)$	III. D. 1
(2)	$AE:EB = AD:P$	hyp.
(3)	$AD:P = (AD:DF).(CF:BC)$	(1), (2)
(4)	$AD:P = (AD:DF).(DF:P)$	II. 2
(5)	$DF:P = CF:BC$	cf. (3) with (4)
(6)	$P:DF = CB:CF$	(5)
(7)	$CF = (DB.CB)/P$	(6).

That is, multiply the sinus of the measured altitude by Sin 90° (= Sin CB), and divide by P where [Sin] AD (the complement of the measured altitude) bears to P the ratio of [the sinus of] the latitude of the region to the [sinus of] the maximum altitude of Aries.[3] The answer is the azimuth of the measured altitude.

[1] Throughout, 'chord' must be taken to carry the connotation 'chord of the double arc', or must be understood as referring to sinus.

[2] *post occidentem.*

[3] The text has the verbal formula of the latter part of the sentence (equivalent to (2) above) the wrong way round. AE is the geographical latitude but BE is simply its complement. There was no compelling need to mention the altitude of Aries.

Sed nota bene quod ista demonstracio probandi arcum elevatum de equinoxiali vel azimuth per Solis vel stelle altitudinem non valet nisi supposito quod centrum Solis vel stelle sit in linea equinoxiali, quod patet sic:

Disponatur enim[a] catha ut prius, et preter hoc protrahatur arcus stelle vel Solis habentis latitudinem meridionalem vel septentrionalem. Et sit linea latitudinis meridionalis *GH*[b] et *KI*[c] linea septentrionalis latitudinis. Dico igitur quod nulla catha arcualis in spera potest fieri ex *AB*, *BH* principalibus et ex *HG*, *AF* reflexis, nec similiter ex lineis *AB* et *BK* principalibus et *KI* et *AF* reflexis, cum nec linea *HG* nec linea *KI*[d] sint de[e] maioribus circulis in spera, quod tamen presupponitur in precedentibus demonstracionibus Ptholomei. Igitur oportet alterum modum invenire determinandi elevacionem porcionis circuli equinoxialis per altitudinem Solis vel stelle, quando declinant ab equinoxiali versus meridiem vel septentrionem, quia in utroque casu arcus eius non potest esse una de lineis in catha, cum sit minor circulo maximo dividente[f] speram.

Et est modus talis, ex quo ita quod non potest fieri catha ad ostendendum arcum equinoxialem elevatum per altitudinem gradus[g] vel stelle declinantis[h] ab equinoxiali, quia arcus eius[i] diurnus quem describit non est unus de maximis circulis in spera, ut predixi, oportet invenire alterum modum determinandi illud. Et primo disponemus catham ad ostendendum medietatem arcus diei talis gradus vel stelle in circulo obliquo, sive habuerit latitudinem septentrionalem sive meridionalem. Sit ergo *DB* quarta equinoxialis et *BA* quarta meridionalis coluri ab equinoxiali ad polum eius meridionalem *A* sub orizonte. Et[j] erunt iste due principales linee in catha. Et sit *AE* depressio poli meridionalis sub orizonte, et quod est equalis elevacioni alterius poli et latitudini regionis; et sit *EB* elevacio capitis Arietis super orizontem, quod est residuum latitudinis, et protrahatur linea *DE*, quod est orizon obliquus; et in linea *AFC*, colurus transiens per centrum stelle, signetur punctus *F*, que est latitudo stelle ab equinoxiali, scilicet arcus *FC* ita quod *F* punctus cadat in orizonte ubicumque volueris,[k] secundum exigenciam latitudinis *FC*. Ex quo ergo isti sunt quatuor quarte maiorum circulorum in spera, vel quatuor sinus quatuor quartarum circulorum maximorum, secundum veritatem[l] cathe disiuncte.

Erit proporcio *AE* ad *EB* composita ex duplici proporcione, ex ea scilicet que est *AF* ad *FC* et ex ea que est *DC*[m] ad *DB*. Si ergo multiplices primum *AE* in quartum *FC*, et productum dividas per tercium *AF*, exibit quiddam ad[n] quod se habebit[o] *AE* primum sicut se habet *AF* secundum ad *FC* tercium. Illud sit *P*. Ergo[p] cum proporcio *AE* ad *P* sit sicut proporcio *AF* ad *FC*, igitur proporcio *AE* ad *EB* fit ex *AE* ad *P* et

[a] M in [b] C +meridionalis [c] CM λ [d] CM *KF* [e] C λ
[f] M dividende [g] M 3λ [h] M declinaciones [i] M est
[j] C 17λ [k] M voluerit [l] C verificatem [m] CM *DE*
[n] CM λ [o] CM +ad [p] C 12λ

Note well, however, that this proof of the derivations of the time[1] or the azimuth by means of the altitude of the Sun or of a star does not hold unless it be supposed that the star or the centre of the Sun lie on the equinoctial. This is explained as follows.

A cata is arranged as before, and in addition an arc is drawn for the Sun or the star on the assumption that they have some northern or southern declination.[2] The line of southerly declination is to be *GH*, and of northerly declination *KI*.[3] I hold, then, that no cata in arcs on a sphere may be constructed from principal arcs *AB* and *BH*, and reflex arcs *HG* and *AF*; nor, similarly, from principal lines *AB* and *BK* and reflex arcs *KI* and *AF*. The reason is that neither the line *HG* nor *KI* is an arc of a great circle on the sphere, [a condition] presupposed in the preceding proofs of Ptolemy. It is necessary, therefore, to find some other way of determining the time from the altitude of the Sun or of a star, when they have some northern or southern declination, for in either case there is an arc which does not qualify as a line in a cata, being less than a great circle dividing the sphere.

Since, as I have said, it is not possible to construct a cata to show the movement of the equinoctial from the altitude of some point or star having a definite declination from it, the diurnal arc not being a great circle on the sphere, it is necessary to find some other way of determining it. First we arrange a cata to show the semidiurnal arc of the point or star above the horizon,[4] whether its declination be north or south. Let *DB* be a quadrant of the equinoctial and *BA* a quadrant of the meridional colure, from the equinoctial to its southern pole *A* beneath the horizon. These will be the two principal lines of a cata. Let *AE* be the depression of the south pole below the horizon, which is equal to the altitude of the other pole and to the latitude of the region. Let *EB* be the [maximum] altitude of the head of Aries above the horizon, that is, the complement of the latitude: and let the horizon *DE* be drawn. A point *F* is marked on the colure *AFC* passing through the centre of the star,[5] such that *FC* is the declination of the star, *F* being at some point or other on the horizon in keeping with the [known] declination.[6] From this it follows that these are four quadrants of great circles on the sphere, or four sinus of four quadrants of great circles, in our theory of disjunct catas. Then

(1) $AE{:}EB = (AF{:}FC).(DC{:}DB)$	III. D. I
(2) $(AE.FC)/AF = P$	def.
(3) $AE{:}P \;= AF{:}FC$	(2), II. I
(4) $AE{:}EB = (AE{:}P).(DC{:}DB)$	(1), (3)

[1] Literally the 'elevation of an arc of the equinoctial'.
[2] *latitudo*; but '*ab equinoxiali*' was evidently understood.
[3] These lines are in fact the semidiurnal arcs.
[4] *in circulo obliquo.*
[5] This does not mean that it does so at the moment the altitude is taken.
[6] *F* is the point at which the object crosses the horizon.

ex *DC* ad *DB*. Sed[a] *AE* ad *P* fit ex *AE* ad *EB* et *EB* ad *P*; hec[b] erit *DB* ad *CD*. Si igitur multiplices *P* medium in *DB* medium, et quod surgit dividas per *EB* primum, surget quantitas *CD*. Sed *CD* est illud quod est differencia ortus *F* in circulo directo et obliquo. Si ergo *F* sit meridionalis, subtrahe[c] *CD* de toto *DB*[d] et remanebit *CB*, quod est arcus stelle meridionalis habentis talem latitudinem; vel adde *CD* super 90, si sit *F* stella septentrionalis, et habebis medietatem arcus diei certissime.

Unde regula est[e] ista:

Pro arcu diei mediato multiplica sinum latitudinis regionis [*AE*] in sinum latitudinis stelle [*FC*], et productum dividas per sinum residui latitudinis stelle. Exibit quiddam quod se habebit ad secundum sicut sextum se habet ad quintum. Illud igitur quod sic exit multiplica per sextum, *DB*, quod est 150, et quod surgit divide per sinum residui latitudinis regionis [*EB*]. Et exibit sinus arcus *DC*,[f] cuius quere circuli porcionem; quod adde super 90 si sit stella septentrionalis, vel minue de 90 si sit meridionalis; et habebis medietatem arcus diei.

IV. 12

Iam ad propositum inveniendi porcionem circuli equinoxialis elevatam cum stella in hora altitudinis accepte accedamus, et est conclusio ista:

Si sinum versum medietatis arcus diei stelle duxeris in sinum rectum altitudinis accepte in hora qua volueris, et productum diviseris per sinum rectum altitudinis meridionalis[g] eiusdem stelle, exibit quiddam quod si subtraxeris de sinu verso medietatis arcus diei et residui queras arcum versum quod subtrahe[h] de medietate arcus diei. Et residuum erit quod elevatur de arcu equinoxiali ab hora ortus stelle usque ad horam altitudinis accepte, quod sic declaratur:

Fiant tres circuli equales vice unius, quia hec figura non potest bene protrahi in[i] plano sicut potest imaginari in spera, quia ibi quelibet corda versa et recta est distincta ab alia, sed in plano necessario una erit pars alterius. Et sint isti circuli *A*, *B*, *C*. Et supponatur latitudo

[a] C 18λ+Sed ex prioribus notis, nota est proporcio *EB* ad *P*.
[b] M 5λ+igitur qualis est proporcio *EB* ad *P*.
[c] M abstrahe [d] CM *CB* [e] C λ [f] CM *DE* [g] M meridionale
[h] C subtrahes [i] C λ

(5) $AE{:}P$	$= (AE{:}EB).(EB{:}P)$	II. 2
(6) $EB{:}P$	$= DB{:}CD$	cf. (4) with (5)
(7) CD	$= (P.DB)/EB$	(6).

But *CD* is the difference between the rising of *F* as between the right and oblique horizons. If, therefore, *F* be southwards,[1] subtract *CD* from the whole of *DB*, and *CB* will remain, this being the arc of the southern star having the supposed declination. If *F* be a northerly star, add *CD* to 90, and you will quite certainly have the semidiurnal arc.

The rule, therefore, is this:

> For the semidiurnal arc, multiply the sinus of the latitude [*AE*] of the region by the sinus of the declination of the star [*FC*], and divide the product by the sinus of the complement of the star's declination [*AF*]. Something results which bears to the second what the sixth bears to the fifth. Multiply the result, therefore, by the sixth (*DB*), that is, by 150, and divide by the sinus of the complement [*EB*] of the latitude of the region. This gives Sin *DC*. Find the arc corresponding to the sinus, and add it to 90 if the star be north [of the equator], or subtract from 90 if the star be to the south. You will then have the semidiurnal arc.

IV. 12

We now come to the proposition on determining what arc of the equinoctial has risen when a star has risen to a certain measured altitude. This is the conclusion:

> If you multiply the sinus versus of the semidiurnal arc of the star by the sinus rectus of the measured altitude at the desired time, and if you divide the product by the sinus rectus of the meridian altitude of the same star, the result is something which, if you subtract it from the sinus versus of the semidiurnal arc and seek the arc which has the difference as its sinus versus, you will have that which is to be subtracted from the semidiurnal arc.[2] The result will be that part of the equinoctial which has risen between the moment of the star's rising and its reaching the altitude measured.

This is explained as follows.[3] Let there be three equal circles [as illustrated] in place of the single circle [of the last chapter], since the present figure is not so easily drawn in a plane as it may be pictured on a sphere, where right and versed chords are clearly distinguished one from another, whereas on a plane one of them will necessarily be part of another. Let these circles be *A*, *B*, and *C*. Let it be supposed that the

[1] That is, south of east, meaning that the star has a southern declination.

[2] He should have continued: 'or added to it, when the star has crossed the meridian'. This qualification is made later (IV. 14), when the theorem is stated again.

[3] But not proved. The remainder of the chapter is difficult to follow, but certainly contains no semblance of a proof of the theorem relating ascensions and altitudes, except in a special case. See our commentary.

regionis esse 45 grad., sicut est[a] Novarie[b] et Wyntonie. Et supponatur una stella cuius latitudo septentrionalis ab equinoxiali sit 15 graduum; igitur altitudo meridiana stelle[c] erit 60 graduum—ibi in primo circulo ponatur *D*, et sit linea *DE* sinus rectus eius altitudinis meridiane. Et pone quod stella elevetur ante meridiem 30 grad. altitudinis, et ponatur ibi in primo circulo *G*, cuius altitudinis corda recta sit *GF*. Iterum, cum latitudo regionis fuerit 45 graduum, et declinacio stelle 15, per proximum capitulum operere et invenies medietatem arcus diei stelle esse 105[d] grad. et 30 min., quem arcum describas in secundo circulo. Incipe[e] superius in linea meridiana a puncto *K*, computando 105[f] grad. et 30 min. versus *A* et ubi arcus terminatur sub *A*, signetur punctus *M*, vel a *K* versus *B* et ibi signetur punctus *N*, ita quod totus arcus duplatus *MKN* sit totus arcus diei stelle, cuius medietatis arcus corda versa signetur *KP*.

Iam ergo in primo circulo inveniatur una corda versa que se habet ad lineam rectam altitudinis 30 grad. (*GF*) sicut se habet corda versa *KP* ad lineam altitudinis meridiane rectam (*DE*), quod sic invenietur:

Protrahatur in isto primo circulo linea altitudinis meridiane *DE* et linea altitudinis accepte *GF*. Deinde producatur linea *DP* ita quod linea *EP* sit equalis linee *KP*, et a puncto *G* producatur altera linea *GL* equidistans linee *DP*. Dico ergo quod linea *FL*[g] se habet ad lineam *GF* sicut se habet linea *EP* ad lineam *DE*, quod patet quia isti duo trianguli *FGL* et *EDP* sunt equianguli, ergo latera respicientia angulos equales sunt[h] proporcionales. Quod sunt equianguli patet eo quod tam *F* quam *E* sunt recti; de racione sinus perpendicularis similiter *P* et *L* sunt equales mediante[i] extrinseco coalterno[j] uni et opposito alteri. Igitur reliqui anguli sunt equales, *D* et *G*. Abstrahatur ergo a corda recta *KP* nota linea *FL* nota et residuum erit notum quod sit *KO*, cuius quere circuli proporcionem versam quod sit *KR*. Dico ergo quod quando stella habet altitudinem 30 grad. ante meridianam, tunc elevatur de eius arcu meridie diei totum *MR*, vel *SN*. Si fuerit altitudo post meridiem

[a] C λ [b] M Novaria [c] M stellet [d] C 150 [e] C incipes
[f] C 150 [g] C *PL* [h] M λ [i] C +*K*; M +*L*
[j] M quo alterno

latitude of the region is 45°, as is the case at Novaria[1] and Wyntonia.[2] Let it be supposed that there is one star, with declination 15° (north), making for a meridian altitude of 60°. There, [where the star crosses the meridian] on the first figure, let *D* be placed, and let the line *DE* be the sinus rectus of the meridian altitude. Suppose the star, before reaching the meridian, to have an altitude of 30°, and let *G* be placed [at that altitude] in the first figure. *GF* is to be the corda recta of the angle. Now since the latitude of the region is 45°, and the declination of the star is 15°, work in accordance with IV. 11 and you will find the semi-diurnal arc of the star to be 105° 30′. Describe this arc on the second figure, beginning from the point *K* at the top on the meridian line, and reckoning 105° 30′ towards *A*. Where the arc ends, below *A*, the point *M* is to be marked, and likewise working from *K* towards *B*, the point *N* is to be placed so that the entire double arc *MKN* is the entire diurnal arc of the star. The corda versa of *MKN* is *KP*.

A corda versa is now to be found in the first figure such that it bears to the straight line of altitude 30° (*GF*) the same ratio as the corda versa *KP* bears to the straight line of the meridional altitude (*DE*). The line is to be found thus:

Let the lines of meridian altitude (*DE*) and of measured altitude (*GF*) be drawn in the first figure. Next, let the line *DP* be produced so that *DP* is equal to *KP*, and from the point *G* let another line *GL* be drawn parallel to *DP*. I say then that

$$(1)\ FL\!:\!GF = EP\!:\!DE,$$

which is evident since the two triangles *FGL* and *EDP* are equiangular, and therefore corresponding sides (with respect to equal angles) are proportional. That they are equiangular follows from the fact that [the angles at] *F* and *E* are right angles; and similarly *P* and *L* are equal, being corresponding [angles between parallels]. The remaining angles, *D* and *G*, are therefore also equal. Then[3]

(2) $KO = KP - FL$	constr.
(3) KO^*	(2), FL^*, KP^*
(4) $KO = \mathrm{VSin}\ KR$	def.
(5) KR^*	(3), (4).

I say then that when the star has an altitude of 30° before crossing the meridian, the whole of *MR* (or *SN*) is the arc elevated, [when the star is approaching] the meridian arc. If the altitude is measured after [the

[1] Modern Novara—the classical spelling (e.g. in Pliny, III. 17) was Novaria. The latitude is 45° 27′, but there are tables of the twelfth and thirteenth centuries giving it as 45° precisely.

[2] Winchester, with a longitude of more than 51°, cannot have been intended. The towns of Winchester and Novara are often found together in lists of terrestrial latitudes, but it is not improbable that the error arises through miscopying the name of Vicenza (lat. approximately 45° 33′), often written 'Vincentia', whence 'Vinconia' and 'Wintonia'.

[3] This is in no sense a justification of the rule stated at the outset of the chapter, but merely a translation of that rule into geometrical terms. Nothing has been added by way of proof. In our earlier notation, $KR = \theta$, $KP = s$, $GF = a$, $DE = m$.

accepta, secundum quod *KR* et *KS*[a] sunt arcus equales de racione *RO* et *OS* sinuum rectorum, pertransitur de eius arcu totum[b] *MS*, cuius demonstracio est ista:

Notum est quod in orizonte recto idem est arcus elevatus stelle non habentis latitudinem ab equinoxiali, et arcus altitudinis et eadem altitudo meridiana et medietas arcus diei. Probacio ergo ibi quod ista proporcio corde recte ad versam tenet et pro hoc patebit in ceteris.

Fiat ergo in tercio circulo *AC* altitudo meridiana stelle 90 grad., et per consequens *AC*[c] est medietas arcus diei ibidem, cuius corda recta altitudinis meridiane sit *CD*, que etiam equalis est corde verse medietatis arcus diei. Assignentur alii[d] duo arcus altitudinis ante meridiem *EA* et *FA*, quorum corde recte sint *FG*, *EH*. Dico ergo quod que est proporcio *CD* ad se ipsum, illa est *FG* ad *FI* et *EH* ad *EK*. Sicut igitur *AC* altitudinem correspondet *AF* arcus elevatus et *FC* pertranseundum cuius corda versa est *CI* residuum tocius corde verse, similiter ad altitudinem *AE* correspondet *AE* arcus elevatus. Et arcus inter stellam et meridianam est *EC*, eo quod corda versa eius est residuum post extraccionem equalis ad *EH*, puncta *DK*. Remanet *KC*, cuius arcus versus est *CE*; et hoc est propositum.

Item arguo ex altera parte circuli, ubi non sunt idem arcus et altitudo. Sit ergo altitudo stelle meridiana *BR*, et medietas arcus diei *QC* cuius corda versa sit *CN*; et sit altitudo accepta *BM* cui inveniatur corda proporcionalis sicut est *CN* ad *RS* que sit *NT*, quo abiecto a toto *NC* remanet corda versa *TC*, cuius arcus versus est *CX*. Dico ergo quod totum *XC* nondum est pertransitum de medietate arcus diei, sed solum *QR*,[e] quod probo sic:

Sicut ad elevacionem totius altitudinis *BR* elevatur totus arcus *QC*, et quodammodo[f] pertransitur corda versa *NC*, ita proporcionaliter ad elevacionem altitudinis *BM* pertransitur pars proporcionalis *NC*; que pars probatur esse *NT*, eo quod *NT* se habet ad *ML* sicut se habet *NC*

[a] CM *KD* [b] C totam [c] CM *A* [d] C alia [e] C *QX* [f] C quomodo

star has passed] the meridian, since *KR* is equal to *KS* (their sinus being *RO* and *OS*), it will have covered the whole of the arc *MS*. The proof of this is as follows.

Suppose that the star has the same ascension when crossing the right horizon as a star on the equinoctial, and [the same] arc of the altitude, and the same meridian altitude and semidiurnal arc.[1] The ratio of corda recta to versa is therefore the same,[2] as will be evident from what follows.

In the third figure, let *AC* be the meridian altitude of a star of [declination] 90° (and in consequence *AC* [may be regarded], in the same figure, as a semidiurnal arc). The chord[3] of the meridian altitude of the star is to be *CD* (which is also equal to the versed chord of the semidiurnal arc). Let two arcs of altitude before the meridian be called *EA* and *FA*, with right chords *FG* and *EH*. I say then that the ratio *CD*:*CD* [may be rewritten in terms of either] *FG*:*FI* or *EH*:*EK*.[4] The arc of the altitude, *AF*, is a part of the altitude *AC*, as is *FC* (the versed chord of which is *CI*, the complement of the whole versed chord). So, likewise, to the arc of the altitude *AE* the altitude *AE* corresponds. And the arc between the star and the meridian is *EC*, since its versed chord is the remainder after the subtraction [from *CD*] of a quantity equal to *EH*, namely *DK*, giving *KC*, whose arcus versus [*sic*] is *CE*; and this is what was proposed.[5]

Next, I argue from another part of the figure,[6] where the arc and the altitude are not the same. Let *BR* be the meridian altitude of the star, and *QC* the semidiurnal arc, whose corda versa is *CN*; and let *BM* be the measured altitude. A chord *NT* is to be found proportional to it, so that

$$(6)\quad NT:[ML] = CN:RS$$
$$(7)\quad TC = NC - NT.$$

The arcus versus of *TC* is *CX*. I say then that the whole of *XC* does not yet encompass the whole of the semidiurnal arc, but only *QR*, which I prove as follows.

Just as at the ascension of the whole of the altitude *BR* the whole [semidiurnal] arc *QC* is risen, and some proportion of the corda versa *NC* is covered, so the latter happens when the altitude is *BM*. This part is proved to be *NT*.

$$(6)\quad NT:ML = NC:RS.$$

[1] It appears that we are to suppose the observer to be on the equator, and the star to be on the equinoctial.

[2] The sentiment is positively redundant!

[3] This 'chord', like those which follow, is in fact a sinus.

[4] The passage is obscure.

[5] The purpose of this extremely obscure paragraph is discussed in our commentary.

[6] That is, the right-hand side of the figure found in the MSS.

ad *RS* et hoc[a] est propositum. Sed dicet falsigraphus,[b] non est proporcio illa inter sinus rectos altitudinis et sinus versos arcuum elevatorum. Convincetur statim. Quia pars linee *NC* excedit in proporcione cordam rectam altitudinis acceptam, vel exceditur, necessario erit arcus[c] diei maior vel minor quam supponitur, quod est contra hypotesim.

IV. 13

Eodem modo, si per Solis vel stelle altitudinem[d] vis cognoscere eius[e] azimuth, accipe[f] cordam versam tocius arcus orizontis, a puncto ortus stelle usque in punctum meridiei, et illum multiplica in cordam rectam altitudinis accepte, et productum divide per cordam rectam altitudinis meridiane; et exibit quiddam quod si extraxeris de corda versa medietatis arcus azimuth et residui queras arcum versum, ille arcus erit arcus omni azimuth inter azimuth stelle in eadem hora et meridiem; quod si extraxeris de toto arcu azimuth, habebis azimuth pertransiens ab ortu stelle, vel si extraxeris illud de 90 habebis distanciam ab oriente.

De declaracione declinacionis cuiuslibet gradus zodiaci ab equinoxiali circulo, quere in primo libro *Almagesti* capitulo 12 secundum modum Ptholomei.

IV. 14

Pro arcu elevato diurno per stelle altitudinem ab eius ortu, quere sinum versum medietatis arcus diei eiusdem, et illum multiplica in sinum rectum altitudinis accepte (ante vel post meridiem), et quod ex multiplicacione provenit divide per sinum rectum altitudinis meridiane; et quod tunc exit subtrahe de sinu verso medietatis arcus diei, et tunc residui quere circuli porcionem, quam subtrahe a medietate arcus diei stelle (si fuerit ante meridiem) vel adde (si fuerit post). Et quod fuerit post addicionem vel subtractionem est illud quod elevatur de equinoxiali

[a] C 2λ+habetur
[b] C philigraphus; M pligraphus(?) (See the note to the translation.)
[c] M alicuius [d] M λ [e] C λ
[f] M 3λ+acope corda versa

But pseudography[1] will tell that there is no ratio between the sinus recti of the altitudes and the sinus versi of the arcs which have risen.[2] It will be demonstrated at once. Since part of the line *NC* exceeds in ratio the accepted corda recta of the altitude, or is exceeded [by it], the diurnal arc will necessarily be greater or less than is supposed, which is contrary to hypothesis.[3]

IV. 13

In the same way, if from the altitude of the Sun or star you wish to know its azimuth, find the corda versa of the whole arc of the horizon, from the point where the star rises to the point on the meridian, and multiply it by the corda recta of the measured altitude. Divide the product by the corda recta of the meridian altitude, and the result will be something which, if you subtract it from the corda versa of half the arc of the azimuth and find the arcus versus of the difference, will be that arc which is the difference for any azimuth between the azimuth of the star at that moment and the meridian.[4] Subtracting this now from the whole azimuth, you will have the azimuth traversed from the rising of the star; or, if you subtract the same thing from 90, you will have the distance from the west.

On the declination of any degree of the zodiac, consult *Almagest* I. 12 for Ptolemy's method.[5]

IV. 14

To determine [that part of] the diurnal arc of a star which has risen, from its altitude, find the sinus versus of the semidiurnal arc of the star, multiply by the sinus rectus of the measured altitude (whether before or after the meridian), and divide the product by the sinus rectus of the meridian altitude. Subtract the result from the sinus versus of the semidiurnal arc, and find the arc of the circle [which has this as its sinus versus]. This you must subtract from the semidiurnal arc of the star (if it is before the meridian), or you must add it to the same (if the star was past the meridian). The result of the addition or subtraction is that

[1] The copyists of our MSS. did not know how to treat this word, which is sometimes written *filigraphus*. All variations obviously arise from a mistaken expansion of a contraction of '*falsigraphus*', a medieval form of Aristotle's ψευδογράφος. See *De sophisticis elenchis*, $171^{b}34$–$172^{a}17$.

[2] That is, in the notation of our commentary, that it is not possible to write

$$\mathrm{VSin}\ \theta/\mathrm{VSin}\ s = \mathrm{Sin}\ a/\mathrm{Sin}\ m.$$

[3] This is a valid, ingenious, but far from obvious demonstration, by *reductio ad absurdum*.

[4] See our commentary. As it stands, the chapter is unacceptable.

[5] Given above, in fact, in chapter IV. 10.

a tempore ortus stelle usque in aliam[a] horam. Quere[b] figuram ad hoc signum: †.

IV. 15

Si arcum planete tocius[c] diei invenire volueris, scias primo que sit latitudo planete ab equinoxiali, et declinacionem gradus cum quo fuerit planeta, et partes utriusque considera si utrumque declinat versus septentrionem vel meridiem; que si fuerint ambo in una parte iunge utrasque; similiter, si vero diverse minue minus de maiori; et quod remanserit erit elongacio planete[d] ab equinoxiali circulo. Quere ergo sinum longitudinis planete et minue longitudinem de 90, et residui similiter accipe[e] sinum. Similiter accipe[f] sinum latitudinis regionis, et sinum eius residui de 90.

Multiplica ergo sinum latitudinis regionis in sinum longitudinis stelle ab equinoxiali, et quod provenit ex multiplicacione divide per sinum residui longitudinis planete. Et quod inde exit idem multiplica in 150, et partire quod inde provenit per sinum residui latitudinis regionis; et exibit sinus unus, cuius quere[g] circuli porcionem, cuius duplum adde ad 180 si fuerit latitudo planete septentrionalis, vel minue si fuerit meridionalis.

Et quicquam[h] habueris post talem addicionem vel subtraxionem erit arcus planete, quem queris sicut patet demonstracione in presenti catha disiuncta, in qua est *BD* linea equinoxialis, *BA* quarta coluri, *AE* latitudo regionis, *EB* residuum latitudinis, *DFE* quarta orizontis obliqui ad orizontem rectum, *CF* latitudo stelle, *FA* residuum latitudinis. Et argue per catham disiunctam: et cognosces *CD* ignotum per *AE, EB, AF, FC, DB*[i] quinque nota. Quo cognito, scies *DF* latitudinem ortus per *DB, BC, DE, AF, AC*[j] quinque nota. Nunc patet ad oculum et hec est figura que sequitur proximo.

IV. 16

Quantitatem arcus altitudinis a cenith capitis ad quodlibet punctum circuli signorum invenire; cuius probacio est hec:

Considerabimus circulum[k] meridiei *ABGD*, et infra eum medietatem orizontis *BED*, et medietatem orbis signorum *ZHTF*. Et sit punctum *H* caput Cancri secundum quodlibet tempus distans a[l] linea meridiana,

[a] C illam [b] C 5λ [c] C 2λ [d] M (et passim) plene [e] M acope
[f] M acope [g] M que [h] C quicquid [i] C *DD* [j] M *AE* [k] C λ
[l] C 2λ+aliam

part of the equinoctial which has ascended from the time of the star's rising up to the other time. See the figure at this mark: †.[1]

IV. 15

If you wish to find the full diurnal arc of a planet, first find the planet's latitude,[2] and the declination of the degree which the planet will have, and consider whether each is a declination north or south. If both are to the same side [of the equinoctial], add them; and similarly, if they are on different sides, subtract; the lesser from the greater. The result will be the distance of the planet from the equinoctial. Find the sinus of this distance, [δ], of the planet, subtract it from 90, and similarly take the sinus of the complement. Likewise take the sinus of the latitude, [ϕ], of the region, and of its complement. Now form the product

$$(150' \text{ Sin } \phi . \text{Sin } \delta)/(\text{Cos } \delta . \text{Cos } \phi),$$

which is equal to the sinus of an arc which you must find. Add double the arc to 180 if the declination of the planet is north, or subtract if south. Whatever you have after such an addition or subtraction will be the arc of the planet required.

What you are looking for is clear in the proof of the disjunct cata of the present work. In this, *BD* is the equinoctial, *BA* a quadrant of the colure, *AE* the latitude of the region, *EB* the co-latitude, *DFE* a quadrant of the [oblique] horizon at the right horizon, *CF* is the declination of the star, and *FA* its complement. Use the argument of the disjunct cata: you will know the unknown *CD* from the five known quantities *AE*, *EB*, *AF*, *FC*, and *DB*. Knowing *CD*, you will know the distance [from east] of rising, *DF*, in terms of the five known quantities, *DB*, *BC*, *DE*, *AF*, and *AC*. It is now plain to see, and the figure is that which follows next.

IV. 16

To find the zenith distance of any point on the circle of signs, the procedure is as follows.

We shall consider the meridian circle, *ABGD*, and below it a half of the horizon, *BED*, and also a half of the circle of signs,[3] *ZHTF*. Let the head of Cancer be *H* at some time when it is distant from the meridian

[1] No mark is visible. Since this chapter is nothing more than a restatement of the rule of IV. 12, this was perhaps originally intended to refer the reader back to the figures of that earlier chapter.

[2] This is presumably the correct interpretation. The planet's ecliptic latitude being small, the error will generally be slight. For an unlikely but possible interpretation see IV. 25 below, and the commentary to it.

[3] *orbis signorum*—used indifferently as an alternative to *circulus signorum*.

et exempli causa distans per unam horam;[a] et punctum *Z*[b] medians celum, et punctum *T* orizontis sit notum per 18am proposicionem, ubi ostenditur latitudo ortus cuiuslibet gradus orbis signorum.

Faciam ergo super summitatem capitis *A* et super caput[c] Cancri *H* transire[d] porcionem circuli altitudinis *AHEG*. Scrutabor ergo quantitatem arcus *AH*. Est itaque sicut premisimus arcus *ZT* notus; et arcus *HT*, cum *H* sit principium Cancri; et arcus *AZ*,[e] propter declinacionem puncti *Z* et altitudinem poli notas notus est; et arcus *ZB* quia[f] est complementum quarte note. Hiis ergo cognitis vides quod proporcio *BZ* ad *BA* aggregatur ex duabus, una scilicet que est *EH* ad *EA*, et altera que est *TZ* ad *TH* per conversam prime de cathis coniunctis; et scias quod loquor de sinibus.

Cum ergo ceteri sint noti, erit et arcus *EH* notus, ergo et residuum quarte *AH* erit notum. Regula operacionis brevis est ista:

> Si sinum arcus meridiani deprehensi inter medium celi gradum et orizontem multiplices in sinum arcus circuli signorum deprehensi inter orizontem et punctum circuli signorum ad quod circulus altitudinis deducitur, et productum dividas per sinum arcus circuli signorum intercepti inter orizontem et gradum medii celi, exibit[g] sinus residui arcus quesiti, quam si a quarta demseris relinquitur arcus altitudinis circuli a[h] summitate capitis ad punctum circuli signorum destinatum, quod est propositum.

IV. 17

Ad demonstrandum gradum cum quo stella oritur,[i] describam orbem meridiei super quem sit *A*, *B*, *D*, *G*, et medietatem orbis equacionis diei, qui sit *AEG*;[j] et sit polus eius punctus *Z*; et medietatem orbis orizontis, supra quem sit *B*, *E*, *D*, et[k] sit elevacio, id est ortus stelle super punctum *H* orizontis. Et describam quartam orbis magni, id[l] est colurum, super duo puncta *Z*, *H*.

Et quia inter duos arcus duorum orbium magnorum, arcum *AZ*, arcum *AE*, protrahuntur arcus *ZT* et arcus *EB*, erit proporcio *ZB* ad *BA* aggregata ex duobus, ex proporcione *ZH* ad *HT* et ex proporcione *ET* ad *EA* (et loquor de sinibus—patet per catham disiunctam). Sed *ZA*, *ZT*, *EA* quilibet est quarta circuli. Sed propter latitudinem stelle sciemus lineam *HT*, et propter gradum medii celi sciemus punctum *T*; et sciemus propter latitudinem regionis arcum *ZB*. Ergo necessario sciemus sextum *TE* per catham disiunctam.

[a] C λ [b] C λ [c] M λ [d] C +per [e] C *HZ* [f] M 2λ+quem
[g] M +residui [h] C 2λ+ad summitatem [i] M +Ptholomeus
[j] CM *ACG* [k] C si [l] M 2λ

—and for the sake of an example [it may be taken as] an hour distant. The point Z is to be mid-heaven, and T on the horizon is to be known by means of IV. [15],[1] where the azimuth of any ascendent point of the sphere of signs is determined.

I shall draw, then, an arc of a circle of altitude $AHEG$ through the zenith,[2] A, and the head of Cancer, H. I shall carefully investigate the size of the arc AH. As promised, the arc ZT is known; and also the arc HT, H being the beginning of Cancer; and the arc AZ is known since both the altitude of the pole and the declination of Z are known, and also ZB, its complement. These quantities being known, you should see that

(1) $BZ:BA = (EH:EA).(TZ:TH)$ conv. III. C. 1

(and you are to know that I am talking of sinus).

Since all the others are known, therefore so is EH known, and thus also its complement AH. There is this short procedural rule:

> If you multiply the sinus of the meridian altitude of mid-heaven by the sinus of an arc of the ecliptic intercepted between the horizon and a point [H] of it on a certain circle of altitude, and then divide by the sinus of the arc between mid-heaven and the horizon, you will obtain the sinus of the altitude of H, that is to say, the sinus of the complement of its zenith distance. And this was proposed.

IV. 17

In order to show with what degree [of the equinoctial] a star rises, I shall describe a meridian circle on which are to be the points A, B, D, and G, and I shall describe AEG, which is half of the circle of the equation of the day.[3] Its pole is to be Z. I shall draw half of the horizon, on which are the points, B, E, and D. Let the star be rising at the point H of the horizon. I shall describe a quadrant of a great circle, that is, a colure, over the two points Z and H.

Since the arcs ZT and EB are drawn between the two arcs of great circles, AZ and AE, therefore (and I am speaking of sinus)

(1) $ZB:BA = (ZH:HT).(ET:EA)$ conv. III. D. 17.

But ZA, ZT, and EA are all quadrants of a circle. We shall know HT from the declination of the star, the point T from the known degree of mid-heaven,[4] and ZB from the latitude of the region. Necessarily, therefore, we shall know the sixth quantity TE, by disjunct cata.

[1] The reference is doubtful. The 'IV. 18' of the text is clearly mistaken.

[2] *summitas capitis.*

[3] *orbis equacionis diei*—the equator (equinoctial).

[4] This remark is superfluous. See our commentary.

Et breviter, sciemus quod si nos acceperimus in eis que occidunt arcum equalem arcui *TE* in eo quod precedit[a] punctum *T*,[b] arcus *TK* erit occasus in arcu equali arcui *BH*, id est angulus qui est in eo quod precedit de orbe meridionali, etiam equalis angulo qui continetur ab *AZ* et *TZ*, in eo quod sequitur secundum hanc figuram.

Et sciemus per illud cuius iam precessit declinacio de numeracione elevacionum in omni climate, cum eo quod elevatur et occidit de partibus orbis signorum, qui oritur cum puncto *T* (qui[c] est punctus in linea orbis equacionis diei) et cum stella[d] et parte[e] que occidit de orbe signorum cum puncto *K* (qui[f] est punctus similiter orbis equacionis diei) et cum stella. Et[g] manifestum est quod cum fuerit Sol in aliquo illorum graduum erit arcus Solis et stelle illius equalis, quia similiter oriuntur et occidunt. Et hec vocantur 'centra[h] altorum vera', hec ibi 8 libri *Algamesti* et conclusio quinta.[i]

IV. 18

Regula commentatoris in primo libro *Almagesti* de arte inveniendi declinacionem cuiuslibet[j] gradus zodiaci per calculacionem—et eius certitudo est in precedentibus determinata.

Regula. Si sinus porcionis ab equinoxiali inchoate (cuius finalis puncti declinacio ab equinoxiali queritur) ducatur in sinum[k] maxime declinacionis, et productum dividatur per sinum quarte circuli, exibit sinus quesite declinacionis.

IV. 19

Regula ascensionum calculandarum[l] in orizonte recto est[m] talis, ibidem proposicione 7:[n]

Si[o] sinus residui maxime declinacionis ducatur in sinum declinacionis illius porcionis inchoati[p] ab equinoxiali linea (cuius porcionis queritur elevacio), productumque dividatur per sinum residui declinacionis illius porcionis, et quod exierit itidem[q] ducatur in sinum quarti circuli, scilicet 150 minutorum, productumque[r] dividatur per sinum maxime declinacionis, exibit sinus quesite elevacionis.

Et hec conclusio prius probatur[s] certissime per catham disiunctam.

[a] M procedet [b] CM +et cum [c] M quod [d] C stellam [e] C partem
[f] M quod [g] C λ [h] C λ [i] C 15ª [j] C alicuius
[k] C +in sinum [l] C +ut [m] C 2λ [n] C λ; M 17 [o] C Sed
[p] C inchoato [q] M ididem [r] M productum quia [s] C probat

And briefly, we know that if we are dealing with the case of a setting [star] there will be an arc equal to *TE* now preceding *T*, [the star],[1] and that the arc *TK* sets below an arc [of the horizon] equal to *BH*, that is, the angle preceding the meridian circle, also equal to the angle between *AZ* and *TZ*. This all follows in accordance with the figure.

And we shall know, concerning a point of given declination, how to enumerate its ascensions at any climate. We shall know those ecliptic arcs which rise and set—that which rises with the point *T* (the point on the equator),[2] and with the star, and that which sets with the point *K* (which is a point similarly on the equator), and with the star. And it is manifest that when the Sun is in any of these degrees, the arcs of the Sun and of the star will be equal, since they rise and set in like manner. And these are called 'true centres of the heights'.[3] (See *Almagest* VIII. 5.)

IV. 18

A rule of the commentator on Book I of *Almagest* on the means of finding the declination of any degree of the zodiac by calculation—and its correctness is assured in what has gone before.

Rule. If the sinus of an incomplete arc [of the ecliptic measured from the] equinoctial (the declination of the end-point of which is sought) be multiplied by the sinus of the obliquity,[4] and the product be divided by the sinus of a quadrant, the result is the sinus of the declination sought.

IV. 19

The rule for calculating right ascensions is this, [which is to be found] in the same place as proposition 7:[5]

If the sinus of the complement of the obliquity be multiplied by the sinus of the declination of the incomplete arc [of the ecliptic measured from] the equinoctial, the ascension of the arc being that which is sought, and the product be divided by the sinus of the complement of the declination of that arc, and if in like manner the result be multiplied by the sinus of a quadrant (namely 150′), and the product be divided by the sinus of the obliquity, the result will be the sinus of the required ascension.

This conclusion is first proved conclusively by means of a disjunct cata.

[1] It precedes, if regarded in the direction of the order of the signs.

[2] *linea orbis equacionis diei.* [3] *centra altorum vera.* [4] *declinacio maxima.*

[5] The '17' of the MSS. seems to be a mistake, and perhaps stems from the accidental elongation of the first of the points (in '·7·') in an early copy. In IV. 7 above there is a proof of the result.

IV. 20

Regula inveniendi arcum diei minimi et maximi in quovis climate per poli altitudinem notam—et est regula hec in principio secundi libri, proporcione prima coniuncti:

Si sinus altitudinis poli ducatur in sinum maxime declinacionis, et productum dividatur per sinum[a] residui maxime declinacionis, et quod exierit iterum ducatur in semidiametrum et productum dividatur in sinum residui altitudinis poli, exibit differencia mediata minimi[b] diei ad equinoxialem diem.[c] Patet per catham disiunctam, ut prius ostensum est.

IV. 21

Regula inveniendi arcum orizontis in quovis climate qui est inter ortum tropici et equinoxialem per assignacionem minimi[d] diei—et est regula per catham[e] coniunctam:

Si ducatur sinus dimidii arcus diei communis in sinum residui maxime[f] declinacionis, productumque dividatur per sinum quarte porcionis circuli, exibit sinus residui arcus orizontis qui est inter ortum utriusque tropicorum et[g] circulum equinoxialem. Similique racione invenire poteris differenciam ortus[h] cuiuslibet signi vel gradus ab equinoxiali. Hec patet per catham coniunctam. Posito igitur quod dies brevissimus[i] 14 horis equalibus, et divisio terminetur ut est in Rodas insula, invenies[j] arcum inter[k] equinoxialem et ortum tropici esse 30 graduum circuli orizontis 360 graduum.

IV. 22

Altitudinem poli per arcum diei minimi notum faciliter poteris sic investigare. Si sinum differencie meridiei diei minimi ad diem equinoxialem ducas in sinum residui quarte orizontis, et productum divides per sinum arcus orizontis qui est inter ortum tropici et equinoxialem, atque quod exierit[l] ducatur in sinum quadrantis, productumque itidem dividatur per sinum arcus medie diei, et exibit sinus altitudinis poli.

[a] CM 2λ+residuum [b] C minimi [c] C diei [d] CM minime
[e] M 2λ+catha coniuncta [f] M 12λ [g] M +arcum [h] M arcus
[i] C brevissima; M brevissimam [j] C invenias [k] M λ [l] M exiret

IV. 20

The rule for finding the least and greatest diurnal [solar] arc, for any climate, from the known altitude of the pole—and this rule is at the beginning of [*Almagest*] Book II,[1] in the first proportion on the conjunct [cata]:

If the sinus of the altitude of the pole be multiplied by the sinus of the obliquity, and the product be divided by the sinus of the complement of the obliquity, and if the result be then multiplied by the semidiameter, and the product divided by the sinus of the co-latitude, the result is [the sinus of the complement of the arc corresponding to] half the difference between the shortest day and the equinoctial day. This is clear from a disjunct cata, as was shown earlier.[2]

IV. 21

The rule for finding, for any climate, the arc of the horizon which lies between the rising of the tropic and the equinoctial, by considering the shortest day—and this rule is [proved by] a conjunct cata:

If the sinus of the semidiurnal arc be multiplied by the sinus of the complement of the obliquity, and if the product be then divided by the sinus of a quadrant, there results the sinus of the complement of the arc of the horizon which lies between the rising point of either of the tropics and the equinoctial circle. And by a similar argument you may find the difference between the rising points of any sign or degree from the equinoctial. This is clearly determined by means of a conjunct cata. It being supposed, then, that the shortest day is of 14[3] equal hours, and that the division is made as for the island of Rhodes, you will find the distance between the equinoctial arc and the rising of the tropic to be 30° of the horizon.

IV. 22

You may easily investigate the altitude of the pole, [that is, the latitude of the region], through the known arc of the shortest day, in this way. If you multiply the sinus of the difference between [half] the shortest day and [half] the equinoctial day by the sinus of the complement of an arc on the horizon, and divide the product by the sinus of that arc of the horizon (between the rising of the tropic and of the equinoctial), and multiply what results by the sinus of a quadrant, and if in like manner the product be divided by the sinus of the semidiurnal arc, the sinus of the altitude of the pole results.

[1] *Almagest* II. 2.
[2] See the final part of IV. 11 above for the proof of a theorem of which this is a special case.
[3] *Almagest* II. 2, the source of the example, has $14\frac{1}{2}$.

Exempli gracia. Sit *Z* polus mundi, et *E* caput Arietis, *BZ* elevacio poli, et *EB* quarta orizontis, et *EH* ortus tropici hiemalis. Igitur *TA* erit arcus equinoxialis minime diei, et *ET* differencia inter ipsum et equinoxialem; patet igitur per catham disiunctam quod proporcio *ET* ad *TA* fit ex proporcione *EH* ad *HB* et ex *ZB* ad *ZA*,[a] quare[b] si ducas primum, scilicet *ET*, in quartum, scilicet in *HB*, et productum dividas per tercium, *EH*, exibit quiddam quod se habebit ad secundum, scilicet *TA*, sicut quintum *ZB*,[c] ad sextum, *ZA*. Sed illorum tria sunt nota ex ypotesi, secundum vero quia *TA* est medietas[d] arcus, et *HB* quia est residuum latitudinis orizontis inter equinoxialem et tropici ortum, et tercium est notum, scilicet *ZA*, quia medietas[e] diametri; ergo notum erit quartum, scilicet *ZB*. Et hoc est quod volebamus declarare.[f]

IV. 23

Arcum orizontis qui est inter tropicum et equinoxialem per altitudinem poli notam[g] declarare non est difficile cuius *regula* est:

Si sinum maxime declinacionis ducas in semidiametrum, et productum dividas per sinum residui altitudinis poli, exibit sinus arcus orizontis qui est inter ortum[h] tropici et equinoxialem.

Similiter idem potest cognosci per altitudinem poli et per notam declinacionem cuiuscumque stelle, videlicet arcus interceptus de orizonte inter eius ortum et equinoxialem. Sic resumatur figura eadem que immediate ante,[i] et sit *H* una stella vel gradus cuius latitudo ab equinoxiali sit nota, scilicet arcus *HT*, et elevacio poli *ZB*. Per hec duo contingit investigare arcum *EB*. Sicut notum est ex catha coniuncta quod conversis proporcionibus quod proporcio *AB* ad *AZ* constitui ex *TH* ad *TZ* et ex *BE* ad *EH*. Sed *TZ* et *BE* sunt equales, quia due quarte, igitur qualis est proporcio *AB* ad *AZ*, talis est *TH* ad *EH*. Sed primum est notum, quia residuum elevacionis poli, et secundum est notum, quia semidiameter, et tercium est notum, scilicet *TH*, quia latitudo stelle, est sinus maxime declinacionis eius; quare notum erit quartum necessario operando per regulam premissam.

[a] C *ZH* [b] M quere [c] M *EB* [d] M medietatis [e] M medietatis
[f] C ostendere [g] C λ [h] C arcum [i] M trium

Proof. Let Z be the pole of the world, E the point [where] the head of Aries [rises], BZ the altitude of the pole, EB a quadrant of the horizon, and EH the rising of the winter tropic. TA will then be the [semidiurnal] equinoctial arc of the shortest day, and ET the difference between it and the [corresponding arc] at the equinox. Then, by a disjunct cata, it is clear that

$$(1)\ ET:TA = (EH:HB).(ZB:ZA).$$

If you multiply the first by the fourth and divide by the third, the result will have to the second what the fifth has to the sixth, that is

$$(2)\ ((ET.HB)/EH):TA = ZB:ZA \qquad (1).$$

But of these, [the first] three quantities [in (2)] are known by hypothesis, the second[1] is known since TA is the semidiurnal arc [of the shortest day], HB is known since it is the complement of the distance along the horizon between the rising points of equinox and tropic, and the third (ZA) is known since it is half the diameter.[2] The fourth, namely ZB, will therefore be known; and this is what we wanted to establish.

IV. 23

It is not hard to determine the arc of the horizon between [the points at which rise] the tropic and the equinoctial, in terms of the altitude of the pole. The rule is that if you multiply the sinus of the obliquity by the semidiameter, and divide the product by the sinus of the co-latitude, the result is the sinus of the arc of the horizon between the rising of the tropic and of the equinoctial.

From the altitude of the pole and the known declination of any star, the same [sort of thing] may be discovered, namely the arc of the horizon intercepted between the point at which it rises and that of the equinoctial. Once again, let the same figure be taken as immediately beforehand, and let H be a star or degree whose declination (HT) is known. Let the altitude of the pole be ZB. From these two things it is left to find the arc EB. From a conjunct cata, by converse proportion,

$$(1)\ AB:AZ = (TH:TZ).(BE:EH)$$

$$(2)\ TZ = BE = 90^\circ$$

$$(3)\ AB:AZ = TH:EH \qquad (1), (2).$$

But the first, [AB], is known, being the complement of the elevation of the pole; and the second is known, being a semidiameter;[3] and the third, since it is the 'latitude' TH of the star, is the sinus of its [declination, which in the case of the Sun you will take to be] the maximum declination. The fourth term will therefore of necessity be known, working with the aforesaid rule.

[1] (2) is here naturally regarded as a proportionality between four terms.

[2] Meaning, of course, that Sin ZA is so.

[3] Meaning that Sin $AZ = R$.

IV. 24

Cuiuslibet porcionis circuli obliqui ascensiones in orizonte obliquo invenire potes per hanc *regulam*:

Si sinum altitudinis poli duxeris in sinum declinacionis porcionis inchoate ab equinoxiali puncto et productum dividas per sinum residui declinacionis,[a] et quod exierit itidem ducas in semidiametrum, et productum dividas per sinum residui altitudinis poli, exibit sinus[b] differencie ascensionum sumpte porcionis in spera recta et obliqua, cuius determinacio est premissa.

IV. 25

Si ecliptica intercipiatur inter stellam[c] [et] equinoxialem, pro latitudine stelle extrahenda, sic operaberis:

Accipe totam declinacionem ad 90 et illius aggregati accipe sinum: sit sinus primus. Item accipe declinacionem veri gradus stelle, quam addes[d] super 90, et illius aggregati quere sinum, qui sit secundus. Deinde ex eodem aggregato subtrahe residuum declinacionis stelle ab ecliptica, et residui quere sinum, qui sit tercius. Deinde multiplica primum in tercium et productum divide per secundum sinum, et exibit unus sinus cuius quere arcum rectum et hec est vera declinacio stelle ab equinoxiali circulo; qua[e] habita, sic invenies gradum cum quo mediat celum.

IV. 26

Accipe ut prius sinum tocius aggregati ex 90 et tota declinacione, et[f] sit primus. Deinde accipe sinum tocius declinacionis et sit secundus. Tercio, accipe sinum declinacionis stelle ab equinoxiali circulo, et sit tercius. Quarto, illam declinacionem stelle ab equinoxiali extrahe de 90, et sinus residui sit quartus. Quinto, cum longitudine veri gradus stelle a principio Capricorni vel Cancri intra tabulam ascensionum in circulo directo inter gradus equinoxiales quam precise poteris, et quod in eius directo fuerit in linea numeri inter gradus equales, accipe et illius arcus quere sinum qui sit quintus. Age igitur sic: multiplica secundum in

[a] C 14λ [b] C λ [c] C λ [d] C addas [e] M quia [f] CM 3λ

IV. 24

You may find the ascensions on the horizon of any arc of the ecliptic by means of this rule:

If you multiply the sinus of the altitude of the pole by the sinus of the declination of the end point of an arc [of the ecliptic] beginning at the equinoctial point, and divide the product by the sinus of the complement of the declination, and then in the same way multiply the result by the semidiameter, and divide the product by the sinus of the co-latitude, the result is the sinus of the difference of the right and oblique ascensions of the assumed arc. The proof of this was set out earlier.[1]

IV. 25

If the ecliptic lie between a star and the equinoctial, to find the declination[2] of the star this is how you will proceed:

Take the sinus of the sum of the obliquity and 90 [degrees]. Let this be the first sinus. Then take the quasi-declination[3] of the ecliptic point with the same longitude as the star, and add it to 90, and take the sinus of the total. This is to be the second sinus. Then from this total subtract the complement of the star's ecliptic latitude,[4] and find the sinus of the result. Let this be the third sinus. Now multiply the first by the third, and divide the product by the second sinus. The result is the sinus [rectus] of the required arc, that is, the true declination of the star from the equinoctial. Having found it, find the mediation as follows.

IV. 26

As before, take the sinus of the sum of 90 degrees and the obliquity, and let this be the first [term]. Then take the sinus of the obliquity, and let this be the second. Thirdly, take the sinus of the declination of the star, and let this be term number three. The fourth is the sinus of the complement of the declination. Fifth, enter the table of right ascensions with the distance of the 'true degree' of the star from the beginning of Capricorn or Cancer, as precisely as you can, among the equinoctial degrees.[5] Take what you find opposite, [which] will be in the line of number amongst the equal degrees, and find the sinus of its arc, which will be the fifth [term]. Then proceed as follows. Multiply the second

[1] See IV. 8, above.

[2] *latitudo*—only later do we see the meaning.

[3] This is the line *IN* in the figure of IV. 10 above (or IV. 26 below). We denote *IN* by the symbol Δ. In the modern sense of the word, *IN* is not a declination.

[4] *declinacio ab ecliptica*. In our standard notation this construction merely leads us to the sum $(\beta+\Delta)$.

[5] This method of using the tables of ascensions in reverse was introduced in IV. 9–10 above. Referring to the figure of IV. 10, the method provides *EN*, given the distance *IG*.

tercium et productum divide per primum, et quod exit, itidem[a] multiplica per quintum et[b] illud quod surgit itidem divide per quartum. Et exeuntis sinus quere arcum, quem arcum[c] addes ad arcum quinti sinus; et cum illo arcu aggregato intrabis iterum tabulam ascensionum circuli directi inter gradus equinoxiales, et quod in directo illius inveneris in linea numeri inter gradus equales, ille est gradus cum quo ipsa stella mediat celum.

IV. 27

Si stella fuerit in ecliptica, eadem est[d] latitudo stelle que est declinacio veri gradus eius, et idem gradus verus et gradus medii celi. Igitur si fuerit stella intercepta inter equinoxialem et eclipticam, sic invenies latitudinem stelle:

Accipe sinum aggregati ex tota declinacione et 90, et sit sinus primus. Deinde adde declinacionem veri gradus stelle super 90 et istius sinus sit secundus. Deinde de isto ultimo aggregato subtrahe[e] residuum declinacionis stelle ab ecliptica, et sit sinus remanentis tercius. Multiplica igitur tercium in primum, et productum divide per secundum; et exeuntis sinus quere circuli porcionem, que est declinacio stelle ab equinoxiali certissime. (Hic nulla est diversitas in operando a prima declaracione.)[f]

IV. 28

Habita igitur latitudine stelle ab equinoxiali, sic invenies gradum medii celi, omnino sicut prius fecisti per illas[g] 5 cordas notas; ergo, stella existente in eadem[h] parte equinoxialis[i] cum vero gradu eius nulla omnino est differencia in operando nec pro latitudine nec pro gradu medii celi.

IV. 29

Si autem stella fuerit in equinoxiali, nullam habet latitudinem, sed pro gradu medii celi sic operaberis:

Accipe arcus veri gradus stelle, computando a Capricorno secundum[j]

[a] M eodem [b] M 2λ+aliquid [c] C λ [d] CM λ [e] M subtrahere
[f] C λ [g] M illa scilicet [h] C λ [i] C equinoxiali [j] C λ

by the third, and divide the product by the first. In like manner multiply what results by the fifth, and in the same way as before, divide, now by the fourth. Find the arc of the resulting sinus, and add it to the arc of which the fifth was the sinus. With the aggregate arc you will again enter the equinoctial degrees in the table of right ascensions. Whatever you find opposite, in the line of number among the equal degrees, that will be the mediation of the star.

IV. 27

If a star be on the ecliptic, the latitude of the star is the same as the quasi-declination of its degree, and the ecliptic longitude is the same as the mediation. If the star lie between the equinoctial and the ecliptic, find its declination as follows:

Take the sinus of the sum of 90 and the obliquity, and let this be the first sinus. Add the quasi-declination of the ecliptic place of the star to 90, and let its sinus be the second. Then from the last sum subtract the residuum[1] of the ecliptic latitude of the star, and let the sinus of this remaining [arc] be the third. Multiply, now, the third by the first, and divide the product by the second. Find the arc corresponding to the resulting sinus. This arc is most certainly the declination of the star. (This method is no different from that explained in the first statement [of IV. 26].)

IV. 28

Having [found] the declination of the star, this is how you will find the mediation. [You will find it] exactly as you found [it] before, from five known chords. If the star is now at the same part of the equinoctial and with the same degree, there is nothing different about the working, whether for declination or mediation.

IV. 29

If, however, the star is on the equinoctial, it has no declination. For the mediation then work as follows:

Take the arc of the true [ecliptic] degree of the star, reckoning from

[1] This word is left untranslated, since it could mean either the result of an earlier unspecified operation (but none is obvious—we require it to be the addition$(\beta+90^\circ)$), or the complement $(90^\circ-\beta)$, in which case we obtain the wrong result, *unless* the fact that the latitude is south is taken into account, as we should take it into account by prefixing a minus sign.

successionem signorum, cum quo arcu intrabis[a] tabulam ascensionum circuli directi inter gradus equinoxiales; et quod in directo illorum fuerit in linea numeri de gradibus equalibus, ipse est gradus medii celi.

IV. 30

Si autem linea equinoxialis fuerit inter stellam et verum gradum eius, pro latitudine extrahenda sic operaberis:

Abstrahe declinacionem veri gradus stelle de 90, et residui quere sinum, et sit primus; a quo extrahe[b] residuum declinacionis stelle ab ecliptica, et illius tunc residui quere sinum, qui sit secundus. Deinde extrahe totam declinacionem de 90, et residui quere sinum, qui erit tercius. Deinde multiplica secundum in tercium et productum divide per primum, et exeuntis sinus quere arcum, qui est latitudo stelle ab equinoxiali[c] certissime, per conversam 18e de cathis coniunctis.

IV. 31

Habita itaque latitudine, sic extrahes gradum[d] medii celi:

Extrahe totam declinacionem de 90 et residui quere sinum qui sit primus, et sinum tocius declinacionis, qui est secundus. Tercio, sinum declinacionis stelle ab equinoxiali qui est tercius, et sinum residui eius de 90, et sit quartus. Deinde cum arcu distancie veri gradus stelle a principio Capricorni vel Cancri intra in circulo directo inter ascensiones inter gradus ascensionum, et arcus qui in eius directo fuerit[e] inter gradus equales tam in signis quam in gradibus accipe sinum, qui erit quintus.[f] Argue igitur per conversam primi de cathis disiunctis sic:

Multiplica secundum in tercium, et productum divide per primum; et quod exit multiplica per quintum, et quod inde surgit itidem divide per quartum; et exeuntis sinus quere circuli porcionem, quam subtrahe de arcu quinti sinus. Et cum residuo intra in circulo directo inter ascensiones, et quod in eius directo inveneris inter gradus equales, ille certissime est gradus cum quo[g] pervenerit ad celi medium.

Ut autem melius et planius predicta omnia et consimilia intelligas, pono hic sex regulas predictas sub aliis verbis.

[a] C intras [b] C trahe [c] CM equinoccio [d] M +medium
[e] C λ [f] M 15us [g] C 2λ+perveni

[Cancer or] Capricorn in the natural order of the signs, and with this arc enter the table of right ascensions, among the equinoctial degrees. What you find opposite will be in equal degrees in the line of number, and that will be the degree of mediation.

IV. 30

If, however, the equinoctial lie between the star and its true [ecliptic] degree, you will work as follows to determine the declination:

Subtract the quasi-declination of the ecliptic place of the star from 90, and find the sinus of the result. Let this be the first [term]. Subtract from [this arc] the complement of the ecliptic latitude, and find the sinus of the result, which shall be the second [term]. Next, subtract the obliquity from 90, and find the sinus of the remainder, which will be third. Now multiply the second by the third and divide the product by the first. Find the arc corresponding to the resulting sinus. This is most certainly the declination of the star, by the conversion of II. C. 18.

IV. 31

Having obtained the latitude, this is how you find the degree of mediation:

Subtract the obliquity from 90 degrees and find the sinus of the difference. Let this be the first. Take the sinus of the obliquity, which is second. Third comes the sinus of the declination. Let the fourth be the sinus of its complement. Then with the arc of the distance of the true [ecliptic] degree of the star from the beginning of Capricorn or Cancer, enter the [table of] right ascension among the degrees of ascension. Take the sinus of the arc which is found opposite among the equal degrees (both in signs and degrees), and this will be the fifth [term]. Then argue by II. D. I (converse) thus:

Multiply the second by the third, and divide the product by the first. Multiply the answer by the fifth, and what next results divide in like manner by the fourth. Find the circular arc of the resulting sinus, and subtract it from the arc corresponding to the fifth sinus. With the result, enter the [table of] right ascension among the ascensions, and whatever you find opposite, among the equal degrees, that is most certainly the degree of mediation.

However, in order that you should appreciate everything I have already said, and similar matters, I explain the six aforesaid rules in different words.

IV. 32[a]

Si sex quantitatum proporcio prime ad terciam[b] componitur ex proporcione tercie ad quartam et quinte ad sextam, et econverso, secunde ad primam et ex sexte ad quintam et quarte ad terciam, fiunt 18 combinaciones utiles directe et 18 converse in coniunctis et totidem in disiunctis; ut *primus* modus sit ille iam suppositus, erit ergo *secundus* talis quod proporcio prime ad secundam componitur ex tercie ad sextam et quinte ad quartam; *tercius* modus erit quod proporcio prime ad terciam fiet ex secunda ad quartam et quinte ad sextam; *quartus* quod prima ad terciam fiet ex proporcione secunde ad sextam et quinte ad quartam, et sic deinceps. Ergo earum[c] quibuscumque[d] quinque notis,[e] erit sextam notam ex radice habita in[f] quinto libro et septimo Euclidis. Si igitur istarum sex quinque fuerint[g] note et sexta ignota, multiplica primam in quartam, productum divide per terciam, iterum duc secundam in quintam, et quod surgit divide per illud quod exit in prima divisione, et exibit sexta; vel sic duc secundam in terciam, et productum divide per quartam, et quod exit iterum duc in quintam; productum divide[h] per primam et exibit sexta. Ex hec est operacio duplex sicut quelibet alie.

Et si quinta sit ignota, duc primam in quartam et productum divide per terciam, et quod exit multiplica in sextam, et productum divide per secundam; et exibit quinta; vel sic—duc secundam in terciam et productum divide per quartam et quod exit serva; itidem duc primam in sextam, et productum divide per servatum, et exibit quinta.

Et si quarta sit ignota . . . [etc.]

* * *

[a] The case-endings of the ordinal numbers in all manuscript copies of this chapter appear to have been decided upon in a purely random way. They are here silently amended, the majority being now feminine, qualifying *quantitas* or *proporcio*.

[b] C secundum; M tercium [c] C +notis [d] M quibus omnibus [e] M λ
[f] M λ [g] C +prima [h] C λ

IV. 32

If the ratio of the first of six quantities to the third is compounded of the third to the fourth and the fifth to the sixth, and conversely, if the second to the first is compounded of the sixth to the fifth and the fourth to the third, there are 18 useful direct and 18 converse conjunct combinations, and the same number of disjunct. Thus if the first mode be that now supposed, the second will be

$$\text{first}:\text{second} = (\text{third}:\text{sixth}).(\text{fifth}:\text{fourth}),$$

the third will be

$$\text{first}:\text{third} = (\text{second}:\text{fourth}).(\text{fifth}:\text{sixth}),$$

the fourth will be

$$\text{first}:\text{third} = (\text{second}:\text{sixth}).(\text{fifth}:\text{fourth}),$$

and so on. Then any five of these [terms] being known, the sixth will be known by the methods of Euclid V and VII. If, then, of these six, five are known and the sixth unknown, [combine the terms thus]:

$$(\text{second}.\text{fifth})/\{(\text{first}, \text{fourth})/\text{third}\},$$

and the answer is the sixth. Or thus:

$$\llbracket\{(\text{second}.\text{third})/\text{fourth}\}.\text{fifth}\rrbracket/\text{first},$$

resulting in the sixth. And this is a double operation, like certain others.

If the fifth be unknown,

$$\llbracket\{(\text{first}.\text{fourth})/\text{third}\}.\text{sixth}\rrbracket/\text{second}$$

gives the fifth. Or as follows:

$$(\text{first}.\text{sixth})/\{(\text{second}.\text{third})/\text{fourth}\},$$

again giving the fifth.

And if the fourth be unknown [etc.]

[The remaining transformations of the equation in the first mode are given in the text in exactly similar words. They are omitted—about 200 words in all—from the Latin version opposite, and are given here, together with those above, only in summary form, the latters A to F being used to denote the six terms:

$$\begin{aligned}
F &= (B.E)/\{(A.D)/C\}\\
F &= \llbracket\{(B.C)/D\}.E\rrbracket/A\\
E &= \llbracket\{(A.D)/C\}.E\rrbracket/B\\
E &= (A.F)/\{(B.C)/D\}\\
D &= (B.C)/\{(A.F)/E\}\\
D &= \llbracket\{(B.E)/F\}.C\rrbracket A\\
C &= \llbracket\{(A.F)/E\}.D\rrbracket/B\\
C &= (A.D)/\{(B.E)/F\}\\
B &= \llbracket\{(A.D)/C\}.F\rrbracket/E\\
B &= \llbracket\{(A.F)/E\}.D\rrbracket/C\\
A &= \llbracket\{(B.C)/D\}.E\rrbracket/F.\ ]
\end{aligned}$$

Racio istarum regularum est hec:

Si proporcio *A* prime ad *B* secundam sit ex *C* tercia ad *D* quartam, et ex *E* quinte ad *F* sextam, dico quod si quinque prime sint note, erit *F* nota, que est sexta.

Racio. Notandum est[a] quod si dividas *A* per *B* exit eorum proporcio, quod vocetur *G*. Similiter, divide *C* per *D* et exibit eorum proporcio, que vocetur *H*. Item, si dividas *G* per *H* exibit eorum proporcio que sit *K*. Dico ergo quod proporcio que est inter *E* et *F* est *K*; sed *E* est nota, et similiter *K*, ergo *F* erit nota necessario: per ductam *E* in *K* necessario exibit *F*, scilicet[b] si *F* sit magis quam *E*, vel per divisionem *E* in *K* exibit *F*, scilicet si sit minus.

Sed nunc probo quod si dividatur *G* per *H* exibit *K*, quod est proporcio *E* ad *F*: ex quo enim per divisionem *G* per *H* exit *K*, certum est quod *G* componitur ex ductu *K* in *H*. Sed cum *G* sit proporcio *A* ad *B* et *H* proporcio *C* ad *D*, *G* componitur ex *H* et proporcione *E* ad *F*; cum igitur componitur [*G*] ex *H* in *K*, erit[c] *K* ex proporcione *E* ad *F*, quod demonstrare volebamus, et sic devenitur in noticiam ignote per solam divisionem.

Item potest idem probari[d] per multiplicacionem sic: ut si *A* prima ducatur in *D* quartam, exibit *P*. Item dividatur per *C* terciam et exit *H'*, ergo ducta *H'* in *C* producit *P*. Sit ergo *A* primam, *H'*[e] secundam, *C* terciam, *D* quartam, ergo quod fit ex ductu[f] *A* prime in *D* quartam equatur ei quod sit ex ductu *H'*[g] secunde in *C* terciam; igitur que est proporcio *A* prime ad *H'*[h] secundam, illa est *C* tercie ad *D* quartam. Sed, ex ypotesi, *A* ad *B* proporcio fit ex *C* ad *D* et *E* ad *F*, igitur *A* ad *B* fit ex *A* ad *H'* et *E* ad *F*. Sed *A* ad *B* fit ex *A* ad *H'* et *H'* ad *B*, ergo *H'* ad *B* est sicut *E* ad *F*. Sed proporcio *H'* ad *B* est nota, quia utraque est nota. Sed *E* est nota, ergo *F* (quia, si proporcio est nota et alterum

[a] C λ [b] C sed [c] CM 4λ; *B* proporcio ex; M +cum *B* proporcio
[d] C λ [e] C *B*; M *P* [f] M dictu [g] C *B* [h] C *B*

This is how these rules are arrived at. If

$$(1)\ A{:}B = (C{:}D).(E{:}F),$$

I hold that if the first five are known, *F*, which is the sixth, will be known. The reason is that if you divide *A* by *B*, their ratio results, which is to be called *G*. Similarly, divide *C* by *D* and their ratio results, which is to be called *H*. Next, if you divide *G* by *H*, the outcome is their ratio, to be called *K*. I say then that

(2) $E{:}F = [1/K \text{ or}]\ K$[1]	(1), defs.
(3) $F = E.K\ (F > E)$	
or $F = E/K\ (F < E)$	(2), II. 1
(4) A^*, B^*, C^*, D^*, E^*	hyp.
(5) G^*, H^*	def., (4)
(6) K^*	(5), (8) below
(7) F^*	(3), (4), (6).

But now I prove that

$$(8)\ G/H = K = E{:}F.$$

From the [supposition] that G/H is equal to *K*, it follows that

$$(9)\ G = K.H \qquad \text{II. 1.}$$

Since

(10) $G = A{:}B$, $H = C{:}D$, and $K = E{:}F$, and *G* is the product of *H* and *K*,

this proves what we wanted to prove, and thus a knowledge of the unknown is arrived at by a single division.

The same thing may be proved by multiplication, as follows. Let

$$(1)\ P = A.D \text{ and } H' = P/C \qquad \text{def.}$$[2]

Then

(2) $H'.C = P$	(1), II. 1
(3) $A.D = H'.C$	(1), (2)
(4) $A{:}H' = C{:}D$	(3), II. 1
(5) $A{:}B = (C{:}D).(E{:}F)$	hyp.
(6) $A{:}B = (A{:}H').(E{:}F)$	(4), (5)
(7) $A{:}B = (A{:}H').(H'{:}B)$	II. 2
(8) $H'{:}B = E{:}F$	cf. (6) with (7)
(9) $(H'{:}B)^*, E^*$	(1), hyp.
(10) F^*	(8), (9).

(If a ratio is known, and either extremum, the other extremum will

[1] The first alternative is included to make the first line of (3), below, intelligible. There was a fiction, seldom expressed, that the number expressing a ratio is greater than unity, which is why the first alternative of (3) is included. In the proof of (8)–(10), the algebraically respectable alternative is considered.

[2] The letter *H* is used here in the original treatise with a meaning distinct from that of the previous paragraph, and the two occurrences are therefore here distinguished as *H* and *H′*.

extremum, necessario erit secundum extremum notum). Et hec est racio et sollercia[a] in cathis.

Explicit quartus tractatus de corda recta et versa quem[b] *composuit frater Ricardus Walingford monachus Sancti Albani, in quo apparent omnia facilia et levia que peritis astrologis usque ad tempus suum fuerant quasi occulta,*[c] *et modernis loquentibus omnino ignorata.*

[a] M solercis

[b] C 30λ; M 8λ (erasure) (On the subject of the termination of different copies, see the note to the translation, opposite, the notes to the MSS., at p. 164 above, and Appendix 14.)

[c] C λ (see previous note); M occultis

necessarily be known.) And this is the underlying rationale and the technique [required in the manipulation of] catas.

Here ends the fourth treatise on corda recta and versa, written by brother Richard of Wallingford, monk of Saint Albans. Everything which to skilled astronomers[1] *before his time had been as though it were concealed, and which in disquisitions of the day was altogether disregarded, seems in this work easier and more straight-forward.*[2]

[1] *astrologi.*

[2] This colophon is taken from MS. M (Digby 190), f. 105^{r}, and certain words which in that copy were erased (for reasons which are difficult to imagine) are restored with the help of MS. T (Trinity Coll. o.9.6), f. 255^{v}. The missing words could have filled the available space, but the same can be said of part of an otherwise largely different colophon in MS. D (Digby 178), which makes Richard '*quondam abbas*', rather than '*monachus*'. Even the ending of *De sectore*, apparently written (see below) when Richard was abbot, makes no mention of his highest office, and refers to him simply as a monk.

TRACTATUS DE SECTORE

Cambridge University Library, MS. 1572 (Gg. VI. 3)

Hic incipit tractatus de sectore

SECTOR est figura in superficie spere ex arcubus circulorum magnorum semicirculo minoribus constituta, cuius sciencia predita nosse querit distanciam cuiusvis stelle aut puncta a quocunque puncto in orbe celi assignato, in longitudine et latitudine, iuxta mensuram per orbem circuli magni. Huius autem noticiam tradit Ptholomeus in primo *Almagesti*, ex combinacione 6 quantitatum proporcionalium, sed obscure. Gebir et Commentarius invicem [*lit.* invicens] undique brevitati sub 4 quantitatibus proporcionalibus eam tradit, neuter tamen minus provectis in mathematica satisfacit, eo quod pauca prevenerunt de hiis, que autem huius figure [ad] noticiam oportet prescire, que vix alicubi simul aut lucide colliguntur, cuius pretexti plerique desidiotam [*sic*] decore veritatis allecti ita (?) eius inquisicioni anxietates non modicas sunt perpessi.

Quid igitur conferat hic tractatus hiis quibus speculacionis huius preciosus est habitus in quo omnia noticia sectoris necessaria compilantur? Soli non erunt qui ad compilacionem presisam (?) ex dispersis ad hanc scienciam proprio studio pervenerunt.

In prima parte huius opusculi tractatur de sinu et arcu satis complete.

In secunda eius parte tractatur quot et quibus modis ex 5 quantitatibus proporcionalibus cognitis devenitur ad sextam ignotam [*lit.* ignotum], et qualiter ex proporcione trium notorum sciri potest quartum ignotum.

In tercia parte huius expresse ponuntur demonstraciones Ptholomei et Gebir, de figura sectore.

In quarta parte et ultima ponitur modus operandi in omnibus ad que necessarie [? *lit.* notioie] huius utilitas se extendit.

I

Cap. 1 (I. Prol.) *De diffinicionibus.*

Add.: Solet autem circulus dividi in 360 partes equales, et diameter eius in 120 sive in 300 partes equales; nihil tamen refert in quot partes diameter dividatur, isti tamen numeri eliguntur quia partes aliquotas plures habent.

Om.: Para. 1.

I. 1. *Cognito sinu recto alicuius arcus sinum versum eiusdem arcus ostendere.*

Subst.: *ABCDEF/CP( )DEF.*

Add.: Hoc etiam patet ex Dulkarnon. Si protrahatur linea *ED* equalis linee *AD*, constat quod quadratum *EF* et quadratum *FD* simul accepta valent quadratum *ED*, et per consequens quadratum *AD*. Ergo cum *AD* linea sit mihi nota cum sit medietas diametri, que erit 60 partes secundum quod tota diameter supponitur esse 120 partes, vel 150 partes si diameter dividatur in 300 partes, accipiam igitur quadratum linee *AD* puncta 3,600 vel 22,500, non curo, a quo subtraham quadratum *FE* linee mihi note, et remanebit quadratum *FD* cuius radix erit quantitas linee *FD*, qua per radicis extractionem inventa, subtraham ipsam lineam *FD* iam mihi notam de semidiametro *AD* mihi nota, et remanebit quantitas linee *AF* nota (*lit.* notum); et hoc fuit propositum, per noticiam sinus recti *FE* datam ostendere quantitatem linee *FA*, que erit sinus versus porcionis eiusdem. Et nota hic pro *corelario*, quod qualitercunque dividatur quarta pars circuli in duos arcus, quadratum sinus tocius quarte valet 2 quadrata sinuum rectorum arcuum corresponden[tium?].

I. 2. *Cognito sinu verso alicuius arcus, sinum eius rectum ostendere.*

Subst.: *ABCDEF/CP( )DEF.*

Add.: Et nota hic pro *corelario* quod corda versa arcus minoris quarta circuli et corda recta residui eiusdem quarte adequant semidiametrum. Et ideo omnis sinus rectus arcus infra quartam circuli cadit super diametrum, quod linea cadens inter sectionem et centrum equalis est corde recte residui eiusdem quarte, quod patet si a puncto *E* producatur equidistans linee *FD*, terminata ad lineam *DC*, que sit linea *EG*; erit quia *EG* equalis linee *FD*, ex 24 primi [Euclidis].

I. 3. *Dato sinu recto arcus cuiuscunque medietatis eiusdem arcus sinum rectum ostendere.*

I. 4. *Dato alicuius arcus sinu recto sinum duplicis arcus ostendere.*

Add.: Et nota hic pro *corelario*, si quadratum equilaterum vel oblongum dividatur per unum latus eius, exibit quantitas alterius lateris.

I. 5. *Cognito sinu recto alicuius arcus, sinum rectum alterius arcus qui tantum deficit a tota quarta quantum est sinus dati ostendere.*

I. 6. *Omnium cardagarum coniunctim et disiunctim sinum rectum ostendere.*

Add.: Verumtamen ita precise operari poteris in extraccione radicis numerorum qui surdi appellantur, per addicionem multarum ciphrarum parum (?)o peracione Indorum quod parmcas (?) (permcas?) in communem sura bellitatis (?) ymaginacionem omnem subterfugiet.

I. 7. *Diversarum porcionum sinibus cognitis unam cordam illis omnibus communiter subtensam ostendere.*

Add.: Et nota hic pro *corelario* quod cognitis quibuscunque duobus lateribus trianguli, tercium latus potest cognosci, si idem triangulus inscribatur circulo, eo quod tercium latus subtenditur utrisque arcubus duorum laterum reliquorum; ut sit triangulus *ABC*, et producto diametro intersecante lineam ignotam *BC*, que sit diameter *AD*, coniunctis *BD* et *CD* habemus quadrangulum diametris suis distinctum, scilicet *BCDA*. Et cum nota fuerit linea *AB*, nota erit linea *BD*. Item, cum nota fuerit linea *AC*, nota erit linea *CD*, quia *C* et *B* anguli sunt recti. Ergo huius quadranguli nota sunt omnia latera, et una diameter; ergo vigilanti, nota erit diameter altera, scilicet *BC*.

Item nota pro *corelario* quod in omni triangulo quodlibet latus est sinus duplatus unius arcus qui duplus est angulo eidem lateri opposito, et ideo cognito latere alicuius trianguli notus erit angulus sibi oppositus, et econverso; hoc enim patet quia omnis triangulus est inscriptibilis circulo. Fit igitur triangulus *ABC*, cuius latus *AB* opponitur angulo *C*, qui subduplus est ad angulum factum in centro circuli super eandem basim, per 25^{am} tercii libri [Euclidis]. Et sic de reliquis lateribus et angulis est probandum.

I. 8. *Maioris arcus et minoris sinibus rectis notis porcionis illius in qua unius arcus excedit reliquum sinum rectum ostendere.*

Add.: Et nota hic pro *corelario* quod licet arcus excedat arcum in duplo, corda non excedit cordam in eadem proporcione, quia *FD* non est duplus ad sinum *GO* . . . ut probat Ptholomeus in primo *Almagesti* universaliter . . .

Subst.: *ABDEFGOI*/*NCOE*( )( ).

I. 9. *Sinum duplatum decime et quinte partis circuli ostendere.*

Add.: Nota tamen hic pro *corelario* quod si coniunxeris quadratum corde duplate 36 graduum, scilicet 10^{e} partis circuli, ad quadratum semidiametri, totius collectionis radix erit corda duplata 5^{e} partis circuli, quod patet per Dulkarnon cum triangulus *BOE* sit rectangulus, et *OE* est corda 10^{e} partis circuli, sicut hic probatur; et etiam *EB* est corda 5^{e} partis, ut probatur alibi; et *O* est angulus rectus.

Item nota pro *corelario* quod si fuerint 2 quadrata, quorum unum est quintiplum [*sic*] ad relicum [*sic*], erit radix minoris quarta pars diametri, et radix maioris continebit eandem quartam partem diametri et mensuram per cordam subtensam decime parti circuli.

Et sic inveniri poterit corda septime partis, per invencionem trianguli habentis 2 latera equalia, cuius unus angulus super basim fuerit triplus ad angulum in prima latera, sicud in triangulo *AOF* angulus *F* est duplus ad angulum *O*. Et similiter procedendum est circa invencionem 9^e^ partis circuli, per angulum quadruplum.

Subst.: *ABCDEFGHKLO/K( )QRDLCNMPE.*

I. 10 (I. 10–11). *Cuiuslibet gradus et omnium unius quarte circuli sinum rectum ostendere.*

Add.: . . . et sic procedam operando donec complevero quartam circuli sive in cordis rectis sive duplatis, quia tabule *Almagesti* habent cordas duplatas et tabule Gebir et Albategni habent cordas mediatas; et hoc est propositum unius quarte circuli sinum rectum ostendere.

Et nota hic pro *corelario* quod, habito uno arcus sinu noto per 3^{am} et per 7^{am} huius, sufficienter inveniri possit sinus cuiuslibet alterius arcus, sine errore sensibili; nec est hesitandum, quia due sexte sint medietas duarum terciarum, quia 3 tercie et 6 sexte sunt equales, ergo 3 sexte est medietas 3 terciarum sicud 6 sextarum, et 2 sexte tercia pars 3 terciarum, sicud est tercia pars 6 sextarum et 2 tercie valent 4 sextas, ergo 2 tercie est duplum ad duas sextas.

I. 11 (I. 12–13). *Cuiuslibet partis unius quarte cognito sinu recto et verso totius semicirculi sinum rectum et versum ostendere.*

II

II. Prol. In secunda parte huius tractatus oportet investigare proporcionem sinus ad sinum . . .

Om.: Defs. 3 and 4.

Subst. & Add.: *Tercia.* Proporcionem componi ex proporcionibus est denominacionem creari ex denominacionibus per multiplicacionem surgendo vel per divisionem exeundo (?).

Quarta. Quando quantitates ex quarum proporcionibus componitur proporcio extremorum continue (?) crescunt vel decrescunt a primo ad ultimum, item componitur proporcio extremorum ex sigillata multiplicacione denominacionum proporcionum intermediorum . . .

Quinta. Proporcio coniuncta quando alia quantitas cum addito se habet ad aliam quantitatem sicud tercia quantitas cum alio addito se habet ad quartam, in eadem proporcione in qua se habent eedem quantitates sine illis additis. . . .

Sexta. Proporcio disiuncta est quando de antecedentibus vel consequentibus 4 numerorum proporcionalium separantur 2 partes proporcionales; adhuc salvetur proporcio inter 4 extrema, ut quia 12 ad 4 sicud 6 ad 2, separatis de 12 et 6 proporcionalibus partibus, ut 1 et 2, adhuc salvatur proporcio 4 extremorum 10 ad 4 sicud 5 ad 2. Hec patent in 14[a] libri 7[i] Euclidis.

II. 1. *Si denominacio proporcionis quorumcunque duorum extremorum multiplicet minus extremum, producitur maius, et si dividat maius, exibit minus extremum.*

II. 2 (—). *Propositis duabus quantitatibus eiusdem generis per secundam multiplicetur tercia quantitas et productum dividatur per primam, quod exit se habebit ad terciam sicut secunda ad primam.*
Ista patent ex precedenti, quia . . .

II. 3 (II. 2). *Propositis duabus quantitatibus extremis et medio eiusdem generis interposito, erit proporcio extremi ad extremum composita ex proporcione primi extremi ad medium et medii ad ultimum.*

II. 4 (II. 3). *Quotlibet mediis inter 2 extrema positis, erit proporcio primum ad ultimum ex omnibus intermediis proporcionibus composita.*

II. 5 (II. 4). *Si aliqua proporcio componitur ex proporcionibus, eius conversa componitur ex eisdem conversis.*

II. 6 (II. 5). *Cum 6 quantitatum eiusdem generis proporcio prime ad secundam composita fuerit ex proporcione tercie ad quartam et quinte ad sextam accident ex combinacione 360 species composicionis, quarum 36 sunt utiles, cetere autem omnes inutiles, id est 324.*

II. 6 cont. (II. 6). Duo iam restant probanda . . .

II. 6 cont. (II. 7). Patet ergo quod quelibet istarum 9 combinacionum . . .

II. 6 cont. (II. 8). Et secundum similitudinem huius deductionis . . .

II. 6 cont. (III. Prol.). Nunc autem restat ostendere omnes modos utiles . . .

III

Incipiunt cathe ostense in coniunctis.

Proposicio 1–Proposicio 18 (III. C. 1–III. C. 18). Si a puncto *A* extrahatur due linee . . .
. . . *Explicit de cathis coniunctis, incipit de disiunctis.*

Proposicio 1–Proposicio 18 (III. D. 1–III. D. 18). Manente igitur eadem figura . . .

Add.: *Tabula* [*que*] *ostendit quot modis in coniunctis* [*vel disiunctis*] *cadit composicio.* [Vide Plate II.]
[*Explicacio tabule*] Si igitur cadant 6 quantitates in figura cathe . . .

[Capitulum] (IV. 32). Causa istius satis habetur ex 7° libro et 15ª Euclidis. Ponam igitur sex regulas operandi, secundum quod primum fuerit ignotum, vel secundum, et sic deinceps, ut pro modis calculator invenire possit quod querit. *Prima regula* . . .

. . . Explicit secunda pars [tercie partis]

Add.: In tercia parte huius tractatus ponentur demonstraciones Ptholomei et Gebir commentarii de figura sectore, quibus prepono 4 conclusiones de primo *Almagesti* ad demonstraciones Ptholomei necessarias, ex quibus consequenter ostendam figuram sectoris secundum 3 modos quibus eius figuram est possibile variari. Deinde addam quasdam conclusiones Gebir suis demonstracionibus necessarias, ex quibus concludam propositum Gebir mirabiliter compendiosum et brevem, ipse quo utitur cordis mediatis (Ptholomeus autem duplatis).

III³ (tercia pars)

III³. 1 (IV. 1). *Cum aliquis semicirculus in duos arcus quolibet modo contigerit dividatur, erit corda dupli arcus unius equalis corde dupli arcus alterius.*

III³. 2 (IV. 2). *Si diameter circuli secaverit cordam duplam alicuius porcionis circuli, erit proporcio unius partis corde ad aliam qualis est proporcio corde dupli arcus unius partis porcionis secte ad cordam duple porcionis alterius partis.*

III³. 3 (IV. 3). *Si ab aliquo puncto semicirculi protrahatur linea que extra circulum concurrat cum diametro, erit proporcio partis linee intra circulum qualis est proporcio corde dupli arcus inter ipsam lineam et diametrum intercepti ad cordam dupli arcus tocius porcionis a principio eiusdem linee ad diametrum.*

III³. 4 (IV. 4). *Cum alicui arcui minori semicirculo subtenditur corda cui a centro circuli protrahitur equidistans, erit corda dupli arcus inter illas intercepti equalis corde dupli arcus compositi ex eo cui corda subtenditur et intercepta.*

III³. 5 (IV. 5–6). (Om. para. 1.) *Si ab aliquo puncto spere duo arcus circulorum magnorum semicirculo minores exierint a quorum terminis 2 alii arcus reflexi sese intersecaverint ad primos arcus, facta erit proporcio corde dupli arcus unius eorum ad cordam dupli arcus alicuius alterius correspondente composita ex proporcione corde dupli arcus alicuius tercii eorum ad quartam, et ex proporcione corde dupli arcus quinti ad cordam duplicis arcus sexti.*

Add.: Patebit etiam hec veritas per omnes modos in proporcione coniuncta . . . nulla est proporcio, quia eadem corda sit. *Explicit Ptolomeus, incipit Gebir.*

Gebir undecimo capitulo libri sui totam scienciam sectoris reducit ad noticiam trianguli rectanguli ex arcubus magnis compositi, et potest triangulus arcualis habere 3 angulos rectos, vel duos, aut unum, vel nullum; sed de eo, qui nullum habet angulum rectum in superficie spere, non est hic consideracio.

Add.: III⁴ (pars quarta)

III^4. 1 (Gebir, *Flores* I. 11). *Omnis trianguli arcualis in quo est unus angulus rectus unusquisque duorum laterum continencium angulum rectum mutuatur angulum sibi oppositum.*

III^4. 2 (loc. cit.). *Omnis trianguli arcualis orthogonii, si aliquod duorum laterum continencium angulum rectum fuerit quarta circuli, latus tercium oppositum ipsi angulo recto est quarta circuli, et econversa.*

III^4. 3 (loc. cit.). *In omni triangulo arcuali rectangulo, si ambo latera includencia angulum rectum fuerint maiora vel minora quarta circuli item latus oppositum recto angulo erit minus quarta circuli.*

III^4. 4 (loc. cit.). *In omni triangulo arcuali rectangulo, si unum duorum laterum continencium angulum rectum fuerit maius quarta circuli, et reliquum latus minus quarta, erit latus oppositum ipsi recto angulo maius quarta.*

III^4. 5 (converse to prop. 2). *In omni triangulo arcuali rectangulo, si latus oppositum angulo recto fuerit quarta circuli, erit unum duorum laterum includencium angulum rectum quarta circuli.*

III^4. 6 (conv. prop. 3). *In omni triangulo arcuali rectangulo, si latus oppositum angulo recto fuerit minus quarta circuli, erit utrumque latus quod includit angulum rectum maius quarta vel minus quarta circuli.*

III^4. 7 (conv. prop. 4). *In omni triangulo arcuali rectangulo, si latus oppositum recto angulo fuerit maius quarta circuli, erunt 2 latera rectum angulum includencia diversi speciei, ita quod si unum illorum fuerit maius quarta, relicum erit minus quarta, neutrumque pars quarta circuli.*

III^4. 8 (conv. prop. 1). *In omni triangulo arcuali rectangulo, utrumque duorum reliquorum angulorum ab angulo recto imitatur latus sibi oppositum.*

III^4. 9 (*Flores* I. 12). *Si fuerint duo circuli magni in spera signati, et non transierit unus eorum super polum alterius, et in circumferencia unius eorum signentur 2 puncta, et producantur a polo alterius circuli 2 arcus transeuntes per illa 2 puncta in circumferencia alterius signata usque in circumferencia illius circuli a cuius polo idem 2 arcus exierint, item proporcio sinus illius arcus qui est inter communem sectionem duorum circulorum et alterum illorum duorum punctorum ad sinum predicti arcus ab eodem puncto ad circulum a cuius polo procedit est sicut proporcio sinus arcus exeuntis ab eadem communi sectione circulorum ad alterum illorum 2 punctorum ad arcum exeuntem ab eodem puncto ad circumferenciam circuli a cuius polo procedit.*

III^{4}. 10 (*Flores* I. 13). *In omni triangulo arcuali circulorum magnorum, proporcio sinus cuiuscunque lateris ad sinum arcus anguli illius cui ille arcus opponitur est sicut proporcio sinus cuiuscunque alterius lateris eiusdem trianguli ad sinum arcus anguli cui ipsum latus secundum opponitur.*

III^{4}. 11 (*Flores* I. 14). *In omni triangulo arcuali rectangulo est proporcio sinus arcus anguli unius duorum reliquorum angulorum ad sinum arcus anguli recti sicut proporcio sinus arcus complementi anguli reliqui ad sinum arcus complementi lateris eidem angulo subtensi.*

III^{4}. 12 (*Flores* I. 15). *In omni triangulo arcuali habente unum angulum rectum proporcio sinus complementi arcus subtensi recto angulo ad sinum complementi unius duorum laterum ipsum rectum angulum includencium est sicut proporcio sinus complementi alterius lateris ad sinum quarte circuli.*

III^{4}. 13 (coroll. prop. 10). *Notis duobus lateribus et uno angulo vel duobus angulis et uno latere trianguli arcualis reliquos angulos et reliqua latera cognoscere sic contingit.*

III^{4}. 14 (*Flores* I. 25). *In omni triangulo rectilineo duobus lateribus et uno angulo notis vel duobus angulis et uno latere, reliquos angulos et reliqua latera cognoscere.*

. . . et econversa notis 2 angulis et uno latere. *Explicit tercia pars.*

IV

In quarta parte huius tractatus, signato sectore in superficie spere, ostendetur alternatim quomodo est operandum per 6 quantitates proporcionales modo quo tradit Ptolomeus tam in sinibus rectis quam in duplatis licet in idem redeunt, et quomodo operandum est per 4 quantitates proporcionales in triangulo arcuali rectangulo, modo quo docet Gebir operari, et hoc penes omnes diversitates tactas in *Almagesti*, cuius gracia hoc presens opusculum maxime compilatur.

IV. 1 (cf. *Quad.* IV. 9). *Declinacionem cuiuslibet puncti circuli orbis signorum a circulo equatore ostendere.*

IV. 2 (cf. *Quad.* IV. 7). *Arcum circuli equatoris ascendentem cum quolibet gradu orbis signorum* . . . (illegib.) . . . *proximo in orizonte directo ostendere.*

IV. 3 (cf. *Quad.* IV. 8). *Quantum elevatur de equatore cum arcu ecliptice dato in orizonte obliquo ostendere.*

IV. 4 (cf. *Quad.* IV. 21). *Arcum orizontis qui est ortum puncti equinoxii et ortum puncti unius tropicorum per nota poli altitudinem, et econverso, ostendere.*

IV. 5 (cf. *Quad.* IV. 10). *Gradum cum quo stella mediat celum et eius latitudinem ab equinoxiali per eius verum gradum et per eius latitudinem ab ecliptica notam ostendere.*

IV. 6 (cf. *Quad.* IV. 17). *Punctum ecliptice et equatoris cum quo stella oritur in orizonte obliquo ex premissis ostendere.*

IV. 7 (—). *Signatis duobus punctis quorsumlibet in spera per notam longitudinem et latitudinem eorum respectu ecliptice eorum distanciam ab invicem demonstrare.*

IV. 8 (cf. *Quad.* IV. 11). *Arcum elevatum ab ortu stelle exeuntis in circulo equinoxiali eius azimuth per notam eius elongacionem a cenit vel orizonte ostendere.*

IV. 9 (cf. *Quad.* IV. 11). *Quantacunque fuerit latitudo Solis vel stelle ab equinoxio medietatem arcus diei eius et altitudinem meridianam cognoscere.*

IV. 10 (=?). *Cognita medietate* . . . (illegib.)

IV. 11 (=?). *Quantitatem anguli et* . . . (illegib.)

IV. 12 (=?). *Quantitatem anguli proven* . . . (illegib.)

IV. 13 (=?). *Quant* . . . (illegib.)

IV. Explicit: *Albyon est instrumentum* . . . 1335 [vide supra].

III

EXAFRENON PRONOSTICACIONUM TEMPORIS

A resonable soule is more worthi than sterres,
and is not harmed of hem but by ignoraunce.
Trin. Cantab., MS. O. 5.26, f. 98v.

SIGLA

A Oxford, Bodleian Library, MS. Ashmole 1796.

B London, British Museum, MS. Royal 12. C. XVIII.

C Cambridge University Library, MS. 1693 (Ii. 1. 1).

D Oxford, Bodleian Library, MS. Digby 67.

R Oxford, Bodleian Library, Digby Roll 3.

T Cambridge, Trinity College, MS. O.5.26.

For full discussion of the manuscripts, Introduction to the text, and Commentary see vol. ii, pp. 83–126. See also vol. iii, p. 17.

The Middle English translation is based on MS. D, and the following conventions have been used:

My own translation, supplementing the original where necessary, is in square brackets.

Where the text is restored from other English MSS., square brackets are also used, and the source indicated in a footnote.

Where a sequence of words is present in the English but not in the original Latin, it is printed in smaller type.

All other amendments or matters of note are indicated in footnotes.

EXAFRENON PRONOSTICACIONUM TEMPORIS

AD PERFECTAM noticiam iudiciorum artis astrologie,[a] que natura regulante ex effectibus planetarum oriuntur, primo oportet scire tempus introitus Solis in primum minutum cuiuslibet[b] signi,[c] eo quod ab[d] introitu Solis in primum minutum Arietis incipit principaliter diversitas annone tocius revolucionis ipsius anni. Et secundario, ab introitu Solis in Tauro incipit diversitas mensis Aprilis; et sic de ceteris signis zodiaci et mensibus solaribus. Et hoc erit tibi signum, quia ut communiter diversificantur tempora in medio mensium quantum ad effectus[e] qui[f] sequuntur Solem. Expedit eciam scire introitum Solis in[g] omnem gradum zodiaci pro particularibus iudiciis que sequuntur Lunam in coniunctione et opposicione, et alios planetas. Et sciencia istius ordinatur ad inveniendum[h] gradum ascendentem[i] super orizontem eadem hora, quia secundum ascendentem debent domus mensurari.[j]

Secundo oportet te scire artem investigandi quis gradus zodiaci pululat[k] super orizontem tuum in momento illius hore in quo vis stabilire iudicium tuum; quia ex[l] quo in alia[m] et alia disposicione[n] firmamenti planete aliam[o] et aliam virtutem[p] secuntur in emisperio[q] tuo, adinvenerunt[r] philosophi artem dividendi totum firmamentum in 12 partes quas partes vocabant domos, incipientes ab ascendente, id est, a gradu orienti,[s] tanquam a principali; quia secundum Philosophum in libro de celo et mundo, virtus deorum, id est planetarum, maxime apparet in oriente; et tunc per illum gradum ascendentem adequabis omnes 12 domos, sicut docebitur[t] in capitulo suo.

Tercio oportet te scire fortitudines substanciales planetarum, quarum quasdam habent ex natura propria unde dicuntur calidi, frigidi, humidi, sicci, ignee nature, vel aeree, vel terree, vel aquatice, et similia, sicut patebit in loco suo. Et alie fortitudines sunt quas habent ex natura signi in quo sunt, et dicuntur dignitates que sunt 5, scilicet domus, exaltacio, triplicitas, terminus, et facies. Quid[u] autem unumquodque istorum est, scies in capitulo suo, cum ad hoc venerimus. Et hic[v] est summa et substancia iudiciorum quia a dignitatibus dependet maxima vis eorum circa[w] quod philosophi anxie[x] laborant.

[a] B astrologice [b] C λ [c] C +Arietis [d] C λ [e] B effectum
[f] AC que [g] B 15λ+signa; C 16λ [h] C videndum [i] C ascendentis
[j] A mensurare [k] BC pulsat [l] C 2λ [m] A aliam tabulam
[n] A disposicionem [o] C naturam [p] C naturam [q] A experio
[r] A adinvenersit [s] C orientali [t] A λ [u] B Quod [v] C hee
[w] A 6λ [x] C maximus

Incipit EXAFRENON PRONOSTICATIONUM TEMPORIS

THO THE PERFECTE KNAWLEGE of the domes of the crafte of astronomye, the which be the rewlynge of kynde ar browth forthe of the effectes of planetes, first it is nedful for to knaw the entrynge of the Sonne in to the first minute of [some sign or other, since from the entry of the Sun into the first minute of] Arietis[1] begynnes principaly the chaungynge of the turnynge of abowte of the hole *y*ere. And aftir that, fro the entrynge of the Sonne into Taurum begynnes the chaungynge of the moneth of Aprile; and so of other signes of the zodiack, and other monethis of the *y*ere. And that is the tokynynge, fore comonly in the myddis of the monethis the tymes ben made dyverse as to effecte of the Sonne. Also it is spedful for to knawe the entrynge of the Sonne in to every gree of the zodiack, fore particular domys that followes the Mone in coniunction and opposition, and other planetis. And the knawlethe of this is ordaned for to knawe the gree ascendente opon the orizont in the same tyme, fore aftir the gree ascendente thu moste mesure thyn howses.

The secunde, thu moste knawe the crafte to seeche what gree of the zodiac rises upon thyn orizont in the moment of the howre, in the which thu wele gyfe thyn dome; fore in dyverse disposition of the mevynge of the firmament and planetis takis diverse effectis, scilicet in thin emisperye. And therfore the philosophirs founde a crafte to departe all the firmament in 12 partes, the which partyes be called howses; and thai began at the ascendent, [that is, from the degree in the east], fore it is principal. Fore aftir the Philosophire in the booke of the heven and of the warlde,[2] be the vertu [of the gods, that is], of the planetis, it shewes moste in the est. And so be the gree ascendent thu salt mesure all that other 12 howses, [as in the appropriate chapter].

The thirde thu most knawe the strenghtis and vertues[3] substancial of all the planetis, of the wich summe have of thair awne kynde, wherfore ar clepid hote and colde, drye and moyste, [of a fiery nature, or airy, or earthy, or watery, and so forth], as it shall show in thaire chapitres. And other strenght is that thei have of the kynde of the signe that thay ben in; and thai ar callid 5 dignytes, the which ben domus, exaltatio, triplicitas, terminus, and facies etc. What iche on of thees ben thu salt see in thare chapitres whan we com therto. And this the substance of domes, for upon the dignytes hanghtes moste fors and strenghtes of them (abowte the which philosofris travelis besily).[4]

[1] D +. For thare   [2] D +be   [3] Not in the Latin.
[4] The words in parentheses are not in the Latin of the best MS. (A).

Quarto oportet te scire fortitudines accidentales planetarum quas accipiunt ex diversitate motus eorum in epiciculo, cuiusmodi sunt stacio, progressio, et retrogradacio; et ex eorum distancia a Sole, cuiusmodi sunt ortus et occasus eorum, accessus et recessus ad cenith capitis in regione tua, secundum quod est in alia et alia parte sui deferentis, qui est ecentricus. Et omnia ista scies cum adequaveris loca vera planetarum per argumenta et per centrum et alia que requiruntur, sicut patet per canones eorum, vel per tabulas manuales[a] que dicuntur almanach.

Quinto oportet te scire disponere in figura domorum planetas et in signis eorum, et scire quid est sextilis aspectus, et quid quartilis, et quid trinus, et quid oppositus, et quid coniunctus; et similiter que est virtus et efficacia cuiuslibet istorum; ut scias quando planeta respicit planetam bono aspectu, et quando respicit[b] eum malo aspectu. Et per hoc debilitatur eius virtus sicut confortatur per bonum aspectum alterius, et scies extrahere dominum anni, diei,[c] mensis, et hore.

Sexto et ultimo, per scienciam omnium istorum debes iudicare et pronosticare futura naturaliter contingencia et circumspecta collacione omnium istorum que predixi tibi, ita tamen quod certitudinaliter non pronostices nisi certus sis nullum predictorum omisisse. Sicut enim dicit Philosophus in *Elenchis*, ad pauca respicientes faciliter decipiuntur unde et a sapientibus merito deridentur, non autem ars illa quam credunt scire sed ⟨non sciunt⟩. Econversa ab ignaris non ipsi, sed ars tanquam frivola habetur in contemptu, sicut dicit Albumazar in maioribus naturalibus suis, libro primo, capitulo quarto, de questionibus[d] in fine. Ex eo siquidem quod plerosque[e] huius artis professores minus peritos[f] in officio suo non nunquam errare vident[g] artificis culpam[h] arti inponunt; qui vero rerum naturalium iura sano iudicio tractant nequaquam huius erroris culpam arti innocue[i] ascribendam[j] censebunt. *Explicit prologus, incipiunt capitula.*

I

Capitulum primum est ad sciendum verum introitum Solis in primum minutum Arietis et in alia signa; quia sicut dicit Albumazar *De floribus*

[a] C eorum [b] A 2λ [c] B λ; C rei [d] BC operacionibus
[e] A plures; C plerique [f] A partes [g] B videntur [h] C λ
[i] C 2λ+immo tue ignorancie [j] A ascribendum

The forthe is thu moste knawe the strenghtis that befallynge to tham be certayne movyngis in thaire differences or ellis in thaire epiciclis, as is station, progression, and retrogradation; and of thaire distance fro the Sonne, as is risynge and goyng downe, and commynge toward or nygh the cenyth and goynge awayewarde [in your region], aftir that it is in dyverse parties of thayre differencis[1] [which is eccentric]. And all those thu salt knawe when thu hast made equacions of the planetis be the argumentis and the centris and other thyngis[2] as it shewis be thayre canouns of tham, and be tables that ar callid armanac.

The fifth chapitre is thu moste knawe for to ordire in figure of howsynge the planetis in thaire signis and to knawe which is the sextile aspecte [and whych quartil, whych trinus, and whych oppositus, and][3] wich coniunction; and also wich is the vertue and the effecte of ich on of thame. And witt thu wele that a planete is comforted by a good aspecte of an other planete, and his vertue is feblid be an evil aspecte, et cetera. Also thu moste knawn to draw owt the lorde of the *y*ere, of the thynges,[4] of the monethes, and of the howre.

The sexte and the laste is thu moste kunne [by the science of all these things], deme, and telle before, thynges that ar to come kyndely be avysement and of the gederynge togedir of all thees that I have sayde, never the les, so that thu saye non thynge before certanely, but thu wytt wele that thu have lefte no thynge behynde that I have spokyn of; as says the Philosofre in *Elenchis*, tho that lookis unto fewe thynges ar lightly begiled, wherfore certanly thei are right wisly scorned, and kan not the science that thai trowe thai kan, and kan it not. But certanly of unkunnynge men thai are not dispised; but the science is haldyn vauyne and but a japerye, as sayes Albumazare and tellis [in his bokis[5] I. 4, on the questions at the end. From this, there are many teachers of the craft who, when they seem ever to err in its practice, put the blame on its author; but those who concern themselves with natural things by the application of sound judgement will certainly not put the blame for any error on the harmless craft itself.]

[*Here ends the prologue, and here begin the chapters.*]

I

The firste chapitre is for to knawe the verray entre of the Sonne into the firste minute of Arietis and into othere signes. For as Albumazar

[1] That is, the deferent circle.
[2] D thyingis
[3] The bracketed passage is supplied from R.
[4] We should here, following the text, have 'of the day'.
[5] The first three words of the addition are taken from R.

in principio, magnus error accidit circa accepcionem domini anni et mensis si hoc negligatur, et ideo quia operacio in tabulis almanach non est verissima ad hoc propter equinoxia et solsticia mobilia sicut patet, quia cum solsticium yemale fuit aliquando in natali Domini nostri Iesu Cristi, nunc retrocessit usque ad festum Sancte Lucie fere pro 11[a] diebus, ideo composui tibi tabulam brevem ad habendum omnia predicta sine errore pro[b] quocumque tempore preterito et futuro faciliter et absque tedio. Et primo ut scias clarius racionem operandi per istas tabulas ostendam tibi causam illius retrocessionis solsticiorum et equinoxiorum. Et quantitas retrocessus quolibet anno inventa est per sapientes compotestas[c] in isto novissimo tempore speculantes, quod Sol non reddit ad idem punctum celi precise in 365 diebus et sex horis, prout supposuerunt fundatores kalendarii Romanorum sed transit ultra per quintam partem unius hore fere. Et propter hoc solsticia et equinoxia et omnia alia festa que ponuntur fixa in Kalendario sub certis Kalendis multum sunt elongata temporibus istis a locis illius temporis in quibus posita fuerint primitus in Kalendario, quia per 11 dies fere ut apparet per diem natalis Domini; fuit enim hec dies secundum scripturas in solsticio yemali a quo fere istis temporibus per 11 dies elongatur a tempore in quo hoc festum celebratur, sicut certissimis machinamentis sapientes Arabes Greci et Latini probant.

Et dixi per quintam partem unius hore fere retrocedit[d] annuatim quia non totaliter recedit per quintam partem, quod est 12 minuta unius hore, sed tamen per 11 minuta unius hore, sicut patet per divisionem[e] longissimi temporis a solsticio invento tempore Abrachis usque ad solsticium inventum tempore Ptolomei; et a tempore suo usque ad tempus nostrum. Et surgunt precise illa 11 minuta addenda vel diminuenda, patet namque quod pro temporibus futuris sunt illa 11 minuta subtrahenda[f] et pro tempore preterito addenda, quolibet anno; quia ex quo solsticia et omnia festa fixa retrocedunt, ad habendum certum locum Solis anno revoluto oportet subtrahere 11 minuta a vero loco in quo fuerat anno preterito eadem hora diei; et pro preterito eadem racione addenda sunt. Unde ad claram noticiam predictorum describam tibi istas duas tabulas quas predixi, quarum prima est tabula mensium, secunda annorum collectorum Cristi, quas feci ad meridiem Londonie pro anno Cristi 1296 anno bisextili hoc modo:

[a] B xiii (and throughout this paragraph) [b] A 2λ+procumque
[c] C componentes [d] B 15λ; C 16λ+scilicet [e] A diem [f] B 11λ+quoad

says in his book *De floribus*, greet arrours may fall abowte the takynge of the lorde of the *y*ere and of the moneth, yf that itt be reklesly loked or takyn, and reson whi, for the werkyn in the tables of the almanac is not verrey, and that is fore the equinoctis and the solstitiis are mevable as it shewis; fore the solsticis of wynter was summe tyme in the feste of the nativite of Criste. Nowe it is gone bakward unto the feste of Sancte Lucie, [changing by about eleven days], and therfore I have made [for you] a table of all the enterynges of the Sonne in to dyvers signes withowtyne erroure, bothe fore tyme passid and fore to come, lightly and withowtyne travayle, and cause of goynge bakward of the solsticye and the equinocis. And the quantite of the goynge bakwarde iche *y*ere is founden be wise countoures in this tyme that the sesoun[1] goyes not agayne unto the same poynte of the zodiac precise in 365 dayes and 6 houris as the fownder[s] of the kalendar [of the Romans] supposid that it shuld doo, but it passid owre the 5 parte of ane howre almoste. And therfore the solsticy and the equinoctii and the festis that are sette fixe in the kalendre [are further removed in these times from the positions assigned to them in the kalendar. For it appears that the birth of our Lord, according to the scriptures having taken place at the winter solstice, is, as it is now celebrated, eleven days removed in time from the solstice], as wise menn of Grece, of Arabie, and of Rome have previdid be certane experience.

[And I said, it goes back through approximately one fifth part of an hour annually, since it does not return by a complete fifth part, or 12 minutes of an hour, but only by 11 minutes of an hour, as is apparent from a division of the long period of time between the solstice as found in the time of Hipparchus and the solstice as found in the time of Ptolemy; and from his time up to our own. And the excess is of exactly those 11 minutes, to be added or subtracted—subtracted for times future and added for times past, for any year. From this, the solstices and all the fixed feasts move backwards. To find the precise place of the Sun in the annual cycle it is necessary to subtract 11 minutes from the true place which it occupied in the previous year at the same hour of the day; and for an earlier year it is necessary to add, by the same reasoning. Wherefore, as a means of clarifying what I have explained, I will describe to you the two tables of which I spoke earlier, the first of which is a table of months, the second of collected years. These I made for the meridian of London for the year of Christ 1296 (a bissextile year), in the following manner:

[1] This should have read 'Sun'.

Primo investigavi[a] diligenter momentum introitus[b] Solis in quodlibet signum pro quolibet mense quatuor annorum continue sequencium, quorum primus annus fuit bisextilis, et ista intitulatur *tabula mensium radicalis*, eo quod in illa[c] ponuntur radices omnium annorum preteritorum et futurorum. Deinde[d] composui tibi aliam tabulam biparcialem habentem duos numeros annorum collectorum cressencium pro 5 et 5 annos, ita quod primus numerus est annorum predictorum et secundus erit futurorum.

Debes igitur cum istis tabulis sic operari: primo, si vis invenire horam introitus Solis in aliquod signum, intra cum eodem signo in tabula mensium radicali contra illum mensem et sub anno tuo, sive fuerit bisextilis sive primus vel secundus vel tercius post bisextum, sicut patet in tabula. Et quod in concursu istorum inveneris sunt dies et hore introitus Solis in primum minutum illius signi pro anno Cristi 1296 si intrasti cum anno bisextili, vel pro anno Cristi 1297[e] si intrasti cum anno primo post bisextum, et sic deinceps usque ad annum Cristi 1299, exceptis illis 11 minutis que annuatim oportet te subtrahere pro futuro tempore vel addere pro preterito, et immo ad hoc faciendum debes intrare tabulam annorum qui post primos quinque annos expansos cressit continue per 5 et 5 annos collectos; et contra annum illum invenies quid addendum est pro preterito et quid fuerit subtrahendum pro futuro. Si vero contingat te intrare in aliquem annum intermedium, scilicet inter 5 et 5, tanto minus subtrahes vel addes quanto ille annus distat a quinto anno sequenti. Et[f] sequitur tabula.

[a] C investigando [b] AC introitum [c] AB illo [d] A 28λ [e] C 15λ
[f] C 3λ+Ut in sequenti tabula patet

First I carefully found the moment of entry of the Sun into every sign for every month of a continuous cycle of four years, the first of which was bissextile. This table is entitled *table of radices for months*, since in it are placed radices, for all years past and future. I then composed for you another table in two parts, having two sets of numbers of collected years increasing five by five, so that the first number is of past years and the second of years of the future.

You must then use these tables as follows. First, if you wish to know the hour of the entry of the Sun into any sign, enter the table of radices for months with that sign and under your year, whether it be bissextile or the first or second or third after the bissextile, as is clear in the table. Whatever you find at the meeting of both ⟦i.e. opposite the one and under the other in the table⟧ will be the day and hours of entry of the Sun into the first minute of that sign for the year 1296 if you enter with the bissextile year, for 1297 if you enter with the first year after the bissextile, and so on up to the year of Christ 1299. The annual 11 minutes must be subtracted for times future or added for times past, and to take them into account you must enter the table of years, which after the first five expanded years increases steadily by five and five collected years. Against the year you will find what has to be added for the past and what will have to be added for the future. If you happen to have to enter with some intermediate year, that is to say, between the multiples of five, subtract or add with the next five-year entry an amount appropriate to the difference between that multiple of five and your year. And here follows the table.]

[Tab. 1.1] *Tabula radicalis mensium anno Cristi 1296 Londonie*

Nomina mensium	Anno bisext.		Anno primo		Anno secundo		Anno tercio		
	d	h	d	h	d	h	d	h	
Januarius	11	10	11	16	11	22	12	4	Aquarius
Februarius	10	3	10	9	10	15	10	21	Pisces
Marcius	12	2	12	8	12	14	12	20	Aries (equinoc.)
Aprilis	12	20	13	2	13	8	13	14	Taurus
Mayus	13	1	13	7	13	13	13	19	Gemini
Junius	13	12	14	18	14	0	14	6	Cancer (solst.)
Julius	14	12	15	18	15	0	15	6	Leo
Augustus	15	2	15	8	15	14	15	20	Virgo
September	14	18	15	0	15	6	15	12	Libra (equinoc.)
October	14	21	15	3	15	9	15	15	Scorpio
November	13	13	14	19	14	1	14	7	Sagittarius
December	12	23	13	5	13	11	13	17	Capricornus (solst.)

[Tab. 1.2][a] *Tabula annorum Cristi collectorum*

tempus preteritum	tempus futurum	dies	hore	minuta
b′ 1296	1296	0	0	0
1291	1301	0	0	55
1286	1306	0	1	50
1281	1311	0	2	45
1276	1316	0	3	40
1271	1321	0	4	35
1266	1326	0	5	30
1261	1331	0	6	25
1256	1336	0	7	20
1251	1341	0	8	15
1246	1346	0	9	10
1241	1351	0	10	5
1236	1356	0	11	0
1231	1361	0	11	55
1226	1366	0	12	50
1221	1371	0	13	45
1216	1376	0	14	40
1211	1381	0	15	35
1206	1386	0	16	30
1201	1391	0	17	25
1196	1396	0	18	20
1191	1401	0	19	15
1186	1406	0	20	10
1181	1411	0	21	5
1176	1416	0	22	0

[Tab. 1.3][b] *Tabula annorum Cristi expansorum*

tempus preteritum	tempus futurum	dies	hore	minuta
b′ 1296	1296	0	0	0
1295	1297	0	0	11
1294	1298	0	0	22
1293	1299	0	0	33

[a] Variant readings are trivial, and not recorded. Note that for the opening entries 1296, 1296, B has 1292, 1200

[b] All MSS. but A have a fifth line: 1176 (in B 1276), 1416, 0, 22, 0

The firste table that I sett here is the table of the monethis of the kalender be 4 *y*ere sewande, and thai ar made on this manere: firste I sowthe the entrynge of the Sonne in til every signe be 4 *y*ere, of the which the firste is the bysexte. And the tables are of myne awne makyng; fore the tables of the Abbot of Sancte Albones [ben] made full of errourues, and that I trowe bee throwgh the vices of writers, and nott of the wirshipfull clerke and prelate that the trectes made.

The 2^{e} table is made to drawe owt of the firste table, and on this manere salt thu werk with this table. If thu wile have the entre of the Sonne in til any signe, entre in with the *y*ere wher it be bissexte, or the firste, or the secunde, or the thirde, anente the moneth that the signe is in. And write owt the day, the howre, and the mynute. And those ar the trewe entres of the Sonne in to the signes, fore the *y*ere that thai ar titelid fore, that is to say, the *firste y*ere was fore the firste *y*ere after the bissexte, the *secunde* for the 2^{e}, the *thirde* fore the 3^{e}, and the *fowrte* was the bissexte, that to say 1388 *y*ere of Criste. And fore thes fowre *y*ere the entres was as thai stonde in the table, and fore 1389 *y*ere in the 2^{e} table is 42 mynutes 57 secunda 5 tercia. And that sal be drawne owt of ich on of the 4 *y*ers sewande, et cetera ut infra. [And than adde the equacioun of the day in minutis therto, and so schalt thu have the verray introite of the Sunne into a signe.][1]

[1] Supplied from T; missing in R.

Notandum[a] quod 15 gradus faciunt unam horam equinoxialem, et 7 grad. et dimidium faciunt 30 minuta unius hore, et unus gradus fecit 4 minuta unius hore, et 30 minuta unius gradus faciunt duo minuta unius hore, et 15 minuta unius gradus faciunt unum minutum unius hore. Et nota quod per spacium temporis unius milliarii in Terra pertransit Sol 5 gradus in firmamento. Cum ergo in una hora equinoxiali sint terquinque gradus, patet quod una hora equinoxialis sit spacium temporis trium milliariorum.

2

Capitulum secundum est ad investigandum quis gradus zodiaci pululat[b] super orizontem in quacumque hora[c] volueris incipere iudicium. Et quia instrumenta arsmetrica certiora sunt ad hoc quam geometrica, licet ista sint faciliora, quia[d] per revolucionem rethis astrolabii secundum exigenciam hore statim patet quis est gradus ascendens super orizontem tuum. Verumptamen si habueris tabulas calculatas pro ascensionibus ville tue, cuiusmodi sunt tabule Maudith calculate ad latitudinem Oxonie, certissime poteris scire ascendens hoc modo:

Primo per aliquod instrumentum capias altitudinem Solis de die, vel stelle de nocte, ut ex hoc scias quot hore transite[e] sunt de die vel de nocte, per astrolabium vel certius per tabulas de corda et arcu, multiplicando cordam versam medietatis arcus diei Solis (vel stelle) in cordam rectam altitudinis[f] accepte[g] Solis vel stelle; et quod surgit[h] exinde divide per cordam rectam altitudinis meridiane Solis vel stelle; et quod per talem divisionem exit erit una corda cuius arcus est distancia corporis luminosi cum quo operatus es ad[i] meridiem, quem arcum si subtraxeris de medietate arcus diei corporis illius luminosi habebis quantum est per ortum[j] vel ab ortu[k] eiusdem usque ad illud tempus si fuerit ante meridiem, vel addendo illud arcum super medietatem arcus diei eiusdem si fuerit post meridiem.

Et sic per istum modum operandi habebis certissime quicquid de die vel de nocte est pertransitum[l] usque ad minuta et secunda, quod non posset haberi per astrolabium licet eius diameter haberet in longitudine duo milliaria. Canonem istius operacionis[m] inveniendi arcum equinoxialem elevatum ab ortu vel occasu Solis invenies satis expresse inter canones quas tibi scripsi super tabulas Maudith, ideo eas hic non pono, nisi quatenus expedit hic facere de eis mencionem.

[a] The entire paragraph found only in A [b] BC pulset [c] A λ [d] BC +cito
[e] BC transacte [f] B altitudines [g] A accipe; B accipie [h] BC 13λ
[i] B λ [j] B arcum [k] B arcu [l] B transactum [m] BC operis

[It is to be noted that 15 degrees make an equinoctial hour, and 7½ degrees make 30 minutes of an hour, and 1 degree makes 4 minutes of an hour, and 30 minutes of a degree make 2 minutes of an hour, and 15 minutes of a degree make one minute of an hour. And note that through a space of time of one mile on the Earth the Sun traverses 5 degrees of the firmament. Since an equinoctial hour is thrice five degrees, it is therefore clear that an equinoctial hour is a space of time of three miles.]

2

THE SECUNDE CHAPITRE is to seke whatt gree of the zodiac liggis upon the orizonte what owre thu welet begyn thi dome. [And since arithmetical instruments are more reliable for this purpose than geometrical ones, the latter should naturally be easier, for by revolving an astrolabe rete in keeping with the hour, the ascendent degree for your horizon is immediately evident. Nevertheless, if you have tables calculated for ascensions at your own town, in the manner in which Maudith's tables were calculated for the latitude of Oxford, you may determine the ascendent most precisely in the following way.

First, by the use of some instrument or other, take the altitude of the Sun by day or of a star by night. From this you will know how many hours have passed by day or by night, by using an astrolabe or, more precisely, tables of chord and arc. Multiply the corda versa of the half arc of the solar day (or of that of the star) by the corda recta of the altitude of the Sun or star, and divide the product by the corda recta of the meridian altitude of the Sun or star; and what results from such a division will be a chord whose arc is the distance from the meridian of the luminous body with which you have worked.[1] If you subtract the arc from the semidiurnal arc of that luminous body you will have the size of the arc through or from its rising and up to that time, if the body had not reached the meridian. If it had passed the meridian, you would have the result by adding the arc to the semidiurnal arc.

Working in this way you will find the angle, no matter what it is, whether by day or by night, very precisely—in fact up to minutes and seconds, something which would not be possible on an astrolabe unless its diameter were a length of two miles. You will find this canon, explaining how to determine the equinoctial ascension measured from the rising or setting of the Sun, among the canons which I wrote for you over Maudith's tables, and therefore I will not repeat them here, unless it is expedient to call them to mind.

[1] There is a twofold error in this account, probably, but not certainly, a textual matter. See the commentary.

Habito itaque arcu equinoxiali elevato ab ortu Solis 1 gradus horarum vel eius occasu quam precise poteris tam[a] in gradibus quam minutis et bene servato, intrabis cum gradu primo Solis in tabulam ascensionum[b] ville tue et extrahes ibidem ascensiones gradus Solis id est contra primum gradum Solis sub quo signo queris. Scribe[c] et illud adde omnibus gradibus extractis de horis[d] si sit de die, vel nadir Solis si sit de nocte, quibus ascensionibus adde arcum equinoxialem elevatum ab ortu Solis si sit de die, vel super ascensionibus nadir[e] Solis si sit de nocte. Deinde cum toto illo numero intra iterum in tabulas ascensionum ville tue, quousque inter gradus ascensionum inveneris consimilem numerum, quem cum inveneris scito quod[f] quiscunque gradus in[g] directo illius fuerit in linea numeri de gradibus equalibus, ille profecto est gradus ascendens eadem hora. Et si precise non inveneris ibi numerum ascensionum[h] quem queris, poteris intrare bis, cum maiori et minori numero, sicut sit in arte equacionis[i] planetarum, et tunc absque omni errore habebis gradum minuta et secunda ascendentis super orizontem tuum; et hec est operacio carens omni hesitacione si Deus voluerit.

Et cum habueris gradum ascendentem sicut predixi, ille gradus est principium prime domus, cuius gradus considera ascensiones in[j] orizonte recto et scribe illas ascensiones extra[k] in tabula tua calculatoria. Deinde accipe quantitatem duarum horarum nadir ascendentis in villa tua sicut reperies in illis tabulis Maudith et ille numerus dicitur quantitas domus;[l] quam quantitatem domus, scilicet duarum horarum, adde super ascensiones ascendentis, quas certe[m] pro directo servasti extra; et cum illo totali numero intra in tabulam ascensionum in circulo directo et vide inter illas ascensiones consimilem numerum, et[n] in directo illius gradus equalis inventus est numerus[o] profecto inicium[p] secunde domus. Deinde illum totalem numerum cum quo invenisti iam inicium secunde domus conserva, et iterum adde eidem illam eandem quantitatem domus, scilicet duas horas inequales ipsius nadir ascendentis, et cum toto collecto intra ut prius tabulam ascensionum in circulo directo; et ubi inveneris consimilem numerum, in directo illius accipe gradum equalem et ille gradus procul dubio erit inicium tercie domus. Scias hoc. Deinde tercio, cum isto numero ultimo cum quo invenisti principium tercie domus, si eidem addideris ut[q] prius eandem quantitatem domus, scilicet[r] quantitatem duarum horarum nadir ascendentis, et intraveris ut prius tabulam ascensionum circuli directi, in directo illius numeri ultimo collecti invenies inicium quarte domus.

[a] A λ [b] C ascensionis [c] BC λ [d] B +deinde [e] B λ [f] C et
[g] A 2λ [h] A ascensionem [i] A equaciones [j] C 7λ [k] A λ [l] BC λ
[m] BC 3λ [n] C 17λ [o] AC numero [p] B λ [q] BC λ [r] C 6λ

Having now found and kept the equinoctial ascension of the Sun, [e.g. for] a step of an hour from rising or setting, and having done so as precisely as possible in degrees and minutes, you will enter with the first degree of the Sun in the table of ascensions for your town, and you will there find the ascensions of the degree of the Sun, that is, against the first degree of the Sun under whatever sign you are looking. Write it down and add it to the calculated angle (in degrees of an hour) if this is by day, or to the nadir of the Sun if by night, [that is], add to the ascensions the equinoctial ascension from the rising of the Sun if by day, or the ascensions of the nadir of the Sun if by night. With the resulting number now enter once more the tables of ascensions for your town until you find that number among the degrees of ascension. When you have found it, be assured that whatever falls opposite it in the equal degrees of the line of number, that will be the degree ascendent at the same hour. And if you fail to find the exact number of the ascension there which you are seeking, you may enter twice, with a greater and a lesser number, exactly as is done in the art of planetary equations, and then you will find a degree, minute, and second for the ascendent at your horizon, without incurring any error; and this is an operation devoid of all uncertainty, if God be willing.

If you have found the ascendent as I have explained, that degree is the beginning of the first house. Consider its ascensions on the right horizon, and write them down on your calculating tablet. Then take a quantity of two [canonical] hours corresponding to [the case where the Sun is at] the nadir of the ascendent for your town, such as you find in those tables of Maudith, and that number is called the 'size of the house'. Add the size of the house, that is, of two hours, to the ascensions of the ascendent, which you carefully kept on one side; and enter the table of right ascensions with that total, looking for that same number among the ascensions, when the beginning of the second house will be the corresponding degree at the beginning of the line. Keep then the total number giving you the beginning of the second house, and once again add to it the same size of house, namely two unequal hours for the nadir of the ascendent, and as before enter with the sum into the table of right ascensions. Opposite where you find the number, take the whole-number degree, which will beyond all doubt be the degree of the beginning of the third house. Remember this. Then thirdly, if you add to that last number (which you found to correspond to the beginning of the third house) the same size of house as before, namely two hours corresponding to the nadir of the ascendent, and enter as before the table of right ascensions, opposite the last aggregate number you will find the beginning of the fourth house.

Deinde vide quantitatem duarum horarum gradus ascendentis, vel[a] supra in tabula tercia Maudith vel qualitercunque volueris, et illam quantitatem duarum horarum adde super numerum ultimo collectum cum quo invenisti principium quarte domus; et cum toto isto collecto intra ut prius tabulam ascensionum circuli directi, et in directo illius numeri inter gradus equales invenies principium quinte domus. Iterum isti toti collecto iam ultimo adde eandem quantitatem duarum[b] horarum gradus ascendentis, et cum collecto intra ut prius eandem tabulam ascensionum in circulo directo et gradus equalis in directo illius est inicium sexte domus.

Habitis itaque iniciis sex primarum domorum[c] per eorum nadir invenies inicia[d] aliarum sex domorum. Sed[e] si[f] dubitaveris ad inveniendum octavam domum, nonam domum, et decimam domum, semper addere debes quantitatem nadir ascendentis et habebis undecima domo et duodecima domo quantitatem domus ascendentis, et sic evades errorem et dubium. Nam nadir ascendentis est inicium septime domus, et nadir principii secunde domus est inicium octave domus, et principium none domus est nadir tercie domus, et principium decime domus est nadir quarte, et principium undecime est nadir quinte, et principium duodecime est nadir sexte domus.

Et priusquam incipias adequare domos facias tibi talem figuram qualis subscribitur[g] ut statim habitis iniciis[h] domorum inscribere poteris ea[i] in locis suis, si iudicium tuum fuerit per domos et non solum per aspectus planetarum, cuius figura est circularis sicut patebit loco suo. Et scribes datam tuam pro quo tempore domos adequasti, et subtus dispones loca omnium planetarum seorsum, non in ipsa figura propter[j] confusionem collectam;[k] et tunc scire poteris quis planeta in qua domo cadit, quia omnis planeta existens in aliquo gradu inter gradum ascendentem et principium secunde domus cadit infra limites prime domus.

Verbi gracia. Annis perfectis Arabum die Dominico 673 9 mensibus 5 diebus 10 horis et 50 minutis hore[l] erat hec disposicio firmamenti in domibus et hec loca planetarum equata pro eodem tempore:[m]

Locus Lune in 7 gra. Geminorum,
Locus Saturni in 11 gra. Libre,
Locus Iovis in 11[n] gra. Arietis,
Locus Martis in 8 gra. Tauri,
Locus Veneris in 3 gra. Arietis,

[a] AB ubi; C ut [b] C quatuor [c] A domuum [d] AC inicium [e] BC 34λ
[f] A λ [g] C describitur [h] A inicium [i] A λ [j] C ad
[k] C tollendam [l] C horarum
[m] All MSS give the planetary positions after their houses, but the words of the text show that this was not intended. [n] C 2

Next find out the size of two hours corresponding to the ascendent degree, whether as above in Maudith's third table or in any other way you please, and add that quantity of two hours to the last aggregate number, with which you found the fourth house begins. Enter as before the table of right ascensions with the new total, and opposite the number, among the equal degrees, you will find the beginning of the fifth house. Once again, add to that last aggregate the same quantity of two hours corresponding to the ascendent degree, and enter with the sum as before into the table of right ascensions. The equal degree opposite is the beginning of the sixth house.

Having thus found the beginnings of the six houses, through the nadir of each you will find the beginnings of the other six. But if you have any doubts about finding the beginnings of the eighth, ninth, or tenth house, you may always add the magnitude of the [house, supposing the Sun to be at the] nadir of the ascendent, and you will have at the eleventh and twelfth houses the size of house [corresponding to the Sun in] the ascendent, and you will thus avoid error and doubt. For the nadir of the ascendent is the beginning of the seventh house, and the nadir of the beginning of the second house is the beginning of the eighth house, and the beginning of the ninth house is the nadir of the third house, the beginning of the tenth the nadir of the fourth, the beginning of the eleventh the nadir of the fifth, and the beginning of the twelfth the nadir of the sixth house.

Before you begin to equate the houses, draw for yourself a figure of the sort copied out below, so that as soon as you have the beginnings of the houses you may write them in at the appropriate place, assuming that your judgement is by houses and not merely by planetary aspects, the figure for which is circular, as will be evident in its proper place. Write in the date for which you equated the houses, and underneath—and separately—set out the places of all the planets. Do not do this in the figure itself, or everything will be confused. You will then be in a position to know which planet is in which house, for each planet lying between the ascendent degree and the beginning of the second house falls within the limits of the first house.

Example. On a Sunday, the Arab year 673 complete, together with 9 months 5 days 10 hours and 50 minutes of an hour, this was the division of the firmament into houses and these the equated places of the planets for the same time:

Moon . . . 7° Gem.,
Saturn . . . 11° Lib.,
Jupiter . . . 11° Aries,
Mars . . . 8° Tau.,
Venus . . . 3° Aries,

Locus Mercurii in 5 gra. Piscium,
Locus Capitis Draconis in 13[a] gra. Geminorum;

ex quibus patet quod Luna est in septima domo, et Saturnus in decima domo, et Iupiter in quarta domo, et Mars in quinta domo, et Venus in quarta domo, et Mercurius in quarta domo, et Caput Draconis in sexta domo. Et hec est perfecta disposicio istius capituli.

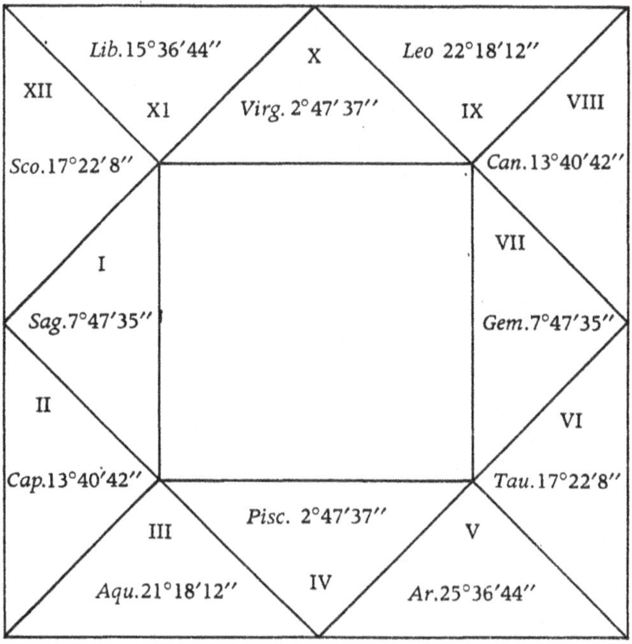

Memorandum[b] quod anno Cristi 1276, [ante diem] 11[c] Kalendas Aprilis, littera domenicali[d] D, aureo numero currente per 4 fuit hec computacio facta, ad extrahendum dominum anni in hora introitus Solis in primum minutum Arietis; et fuit ascendens tunc ut patet 7 gra. 47 min. 35 sec. Sagittarii, et cetere domus patent et annis arabum.

[a] C 3
[b] B and C omit the final paragraph, but C adds it to the end of chapter 3—whence the variant readings of the following notes.
[c] C 14
[d] AB domenicalis

Mercury . . . 5° Pisc.,
Caput Draconis . . . 13° Gem.

From this it is clear that the Moon is in the seventh house, Saturn in the tenth, Jupiter in the fourth, Mars in the fifth, Venus in the fourth, Mercury in the fourth, and Caput Draconis in the sixth[1] house. And this is the finished plan of the chapter.

It is to be remembered that this calculation was made [for] the year of Christ 1276, the eleventh day before the Kalends of April [= Sunday, March 22], dominical letter D, golden number 4, and it was made to find the lord of the year at the hour of entry of the Sun into the first minute of Aries. The ascendent was then 7° 47′ 35″ of Sagittarius, and the other houses are evident [being no different in?] Arab years.]

And that salt thu doo lightly one this manere. I suppose I will have the gree ascendent the therdd howre aftir none and 36 mynute of an howre, the 17 day of Auguste, fore that tyme I wrot this canon. Three houres contenes 45 grees of the equinoctiale, for 15 grees makis evermore an howre in rysinge of the equinoctial, and 45 ar thris 15, therfore I write in my stone 45 grees; and fore I have more be 36 minutis I adde therto 9 grees, for 4 sithys 9 makis 36, and ich gree awnsweris to 4 minutis of an houre—and therfor I adde thes 9 grees to 45. And therof rises 54 grees, the which I adde unto the ascensioun of the 3 gree of Virginis in the cercle directe. And that is 244 grees and 58 mynutis, and aftire that addition I have 298 grees and 58 mynutis. And that nowmbre I fynde aganys the firste gree of Scorpioun, and that wate I wele is the tyme in the myddis of the firmament eveyn in the Sowthe, and that gree is the begynnynge of the 10 hows. The same nowmbre I seek in the table of the ascension of Oxonforde, and I fynde it in the 26 gree of Sagittarys. And that gree wayte I wele is the ascendent that same tyme in the orizonte of Oxenforde. And that the begynnynge of the firste hows, and his fadir[2] is goynge down in the Weste. And that is the begynnynge of the 7 hows. The fadir of the 10 hows is eveyn under the Erthe in the north, and that is the begynnynge of the fowrt, id est 4, hows; et cet.

And thus may I have the 4 principale anglis and the other howses be table of equation of howses, as ar the tables of Mayster Iohn Walters—fore thos ar the beste that are nowe usid, and the moste redy; or else[3] be the turnynge of the astrolabye thu may have the same. *Y*if thu have the ascendent aftir midnyght, [do for each hour after midnight][4] 15 grees, and fore 4 mynutis 1 gree, unto the ascension of the fadir of gree of the forme. And looke that nowmbre that comes there of bothe in the tables of the ascension, and in the tablis directis and cercles croked, and thu salt have the sarcles directe the 10 hows; and be the cercles crokid, the firste hows as thu had before, and all the othere howses.

[1] Either this, or the position (13° Gemini) assigned to Caput Draconis is a mistake.

[2] T nadir. The word 'fadir(e)' of D is used frequently, as a synonym of 'nadir', but is not likely to be a mistake for that word, which is also much used in the translation. 'Fader' is an old word for 'father', and is still found as a dialect word. It seems clear that 'fadir' and 'fadire' are equivalent to the modern 'farther', here used as a noun to mean 'point farthest away'. In the handful of later cases where T and D differ as between 'nadir' and 'fadir', no note is made of the fact.

[3] D ebb.

[4] From T.

3

Capitulum tercium est de fortitudine et debilitate planetarum secundum naturam et qualitates eorum. Scias igitur primo quod tam signa quam planete sequuntur naturam quatuor elementorum; quia Aries, Leo, et Sagittarius sunt ignee nature, calida sunt et sicca; Gemini, Libra, et Aquarius sunt aeree nature, calida et humida; Cancer, Scorpio, et Pisces sunt aquatice nature, frigida et humida; Taurus, Capricornus, et Virgo sunt terree nature, frigida et sicca. Consimilis natura planetarum in calido et sicco: quia Saturnus est summe frigidus et siccus effective, et quantum est ex se omni nature generabili contrarius, et ideo vocatur malivolus. Iupiter vero temperate calidus est et humidus omni[a] nature transmutabili amicus, et ideo a philosophis vocatur[b] benevolus; plus tamen habundat in eo calor quam humiditas. Mars autem superflue est calidus et siccus vel ut verius[c] dicam inter omnes planetas plus utitur calore et sic est vite contrarius, et ideo dicitur malivolus. Sol vero temperate[d] calidus est siccus in[e] virtute, et inter omnes planetas plus temperatus est et equalis, ideoque dicitur[f] benevolus. Venus vero temperate calida est, superflue tamen humida[g] unde dicitur habere dominium super actus[h] venereos et hoc quantum ad disposicionem corporis ut patet in hominibus carnalibus, qui magis ducuntur[i] sensualitate quam racione. Mercurius habet inclinacionem per naturam suam ad illum planetam cui coniungitur, et cui signo, tamen dicitur quod in se est calidus et siccus, mediocriter tamen; ideo si coniungatur cum frigido planeta commutat[j] calorem suum in frigus, et ideo maxime apparet eius effectus cum coniungitur cum alio planeta. Luna vero habet naturam[k] temperate frigidam et summe humidam, unde dicitur fons humiditatis sicut Sol caloris, licet[l] sit temperate calidus. Caput Draconis est calidus et humidus, et Cauda frigida et sicca.

Sunt eciam in predictis signis fortitudines quedam, que[m] substancialiter attribuuntur planetis ex locis suis, et dicuntur potestates sive dignitates vel testimonia, cuiusmodi sunt *domus*, *exaltacio*, *triplicitas*, *terminus*, et *facies*. Unde a philosophis signa zodiaci comparantur corporibus et planete animalibus, unde sicut corpora sine anima nichil possunt facere sic nec signa sine planetis.[n] Et sicut anima cum[o] est in corpore operatur secundum complexionem[p] corporis cui alligatur quia aliter in colerico aliter in malencolico, sic planeta cum est in signo

[a] A cum [b] C +amicus sive [c] AC vertus [d] C temperatus [e] AC 2λ
[f] A λ [g] C humidus [h] AC usus [i] AC ducatur
[j] AC coniungat [k] C +naturaliter [l] A dicet [m] C 2λ [n] A planetas
[o] C dum [p] C complecionem

And before thu begyne to equare domus make suche a figure as folows,[1] and sett ilke planete in his hows as it fallis be the almanac. And thu sal have thi proposite.

3

THE 3 CHAPITRE is of the strenght and the febilnes of the planetis, aftir the kynde and the qualites of tham. Knawe thu wele that the planetis folowes the kynde of the 4 elementis, [as do also the signs]; for Aries, Leo, Sagittarys are of kynde of fires, hote and drye; Gemini, Libra, Aquarie ar of kynde of aeyre, hote and mo[y]ste; Cancer, Scorpio, Pisces ar of the kynde of watir, calde and moyste; Taurus, Virgo, Capricorne ar of kynde of erthe, calde and drye, et cetera. One the same manere are the planetis, in hete and drowthe: and fore Saturne is most calde and drye in effecte, and til all kynde abil to gendrid contrarie, and therfore he is callid evil willid. Jupiter is a temprid hoote and moyste till alkynde chaungeable freende, and therfore the philosofres callis hym wele willid; [nevertheles, heete in hym aboundeth more than moysture].[2] Mars is owtragely drye and hote, and certis emonge all the planetis he is moste hote, and therfore he is callid evil willid, [for he is contrarious to al lyf].[3] The Sonne is temprid hete and drye in vertu, and emonge all planetis he is moste[4] temprid and evenne, and therfore he is callid [of][5] gode will. Venus is a tempred hoote [but][6] owteragely moyste, wherfore she [is seide][7] to have lordship of dedis of lecherye, and that as enent the disposition of the body as it shewis in fleshly men of lykynge, [who are led more by sensuality than by reason]. Mercurie has enclynynge be his kynde unto the planete that he is joynet with, and to that signe that he is in. Nevere the les he is hote and drye in his awne kynde, but mesurablely; and therefore yf he be joyned with a calde planete he changis his hete in to calde, and therfore his effecte shewis mostes whan he is joyned with an other planete. The Mone has his kynde a temprid cald and alther hieste moyste, wherfore it is callid the well of moystoure, as the Sonne is callid the well of heete, th[ough] of all it bee but a temprid heete. Caput Draconis is hoote and moyste. Cauda Draconis is cald and drye.

Othere vertuys are in the signes aforesayd, the whilke ar gyvyen substancialy unto the planetis, and thay are callid dignites or witnessyngis, that is to saye, *hows*, *exaltation*, *triplicitas*, *terminus*, and *facies*. Wherfore of the philosophers the signes of the zodiac are likned til a body, and the planetis unto the sawle; fore as longe as the sawle is in the body it wirkis in the body aftir the complexion of the body that it is in, fore otherwise it wirkis in colorike and otherwise in malencoly.[8] So the

[1] The usual figure of the houses is here drawn (see the Latin text, p. 198), but not filled with numerical data. [2] From T. [3] From T. D has 'fore he is contrarie all that life'.
[4] Damaged MS. (D) possibly has another word here. [5] Damaged MS. My addition.
[6] Damaged MS. From T. [7] Damaged MS. From T.
[8] The translator changes the order of sentences at this point.

operatur secundum naturam signi.[a] Et ideo secundum quod planeta est in signo sibi convenienciori, secundum hoc habet ibidem plura[b] testimonia. Unde ubi planeta habet domum, ibi dicitur habere 5 testimonia; ubi habet exaltacionem, ibi habet quatuor testimonia sue virtutis; ubi habet triplicitatem, ibi habet tria testimonia; ubi vero habet terminum, ibi habet duo testimonia; et ubi habet faciem, ibi habet unum testimonium, sive unam virtutem. Quando autem est in signo in quo creatus est tunc dicitur habere domum, quia tunc ille planeta a philosophis comparatur viro in domo sua et in dominacione sua, et dicitur domus gaudii. Si vero fuerit in signo convenienti illi[c] signo in quo creatus fuerat, tunc est in domo accidentali; habet tamen ibi 5 testimonia sicut in priori, non[d] est tamen ibi[e] fortis ita ut in 5 virtutibus suis sicut[f] in primis. *Verbi gracia.* Sol creatus fuit in Leone, ideo habet ibi domum gaudii.

Ubi vero planeta habet exaltacionem, ibi comparatur[g] viro exaltacio in regno suo et in gloria sua. Planeta habens triplicitatem in signo comparatur viro existenti in honore suo et inter auxiliatores suos. Planeta vero habens terminum[h] comparatur viro inter cognatos suos et genus suum. Ubi autem habet faciem comparatur viro existenti in magisterio suo.

Est igitur Leo domus Solis, Cancer Lune, Virgo Mercurii, Libra Veneris, Aries Martis, Capricornus Saturni, Sagittarius domus Iovis. Sed notandum quod 5 planete habent alia domicilia accidentalia, ut Saturnus in Aquario, Iupiter in Piscibus, Mars in Scorpione, Venus in Tauro, Mercurius in Geminis.

Exaltaciones planetarum sunt hee: Sol exaltatur[i] in Ariete, Luna in Tauro, Saturnus in Libra, Iupiter in Cancro, Mars[j] in Capricorno, Venus in Piscibus, Mercurius in Virgine. Et nota quod sicut in Ariete exaltatur Sol secundum suam virtutem, sic in Libra deprimitur eius virtus,[k] cum sit in signum oppositum, et sic intelligendum est de ceteris, quod sicut[l] in Leone habet domum, sic in Aquario habet exilium; et sic de ceteris.

Triplicitatem dicitur planeta habere cum fuerit in signo eiusdem nature cum signo in quo creatus est, ut cum Sol fuerit in Sagittario vel in Ariete et que sunt signa[m] calida et sicca sicut Leo, in quo creatus est, dicitur habere triplicitatem.

[a] B λ [b] B pluria [c] C 2λ [d] A 2λ+quinto [e] C illa
[f] BC 3λ [g] A caperatur [h] A termina [i] C λ [j] C 3λ
[k] AB virtutem [l] B Sol [m] C λ

planete ever whilis he is in a signe he wirkis aftir the kynde of the signe; wherfore, as a body may nothynge doo withowtyn a sawle, on the same maner the signe may nothynge do ne wirke withowte a planete. And where a planete has a hows, there he has 5 dignytes, and where an exaltation 4 dignytes, a triplicite 3 dignites, a terme 2 dignites, and a face 1 dignyte. When a planete is in his hows[1] where he was made, he is likned [by the philosophers] til a man that is in his hows and in his lordshipe, and it is clepid his hows of joye. And yf he bee in a signe acordante unto the same signe where he was made, than he is in his secundarye hows,[2] (and he is not so strange there as he was in the firste hows).[3] Never the les he has there 5 dignytes. [*Example.* The Sun was created in Leo, and so there is its house of joy.]

And when a planete is in his exaltation, he is likned til a kynge in his see and in his joye emonge his men.[4] When a planete is in his triplicite, he is likned till a man in his wirshipe emonge his helpis. When a planete is in his terme he is likned til a man emange his kynn and frenchep. When a planete is in his face, he is likykned [*sic*] til a man emange his maystrys; et cetera.

Leo is the hows of the Sonne, Cancer of the Mone, Virgo the chef hows of Mercurie, Libra of Veneris, Aries of Martis, Capricorne of Saturni, Sagittarie of Jupiter. Also thai have other howses secundarye, that is to say Mercurie Gemini, Venus has Taurum, Mars Scorpion, Saturnus Aquarie, and Jupiter Pisces.

Thee Sonne is exaltid in Ariete (19 gree),[5] the Mone in Tauro (3 gree), Saturne in Libra (the 21 gree), Jupiter in Cancro (the 15 gree), Mars in Capricorne (the 28), Venus in Piscibus (the 27 gree), Mercurie in Virginis (the 15 gree). And as the Sonne is exaltid in Ariete [in keeping with his virtue], so is he put downe in Libra in like gree (id est 19), the whilke is his fadir. And so it is of all othere planetes. And as the Sonne has his hows in Leone, so he has his outlawrye[6] in Aquarye; and so of all the planetis.

A planete is sayd to have a triplicite when he is in a signe of his awn kynde with the signe that [he] was formed in, as when the Sonne is in Sagittarie or in Ariete, that are hote signes and drye, so is Leo that is his hows; and thes 3 are his triplicitees.

[1] 'hows' here translates 'signum'.

[2] *in domo accidentali.*

[3] Occurring later in the Latin.

[4] 'joye emonge his men' translates 'gloria'.

[5] None of the bracketed data of this paragraph occurs in the Latin or in R or in T. See commentary to cap. 3, subheading (ii).

[6] Note that this is not the same as a dejection.

De terminis diversi diversimode sentiunt; famosiores tamen et certiores sunt termini Egipciorum qui dicunt quod Iupiter habet pro termino primo sex gradus Arietis, et dum moratur in primis 6 gradibus Arietis[a] habet terminum, id est duas dignitates quas non est habiturus cum exierit et fuerit introitus[b] in signo. Item Venus habet proximos sex sequentes pro termino suo, et sic de ceteris sicut[c] in presenti tabula subscripta.

Facies autem signorum distinguuntur per decanos; unumquodque enim signum distinguitur in tres partes equales, et unaquaque pars continet 10 gradus et ideo appellantur decani quorum inicium[d] est a primo gradu Arietis usque in 10 gradus et dicitur facies Martis, quia Mars existens infra illas 10 gradus habet unum testimonium sue[e] virtutis. Secunda facies incipit a 10° gradu et terminatur in 20° gradu, et dicitur facies Solis, quia Sol succedit Marti in ordine planetarum.[f] Tercia facies incipit a 20° gradu Arietis et terminatur in 30° gradu et dicitur facies Veneris. Quarta facies incipit in[g] principio Tauri et terminatur in 10°[h] gradu, et dicitur facies Mercurii sicut patet in presenti[i] tabula;[j] tot enim et tanta miracula operatus est Deus in nona spera, in[k] qua describitur zodiacus, quod non est gradus in ea quin aliam et aliam influenciam operatur in planetis et super omnia transmutabilia, quorum[l] magis palpabilia ponuntur in presenti tractatu, que omnia extrahuntur a maioribus naturalibus[m] Rabyn Albumazar.

Aries	Jupiter	6	Venus	6	Mercur.	8	Mars	5	Saturn	5
Taurus	Venus	8	Merc.	6	Jupiter	8	Saturn	5	Mars	3
Gemini	Merc.	6	Jupit.	6	Venus	5	Mars	7	Saturn	6
Cancer	Mars	7	Ven.	6	Merc.	6	Jupit.	7	Saturn	4
Leo	Jup.	6	Ven.	5	Sat.	7	Merc.	6	Mars	6
Virgo	Merc.	7	Ven.	10	Jup.	4	Mars	7	Sat.	2
Libra	Sat.	6	Merc.	8	Jup.	7	Ven.	7	Mars	2
Scorpio	Mars	7	Ven.	4	Merc.	8	Jup.	5	Sat.	6
Sagitt.	Jup.	12	Ven.	5	Merc.	4	Sat.	5	Mars	4
Capric.	Merc.	7	Jup.	7	Ven.	8	Sat.	4	Mars	4
Aquar.	Merc.	7	Ven.	6	Jup.	7	Mars	5	Sat.	5
Pisces	Ven.	12	Jup.	4	Merc.	3	Mars	9	Sat.	2

Aries	Mars	10	Sol	20	Ven.	30
Taur.	Merc.	10	Lun.	20	Sat.	30
Gem.	Jup.	10	Mars	20	Sol	30
Canc.	Ven.	10	Merc.	20	Lun.	30
Leo	Sat.	10	Jup.	20	Mars	30
Virgo	Sol	10	Ven.	20	Merc.	30
Libr.	Lun.	10	Sat.	20	Jup.	30
Scorp.	Mars	10	Sol	20	Ven.	30
Sag.	Merc.	10	Lun.	20	Sat.	30
Capr.	Jup.	10	Mars	20	Sol	30
Aquar.	Ven.	10	Merc.	20	Lun.	30
Pisc.	Sat.	10	Jup.	20	Mars[n]	30

[a] C λ [b] B interius
[c] B ends: sicut patet in presenti tabula parva quam hic pono. C ends: sicut in presenti tabula subscripta. [d] B inicio [e] C 2λ+sive virtutem [f] C signorum
[g] BC a [h] A λ [i] C precedenti [j] B +proximo sequenti; C λ
[k] A 37λ [l] AB quarum [m] AB nullalibet; C λ [n] B Oxis (!)

Of termes diverse men dyversly spekis, never the les the moste certane are the termes of Egypcianes, [who say that Jupiter has for its first term six degrees of Aries, and when it is in the first six degrees of Aries it has a term, that is, two dignities which it will lose as soon as it leaves and goes out into the sign. Venus has the next six degrees for its term, and so for the others], the whiche I have sett in this table that folowes nexte in figure.

The facies of the signes are departid be 10 and 10 (as thus: 10–20–30), [and the parts are therefore called *decans*, the first of which is] the firste 10 gree of Aries [which] is the face of Mars, [since when Mars falls within those ten degrees he has a testimony or virtue]. And froo 10 unto 20 is the face of the Sonne, [since the Sun follows Mars in the order of the planets]. And fro 20 unto 30 is the face of Venus. (And so forthe be the ordere of planetis.) Mercurie has 10 grees of Tauri, fro 10 unto 20 the Mone, and fro 20 unto 30 Saturne. And so forthe as it shewis in the table next folowynge. Godd has wrought so many merviles and so many wondris in the [ninth][1] spere, in the which the zodiac is sett, that thare is noo gree but it has grete influence in dyverse thynges,[2] [on the planets and all that is changeable. The most palpable of these effects are set out in the present treatise, all taken from Albumasar's *Great Introduction*.[1]]

[1] D 8; T 9.

[2] D adds 'et cetera', and then follows with a blank page (f. 8^r^), which is, however, ruled, presumably to take the table promised.

[3] *extrahuntur a maioribus naturalibus Rabyn Albumazar*. It is conjectured that this refers to the *Great Introduction*. See commentary.

Ponam eciam[a] hic tabulam generalem in qua dignitates et nature omnium planetarum patent, simul cum natura signorum cum omnibus gradibus eorum, puta qui gradus sunt lucidi, qui tenebrosi, et qui masculini, et qui feminini et que sunt signa terrea, que ignea, que aerea, et que aquatiqua,[b] ut patebit hic.

Tabula ad cognoscendum cuius virtutis est quilibet gradus zodiaci et quam potestatem habet in signo

Aries[c]	Domus est Martis. Exaltacio Solis in 19° gra. Prima triplicitas Solis in die, Iovis autem in nocte, horum particeps est Saturnus. Terminus Iovis 6, Veneris 6, Mercurii 8, Martis 5, Saturni 5. Facies Martis, Solis, et Veneris. M 7; F 2; M 6; F 7; M 8. T 3; L 6; F 8; T 4; L 5; U 1. Pleni 12, vacui 18. Masculinum, igneum, orientale.
Taurus	Domus est Veneris et exaltacio Lune 3° gradu. Prima triplicitas Veneris in die, Lune in nocte, horum particeps est Martis. Terminus Veneris 8, Mercuri 6, Iovis 8, Saturni 5, Martis 3. Facies Mercurii, Lune, et Saturni. M 7; F 8; M 15. T 4; L 4; U 5; L 4; U 5; L 7; T 2. Pleni 20, vacui 10. Femininum, terreum, meridionale.
Gemini	Domus Mercurii. Exaltacio Capitis 3° gra. Prima triplicitas Saturni in die, Mercurii in nocte, horum particeps est Iupiter. Terminus Mercurii 6, Iovis 6, Veneris 5, Martis 7, Saturni 6. Facies Iovis, Martis et Solis. F 5; M 11; F 6; M 4; F 4. L 4; T 3; L 5; U 4; L 6; T 5; U 3. Pleni 21, vacui 9. Masculinum, aereum, occidentale.
Cancer	Domus Lune. Exaltacio Iovis 15 gra. Prima triplicitas Veneris in die, et Martis in nocte, horum particeps Luna. Terminus Martis 7, Veneris 6, Mercurii 6, Iovis 7, Saturni 4. Facies Veneris, Mercurii, Lune. M 2; F 6; M 2; F 2; M 11; F 4; M 3. L 12; T 2; U 4; F 2; L 8; U 2. Pleni 19, vacui 11. Femininum, aquaticum, septentrionale.

[a] BC 2λ+tibi [b] C aquea

[c] There are very many trivial variant readings in the tables, but since most of these may be removed by reference to the main text, they are not for the most part recorded here. This does not apply to the division of degrees by sex, by vacuity, or by lucidity (etc.). In very many instances these do not, for a given classification, total 30 degrees, as they obviously should. The numerical data of the following table is prepared on the basis of the English translation (MSS. D, R, T), in connection with which variants are recorded, corrected where necessary from the somewhat inferior Latin text of A, B, and C. (One frequent error in these MSS. is where the number 21 is written as 12.)

Note that the order of the signs in this table is the natural order in MSS. A and B, but in C the sequence is: Aries, Leo, Sagittarius, Taurus, Virgo, etc., namely an ordering by triplicities.

Here before abowve I have sett a table general, in the which the dignites and the kyndes of all the planetis shewis with the kyndes of the signes with all thaire grees, as lucidi, tenebrosi, (fumosi and vacui); and which benne masculyne and which femynyne; and so which signe is firey and which erthy, and which is aery and which is watery.

Aries is the hows of Mars. Exaltation of the Sonne in the 19 gree. The first triplicitee is of the Sonne in the day, of Jupiter in the nyght, and parcenel is Saturn. Terme and face as in the firste table before. Masculyne grees in Ariete are the first 7, feminyne 2, masculyne 6, feminyne 7, masculyne 8. Tenebrosi 3, lucidi 6, fumosi 8,[1] tenebrosi 4, lucidi 5, umbrosi 1. Pleni 12, vacui 18. Masculyne, firye, orientale.

Taurus is the hows of Veneris. Exaltation of the Mone in the 3 gree. The firste triplicite is of Veneris on the day, of the Mone in the nyght, and the parcenel is Mars. The termes and the faces as in the tables befor. Masculyne grees in Tauro firste 7, feminyn 8, masculyne, 15. Tenebrosi 4, lucidi 4, umbrosi[2] 5, lucidi 4, umbrosi 5, lucidi 7, tenebrosi 2. Pleni 20, vacui 10. Feminyne, erthye, meredionale.

Gemini is the hows of Mercurie. Exaltation of the dragon heed in the 3 degree. The firste triplicite of Saturne on the day, of Mercurye in the nyght, and particepe is Jupiter. The termes and faces as in the table before. Feminyn 5 grees, masculyn 11, feminyn 6, masculyn 4, feminyn 4. Lucidi 4, tenebrosi 3, lucidi 5, umbrosi 4, lucidi 6, tenebrosi 5, umbrosi [3].[3] Pleni 21, vacui [9]. Masculyne, aerye, and occidentale.

Cancer is the hows of the Mone. Exaltation of Jupiter in the 15 gree. The firste triplicite is of Veneris in the day, of Mars in the nyght; of thees is parcenel the Mone. Termes and faces as in the table before. Masculyne 2 grees, feminyne 6, masculyne 2, feminyne 2, masculyne 11, feminyne 4, masculyne 3. Lucidi 12, tenebrosi 2, umbrosi 4, fumosi 2, lucidi 8, umbrosi 2. Pleni 19, vacui 11. Feminyne, cald, waterye, septentrionale.

[1] Correct figure in T: 6 elsewhere.

[2] D has *vacui*.

[3] Last three data omitted in D. Final figure corrected from 4, a mistake of the Latin text, judging by surviving MSS.

Leo Domus Solis. Exaltacio nulla. Secunda triplicitas Solis in die, Iovis in nocte, horum particeps Saturnus. Terminus Iovis 6, Veneris 5, Saturni 7, Mercurii 6, Martis 6. Facies Saturni, Iovis, Martis.
M 5; F 3; M 7; F 7; M 8.
T 10; F 10; L 5; U 5.
Pleni 21, vacui 9. Masculinum, igneum, orientale.

Virgo Domus Mercurii. Exaltacio Mercurii 15 gra. Secunda triplicitas Veneris in die, et Lune in nocte, horum particeps Martis. Terminus Mercurii 7, Veneris 10, Iovis 4, Martis 7, Saturni 2. Facies Solis, Veneris, Mercurii.
F 8; M 4; F 8; M 10.
T 5; L 3; U 3; L 5; F 6; U 5; T 3.
Pleni 14, vacui 16. Femininum, terreum, meridionale.

Libra Domus Veneris. Exaltacio Saturni 21 gra. Secunda triplicitas Saturni in die, et Mercurii in nocte, horum particeps Iupiter. Terminus Saturni 6, Mercurii 8, Iovis 7, Veneris 7, Martis 2. Facies Lune Saturni, Iovis.
M 5; F 10; M 5; F 7; M 3.
F 5; T 5; L 8; T 3; L 7; U 2.
Pleni 19, vacui 11. Masculinum, aereum, occidentale.

Scorpio Domus Martis. Exaltacio nulla. Secunda triplicitas Veneris in die, Martis in nocte, horum particeps Luna. Terminus Martis 7, Veneris 4, Mercurii 8, Iovis 5, Saturni 6. Facies Martis, Solis, Veneris.
M 4; F 10; M 4; F 8; M 4.
T 3; L 5; U 6; L 6; F 2; U 5; T 3.
Pleni 16, vacui 14. Femininum, aquaticum, septentrionale.

Sagittarius Domus Iovis. Exaltacio Caude 3 gra. Tercia triplicitas Solis in die, Iovis de nocte, horum particeps Saturnus. Terminus Iovis 12, Veneris 5, Mercurii 4, Saturni 5, Martis 4. Facies Mercurii, Lune, Saturni.
M 2; F 3; M 7; F 12; M 6.
L 11; T 3; L 5; F 4; L 7.
Pleni 30. Masculinum, igneum, orientale.

Capricornus Domus Saturni. Exaltacio Mercurii 28 gra. Tercia triplicitas Veneris in die, Lune in nocte, horum particeps Mars. Terminus Mercurii 7, Iovis 7, Veneris 8, Saturni 4, Martis 4. Facies Iovis, Martis, Solis.
M 11; F 8; M 11.
T 7; L 3; F 5; L 4; T 3; U 3; T 5.
Pleni 30. Femininum, terreum, meridionale.

Leo is thee hows of the Sonne. The secunde triplicite in the day of the Sonne, of Jupiter on the ny*gh*t, of thame parcenel is Saturne. The termes and faces as in the tablis. Masculyn grees 5, feminyn 3, masculyn 7, feminyn 7, masculyn 8. Tenebrosi 10, fumosi 10, lucidi 5, umbrosi 5. Pleni 21, vacui 9. Masculyne, firye, orientale.

Virgo is the hows of Mercurie. Exaltation of Mercurie in the 15 gree. The 2 triplicite in the day of Veneris, of the Mone in the nyght, and of thame is Mars parcenel. The termes and faces as before in the table. Femynyn grees [8],[1] masculyne 4, feminyn 8, masculyn 10. Tenebrosi 5, lucidi 3, umbrosi 3, lucidi 5, fumosi 6, umbrosi 5, tenebrosi 3. Pleni 14, vacui 16. Feminyne, erthye, meredionale.

Libra is the hows of Veneris. Exaltation of Saturne in the [21][2] gree. The 2 triplicite Saturne on the day, Mercurie on the ny*gh*t, of them parcenel is Jupiter. Masculyne grees 5, feminyne 10,[3] masculyn 5,[4] [feminyne 7, masculyne 3].[5] Fumosi 5, tenebrosi 5, lucidi 8, tenebrosi 3, lucidi 7, umbrosi 2. Pleni 19, vacui 11. Masculyne, aerye, occidentale.

Scorpio is the hows of Mars. The 2 triplicite of Veneris in the day, of Mars on the ny*gh*t, of thame parcenel is the Mone. Masculyn grees 4, feminyne 10, masculyne 4, feminyn 8, masculyne 4. Tenebrosi 3, lucidi 5, umbrosi 6, lucidi 6, fumosi 2, umbrosi 5, tenebrosi 3. Pleni 16, vacui 14. Feminyne, aquatice, waterye, septentrionale.

Sagittarie is the hows of Jupiter. Exaltation of the taile of the Dragon in the 3 gre. The 3 triplicite is of the Sonne on the day, of Jupiter on the ny*gh*t, and of tham is Saturne parcenel. Masculyne 2, feminyne 3, masculyne 7, feminyne 12, masculyne 6. Lucidi 11, tenebrosi 3, lucidi 5, fumosi 4, lucidi 7. Masculyne, igneum, oriental. Pleni all.

Capricorne is the hows of Saturne. Exaltation of Mars in the 28 gree. The 3 triplicite of Veneris on the day, of the Mone in the nyght, and of tham is Mars parcenel. Masculyne grees 11, feminyne 8, masculyne 11. Tenebrosi 7, lucidi 3, fumosi 5, lucidi 4, tenebrosi 3, umbrosi 3, tenebrosi 5. Feminyne, terreum, meredionale. Plenum per totum annum.

[1] Text and translation have 7 in all MSS.
[2] T alone correct. Text and R and D have 13. [3] D 7. [4] D 3. [5] From T.

Aquarius Domus Saturni. Exaltacio nullius planete. Tercia triplicitas Saturni in die, Mercurii in nocte, horum particeps Iupiter. Terminus Mercurii 7, Veneris 6, Iovis 7, Martis 5, Saturni 5. Facies Veneris, Mercurii, Lune.
M 5; F 10; M 6; F 4; M 5.
F 4; L 5; T 4; L 8; U 4; L 5.
Pleni 30. Masculinum, aereum, orientale.

Pisces Domus Iovis. Exaltacio Veneris 27 gra. Tercia triplicitas Veneris in die, Martis in nocte, horum particeps Luna. Terminus Veneris 12, Iovis 4, Mercurii 3, Martis 9, Saturni 2. Facies Saturni, Iovis, et Martis.
M 10; F 10; M 3; F 5; M 2.
T 6; L 6; T 6; L 4; U 3; L 2; T 3.
Pleni 30. Femininum, aquaticum, septentrionale.[a]

4

Capitulum quartum est de fortitudinibus planetarum que accidunt eis ex diversitate locorum et motu eorum in epiciculis, et distancia eorum a Sole, cuiusmodi[b] sunt stacio, progressio, retrogradacio, elevacio in deferente, ortus, occasus. Sed et ex naturis suis habent quasdam proprietates accidentales, que[c] sunt[d] masculini, feminini, diurni, nocturni, lucidi, tenebrosi, fortunati, infortunati. Tunc autem planeta dicitur stationarius quando argumentum epiciculi eius est tria signa vel prope, plus vel minus; et tunc debilitatur in effectu[e] eius, quia non habet omnem[f] motum sibi convenientem, cum motus sit vita celestium corporum, secundum philosophos.

Sol vero non dicitur *stacionarius*, quia non habet epiciculum, nec Luna propter velocitatem motus eius in epiciculo. Sed dicitur velocior cursu vel tardior secundum quod fuit in illa parte epiciculi que[g] est super eius deferentem vel sub deferente, quod patet per eius argumentum. Semper tamen maioris virtutis est quando est velocior cursu quam quando est tardior.

Tunc eciam dicitur planeta *retrogradus* quando exit a signo et movetur contra successionem signorum cum motu firmamenti; et quia ille motus est nature eius contrarius, ideo in retrogradacione multum debilitatur eius virtus, in tantum quod dicitur tunc a philosophis infortunatus.

[a] At the end of cap. 3 in MS. A (f. 22[v]), and seemingly unconnected with the main text, is a passage of about 220 words, beginning 'In signis: Saturnus in Aquario, Iubiter in Sagittario . . .', and ending 'sicut crescens lumine'. [b] A cuius et; B cuius [c] AC quod [d] A sicut; C sint [e] AC effectum [f] C λ [g] AC quod

Aquarie is the hows of Saturne. The 3 triplicite of Saturne in the day, of Mercurie over the nyght, of tham is Jupiter parcenel. Masculyne grees 5, feminyne 10, masculyne 6, feminyne 4, masculyne 5.[1] [Fumosi 4, lucidi 5, tenebrosi 4],[2] lucidi [8],[3] umbrosi 4, lucidi 5. Plenum totum. Masculyne, aereum, occidentale.

Pisces is the hows of Jupiter. Exaltation of Veneris in 27 gre. The 3 triplicite of Veneris on the day, of Mars in the nyght, the Mone is parcenel. Masculyne 10, feminyne 10, masculyne 3, feminyne 5, masculyne 2. Tenebrosi 6, lucidi 6, tenebrosi [6],[4] lucidi 4, [umbrosi 3, lucidi 2],[5] tenebrosi 3. Feminyne, aquatice, septentrionale. Plenum per totum (annum).[6]

4

THE 4 CHAPITRE is of strenghtes of planetis that fallis to thame of dyversite of stedis and movynge of thame in thare epiciclis, and thaire distance fro the Sonne, of what manere is thaire goynge forth, goynge bakwarde, rysynge up in thaire differente, risynge, goynge downe. But of thaire kynde thai have summe proprites accidentale, id est, fallynge to, the which ar masculyne and which ar feminyne of dayes and of nyghtes; lighte, dirke; fortunat and infortunat. That tyme the planete is clepid stationarye when the argument of his epicicle is 3 [signs] or ellis nyghande, [whether ahead or behind]; and than he is febleid in his effecte, fore he has not mevynge acordynge til hym, senthen that mevynge is the life of the bodyes that are of hevyn, aftir al philosofris.

The Sonne is not clepid *stationarie*, for he has non epicicle; ne the Mone, for the swyftnes of hir movynge, or ellis fore slawnes aftir that she is in hir epicicle. [She is said to move faster or slower according as she is on that part of the epicycle beyond the deferent or below it, which is clear from her argument.] And she is evermore of more vertue when she is of smert movynge than when she is of slawe mevynge.

Than the planete is clepid *retrograde*, id est goynge bakwarde, when it mevis agayne the signe, (as froo Aries to Pisces, or ells froo 6 grees to 5, or fro 5 to 4, and so forth); for that mevynge is til kynde contrarius, therfore in retrogradation his vertewe is mekil feblid, in as mekil that after the philosophris the planete is clepid seeke and unhappy.

[1] D +feminyne 4.
[2] D Lucidi 5, tenebrosi 8.
[3] Text and translation 9 in all MSS.
[4] D 5.
[5] Translation omits, in all MSS.
[6] This addition is meaningless.

Progressus autem est planeta quando movetur cum successione signorum et contra motum firmamenti,[a] et tunc est planeta bene prosperatus in hoc.

Et hec tria per eorum argumenta sciri poterunt vel per almanac; est[b] autem hic[c] brevis tabula ad hoc cognoscendum de omnibus quinque planetis, cuius ars est hec: vide si argumentum planete concordat aliquo[d] numero in directo eius posito, vel prope, et caput tabule dicet si fuerit stationarius vel retrogradarius aut progressivus quia eorum inicia ponuntur in tabula.

[Tab. 4.1][e]

Argumenta	Saturni			Iovis			Martis			Veneris			Mercurii		
Prime stacionis inicium	3	22	44	4	4	5	5	7	28	5	15	51	4	24	42
Retrogradacionis inicium	3	25	30	4	7	11	5	19	15	5	18	21	4	27	14
Retrogradacionis finis	8	4	30	7	22	49	6	10	45	6	11	39	7	2	46
Stacionis secunde inicium	8	7	16	7	25	55	6	22	32	6	14	9	7	5	18
Tituli, et cetera	s	g	m	s	g	m	s	g	m	s	g	m	s	g	m

Quando eciam planeta est in auge sui deferentis, fortior est quam quando est in opposito augis, quia ex quo tunc est remocior a Terra est velocioris motus, et tanto est fortioris impressionis et quanto est propinquior Terre tanto tardioris motus et debilioris impressionis.[f] Hoc patet in Sole sensibiliter, quia dies sunt calidiores ipso existente in Geminis, Cancro, et Leone, quam quando est[g] in eorum nadir. Similiter de Luna, quando est in auge deferentis sunt flumina et fluctus maris multo maioris inundacionis quam in opposito. Similiter pisces omnes fecundiores[h] sunt carnis,[i] et omnes arundines aque plenioris sunt pulpe, et breviter omnia in quibus est natura aquea substancialiter. Hoc quilibet in se experitur, quia magis sensibiliter habundat sanguine in humoribus, Luna existente in auge eius quam in opposito.

Similiter de ceteris planetis, hec materia satis est diffuse tractata in tercio[j] libro maiorum naturalium Albumazar per totum de sensibilibus et mirabilibus effectibus planetarum, et libro quarto. Unde capitulo septimo nunc itaque et infra quoties *Saturnus* a medio absidis sue ascendit in frigido et sicco multum viget, qui[k] si tunc fuerit in signo frigido et[l] quadrante anni frigido, ut autumpno vel hyeme, cum tunc omnia conveniuntur,[m] firmus erit et efficax in frigore. Si vero fuerit in

[a] C 3λ [b] C hec [c] C λ [d] B alicui

[e] C differs in seven readings and B in one of these. It is not difficult to select the author's intended readings (see commentary), and variants are not recorded.

[f] B +et quanto est propinquior Terrus tanto debilioris impressionis et tardioris motus

[g] C 3λ [h] B fecundioris [i] C carne [j] C λ [k] A quia [l] C 4λ

[m] C conveniunt

Progression is when the planete gothe forth aftir the ordre of the signes and agayne the mevynge of the firmament, and than is the planete in wele fare and prosperite, (as says the philosophres).

[These three may be known from their arguments or from an almanac; however,] here is a good shorte table for to knawe when all the [five] planetis are stacionarye or retrograde or progressive, et cetera, ut patet in tabula. [Its method of use is this. See whether the argument of the planet agrees with, or approximates to, any number in the column under the planet's name. The beginning of the row where it is found will say whether the planet be stationary, retrograde, or progressive, for the arguments at which they occur are put in the table.]

Also when a planete is in the auge of his difference, he is strongere than when he is in the opposite, fore ever the ferther he is fro the Erthe, the swithter he is of mevynge and of stronger enprentynge; and ever the nerther the Erthe, the febleere he is, and slawer in mevynge. And showies felyngly be the Sonne, for the dayes ar hoteer when the Sonne is in [Gemini, in] Cancro, or in Leone than when he is in thar nadir. On the same wise of the Mone, when she is in the auge of hire difference the flodes of the see are of more waxynge than in the opposite. Also of fishes, all manere of fishes are fullere of substance, [and al redis and rochis er fuller of pyght],[1] [and in short, everything of an aqueous nature flourishes]. And the bloode of every beste abundis more in the fulle of the Mone, when the Mone is in hir auge, than when she is in the opposite.

[And similarly for the other planets, the whole problem is sufficiently well covered by Albumasar in his longer books of the natures of things [i.e. *Introductorium*], in the third book, for all perceptible and wonderful effects of the planets; and in book four.][2] Morowre, [in book four chapter 7 onwards it is said that] also ofte as *Saturne* ascendis fro the myddis of his aux in colde and in drowthe he faris wele, for than yf he be in a colde signe and in a colde quarter of the *y*ere, as herveste or wyntir, senthen that tyme all thyngis accordis togedre, it sal be stronger in colde.

[1] From R. [2] D has merely 'And all this is allegide in the bokes of Albumazare'.

signis calidis et siccis, et quadrante anni consimili, minuetur quidem frigus, sed[a] siccitas invalescet; et sic de ceteris[b] combinacionibus. Si vero descendat a sua abside, et fuerit in signis frigidis et humidis, quadrante anni convenienti erit planeta nimis frigidus et humidus, et congelans in nivem. Si vero fuerit in opposito absidis, id est augis, descendit et minuitur eius frigus et variatur secundum humidum et siccum ut prius. Et tanto[c] amplius quanto[d] fuerit in signo[e] vel quadrante calidiori, frigus[f] eius non commutabit in calorem.

Mars eciam residens in abside sua, vel prope, cum fuerit ipse effectissime et intemperate calidus, et per consequens siccus, intenditur calor eius et fervor, et fit intolerabilis cum fuerit in signis et quadrante anni calidis[g] et siccis,[h] secundum proporcionem quam prediximus de Saturno. Et in opposito augis deferentis, si fuerit in signis frigidis et quadrante, poterit eius calor multum mitigari et siccitas converti in humiditatem propter minucionem caloris.

Iupiter vero, cum ascendat versus augem accenditur calor eius, et humiditas aerea; et tanto amplius quanto[i] fuerit in hoc signo calido et humido, et quadrante anni consimili. Unde sensibiliter incentiva[j] carnis cressit in prurigine libidinis, propter convenienciam quam habet[k] ad generacionem. Et cetera animancia et vegetabilia vivificantur in radice. In opposito vero absidis[l] et locis et signis contrariis proportionaliter variatur in effectu.

Sol autem in circulo suo ascendens calidus est et siccus siccitate maturificante; descendens vero calidus est et humidus. Et si tunc fuerit in loco humido, humiditas prevalet. Et tanto magis inundat humor aque quanto magis confortatur ab aliis planetis aqueis, ut a Luna in suo bono esse; et econversa in parte contraria—nam de calidis et frigidis altera calida et altera frigida favent.

Venus cum sua natura est calida et habundanter humida cum ascendit[m] in circulo suo versus absidem, confortata ex conveniencia signi et temporis, et non impedita aliunde; multum inclinat appetitus[n] bestiales ad actus venereos nisi[o] appetitu racionali cohibeantur. Et econversa in opposito augis potest[p] convertere calorem eius in tepidam frigiditatem, et humor tunc erit fleumaticus.

Mercurius vero, quamdiu ascendit et tenet se in parte absidis, multum est siccus parum autem frigidus; descendens vero fit humidus cum modico frigore. Set quia omnium quatuor qualitatum est capax, ideo per convenienciam loci et temporis faciliter ad quascunque qualitates[q] permutatur.[r]

[a] C et [b] C consimilibus [c] C tantum [d] C quantum
[e] B signo [f] ABC +cum [g] ABC calidus [h] AC siccus [i] C quantum
[j] C 2λ [k] A habent [l] C calido [m] A ascendent [n] C appetere
[o] A ubi (!) [p] BC posset [q] A qualitate [r] C immutatur

Yf it be in hoote signis and drye, and the quarter of the *y*ere acorde therwith, the cald shal be lessid, and drowghte and heete sal be more; and of othere. *Y*if he come down fro his aux and be in cold signis and moyste, and the quarter of the *y*ere be acordaunte therto, the planete sal be ful colde and wete, and fresynge in to snawe. *Y*if he be in the fadir of his aux, and come downward, he is made less colde, and is diverse aftir hote and drye as before. And in as mekil as he is in a signe or ellis in a quarter of the *y*ere hotere, the calde shal not dryve away the hete.

Mars in his auge, or nythande, is in effecte intemperate, hoote, and drye, and unsufferable he is in hete when he is in hotes signis and in an hote quarter of the *y*ere, and a drye, aftir the proportion that we have sayed of Saturne. And in the nadir of his aux of his differente, yf he be in colde signis and in the quarter that is cold, the heete of hym may be mekile abatid and the drynes be turned in to wete, fore the losynge of heete.

Jupiter, when he ascendis towarde his aux, his heete is more and[1] his moystnes; and ay the more that he be in a hoote signe and in a hoote quarter of the *y*ere, and a moyste. [Wherfore][2] ful ably [comen ofte tyme][3] risynge in til luste of fleshe, fore the acordynge that he has in gendrynge. [Things come to life and vegetation stirs at its roots.] In the fadir of his aux and signis (and quartere) acordaunte unto the [contrarie][4] is dyversid in effecte.

The *Sonne* in his cercle ascendaunte is hote and drye [as of ripening]; descendant he is hoote and wete. And yf he be than in a moyste signe, the wetnes sal owrecome the drynes. And in as mekil sal his moysture abunde as he is comforted of other moyste planetis, as of the Mone [in her kindness]; and [a*g*enward in][5] econtrarye party—[for heat and cold are well disposed towards other heat and cold].

Venus, that is hote and owere moyste when he[6] ascendis toward his aux, and is confortid of his signe and of the tyme, and is not lettid on other parties; she bowes mekil appetid or bestis unto dedis of lechorye, but thai be reformyd be reson. And econtrarye in the nadir of hir cercle of hir auge, for than she may turne hir hete in to a softe calde, and her wetnesse than shal be fleumatike.

Mercurie, as long as he ascendis toward his aux, he is mekil drye and lityll colde. Comynge downward, she[7] is moyste with a litil calde. But fore that she is takande of all the 4 qualites, therfore be the recordynge of the stede that she is in, and the tyme, lightly she is convertid til her qualite.

[1] D than. [2] From T; D λ. [3] From T; D λ. [4] From T; D centris.
[5] From T; D λ. [6] *Sic*. [7] *Sic*.

Luna autem, cum[a] in circuitu[b] suo quatuor solaris circuitus tempora quadam mutacione renovat, quia in primo lunacionis sue quadrante[c] vernalis est nature, in secundo estivalis est nature, in tercio autumpnalis, in quarto hyemalis. Si tunc, dum fuerit in primo eius quadrante, fuerit in eius ecentrico ascendens,[d] calor eius humori[e] prevalebit. Si descendens, econtra experieris.[f] Item, in secundo quadrante eius lunacionis, cum tunc fuerit estivalis, si ascenderit in ecentrico, calor per eam invalescet. Si descendat, calorem temperabit. In tercio autumpnali[g] quadrante,[h] si ascenderit in ecentrico siccitas frigori prevalebit. Si descenderit, econtra erit. In quarto hyemali[i] ascendens, frigus prefert humori. Descendens, econtra reperies.

Ortus stelle dicitur cum aliqua stella fuerit iuxta Solem, tunc eius splendor victus splendore Solis tegitur,[j] et cum inceperit[k] elongari a Sole in tantum quod inceperit[l] videri. Dicitur tunc oriri, et incipit vigere eius virtus, quia radii luminosi quibus influebat in ista inferiora erant refracti et obtecti fontano lumine Solis. Quando vero incipit appropinquare Soli, quod amplius non poterit videri, tunc dicitur planeta habere occasum; et dicitur infortunatus ad iudicia. Et ideo, quia frequenter Mercurius latet et in maiori parte[m] sub alis Solis, non de se reperitur fortunatus nisi raro, et in longissimo tempore, nisi ex coniunctione cum aliis planetis. Et ideo in eorum occasu confortare quidam possunt Solem, vel debilitare; non tamen debet principalis effectus ascribi illis, sed Soli.

Ad extrahendum autem faciliter ortum et occasum matutinum et vespertinum, fecit[n] Ptholomeus tabulas in *Almagesti*, et Arzachel. Quia igitur de his multum curant periti astrologorum, ideo ponam tibi hic tabulas Arzachelis de his, cum canone earum, quia tabule leves sunt et valde necessarie. Prima tabula est ad extrahendum ortum matutinum et occasum vespertinum trium superiorum. Secunda vero tabula est de duobus[o] planetis inferioribus,[p] quia habent duplicem ortum et occasum, si Deus voluerit.

Cum volueris per tabulas alicuius trium superiorum planetarum scire ortum vel occasum considera argumentum eiusdem planete equatum, quod cum fuerit prope 20 grad. erit incipiens apparere quasi exiens de[q] radiis Solis. Occultari[r] autem incipiet atque regi a radiis Solis cum fuerit idem argumentum prope 360[s] gra. Veneris vero et Mercurii

[a] AC λ [b] A circulo [c] C quadrantis [d] AB ascendente [e] A humor
[f] C reperies [g] AB autumpnalis [h] A λ [i] AB hyemalis [j] AB tegebatur
[k] A incipit [l] A incipit [m] C +orbis [n] C ponit [o] C tribus
[p] C superioribus [q] ABC +sub [r] C 8λ [s] C 380

Luna, sythen she is goynge abowte 4 tymes of the *ȝ*ere, that is to say 4 quarteris of the *ȝ*ere of the circuit of the Sonne she makis in iche moneth, she newis the tymes be a maner of the chawngynge. Fore in the firste quarter she is of the tyme that is callid veere, in the secund she is of the kynde of somere, in the 3[e] of the kynde of harveste, and in the 4[e] of wynter. And when she is in hir firste quarter fro the chaunge, *ȝ*if she ascendit in hires cercle, hir hete shal passe her weetnes; and yf she descende, the contrarye, as it provis. And in the secunde quarter, [or summer], yf she ascende, hir hete evenely sal wax stronge. Yf she desconde, her hete sal be temperate. In the 3[e] [or autumn] quarter, yf she ascende, hir drynesse shal be more than hir colde. Yf she descende, the contrarye. In the 4[e], [or winter], quarter, yf she asconde, the colde shal owrcome the wetnesse. Comyng down, the contrarye.

The risynge of a planete is when hee has been hidd undir the bemys of the Sonne, and begynnes to goo a wayward fro undir the bemys, that it may be seen. Than it is sayde that it is risynne, and his vertewe begynnes to encrece, [for the luminous rays by which he influences things below are then broken and hidden in the fountain of light from the Sun]. When he begynnes to com towarde the Sonne, so that he may not be seen (fore owr mekill lighte of the Sonne), the planete is sayde to have his fallynge downe; and he is calid unhappy till domis. And therfore [since] Mercurye [often lies hidden and] lourkis fore the moste party under the wenges of the Sonne, wherefore he is not calid happy of his awne kynde but ful seldowme, and in longe tyme, and but fore his conjunction with ane other planete. And therfore in thaire fallynge down thai may comforthe the Sonne, and feble hym; never [the less] the principale effecte has noot ben writyn to thame, but to the Sonne.

Fore to drawe owte lightly the risynge that is calid ortus, and the fallynge that is clepid occasus, of the morowtide and of the evenynge, (that is clepid in the tablis 'ortus matutinus' and 'vespertinus'), and this techis Ptolome in his *Almageste*, and Arzachel, fore to this wise astronomeris takis mekill heede (and greete entent), et cetera. Here is a table of Arzachel with here canouns, for the tablis are lighte and ful necessarie.[1] [The first table is to find the morning rising and evening setting of the three superior planets. The second table is for the two inferior planets, if God be willing, for they have two possible risings and settings.]

When thu wilt witte the rysynge or the goynge down of any planete [of the three which are superior] be thes tablis above, luke firste what is the argumente (of Saturne, Jupiter, or Mars), and [if] it be nere [20][2] grees it shal begyne til apere and go owte of the bemes of Sonne, and it shal begyne to be hide under the bemys of the Sonne [whan the same argument is neerhand 360 gra. Of Venus and Mercure, the oriental

[1] After the word 'he(e)de', T completes this paragraph differently: 'And here after on that otherside of the lefe foleweth Arzachels tables with her canouns or rules, for the tables beth li*gh*te and ful necessarie to knowe, and namely to alle hem that han likinge and lust in the science of astronomye.'

[2] From T; D 12.

orientalis[a] apparicio erit cum fuerit argumentum prope[b] 18 grad., et occultari cum fuerit prope[c] 360.[d]

Ad investigandum per tabulas quando planeta debet apparere vel occultari,[e] pro tribus superioribus, loca eorum equata et similiter locum equatum Solis considera per almanach, et que sit utriusque differencia et cuius fuerit illa differencia. Considera videlicet an sit excessus Solis vel planete; quod si fuerit planete, apparens in oriente mane vel in occidente in vespera vel tectus[f] sub radiis Solis, intra igitur cum signo in quo fuerit planeta in tabulam priorem; et gradus quos ibi in directo signi et planete inveneris et minuta considera. Quod si fuerint[g] pauciores quam differencia tua que est inter Solem et planetam, erit tunc apparens in occidente. Si vero plures fuerint, erit obtectus nec videbis eum. Si autem supra differencia fuerit Solis, tunc erit planeta apparens vel occultatus in occidente. Ingredere igitur tabulam ut prius cum signo in quo fuerit planeta, et tabula dico[h] ortus matutini eiusdem, et quos ibi gradus et minuta inveneris, si fuerint[i] plures gradus[j] et minuta[k] differencie Solis, erit planeta occultus, nec labora ad videndum eum. Si vero fuerint[l] pauciores, erit apparens in oriente, et tanto clarius quanto differencia fuerit maior. Hoc quantum ad tria superiora.

Veneris[m] vero et Mercurii,[n] propter eorum velocitatem est ortus et occasus vespere et matutine; ut igitur hec eadem scias per tabulas extrahe cuius vis eorum differenciam ad locum Solis ut supra, et cuius sit differencia nota. Post hec considera si fuerit argumentum Veneris minus 137 gra., et si sic, erit Venus in ortu apparens per[o] vesperam vel occultatus. Intrabis igitur tabulam ortus vespertini ipsius Veneris cum signo in quo fuerit Venus; et gradus ibidem inventi, si fuerint pauciores quam illa differencia inter ipsum et Solem, erit apparens; si plures, erit occultata.[p] Si autem fuerit argumentum eius 138 gradibus in 180, erit in occasu vespertino apparens vel occultata,[q] et erit retrograda.[r] Tunc ingredere tabulam eius occasus vespertini, et scies per gradus ibi inventos contra signum in quo est, et per differenciam, sive[s] fuerit Solis vel stelle, utrum fuerit apparens vel occulta.[t] Si autem fuerit idem[u] argumentum a 224 gra. in 360 gra., erit in occasu matutino apparens, vel occultata.[v] Intrabis[w] igitur cum signo in quo fuerit tabulam occasus matutini[x] et per differenciam certe[y] scies utrum sit apparens vel occultata. Si enim

[a] C λ [b] C λ [c] BC λ [d] C 380 [e] C occultare [f] BC tecta
[g] In this and the next sentence all MSS. have 'fuerit' and 'pauciora'; A has 'plura'.
[h] A docet [i] ABC fuerit. Cf. n. *g* [j] B gradus [k] BC minutis
[l] ABC fuerit. Cf. n. *g* [m] A Veneri [n] C Mercurio [o] C propter
[p] ABC occultatus. (Note the random genders implied for *Venus* by the adjectives of this paragraph). [q] A occulta [r] C retrogradus [s] C 5λ
[t] AB occultus; C occultatus [u] C ibidem [v] A occultum; C occultatus
[w] C 20λ [x] A vespertini [y] A λ

apparicioun schal be][1] when the same argument is neere hande 18 grees, and for to be hidd when he is nere 360 grees.

For to witt be the table before when the [planet][2] shall appere or ellis be hidd, fore the 3 hieste planetis, luke wele thaire stedis in the zodiac be the almanac, and also the stede of the Sonne, and luke what different is betwixe thame, and whose is the differente. [Consider, that is to say, whether the excess be of the Sun or planet.] And yf that the excesse be of the planete, apperante in the este in the mornynge, in the weste in the eveninge, [or covered by the rays of the Sun], entre with the signe that the planete is in, in to the firste table; and the grees that thu findest evenfurth of the signe [and of the planet], behalde yf thai be fewer than the differente that is betwixe the planete and the Sonn, than is the planete shewynge in the weste. *Y*iff thai be moo, that it is hidd, [and you will not see it]. It is tokenynge yf the forsaide difference be of the Sonne, than is the planete shewynge or ellis hidd in the este. Entre therfore as thu dede before with the signe that the planete is in, the table I say that is tiled *ortus matutini* of the same, and the grees and the mynutes that thu fyndist thare, yf thai be moo than the differente is of the Sonne, than the planete is hidd. Travayle [not][3] abowte for to shewe hym. Yf thai be fewer, he sal be suande in the este, and ever the clerere the more the differente is. And this is fore the 3 overemoste planetis.

Fore to knawe the same of Venus and Mercurie, [on account of their velocity, rising and setting morning and evening], be the tablis drawe owte the difference of that on be the differente of the Sonne, as of the othere 3, and luke whos is the differente. Afterwarde luke yf the argument [of Venus] be [less than] 137 greis, and yf it be so, Venus is shewynge at evyne, or ellis hidd. Entre therfore the table *ortus vespertini* of Veneris with the signe that she is in; and yf the greis that thu fyndest there be fewere that different that is betwyxte and the Sonne, than Venus is shewynge; and yf thai be moo, than she is hidd. And yf here argument be froo [138][4] in to 180, it is in *occasu vespertino*,[5] or ellis hidd, and it is retrograde. Than entre in till hir table that is *occasus vespertini*, and salt thu witte be the grees that thu salt fynde there agayns the signe that Venus is in, and be the differente, whedir he be of the Sonne or of the planete, whedir he be apperande or hidd. Yf the same argument be fro 224 greis into 360, it shal be in *occasu matutino* apperande, or ellis hidd. Entre therfore with the signe that Venus is in, in to the tablee *occasus matutini*, and be the differente [you will know] certanly whedir it be apperande, other hidd. *Y*if the grees that thu fyndest there be [more

[1] From T; D λ.
[2] D table.
[3] MSS. more.
[4] From T; D 134.
[5] D be sepertinu (?).

gradus ibidem reperti fuerint plures gradibus ipsius differencie, omnino non apparebit propter claritatem radiorum solarium, nisi ab homine perspicui visus, et hoc propter eius magnitudinem; quia vidi eam in meridie estivalis serenitatis. Si autem ipsi gradus fuerint pauciores, differencia ipsa bene videbitur, et causabit umbram quod non accidit de alio planeta nisi[a] de Sole et Luna.

Mercurii autem argumentum equatum, si fuerit ab uno gradu in 152 grad., tunc erit in ortu suo vespertino; et[b] a 153 gra. in 180 gra. erit in occasu vespertino; et a 181 in 248 erit in ortu matutino; et a 249 in 360 gra. erit in occasu matutino. Utrum autem tunc sit apparens vel occultatus in quovis istorum locorum scies, si feceris cum eius differencia et introitu signi, quemadmodum operatus es ad operacionem[c] Veneris. Et fuerint iste tabule preciose Arzacheli a tempore suo.

Quo ad proprietates masculinas et femininas generaliter nota sicut dicit Albumazar in maioribus[d] naturalibus, libro secundo, capitulo octavo: cum ignis et aer sint elementa naturaliter activa, aqua et terra naturaliter passiva. Activa[e] autem competunt in masculinis, passiva vero femininis.[f] Racionabile erat philosophis quod[g] signa, planete, et gradus ignee nature et aeree vocare ea masculina; que vero sunt terree nature et aquatice[h] feminina.

Item pro diurnis et nocturnis subdit capitulo nono cum dies naturaliter sint calidiores suis noctibus, et noctes frigidiores diebus[i] quibuslibet[j] precedunt suas noctes, consequenter vocabat[k] signa scilicet Arietem diurnum, Taurum nocturnum, Gemini diurnum, Cancrum nocturnum, et sic per ordinem. Et de planetis consequenter[l] sequuntur proprietates diei et noctis secundum suas qualitates in calido et frigido, ut sit Saturnus nocturnus et Mars diurnus et cetera.

Similiter de lucido et tenebroso intelligendum est, ut[m] dicit Albumazar, libro quarto, capitulo quinto, stelle que ad temperiem aeris et vite generacionem faciunt, et ad vitalia opera suo motu ducunt; et gradus zodiaci fortunati dicuntur. Que vero ad discrimen[n] nature et generacionis impedimentum et corrupcionem procurant infortunati dicuntur, quorum passiones et proprietates sunt calide,[o] frigide, lucide, sicce, masculine, feminine, diurne, nocturne, lucide, tenebrose, plene, vacue; non quod in se tales[p] sunt, set quia naturaliter[q] et eminenter tales in se continent effectus. Qui vero sunt gradus obscuri, qui lucidi, qui pleni, et qui vacui, patet per tabulam generalem premissam.

[a] A 2λ+quam; B λ+quam [b] C 11λ [c] C apparicionem [d] C +suis
[e] AB activum (These MSS. put the sentence in the singular.) [f] AB femellis
[g] C quia [h] A +vocari [i] ABC dies [j] ABC quolibet [k] C vocabant
[l] A 2λ+sequenter [m] A unde [n] C destruccionem
[o] The case endings of this and the following eleven adjectives vary irrationally in all MSS., being generally -e for the first three (or -i in C) and -um (occasionally -i) for the rest.
[p] ABC talia [q] C nisi taliter

than] the grees of the differente, in no manere it sall shewe fore clernesse of Sonne bemys, but of a man that has full clere sight, and that fore hir mekilnes; fore I have seen hir at noon in somerys tide in a clere daye. Yf the greis be fewere than that differens, it sal be seen so, [and][1] it shal cause a shadowe, [a thing which can happen only in the case of the Sun and Moon among the other planets].[2]

The argument of Mercurie, verey, [if it be from 1 to 152 degrees, the planet will rise in the evening]; yf it be fro 153 greis in to 180, it sal be in *occasu vespertino*; and fro 181 in to 248 it sal be in *ortu matutino*; [and fro 249 into 360 it shal be in *occasu matutino*].[3] Whether it be than apperynge or ellis hydd in on ther of this placis thu salt fynde, and wyte thu do with this difference, and with the entre of the signe as thu has don with Venus. And the tablis was [preciouse],[4] id est namly in the tyme of Arzachel.

As anente the propriteis of masculyne and feminyne generaly, note thu as Albumazar says [in his longer books on natures, II. 8]: sithen that fire and air are elementis kyndely actyfe (id est doynge), the water and the erthe passyve (id est sufferande). Actyve, that is doynge, acordis to masculyne, and passyve, that is sufferynge, to feminyne. [By the reasoning of philosophers it was decided that they would call][5] hote signes, hote planetis, and hote greis, masculyne, colde signis, colde planetis, and colde greis, feminyne.

Also, [as he set down in chapter 9], fore days are hoter than nyghtis, and [nights are colder than the days which precede the nights],[6] therfore thai[7] callid (the firste) signe—Aries—diurnus, id est of the day, Taurus nocturnus, id est of the nyght, [Gemini diurnal, Cancer nocturnal], and so all the signes be ordre. And the planetis folowis the proprites of the day and of the nyght aftir thare qualites in houte and in colde, as Saturnus is of the nyght, Mars of the day, et cetera.

And of the same manere *de lucido et tenebroso*, [as Albumasar explains in IV. 5]. Also goode planetis that wirkis [towards a tempering of the air and the generation of life, and which lead by their motion to life-giving works. And it is said that these degrees of the zodiac][8] are fortunat. And thoo that wirkis corruption and lettynge and perile to generation, thai are infortunat, [and of them the passions and properties are hot, cold, and lucid . . . dry, masculine, feminine, diurnal, nocturnal, lucid, dark, full, empty. Not that these are so in themselves, but because of their nature they manifestly produce such effects. Which degrees are obscure, which lucid, which full, and which empty, is clear from the general table given earlier.]

[1] D that.

[2] D 'that fallis to the Mone' [a gap of about three words' length follows, and then] 'other planete, but to the Sonne and to the Mone'.

[3] From R.

[4] From T; D precise.

[5] D has simply 'and so are'.

[6] D has simply 'days gois before the nyghtis'.

[7] The Latin would have this singular.

[8] D to the life and generation.

[Tab. 4.2 (i)][a]

	Visio Saturni		Occultacio Saturni		Visio Iovis		Occultacio Iovis		Visio Martis		Occultacio Martis	
	Ortus matutinus		Occasus vespertinus		Ortus matutinus		Occasus vespertinus		Ortus matutinus		Occasus vespertinus	
Signa	gra.	min.	gra.	min.	gra.	min.	gra.	min.	gra.	min.	gra.	min.
Aries	29†	28	13	46†	19	33	9	28	29	0	14	12
Taurus	26	26	14	7	18	21	9	38	27	11	15	8
Gemini	22	10	15	5	14†	15	10	16	22	14	16	7
Cancer	17	18	17	9	11	41†	11	44†	18	15	18	15
Leo	14	8	19	48†	9	44†	13	32	16	7	22	14
Virgo	13	8	22†	0	9	7	15	23	15	8	27	11
Libra	12	15*	22	32	9	0	16	7	14	12	29	0
Scorpio	13	1	21	20	9	7	15	23†	15	8	27	11
Sagittarius	13	47	18†	35	9	44‡	13	32	16	7	22	14
Capricornus	16	36	16	36	11	44‡	11	44	18	15	18	15
Aquarius	21	16	14	40†	14	44‡	10	16	22	14	16	7
Pisces	26	46	14	0	18	11	9	38	27	11	15	8

Tabula Veneris [4.2 (ii)][a]

Nomina Signorum	Visionis ortus eius vespertinus a 1 gra. usque ad 137 gradus		Veneris occasus matutinus a 224 gra. ad 360		Ortus eius matutinus ab 180 gra. usque in 223 gra.		Occasus vespertinus ab 138 gra. usque in 180 gra.	
	gra.	min.	gra.	min.	gra.	min.	gra.	min.
Aries	15†	31	4	25	3†	36	2	27
Taurus	13	48	4	29*	4	9	3	30
Gemini	10*	39	7	38	5	14	8	47
Cancer	8	38	8	58	10	12	10	44
Leo	7	5	8†	59	17	45	11	30
Virgo	6	53	10	46	23†	40	7	43
Libra	6	57	11	9	223	27	6	40
Scorpio	7	11	11	26	15	14	6	17
Sagittarius	7	56	12	27†	7	1	5	12
Capricornus	9	18	9	18†	2	18†	2	18
Aquarius	12	47	8	29	1	36	1	14
Pisces	15	20†	7	43	2	43	1	31

[a] The aim has been to establish the most likely Toledan prototype used by Richard of Wallingford, and hence, presumably, his original tables. For our method, and for the meanings of the asterisks and daggers, see the end of the commentary to cap. 4.

Note that in tables 4.2 (ii)–(iii) exchanging the first heading for the second and the third for the fourth gives a correct, but unhistorical, table. See our commentary.

Tabula Mercurii [4.2 (iii)][a]

	Visionis ortus vespertinus a 1 gra. usque in 152		Mercurii occasus matutinus a 249 gra. in 360 gra.		Eius ortus matutinus a 180 gra. in 248 gra.		Occultacionis occasus vespertinus a 153 gra. usque in 180 gra.	
	gra.	min.	gra.	min.	gra.	min.	gra.	min.
Aries	24	10	12	24	22	43	12	9
Taurus	21	15	12	18	24	23	12	12
Gemini	17	10	12	37	22*	28	14	44
Cancer	14	9	14	9	18	48	19	48
Leo	12	53	16	39	15*	8	23	25
Virgo	12	8	20	23	13	15	26	37†
Libra	12	10	23	50	12	29†	25	38
Scorpio	12	41	23	49	12	10	20*	35
Sagittarius	14	3	20	44	11	16	17	41
Capricornus	16	19	16	19	12	15	12	30
Aquarius	20	15	14	7	14	25	11	32
Pisces	24	18*	12*	14*	18	22	11	47

[a] See p. 222 n. *a*.

Est autem in zodiaco via que dicitur conbusta a 20[a] gradu Libre usque ad tercium gradum Scorpii, in quibus 12 gradibus omnis planeta mirabiliter impeditur a suo effectu. Et quicquid potest planeta vertitur in detrimentum nisi aliunde confortetur. Item a principio Capitis usque in 12 gradus confortatur planeta, et infra 12[b] a Cauda debilitatur. Et nota quod quociens[c] planeta fuerit in loco vel aspectu racione cuius impeditur, dicitur infortunatus; cum vero fuerit in loco[d] salubri, dicitur fortunatus.

5

Capitulum quintum est ut scias[e] describere figuram in qua omnes gradus zodiaci appareant ut deinde scias aspectus planetarum ad invicem, et erit propinquacio[f] ad opus quod intendis, cum simul habueris figuram domorum per quam extrahes dominum anni et mensis, quam figuram loco suo preposui. Descripcio vero prioris figure aspectuum est ut facias[g] super idem centrum 12 circulos ita quod primi duo circuli coniungantur simul[h] fere ut inter eos describantur gradus zodiaci. Secundus et tercius parvo laciores ut ibi possint describi numeri graduum; tercius et quartus similiter, ut ibi scribantur nomina signorum. Deinde a quarto usque ad infimum circulum distinguantur omnes circuli per 13[i] spacia equalia, ita quod 12 spacia directe sedeant[j] contra[k] 12 signa in limbo et 13[m][l] spacium debet servari pro nominibus planetarum, ita quod in illo spacio quod est inter finem Piscium et principium Arietis inter circulum quartum et quintum scribatur *Saturnus*, et inter quintum et sextum scribatur *Jupiter*, inter sextum et septimum scribatur *Mars*, inter septimum et octavum scribatur *Sol*, inter octavum et nonum scribatur *Venus*, inter nonum et decimum scribatur *Mercurius*, inter decimum et undecimum scribatur *Luna*, inter undecimum et duodecimum *Caput Draconis*. Et tunc in quolibet spacio scribantur[m] sub signo et planeta eiusdem, testimonia quot[n] habet[o] in illo signo, ita quod si habuerit ibi domum, inscribantur ibi 5, quia domus 5 testimoniis testatur planete quia virtus eius ibi confortatur. Si autem habuerit ibi exaltacionem scribantur 4; si habet ibi triplicitatem, scribantur 3; si terminum 2; si faciem 1. Cum igitur equaveris planetas tuos ad vera loca eorum, statim scire poteris per ipsos quot testimonia habet quilibet in loco suo, et tunc videbis per quem aspectum respiciunt se invicem. Et hec est disposicio[p] figure exposite:

[a] C 28 [b] A +ad ultimum; B +ad vel; C +vel [c] A quocienscunque [d] C λ
[e] A λ [f] C applinquacio [g] A scias [h] AC λ [i] C 12 [j] A sedeat
[k] C cum [l] C 3 [m] A scribatur [n] A quid [o] B habuerit
[p] B ends: istius figure presentis; C ends: istius figure

Also ther is in the zodiac a way, and that is callid the brent way,[1] and that is fro the 20 gree of Libra unto the 3 gree of Scorpion, in the which way [of 12 degrees] yf a planete bee he is wondirful lettid of his vertew. And what so the planete may doo is turnyd into harme but[2] it be otherwise comfortid. Also fro the begynynge of the hede of Dragone unto [12][3] greis, a [planete][4] is comfortid, and within 12 greis of the tayle he is feblid. And nota also when a planete is in a goode stede or in a good aspecte, he is clepid happy; and when he is in a stede where he is lettid, he is unhappy.

5

THE 5 CHAPITRE is to make a figure or a table (that[5] is better than I myself has ymagyned, the which is the table of all the dignites of all the planetis that thei have in the 12 signis). [In this, all the degrees of the zodiac appear in such a way that you may know the mutual aspects of the planets, and it will help in the work you have in hand if at the same time you have the figure of the houses by which you extract the lord of the year and of the month, which figure I set down earlier in the appropriate place. The first figure of aspects is to be described as follows. Draw twelve circles about the same centre, so that the first two circles almost touch, and so that between them the degrees of the zodiac may be marked. The space between the second and third is a little more, so that the numbers of the degrees may be written there; and the space between third and fourth likewise, so that there the names of the signs may be written. From the fourth to the innermost circle, all are divided into thirteen equal spaces, so that twelve of the spaces are directly opposite the twelve signs on the limb, and the thirteenth may serve for the names of the planets, and so that in the interval between the end of Pisces and the beginning of Aries between the fourth and fifth circles is written *Saturn*, between the fifth and sixth *Jupiter*, between sixth and seventh *Mars*, between seventh and eighth *Sun*, between the eighth and ninth *Venus*, between the ninth and tenth *Mercury*, between the tenth and eleventh *Moon*, and between the eleventh and twelfth *Head of the Dragon*. Next, the number of dignities which a planet has is written under the sign and planet in question, so that if this be its domicile, 5 is written, this being the number of dignities of the planet, its virtue being strengthened. If it has an exaltation there, 4 is written; for a triplicity, 3; for a term, 2; for a face, 1. As soon, therefore, as you have found the true places of the planets, you will be able to say how many dignities each has in its place, and you will then see in what aspects it is with the others. And here is the arrangement of the figure described.]

[1] T *via combusta*. [2] Note that this has the sense of 'unless'. [3] D, R 20.
[4] T, R; D λ. [5] T 9λ.

[Vol. iii, p. 17]

Est autem aspectus planete ad planetam unus[a] bonus et alius malus. Per bonum aspectum accipit unus planeta ab alio fortitudinem et confortatur ab eo si conveniat cum illo, et si fuerit contrarius ei non multum nocet, quia non potest esse in tali aspectu bono ubi[b] fuerit in loco sibi disconvenienti, cum aspectus bonus accipiatur a conveniencia graduum. Aspectu autem malo debilitatur[c] econversa planete a se invicem. Sunt[d] autem aspectus 5: *coniunctus*, *sextilis*, *quartilis*, *trinus*, et *oppositus*. Aspectus *coniunctus* est cum fuerint duo planete vel plures in eodem gradu vel prope, et est fortis in convenientibus et multe turbacionis in contrariis; impeditur tamen turbacio, quia alter eorum debilitatur a loco sibi disconvenienti, et ideo dicitur bonus aspectus; non tantum potest impediri ibidem a suo contrario quantum[e] potest promoveri a suo simili.

Sextilis aspectus est quando inter duos planetas sunt duo signa, quod est sexta pars zodiaci, ut si fuerit Sol in 10 gradu Arietis et Venus in 10 gradu Geminorum, vel prope. Et iste aspectus dicitur mediocriter bonus propter convenienciam signorum in altera qualitate, quia Aries est igneus et Gemini aereus, igitur uterque calidus.

[a] A λ   [b] AC nisi   [c] C debilitantur   [d] ABC est   [e] AB λ

(Over the lefte hande are all the signes, and in the hede of the table are the namys of the 7 planetis. And in the metynge of the signe and of the planete are the dignyteis wrytyn. And nota that 5 is writyn fore an hows of a planete; and yf it have beside it a poynt upwarde, thus—✓, it is the principale hows of the planete; and yf it be with a poynt downward, as thus—, it betokyns the hows secundarye. And wher it is turned upsodown thus—ς, it betokyns that the planete is ther exilid. And wher this figure is 4, it signyfies that the planete has there his exaltation. And undir the same figure is writyn the gree of his exaltation. And yf it be turnyd upsodown, as thus—ᔭ, it betokyns his confusion, and that figure evermore the fadir of the exaltation. And where thu seis this figure 3, ther has the planete a triplicite. And yf it have a poynte upward, as thus—3✓, it betokyns that the planete regnis ther on the day. And when the poynte is downward, as thus—3✓, it regnis ther on the nyght. And yf itt have a single poynte as thus—3•, it is parcenel bothe on the nyght and on the day. This figure 2 betokyns a terme, and undir it is written the greis of the terme. This figure 1✓ with the poynte upwarde betokyns the firste face; and with the poynte upward the secund face; and a single poynte thus—1• betokyns the thrid face. The figure of all thes are in the nexte table folowynge, et cetera.)[1]

As enente the aspectis, summe are gode and summe are evil. Be a gode aspecte, a planete of another takis grete strenght and comforth [*y*if thei acorde in qualite],[2] and yf thei acord not, it sal noye hym. [For it may not be in such a good aspect where it is in a place unfriendly to it, good aspects being judged by the extent to which the position is accommodating.] In an evil aspecte a planete may feble an otheris strenght, and harme hym. And knaw thu wele that there bee 5 aspectis, that is to say, *conjunctione*, *sextilis*, *quartilis*, *ternus*, and *oppositus*. *Conjunction* is when 2 planetis or moo ar in on gree or ellis nere hand, and that is stronger in them that are acordid, and mekil tribulation in contrariis. Never the les, the turbation of that on is, fore that on is a stede that is not acordant til hym, and so he is feblid, and therfore he is callid a gode aspecte, and it may not bee [feblid][3] so mekil off hym [that is contrarie to hym][4] as he may be comfortid of hym that is like til hym.

Sextilis aspecte is a gode aspecte menely, and that is 2 signes [or a sixth part of the zodiac] betwene, as Aries and Gemini, (or Aries and Aquarie), [the Sun for example being in 10° of the former and Venus in 10° of the latter, or near by. And this aspect is said to be moderately good, on account of the fact that the qualities of the signs conform, since Aries is fiery and Gemini airy, and both therefore are warm.]

[1] This ingenious system whereby the original diagram of *Exafrenon* is, in the translation, exchanged for one giving more information, is not actually used in D or R, where the spaces for the tables are left blank, while T uses a somewhat different system, requiring red and black inks rather than the auxiliary pointer to register the same distinctions. The information which was to be provided may all be found from our diagram of the commentary to cap. 3.

[2] From T. [3] From T; D felid. [4] From T; D λ.

Quartilis aspectus est cum fuerit inter quoscumque planetas tria signa que sunt quarta pars zodiaci vel prope. Et dicitur iste aspectus malus, propter contrarietatem in utraque qualitate; nam si fuerint planete contrarii undique patebit contrarietas. Si vero fuerint planete similes[a] unus impeditur[b] a confortacione[c] alterius propter loci contrarietatem.

Trinus aspectus est que inter duos planetas fuerint quatuor signa in gradibus, quod est tercia pars firmamenti, ut si fuerit Sol in Leone et Venus in Sagittario consimili gradu. Et est iste aspectus fortissimus et optimus, propter convenienciam signorum, quia si planete fuerint similes multum adiuvant se invicem, si[d] contrarie infortunata non multum impedit stellam fortunatam.

Aspectus *oppositus* est cum fuerit planeta in nadir alterius planete, ut in septimo ab eo. Et est iste aspectus multum impeditivus[e] in contrariis, et minus confortatus[f] in similibus, ut probatur per[g] experiencia magna, et maxime si planeta impediens fuerit in signo in quo est fortunatus.

Et hec est perfectio aspectuum generalium, que[h] per disposicionem figure presenti probari possunt.

Item per istud[i] capitulum oportet te scire extrahere[j] dominum anni, mensis, et hore, et scire potestates eorum quia ex hoc[k] dependet multum sciencia istius operis. Ut igitur scias primo[l] *dominum anni* oportet te primo scire momentum introitus Solis in primum minutum Arietis a quo incipit naturalis revolucio anni; quia magnus error[m] accidet[n] in opere tuo si hoc negligatur secundum Albumazar in *Floribus* et in *Maioribus*, et secundum omnes philosophos de hoc loquentes. Deinde pro[o] eodem momento adequabis domos per invencionem gradus et minutorum ascendentis quam precise volueris; et similiter pro eodem tempore adequabis omnes planetas residuos Sole satis equato per introitum eius in primum minutum Arietis, et scribe illos seorsum iuxta figuram domorum sicut docui te capitulo secundo. Deinde videas quis planeta accidit in qua domo, nam si fuerit planeta aliquis in prima domo, que est domus ascendentis, et si in illa domo habuerit aliqua testimonia sue virtutis, quod scies[p] per gradus qui cadunt in illa domo, tunc noli respicere ad alios planetas in aliis domibus, quia profecto ipse est dominus et dispositor illius anni,[q] et quasi anima illius domus[r] et mirabili influencia influit virtutem ceteris planetis ut[s] sapiant et inclinent ad naturam eius per totam revolucionem illius anni, quod est unum de occultis operibus nature.[t]

[a] C nature similis [b] AB impeditus [c] C fortitudine [d] B scilicet
[e] A impeditivis [f] A confortativus; C fortunatius [g] B λ [h] BC 7λ
[i] C 2λ+illum aspectum [j] AC λ [k] BC huius [l] C post [m] B horror
[n] C accideret [o] A primo; BC post [p] A scieris [q] A λ [r] AB dominus
[s] B insapiant; AC 2λ+incipient (Evidence of dictation?) [t] BC +operose

Quartilis aspecte [is when between any two planets there are three signs, which] is a quartir of the firmament, (as Aries and Cancir), [or when the separation is approximately this]. And that is an [evel][1] aspecte, fore the contrarius in boythe qualiteis. *Y*yf the planetis bee like that on is lettid for the comforth of that other, fore the contrariuste of that stede. *Y*if thei bee discordant, in every partey is contrariuste.

Trinus aspectus (is gode), [and that is when there are between two planets four signs], and that is the 3 parte of the zodiac, (as Aries and Leo), [as when the Sun is in Leo and Venus in the same degree of Sagittarius]. And this the beste and strangeste, fore thei [the signs] acorde in all qualites. [If the planets be like, each assists the other, and if contraries, the evil star does not greatly hinder the good.]

Oppositus aspectus is the fadir of on signe til an other, [as in the seventh from it],[2] (as Aries and Libra). And this the werste, fore it lettis mekil in contrarye planetis, and it comfortis not in them that are like, as it previs oftyn tyms be experience, [and there is greatest comfort if the hostile planet be in a fortunate sign].

[And this concludes the discussion of general aspects, which may be found from the arrangement of the accompanying figure.]

Be thees thu salt knawe to drawe owte the lorde of the *y*ere, of the moneth, and of the howre, and to knawe the poweris of them, therfore ther hyngis mekil in the science of this crafte. For to knawe the *lorde of the yere*, take gode hede to the entre of Sonne in til the first mynute of Arietis, of the which the kyndely revolution of all the *y*ere begynnes; fore aftir Albumazar [in his *Great Introduction* and in his *Flowers of Astrology*], grete errore shal befalle in the wirkynge yf thu forgete that, as says all philosofris that spekis therof. Aftir, thu salt make a figure fore the same tyme as precise as thu kan, and sett the planetis in thare howses aftir that thei are, [at the time of the Sun's entry into the first minute of Aries]. And howe this sal be done it shewis opynly in the 2 chapitre[3] of this treteis. Aftirwarde, luke what planete is in the firste hows, that is, in the firste hows ascendente, and yf he have thare any dignyte that thu may knawe (be the table of dignyteis) [from the degrees which comprise that house, then pay no attention to the other planets in the other houses, for you may] than make hym lord [and controller] of that *y*ere before all othere; and he is as the sawle of that hows, and wondirfully he shal sende vertewis in to all othere planeteis, that thei be buxum and inclinant, and for to folowe his kynde [throughout the course of the year. And this is one of the hidden works of nature].

[1] From T; D λ.

[2] Seventh, that is, counting the first.

[3] That is, cap. 1, but the second chapter if we count the Prologue.

Si vero fuerit in domo ascendentis plures planete, vide ibi quis illorum habuerit plura testimonia influencie[a] virtutis sue, et, omissis ceteris, ipse est dominus quem[b] queris. Si vero in domo ascendentis non fuerit planeta, vide si aliquis in domo decima acciderit, et si tunc habuerit ibi testimonium sue virtutis cum in prima domo non fuerit planeta, sed fuerit quasi corpus sine[c] animo,[d] certus esto quod in domo decima ipse est dominus anni. Eciam si fuerit in domo ascendentis planeta carens testimonio et alius in domo decima cuius virtuti multa contestantur, ipse erit dispositor anni quem queris, et maxime si habuerit[e] ibi exaltacionem.[f] Tante enim dignitatis et eficacie est influencia domus ascendentis quod magni testimonii est planete esse in illa hora revolucionis, ita quod famosi philosophi preeligunt planetam in domo ascendentis[g] carentem[h] alio testimonio quam[i] fortiter ex nobilitate loci[j] quam planetam in decima domo cum facie et termino, vel solum cum triplicitate.

Et sicut predixi de domo ascendentis,[k] ita sencias de domo decima, quod si ibi fuerint plures planete, qui planeta habet testimonia, ipse[l] est dominus quem queris. Si autem non fuerit in domo decima planeta, nec in ascendente, tunc preeligas domum septimam ceteris et idem intelligas ibi quantum ad eleccionem domini anni quod dixi in ascendente[m] vel[n] domo decima; nec queras ulterius, si ubi fuerit planeta superans in testimonio. Si autem nec ibi inveniatur planeta, vide tunc si fuerit aliquis in quarta domo, et eodem modo fac sicut predixi tibi in ceteris tribus precedentibus. Et isti sunt quatuor anguli celi principales; et est angulus orientis[o] principalior angulo medii celi, qui est inicium[p] decime domus, et angulus occidentis principalior angulo terre, qui est angulus domus quarte; et sic per ordinem succedunt sibi[q] iste quatuor domus principales.

Si igitur in nullo istorum quatuor inveneris planetam hora introitus Solis in primum minutum Arietis, vide si fuerit aliquis planeta in undecima domo, et si ibi habuerit[r] aliquod testimonium preeligas eum[s] ceteris ut sit dominus anni, vel si plures ibi fuerint ut prius qui plura habet testimonia ipsum preeligas. Si autem ibi non fuerit planeta,[t] vide si fuerit in octava domo aliquis, et ipsum preeligas pre ceteris. Si autem nec ibi fuerit aliquis, vide si in quinta domo fuerit aliquis, et tunc respicias eum pre ceteris. Si vero nec ibi fuerit tandem vide in secunda domo, et si ibi fuerit adhuc non privabitur revolucio anni principio, sed ipse debet preesse anno.

Si autem in nullo istorum fuerit planeta, scias quod Sol erit rector anni. Sed erit multe variacionis et tempus carens stabilitate, nisi Sol fuerit in aliquo quatuor istorum angulorum, vel in quatuor domibus sequentibus, quia residui quatuor sunt *domus cadentes* nec unquam fuit

[a] C influencia [b] A quod [c] ABC ex [d] ABC animum [e] C fuerit
[f] C exaltacio [g] C ascendente [h] C carente [i] C +quod [j] A Leo
[k] C ascendente [l] C plura [m] C ascendentis [n] A de [o] C qui oritur
[p] BC in manu [q] AB secundum [r] C fuerit [s] BC cum [t] B λ

And yf there bee in the hows ascendant moo planetis than on, louke which of theme has moste dignyteis, fore he shal be lord befor all othere. And yf there bee no planete in the first hows, take hym that is in the tente hows, and yf he have any stede there, it sal be lorde, sythen the firste hows was a body withowtyn sawle. Yf there bee a planete in the firste hows withowtyn dignyte, and an other in the 10 hows with many digniteis, he shal be lorde [of the year which you seek], and namly yf he have theron exaltation. Never the les, the ascendent is of so mykil vertewe, that if a planete be therin withowtyn dignitie, thees wise philosofris chosyne hym the lorde before all othere, for the wirthyeste of the place more than any other in the 10 hows, havynge there a triplicitie, or ellis a terme and a face.

And as I have sayde of the first hows, so take kepe of the 10 hows, that yf there bee moo planetis, he that has (moste) digniteis shal be lorde of the *y*ere. And if there bee noon in the 10 hows, [and none in the ascendent], take hym that is in the 7 hows and do there as I have sayde of the firste hows, or of the 10 hows; and seke no ferthere, (and make hym lorde of that *y*ere) that has moste dignyteis there. And if there bee non there, louke in the 4 hows, and do there as thu didd in the othere 3 howses beefore. Fore these are the 4 *principalle anglis*; [and the eastern angle is more important than the mid-heaven, which is the beginning of the tenth house, while the western angle is more important than the angle of the Earth, which is the angle of the fourth house. And so the four principal houses follow this order.]

And if there be noon planete in non[1] of these 4 anglis in the entrynge of the Sonne in to the firste minute Arietis, loke where there bee any in thee 11 hows, (and do there as I have sayd in the othere 4 howsis). [If any there have any dignity, take it before the others as lord of the year, and if more than one be there, take the one with most dignities, as before.] And yf there be non there, louke hym that is in the 8 hows, (et cetera ut supra), [and take him before the others]. And yf there be non there, take hym that is in the 5 hows (et cetera) [and select him before the others]. And yf ther be non in the 5 hows, take hym that is in the 2 hows, and he shal be lorde of that *y*ere.

And yf there be non in any of thees 8 howsis, than sall the Sonne be principale lorde of the *y*ere. And it sal be a *y*ere offe mekil unstablynesse and of chawngeynge, [if the Sun be in none of the four angles or the four houses following], (and if the Sonne be in non of these 8 howsis). Fore the othere 4 are callid *fallynge*, fore in non of them was never

[1] An illogical but quite typical use of the double negative. There are other examples in the following paragraph.

in eis planeta bene fortunatus, nec in principio revolucionis anni, nec in alio tempore; sed erunt in illo anno infirmitates et pestilencie notabiles, nisi aliunde impediatur, sicut scies post ea. Unde Albumazar, libro quarto, capitulo quinto, dicit.

Saturnus quociens fuerit dominus anni et principatum eius tenet, si non respiciat eum planeta qui poterit eum debilitare malo aspectu, ut Mars trino aspectu vel alius planeta contrarius a quo poterit impediri, per universas boree vicinias[a] hyemale frigus exasperat in tantum quod magnam partem animalium illius regionis exterminat, et germen sicco frigore impedit. Et magis in circulo ecentrico ascendens australibus vero partibus estu Solis fervidis in tantum infrigidat auras quod estus earum convertitur in temperiem [ut] animalium[b] saluti et germinis ubertati non modicum adiciat.

Mars vero, cum habuerit dominium anni, si non respiciatur a Saturno vel ab aliis stellis contrariis, boreales[c] yemes mitigat, australem[d] exasperat. Ut autem supradictum est, solares cursus per circulum obliquum[e] anni tempora sequuntur, unde cum estas estati, hyems hyemi fuerit dissimilis, causa huius est participacio planetarum cum Sole. Unde *Saturnus* in quibusdam signis boreales flatus exasperat et hyemale frigus auget, estivum calorem minuit.[f] Mars quidem econtra in quibusdam signis amicis boreale frigus mitigat et estivales estus accendit.

Iupiter cum fuerit dominus anni Soli permixtus bono aspectu nec aliunde impeditus elementa mundi ex calido et humido[g] ad rerum generacionem et ad animancium propagacionem sanam temperiem procurat.

Venus cum fuerit dominus anni et Soli permixtus nec aliunde impeditus suis anni temporibus et maxime in hyeme et vere annum[h] equa humectacione temperat, sed et estatis et autumni siccitatem revelat,[i] nec aliter quam[j] in bonam partem potest uti suo dominio sicut nec Iupiter.

Mercurius autem cum[k] fuerit dominus anni, si Soli sit bono aspectu coniunctus nec aliena contrarietate impeditus, totum tempus sui dominii variis qualitatibus ventisque diversis exagitat, nec aliter dominium anni regere potest; ideo dicitur meretricula planetarum.

[a] A vicinas; BC animas [b] A +que; B et animalium; C et alium [c] C borealibus
[d] B australibus estantem; C australibus [e] A aliquando [f] C λ [g] C frigido
[h] A anni; B aurem; C auram [i] C relegat [j] C quod [k] C 2λ

planete wele disposide in no tyme of the *y*ere, or of the day. (And yf it befalle that no planete be foundyn in any of thes 4[1] howsis aforesayd in the begynnynge of the *y*ere), than there sall fale pestylence and many othere seknessis, but[2] it be lettid on othere sides, [as you will learn in due course]. Wherefore Albumazar says in his 4 bouke, the 5 chapitre.

Whan *Saturne* is the lorde of the *y*ere principaly, if no planete lett hym with an evil aspecte, as Mars with a (contrarye)[3] aspecte, or any planete that is contrarye til hym that may lett hym, he sal mak so colde a wynter in the northe cowntree that sal kill up wele nythe all the bestis of the lande, and he shal fore doo all the cornes.[4] And moreowre, ever the nere that hee commys toward his aux unto the sowthe partie, the hete of Sonne he kelis, and makis it a temperate that it sal not be harmynge to beesties [and that it shall contribute no little to new growth].

Mars, when he is the lorde of *y*ere, yf he be not lettid with an evil aspecte of Saturne, or ellis of an othere planete that is contrarye til hym, the wynter til them of the north he temperis, and the somere to them in the sowthe he makis owre hote.[5] And as it is sayd before, the 4 tymes of the *y*ere folows [the] cowrse of the Sonne in the zodiac, wherfore sithen the somere [to somer and the wynter][6] be unlike, the cause therof is the medlynge of the planetis with the Sonne. Wherefore *Saturne* in sum signis in the northe makis colde wedir [in winter, increases the wind], and he makis [lesser][7] the hete of the somers. And Mars wirkis contrarye, fore in sum signis the colde wyndis of the northe he temperis, and the hetis of the somere he ekis.

Jupiter, when he is lorde of the *y*ere and mellid with the Sonne in a good aspecte, and not lettid on non othere partie, the elementis he disposis unto genderynge of thyngis, and unto heleful sesone.

Venus, when she is [governynge][8] and medlid with the Sonne, and noghere ellis lettid, in wynter principaly, and in ver, she temperis the *y*ere, in moysture, and the drowght of somer and harveste she schiwes, fore she manote use hir ladyshipe but in gudenesse namore than Jupiter.

Mercurius medlid withe the Sonne, yf she[9] be lorde of the *y*ere and notwar lettid all the tyme of hir reynynge with dyvers qualities and dyverse wyndeis she governys, fore othere wise may she not do; and therefore she is clepid meretrix[10] planetarum, (id est, strumpid of planetis, fore she folows the kynde of every planete that she is medlid with in any aspecte).

[1] T 8.
[2] In the sense of 'unless', of course.
[3] The text has *trinus*.
[4] The text has *germen*, 'bud'.
[5] For 'makis owre hote' the text has simply *exasperat*, 'makes worse'.
[6] From T. D has 'sithen the somere to the wynter and the wynter to the somere be unlike'. This is true but trivial, and does not follow the Latin.
[7] From T; D λ.
[8] From T; D governyd.
[9] Cf. cap. 4, p. 215 notes 6, 7, and the following note, 10.
[10] The Latin is actually *meretricula*, an office rather lower on the social scale than that of *meretrix*.

De *Lune* dominio dictum est ut prius et post dicetur, quia transit circulum Solis duodecies in anno, ideo tociens potest eius dominium variari[a] et in quolibet eius transitu—4, scilicet, singulis lunacionibus—ideo preter dominium annuale recipit dominium aliorum planetarum in qualibet[b] quarta sue lunacionis. Quia eciam quarta prima sue lunacionis est calida et humida, secunda vero calida et sicca, tercia frigida et sicca, et quarta est frigida et humida, ut predictum est, ideo cum Luna fuerit dominus anni et admiscetur Soli nec aliunde impedita quatuor partes anni disponet[c] secundum quatuor qualitates sue lunacionis, ita quod primam partem anni faciet solito calidiorem et humidiorem, et hoc humiditate fecunda et serotina. Secundam partem anni faciet calidam et siccam notabiliter, et sic deinceps.

Quia igitur Saturnus et Mars in calore et frigore sunt intemperati et causa corrupcionis, ideo *infortunati* dicuntur; et quia Saturnus frigidus est et siccus ideo in utraque qualitate contrariatur[d] rerum generacioni,[e] et ideo dicitur magis infortunatus quam Mars; et sic de ceteris dicit[f] Albumazar ibidem, et multa alia bona dicit, que hic inseri[g] non possunt, cum tamen perfecta sciencia hoc requirit.

Eodem modo cum volueris habere[h] *dominum mensis* pro[i] diversitate mensis, accipies primum minutum in quo intrat Sol signum eiusdem mensis, et pro eodem minuto hore adequabis domos, et scies dominum mensis sicut docui te invenire dominum anni, per principaliorem domum et testimonia planete in eodem.

Item *dominus hore* est planeta que accidit in illa hora diei, secundum[j] denominacionem[k] diei et successionem planetarum, ut Sol in prima hora diei Dominice, Venus in hora secunda, Mercurius in tercia, Luna in quarta, Saturnus in quinta, Iupiter in sexta, et sic deinceps. Et hec est perfectio istius capituli.

6

Capitulum sextum est de doctrina ordinandi in opus omnia capitula precedencia per debite[l] experiencie[m] capitula. Cum igitur noveris dominum anni vel mensis, et virtutes planetarum in locis suis et domibus, et adequaveris omnia predicta, et volueris pronosticare de aeris serenitate et fructuum ubertate, considerabis aspectus et habetudines,[n] et apparebit tibi racionabiliter quid[o] futurum est de illis.

[a] C +tociens [b] C λ [c] AB disponent [d] AB +cum
[e] A generacionum contrariatur; B +contrariatur [f] BC hec [g] B misceri
[h] C habetur [i] B 3λ [j] A 12λ [k] C denotacionem [l] C debita
[m] C experiencia [n] C +eorum [o] C +facturum

The *Monee*, fore she passis abowte the cercle of the Sonne 12 tymes in on *y*ere, and therfore hir reynynge may be also ofte dyversid, and in every circuite—that is to say 4 tymes in ilke mone—dyversly governys; therfore withowtyn governynge that she takis be *y*ere of hir awne kynde, she takis the lordeshipe of uthere planeteis in ilke qwarter of hir lunation. Fore in hir firste quartir she is hote and moyste, [and in the 2 hote and drie, and in the 3 colde and drie, and in the 4 colde and moyst],[1] [as was explained before], and therfore when the Mone have principale governynge of the *y*ere, and is medlid with the Sonne and nowther lettid, the 4 parties of the *y*ere she shal ordeyne after the 4 qualites of hir lunation, so that the firste partie of the *y*ere she shal make mor hote than it was wonnt to be, and mor moyst, and that with a moysture that she sal be goode and tyedy. The 2 partie of the *y*ere shee shale make more hote and drye than it wase wonnt to be, and so forthe after hir qualites.

And fore Saturne and Mars in hote and colde are untemperat and the cause of corruption, thei are callid *infortunat*. And fore Saturne is colde and drye in bothe his qualites, he is contrary til generation, and therfore he is callid more infortunat than Mars; and so [of][2] othere. This says Albumazar [in the same place], and many moo, [which cannot be included here, although a perfect knowledge requires it].

And on the same manere, yf thu wilt have the *lord of the monethe*, louke wele the minute of the houre when the Sonne entreis in til the firste minute of the signe of the moneth that thu wilt wirk fore, and make there a figure (as thu did in the entrynge of Aries), and take that planete to be (principale) governore of that moneth as thu did before in the takynge of the lorde of the *y*ere (in all manere of degreis), [by means of the principal house and the dignity of the planet in the same].

Also the *lorde of the howre* is the planete that fallis in the howre of that day aftir the namynge of the day, and the ordre of the planeties, as the Sonne in the firste houre on the Sonneday (at the Sonnyes rysynge begynnis the Sonne for to regne), Venus the 2, Mercurye in the 3, the Mone in the 4, Saturne in the 5, Jupiter in the 6, (Mars in the 7), et cetera. And this is the perfection of 5 chapitrie.

6

THE 6 CHAPITRE is til ordeyne in the werke all the chapitreis before sayd be experience. When thu knawis the lord of the *y*ere, or of the moneth, and the vertuis of the planetis in thare stedys and in the howsis, and thu wilt proficye of wedeirs and of the plentie of fructe, take hede to the aspectis and the begynnyngis of thame, and thu salte resonablye knawe what is for to doo or to come.

[1] From T; D λ. [2] From T; D λ.

Verbi gracia. Quero loca planetarum anno arabum 646 completo, quod est anno Cristi 1249, mense quarto die eiusdem mensis 15, hoc est 17 kalendas May, quo anno erat Iupiter dominus et rector anni, quod patet per introitum Solis in primum minutum Arietis die 14 Marcii, hora tercia, minuta 25ª et[a] secunda hore 38ª. Et erat in domo septimo excedens loco et dignitate residuos planetas. Et pro loco et tempore predicto adequato planetas invenio Solem in 22°[b] gradu Arietis, Lunam in 21° gradu eiusdem, Saturnum[c] in 10° gradu Scorpionis, Iovem in primo gradu Aquarii, Martem in 28° gradu Aquarii, Venerem in 17°[d] gradu Tauri, Mercurium in 14° gradu Aquarii.

Procede igitur sic: Iupiter quantum ad principatum anni, cum in natura sua calidus[e] et humidus[f] fuerit, omnes planetas consimiles multum inclinabit per suam influenciam ad disponendum omnia transmutabilia secundum calidum et humidum. Contrarios vero planetas impediet ab effectu frigiditatis, ut non ita violenter frigus exasperant, sed ut magis faveant calido. Item, cum Sol in natura sua sit temperate calidus et siccus, et tunc fuit in 22° gradu Arietis et sic in exaltacione sua et triplicitate et facie, et sic in universo habuerit octo testimonia sue virtutis, nec est impeditus per aspectum contrarii, ut patet in hac figura quam habeo[g] pre oculis, igitur aer sequetur eius disposicionem et naturam.

Item Venus est calida et humida in 17° gradu Tauri, ubi habet domum et triplicitatem et terminum, et per consequens decem testimonia. Sed Mercurius in natura frigidus[h] est enim in 14° gradu Tauri et ideo est quasi coniunctus[i] Veneri, et habet ibi terminum et faciem, et ideo per tria testimonia sue virtutis aufert a Venere tria testimonia, cum sit ei contrarius, et sic non remanent Veneri nisi septem testimonia. Item Saturnus frigidus et siccus est, et in decimo gradu Scorpionis, et sic quasi opposito[j] aspectu respicit Venerem forti oppugnacione, sed est ibi retrogradus. Habet tamen ibi duo testimonia, et ideo aufert a Venere duo testimonia.[k] Item exaltatur in suo deferente quod est ei iudicium magni testimonii, et sic non permittet testimonia Veneris in aliquo fructificare in adiutorium[l] Solis. Est enim Saturnus prope eius augem quod prevalet multis testimoniis.[m]

Iupiter autem, preter quod est dominus anni, cum sit calidus et humidus, est tunc in primo gradu Aquarii habens ibi triplicitatem et terminum et est in ortu suo et progressivus, et sic est quasi puer de quo est magna spes profectus, et sic Soli est quodamodo favorabilis, preter

[a] C 4λ [b] C 22, but 23 in margin [c] A Saturni [d] BC 14, but C 17 in margin
[e] A calidum [f] A humidum [g] C habes [h] C +et humidus
[i] C 2λ+vernus [j] A oppositum [k] A λ [l] B adventorium
[m] A testimonio

Ane example as thus: I sowth verily the stedis of the planetis in the *y*ere of Criste 1249, [arab year 646 complete], in the 4 moneth and the [15][1] day of the same moneth, that is to saye the 17 kalend of May,[2] in the which *y*ere was Jupiter the lorde and governowre of all the *y*ere, and shewis be the entre of the Sonne in to [the firste minute of] Aries the 14 day of Marcie, the 3 howre, the 25 mynute and 38 secunde; and he was in the 7 hows passynge all othere planetis in stede and in dignitie. The Sonne was in the firste[3] gre of Arietis, Saturne in the 10 of Scorpion, Jupiter in the firste gree of Aquarie, Mars in the [28] Aquarie, Venus in the 17 Tauri, Mercurie in the 14 Aquarie, the Mone [21 degrees of Aries].[4]

[Go forthe thus.][5] Jupiter as enente the governale of the *y*ere, et cetera, sithen that it is hote and moyste, all othere planetis shall bowe unto his kynde fore to dispose all thyngis chawngeable after hete and mostures. Contrarius planetis he sal refreyne frome the effecte of coldnes, so that thei shuld not sharpe thare violente colde, but that thei drawe more to hete. Also, sithen that the Sonne is a temperride hete and drye in his kynde, [and was in 22 degrees of Aries], and [thus][6] in Ariete in his exaltation and his triplicite, and in his face, and so in all he has there 8 digniteis, and he is note distrablid be the aspecte of any contrarie planete as it shewis be the figure [which I have in front of me], and therfore the ayere shall folowe his disposition and kynde.

Also Venus is hote and moyste (and is) in [17 degrees] Tauro where she has a hows and a triplicite [and a terme],[7] and that is 10 dignytes. Mercurie in kynde is cold in [14 degrees] Tauro, therefore he is as conjuncte til Venus, and Mercurie has there a terme and a face, and so he has 3 dignites sen that thei are contrarius, and so to Venus doust (?)[8] but 7 dignytes. Also Saturne is colde and drye, and [is in 10 degrees] Scorpion, and so as in the nadir of Venus and opposit til hire; but he is retrograde. Never the les, he has there 2 dignites, and therfore he revis fro Venus 2 digniteis. Also he is hyngid in his cercle different that he is a toknynge of greet dignite, and so he lettis Venus that she may not perfecte[9] in helpynge of the Sonne. [Saturn is near his aux, which is influential in providing many dignities.]

Jupiter, withowten that he is lorde of the *y*ere, sithen that he is hote and[10] moyste and is in the firste gree of Aquarie, hee has a triplicite and a terme, and he is in ortu suo, id est in his resynge, and directe, as a *y*onge childe of whame is greet hope of profite, and so to the Sonne

[1] D 5. [2] 15 April is correctly given as 17 kal. May.
[3] The Latin has 22° and 21° for Mars, but the confusion is understandable. See commentary.
[4] T λ; D breaks the line short of stating the position. [5] From T; D λ.
[6] D is. The translation changes tense indiscriminately. [7] From T; D λ.
[8] T dwellen. [9] T profite. [10] D +drye!

hoc quod a principio anni influebat[a] Soli influencia mirabili;[b] per quam influenciam multum impedit Saturni ne omnino possit interimere efectum Veneris.

Item Mars est superflue calidus et siccus in 28° gradu Aquarii, habens solum terminum, sed est in ortu suo et progressivus, et in aspectu sextilis ad Solem. Soli ergo in calido et sicco prebet auxilium. Item Luna[c] in natura est frigida et humida in 21° gradu Arietis, sed carens testimonio; est tamen prope augem, quod est sibi magni testimonii. Remanent ergo testimonia Solis inconcussa, quare dispositor erit illius temporis et disponet tempus[d] in temperamento secundum[e] calidum et siccum temperatum;[f] tamen propter Lune vacuitatem et maxime erit temperata siccitas propter regimen Iovis in calido et humido, quod et factum[g] est sicut experti sumus observantes.[h]

Hec est generalis forma secundum quam procedes in iudiciis de aeris serenitate, vel de fructuum ubertate et aliis naturaliter accidentibus circa rerum transmutacionem. Cum eciam calorem excellentem Martis pronosticare volueris, considera per almanac quando plures planete calidi concurrunt vel bene se aspiciunt in trino aspectu in anno in quo Mars est dominus anni, vel alius calidus planeta, et hoc in signo calido et mense vel quadrante anni calido, quia sicut dicit Ptolomeus in *Centilogio*, oportet naturam temporis commisceri cum naturis planetarum, quia si[i] omnes planete significantes[j] humorem[k] in signo humido concurrant in quadrante anni humido, ut in hyeme, cum parva significacione humoris, erit habundancia pluviarum. Si[l] autem acciderit in estate, erunt quidem venti humidi non tamen forte multum pluviosi. Istis igitur concurrentibus ad propositum fiat pronosticacio de calore fervido.

Verbi gracia. Anno Cristi 1249 fuit forte[m] Mars dominus anni. Cum igitur Sol 16 die Iulii, hora 13ª inequali et 45° minuto, ingressus est signum Leonis, tercio gradu Piscium ascendente. Unde tunc per equacionem domorum inventus est Mercurius dominus mensis quia accidebat in septima domo ex aspectu ad Solem habens[n] tria testimonia; quia igitur Sol tunc fuit in domo sua et in triplicitate sua, disponet tempus quantum in se est in calido et sicco. Et hoc regimine[o] domini anni qui est Mars, et similiter regimine[p] Mercurii qui est dominus mensis, nec aliunde

[a] C λ [b] A ynabili [c] A λ [d] C λ [e] A set [f] C temperat
[g] A fons [h] A λ [i] A sicut; C λ [j] A signantes [k] B 2λ; C 3λ+humidum
[l] C 14λ [m] C fortis [n] A +testimonia et habens [o] C regimen
[p] C regimen

he is swmwhat favorable withowtyn that, that fro the begynnynge of the *y*ere he gafe vertew unto the Sone wondirfully; throwgh the whilke infernente (?)[1] he lettis mekil Saturne that he may not fordo the effecte of Veneris.

Also Mars is owteragely hote and drye in the 28 gre of Aquarie, havynge there only a terme, but he is in his risynge and goynge forthwarde and in aspecte sextili unto the Sonne contrarie. Fore he [geveth to the Sunne helpe in drught and heete. Luna in hir][2] kynde is colde and moyste in [21 degrees] Ariete, withowtyne dignite; but she is nere hande hir auge, and that is wowrth a grete dignite til hire. The digniteis of the Sonne dwellis therefore withowtyn harme, wherfore he shal be dispositoure of this *y*ere, and dispose it in hote and drowght atemperate. Never the les, fore the ne*gh*yngis nere[3] of the Mone he sal be full atemperid in drowght fore the governynge of Jupiter in hete and moystoure; and so it was previd.

This is the generale forme after whome thu sall procede in domyes and clernesse of the ayre, or of plente of fructis, and of thyngies that fallis kyndely abowte the chawngeynge of all thyngies. When thu wilt make profecy of a greet hete of Mars, looke when many hote planetis comys togedir, be the almanac, are as goode aspecte togedir as in trino aspectu in the *y*ere that Mars is lorde of the *y*ere, or ellis an othere hote planete, and that in an hote signe, and in on hote moneth, or in an hote quarter of the *y*ere; fore as Ptolome sayes in his *Centiloge*, it behovis the kynde of the tyme, to be mengid with the kynde of the planetis. Fore *y*ofe all the planetis that are moyste came togedir in a moyste signe, in a moyste quarter of the *y*ere, as in wynter, with a litil signification of humore, it sal be grete plente of rayne. *Y*if it be in somere, there sal be weite wyndies but not full rayney. [In like manner you may make a prognostication concerning burning heat from a suitable meeting of planets.]

Ensample as thus. In the *y*ere of Criste 1249, Mars was, [let us suppose], the lorde of the *y*ere, and in the 16 day of Julii, the 13 houre and the 45 mynute, the Sonne entrid in to Leonem. The 3 gree Piscium ascendent. And Mercurie was fuundyn the lorde of that moneth, and she was in the 7 hows of the aspecte to the Sonne havynge 3 dignites; and fore that the Sonne [was] in his [house and his] triplicite, he shal dispose the tyme in hete and drowghte. And [Mars][4] because of governynge, that is lorde of the *y*ere, and also the governynge of Mercurie that is the lorde of the moneth, and nowhere lettid, and therfore itt was an hete of huge

[1] T λ. D could be read 'infervente'. The Latin is *influencia*.

[2] From T; D agance.

[3] Latin: *propter vacuitatem*.

[4] MSS. read 'the more'. The replacement is conjectural, but Mars must certainly be mentioned in this sentence.

impeditus, erit igitur calor excellentis fervoris quod accidit, unde et homines australes multo langore interierunt.

Item cum excellens frigus pronosticare volueris, unde et fructus terre mortificentur, vide similiter quando planete anno vel mense concurrunt qui sint naturaliter frigidi, sicut accidit anno Cristi 1255, Saturno existente in Capricorno in domo sua. Et quinque annis continuis disposuit quinque hyemes in magno frigore et estates gelidas fecit propter aspectum oppositum ad Solem per estatem. Et ita impedivit fructuum maturitates, et erant autumpnorum tempora pervitiosa per 3 Lune cum eo coniuncciones circa plenilunium. Tunc etiam lesit vineas[a] in primo flore, et fructu arbores privavit; et magnas pestilencias per maria generavit per quartilem aspectum ad Martem et Mercurium.

Item cum habundanciam pluvie pronosticare volueris per Lune coniunccionem et per totam lunacionem, primo investigabis horam vere coniunccionis, et pro eadem hora vide quis est ascendens; et adequabis domos pro eodem, et extrahes domum[b] lunacionis. Et si habuerit testimonium in humiditate, et concurrant planete[c] humidi, et in tempore anni humido, scias quod multo habundancius accidet quam putabis. Et si has proporciones noveris,[d] multa proficua et mirabilia poteris pronosticare. Et[e] eodem modo ad videndum si secunda quarta lunacionis erit pluviosa, et[f] equalis erit 3ª vel 4ª pars lunacionis, scire poteris.

Et sic per hanc artem, sicut probat Ptolemeus in *Centilogio*,[g] multa pericula vitare possunt et incomoda, et magnum fructus emolumentum reportare, si non[h] soli litere insistant, sed eciam propriis experienciis. Unde in principio *Centilogii* dicit sciencia astrorum ex te erit, et ex illis, scilicet non solum ex tradicione philosophorum sed eciam ex propria experiencia. Et tunc sicut dixit Albumazar, licet fuerint mundi[i] contemptores diviciarum, habebunt tamen mundi divicias in sua potestate et arbitrio lucrandi. Unde pulchre narrat Aristotiles primo *Politice*, capitulo decimo, de Thale philosopho et astrologo certissimo qui, cum[j] esset ex stirpe regia oriundus, omnes tamen mundi divicias propter amorem philosofandi quasi nichilum contempsit, quem dum amatores mundi tanquam infatuatum et inutile de sua astrologia reputarent, eo

[a] B vintas [b] C dominum [c] A planeta [d] C λ [e] A ut [f] C 3λ
[g] BC +multi [h] AC in [i] B λ [j] A omni

brynnynge, wherfore [the men][1] of the sowthe contre thei deyid of grete langore.

Also, when thu wilt make prophecye of a grete colde, wherthrowghe the fructe of the erthe may be foredowne, [see when during the year or month those planets conjoin which are naturally cold], as befell in the *y*ere of Christe 1255, Saturne beynge in Capricorne in his hows. And so 5 *y*ere contynualy togedir he made 5 wynteris in grete colde, and someris he made frostye fore the opposite aspecte of the Sonne in the somere. And so he lettid the fructe that itt myght not waxe ripe, and the hervest wikkide be 3 monethis ever abowte the full Mone. And than he loste the wynes in ther firste flowrynge, and deprevid treis of there fructe; and he made grete pestilence be the se throwgh the [quartil][2] aspecte to Mars and to Mercurie.

Also, to tel before of gretes raynes be the conjunction of the Mone, and be all the lunationes, firste loke the tyme of the verrey chawnge, and loke what gree is ascendent;[3] and make a figure and take kepe what planete is the lord. And yf he be a moyste planete,[4] and othere moyste planetis come togedir in a moyste tyme of the *y*ere, there shal fall more rayne than men wald trowe. And yf thu kan knawe these proportiouneis, of many prophetable[5] thyngeis and wonderful thu may make prophecye. [In the same way you may discover whether the second quarter of the lunation will be rainy; and the same goes for the third and fourth.]

And so be this arte, as Ptolome tellis in his *Centiloges*, men may flee many perilies, and have mekil prophite therebye, yf thei gyfe not entent alonly to the letter, but to thair awn experience. Wherefore Ptolome sayes in the begynnynge of his *Centilogis*: The science of the sterries is of the, and of them, that is to say, not onlyne of the techynge of philosophris but also of thin awn experience.[6] And as says Albumazar: all tho*ghe*, thei be dispiseris of riches, never the les, thei shal have the riches of the warld in thare powre, and in thaire dome for to wynne, (et cetera). Wherfore tellis Aristotil in [chapter 10 of] the firste of his *Politike* of [Thales], a philosophre, and a certayne astronomyer, the which, yf all he ware borne of kyngis kynde, never the les fore love of philosophye all the riches of the warlde he fore souke, and therefore loveris of warldis riches heldyn hym as a fole, [and held his astrology useless because he

[1] From T; D λ. [2] D sextili.
[3] In D, the words 'all the', 'verrey', and 'ascendent' are on crumbling paper.
[4] Lit. 'if he have dignity in moistness'.
[5] (Despite the spelling) 'profitable'.
[6] The opening words, in fact. Compare with the Middle English translation of Trinity College, Cambridge, MS. O.5.26, f. 93va: '*Here bigynneth the Centilogie of Ptolomei*. The science of sterres of thee and of hem, and it bihoveth not hym that is wise of hem to deme after the forme of particuler effectes as nere it bihoveth a kunnynge man to resseyne forme of singuler sensible thinges.'

quod amplexus mediocritatem divicias mundi contempserit.[a] Volens falso iudicio eorum obviare et ostensa magnanimitate philosophorum avaros confundere,[b] quodam tempore pronosticavit apud se ubertatem[c] olivarum post[d] longam earum sterilitatem, per adequacionem domorum et naturam planetarum et aspectus eorum. Et conduxit omnes olivetas illius provincie pro anno illo, unde in tempore vindemie tota provincia confluebant[e] ad eum pro oleo rediviendo, ex quo ipse ditissimus effectus est. Stetit quodam die in capitolio[f] civitatis et exclamavit, dicens: Ecce incole stulti. Ecce quod philosophi divites esse possent si vellent. Et statim distributis omnibus adquisitis ad priscinam se transtulit paupertatem.

Explicit Exafrenon pronosticacionum temporis.

[a] A contempserint [b] C +qui [c] A utilitatem [d] C 14λ [e] C confluebat [f] C capitologo

refused to enrich himself by it. Wishing to avoid their mistaken judgement, by a high-spirited gesture], than he thowght for to shame [those men of greed],[1] and so be prognostication he sawe a grete plentie of oley commynge, and after that a grete faute of oyle folowande, be makynge of a figure and equatiouns of howses, and the kyndis of the planetis, [and their aspects]. Wherfore he hirid all the oyle makers[2] of that contre fore that *y*ere, wherfore in the tyme of (derthe, id est, scarsnes of the oylis),[3] all the contre come to hym to [bye][4] oyle, fore he was the richeste man of oyle (in all that contre. And so be oyle sellynge he was riche. And when he had gederid togedir a grete sowme of golde), he stode up in a hie towre of the citie, and cryed, and sayde: Behalde, folies, howe riche filosophries may be, yf thai wile! And anone he delte awaye that he had getyne, and went agayne untill his olde poverte. Amen.

Explicit Exafronon pronosticationum temporis, et cetera.

[1] D them.

[2] The classical meaning of *oliveta* given by Lewis & Short makes far better sense here: 'rented all the olive-yards'; and see the usual versions of the story in Aristotle's *Politics* and Cicero's *De divinatione*.

[3] The Latin simply has 'vindemia', 'harvest'.

[4] D be.

IV

TRACTATUS ALBIONIS

SIGLA

C Oxford, Corpus Christi College, MS. (D) 144.
A Oxford, Bodleian Library, MS. Ashmole 1796.
H London, British Museum, MS. Harley 80.
L Oxford, Bodleian Library, MS. Laud Misc. 657.

For full discussion of the manuscripts, Introduction to the text, and Synopsis and Commentary see vol. ii, pp. 127–286. The diagrams and tables are printed in vol. iii, pp. 19–51 and 75–107. See also Plates V–XV in vol. iii.

On matters of editorial principle see pp. xi–xx above, where examples from *Albion* are used to illustrate our general procedure. The geometrical parts of *Albion* are interspersed with so many descriptive clauses that it was impracticable to replace the purely verbal proofs by demonstrations in the form used in the translations of *Quadripartitum*. A synopsis of *Albion* in modern notation is given with the Commentary in vol. ii.

TRACTATUS ALBIONIS

ALBION EST geometricum instrumentum, almanac autem arsmetricum; utrumque tamen eisdem usibus licet inparibus adinventum. Non quod hic novum aliquid adiciatur inventum,[a] sed tamen pauca de disciplinalibus in quibus solent homines studiosi mensuras motuum supracelestium speculari hic ad opus, sed utinam utiliter, applicantur.[b] Ideoque ad detegendum[c] applicacionis huius motiva immo verius ut detur subtilibus materia cogitandi. Huic opusculo non tam correctioni quam detractioni exposito, conclusiones illas, in quibus fundatur huius veritas instrumenti, permittere servitum fore censebam.

CONCLUSIO PRIMA

Distanciam centri deferentis solaris a centro Terre una cum elongacione Solis ab auge circuli deferentis experimentaliter demonstrare.

Hec conclusio prima equivalenti ponitur in *Almagesti* libro 3° capitulo 4°, et in *Almagesti abbreviato* libro 3° capitulo 7°. Et ostenditur isto modo:

Fiat circulus *ABCD*, ductis diametris orthogonaliter *AC*, *BD*. Deinde, cognito tempore quod est inter verum equinoxium vernale et[d] equinoxium hyemale, per rectangulum vel turketum accipiatur medius motus Solis tanto tempori congruens, qui maior erit 180 gradibus, sicut experiencia satis docet; cuius arcus una medietas sit *BE*, et secunda medietas *BF*, producta linea *FE*. Observetur eciam tempus ab equinoxio vernali, quod est punctus *A*, usque ad solsticium estivale, cuius extrahatur medius motus Solis qui sit arcus *ABG*; ducaturque perpendicularis a puncto *G* usque ad punctum *H* in linea *FE*. Deinde ab *H* ducatur linea per centrum *O* circuli *ABCD*,

[a] CAH inventis [b] A multiplicantur [c] CAH deticendum [d] CA ad

TREATISE ON THE ALBION

PART I

The Albion is a geometrical instrument, whereas the almanac is an arithmetical one; both are for the same purposes, however, even though they are differently designed. It must not be thought that any new discovery is set down here, but rather do we apply, I hope usefully, a few of the ideas to be found in those disciplines in which learned men are wont to speculate on the measurement of celestial motions. My purpose is to bring all this into the light, and indeed, to provide subtle minds with more precise material for thought than they would otherwise have. I thought it would be useful if I set down in this little work—which I have written with a view to brevity rather than exactitude—the basic propositions on which the truth of this instrument rests.

PROPOSITION I

[I. 1]

To represent[1] *the distance of the centre deferent of the Sun from the centre of the Earth, and also the elongation of the Sun from the aux of the deferent circle.*

This first proposition is equivalent to that which is found in *Almagest* III. 4 and *Abbreviated Almagest* III. 7. It is shown in the following way.

Draw a circle *ABCD*, *AC* and *BD* being orthogonal diameters. Then, having ascertained the time between the true vernal equinox and the autumnal equinox,[2] let the mean motus of the Sun in the interval be worked out, by means of the rectangulus or the torquetum. It will prove to be greater than 180°, as we know well from experience. One half of this arc is to be *BE* and the other *BF*, the line *FE* being drawn in. The time is also to be found between the vernal equinox, which is the point *E*[3] and the summer solstice. The corresponding mean motus of the Sun, the arc *EBG*, is to be worked out. A perpendicular is to be drawn to the line *FE*, from the point *G*, meeting *FE* in *H*. Let a line then be drawn from *H* through *O*, the centre of the circle *ABCD*, that being the line *HI*. Also let the arc *EBC* be the

[1] *demonstrare*. What are here called 'propositions' ('*conclusiones*') are, as it were, an astronomical counterpart of Euclid's 'constructions'.

[2] Literally 'winter equinox'. The John of Gmunden MSS. have *equinoctium autumpnale*.

[3] Note the confusion over the labelling of this point, here and subsequently.

que sit linea *HI*. Sit eciam longitudo Solis ab *A*, capite Arietis, arcus *ABC*. Dico igitur quod *HO* est distancia centri Terre a centro deferentis solaris, et arcus *IBC* est distancia Solis ab auge deferentis; et per consequens arcus *AI* erit distancia augis a capite Arietis, quod sic patet:

Ductis tribus cathetis, *GK*, *FL*, et *EM*, constat quod totus arcus *BE* est notus et arcus *AB*, qui est 90 partes; et ideo notus erit arcus *EA*; et per consequens eius sinus *EM* similiter erit notus, cui est equalis *PH*, ergo nota per 33 primi Euclidis.

Item notus est arcus *AG* ex hypotesi, ergo similiter erit arcus notus *BG*, cum sit notus arcus *AB*, cuius sinus *GK* similiter erit notus, cui est equalis *PO*; huius ergo trianguli *HPO* rectanguli duo latera *HP* et *PO* sunt nota. Ergo ex resolucione Dulkarnon nota erit linea *HO*; quod est prima pars propositi.

Item duo trianguli *PHO* et *QOI* sunt equianguli, ergo que est proporcio *HO* ad *OP*, illa est proporcio *OI* ad *IQ*. Sed note sunt 3 linee *HO* et *OP* et *OI*, de partibus quibus est *OI* 60 partes. Ergo ex regula 4 numerorum proporcionalium, nota erit linea quarta *IQ*; ergo notus erit eius arcus *IB*; et per consequens notus erit arcus *AI*, que est distancia augis a capite Arietis. Et quia notus est totus arcus *ABC*, non erit ignotus arcus *IC*, que est distancia Solis ab auge; quod est propositum.

CONCLUSIO 2ª

Quamlibet differenciam medii motus Solis ad verum per datam distanciam Solis ab auge vel eius opposito et[a] *notam distanciam centri ecentrici a centro orbis signorum ostendere.*

Hec conclusio similiter ostenditur in *Almagesti* libro 3° capitulo 4°, in virtute, et in *Almagesti abbreviato* libro 3° commento 5°, et Albategni capitulo 28°, sic:

[a] H +per

longitude of the Sun from *E*, the first point of Aries.[1] I maintain, then, that *HO* is the distance of the centre of the Earth from the centre deferent of the Sun, and the arc *IBC* is the [angular] distance of the Sun from the aux of the deferent; and, in consequence, the arc *EI* will be the distance of the aux from the first point of Aries. This will be clear from what follows.

The three perpendiculars having been drawn, *GK*, *FL*, and *EM*, it is agreed that the arc *BE* is known, and likewise *AB* (which is 90°); and thus the arc *EA* will be found.[2] Its sinus,[3] *EM*, will therefore be known, and this is equal to *PH*, by Euclid I. 33.

Next, the arc *EG* is known by hypothesis, and since so also is *EB*, therefore, as before, the arc *BG* and its sinus *GK* (equal to *PO*), will be known. The two sides *HP* and *PO* of the right-angled triangle *HPO* are known, and thus, by Dulkarnon,[4] the line *HO*. This is the first part of the proposition.

Next, the two triangles *PHO* and *QOI* are equiangular, and therefore the ratio *HO* to *OP* is equal to the ratio *OI* to *IQ*. But the lengths of the three lines *HO*, *OP*, and *OI* are known, in units in which *OI* is 60 parts. The fourth line, *IQ*, will be known from the rule of four proportional numbers; and hence its arc *IB*; and, consequently, the arc *EI*, which is the distance of the aux from the first point of Aries. Since the whole arc *ABC* is known, the arc *IC* will not remain unknown. This is the distance of the Sun from the aux, which is what we proposed to find.

PROPOSITION 2

[I. 2]

To determine the difference, whatever it may be, between the mean motus of the Sun and its true [place], given the distance of the Sun either from the aux, or its opposite, and knowing the distance of the eccentric centre from the centre of the sphere.

This proposition is effectively demonstrated in a similar manner in *Almagest* III. 4, in the *Abbreviated Almagest*, Book III, commentary 5,[5] and in Albategni, chapter 28, as follows.

[1] Even if *C* stood for two points, namely the point diametrically opposite *A* and the position of the Sun in general, this sentence seems purposeless.

[2] Use of the sentence pattern '*X*, *Y*, *Z*, . . . being known, *P*, *Q*, *R*, . . . will also be known' is the standard way of expressing a functional relationship, or a calculation sequence. Cf. *Quadripartitum*, especially Part I.

[3] On the interpretation of this and other trigonometrical terms see Appendix 20.

[4] The common medieval name for Euclid I. 47. Chaucer's *Troilus and Criseyde* III, ll. 930–3 contains this plaint by Criseyde: 'I am, til God me bettre mynde sende, / At Dulcarnoun, right at my wittes ende.' (Globe ed. W. S. McCormick (1898), p. 495; F. N. Robinson, *The Works of Geoffrey Chaucer*, 2nd ed. (1937), p. 431.) The word is explained as coming from *dhu'l-qarnain* (Ar.), or 'two-horned', originally an Arabic epithet for Alexander, who claimed descent from Jupiter Ammon, the horned god; but also an apt description of Euclid's diagram.

[5] This is identifiable as the so-called *Parvum Almagestum*. See our commentary.

Fiat circulus *AB* super centrum *C*, producto diametro eius *AB*, et centrum Terre *D*. Sit eciam distancia Solis ab auge *A* arcus *AE*, coniunctis *CE* et *ED*, productaque equidistante linee *CE* a puncto *D*, que sit linea *DF*. Dico quod angulus *EDF* est angulus differencie medii motus ad verum, cuius noticia queritur, quod sic patet:

Angulus *D* et angulus *E* sunt equales, per 29^am primi Euclidis. Sed angulus *E* est differencia inter angulum *ACE*, qui est angulus medii motus computando motum ab auge, quia iam per premissam nota est distancia augis a capite Arietis, et angulum *ADE*, qui est angulus veri motus, quia centrum Terre est *D*, per 32^am primi Euclidis. Qui angulus erit notus, habita noticia linee *CD* 60 parcium linee *CA* semidiametri deferentis, et noto angulo *ACE*. Et econverso, cognitis[a] duobus angulis *ACE* et angulo *EDF*, nota erit linea *CD* de 60 partibus linee *CA*.

Probacio. Producta linea *EC* in continuum, supra quam cadat cathetus a puncto *D* in punctum *G*, erit triangulus *DGC* rectangulus et angulus *GCD* notus, per 15 primi Euclidis. Ergo imaginato circulo supra centrum *C*, secundum quantitatem semidiametri *CD*, erit arcus sinus recti *DG* notus de 360 partibus eiusdem circuli. Ergo sinus erit notus qui est *DG*, per cuius noticiam cognitus erit tercius sinus *CG*, quia ipsa est corda residui ad perficiendum quartam circuli semper. Sunt igitur omnes tres linee trianguli notarum parcium respectu 60 parcium semidiametri *CE*. Ergo nota erit tota linea *EG*. Cum igitur triangulus *EGD* est rectangulus cuius duo latera *EG* et *GD* sunt nota, erit tercium latus quod substenditur angulo recto, puta[b] *ED*, notum per deductionem [47^i] primi Euclidis.

Imaginato igitur circulo super centrum *E* secundum semidiametralitatem linee *CD*, et posito *ED* 60 parcium, erit secundum illas partes nota tam linea *EG* quam *GD*, quia si nota fuerit proporcio *ED* ad *DG* vel *GE*,[c] sub aliquo certo numero erit similiter eorum proporcio nota, sub quocumque numero alterum illorum ponatur. Sed cognito sinu *DG*, cognoscitur eius arcus, qui est arcus anguli *GED*; ergo notus erit angulus *GED*, qui est angulus differencie, ut probatur. Et per consequens sibi equalis erit notus, scilicet angulus *EDF*. Et hoc est propositum.

Variatur tamen hec figura per appropinquacionem Solis ad[d] augis oppositum *B*. Sit igitur arcus *BH* notus, et coniungantur *CH* et *DH*,

[a] A cognoscitis [b] A puncta [c] CAH *CE* [d] A λ

Describe a circle *AB*, with centre *C*, *AB* being a diameter, and let *D* be the centre of the Earth, the Sun's distance from the aux (*A*) being the arc *AE*. *C* having been joined to *E*, and *E* to *D*, a line *DF* is produced through the point *D*, parallel[1] to the line *CE*. I maintain that the angle *EDF* is the angular difference sought, namely that between the mean and true motus. This is evident from what follows.

The angles *D* and *E* are equal, by Euclid I. 29. But the angle *E* is, by Euclid I. 32, the difference between the angle *ACE* (that is, the angle of mean motus, reckoning from the aux, because by Proposition 1 the distance of the aux from the first point of Aries is known) and the angle *ADE*, which—since *D* is the centre of the Earth—is the angle of true motus. This angle will be known, since the line *CD* (in sixtieths of *CA*, the semidiameter of the deferent) is known, as well as the angle *ACE*. Conversely, *CD* will be known in sixtieths of the line *CA*, since the two angles *ACE* and *EDF* are known.

Proof. By Euclid I. 15, when a perpendicular is dropped from *D* upon *EC* produced, which it meets at *G*, the right-angled triangle *DGC* and the angle *GCD* will be known by Euclid I. 15. Imagining, then, a circle centre *C* and semidiameter *CD*, the arc of sinus rectus *DG* will be known in terms of 360th parts of that circle.[2] Its sinus, *DG*, will therefore be known, from which information the third sinus *CG* will be known, this being the sinus of the complementary angle. All three sides of the triangle are therefore of known length, in sixtieths of a semidiameter *CE*, and the total length of *EG* will then be known. Since, therefore, the triangle *EGD* is right-angled, and its two sides *EG* and *GD* are known, the third side subtended by the right angle, taken as *ED*, will be known by the application of Euclid I. 47.[3]

Supposing, therefore, a circle of semidiameter *CD* and centre *E* and making *ED* 60 parts, the lengths of both *EG* and *GD* will be known in terms of those parts. For if the ratio of *ED* to either *DG* or *GE* is known, their ratio will likewise be known as a definite number, depending on which ratio is taken.[4] But the sinus *DG* being known, its arc is known, namely the arc of the angle *GED*; and therefore the angle *GED* will be known, that is, the angle of the difference [between mean and true motus], as is proved. The equal angle *EDF* will consequently be known, as was proposed.[5]

The figure changes, however, when the Sun approaches *B*, the opposite of the aux. Suppose, then, that the arc *BH* be known.[6] Both

[1] There are four ways of expressing the idea of parallelism in *Albion*, and these may be illustrated by four quotations from the first Part: (i) '. . . productaque equidistante linee *CE* a puncto *D* . . .' (I. 2; cf. I. 3, and *Rectangulus* I. 2); (ii) 'producatur linea equidistans linee *HG* . . .' (I. 4); (iii) '. . . protrahatur eidem equidistans . . . in linea *ACB* . . .' (I. 6); (iv) '. . . quorum fuerint semidiametri equidistanter protracte . . .' (I. 8).

[2] The arc is simply the mean motus.

[3] The MSS. have I. 48, which is the converse of I. 47, and cannot have been intended.

[4] *sub quocumque numero alterum illorum ponatur.*

[5] At this juncture Simon Tunsted reminds the reader of Pythagoras' Theorem.

[6] *H* is understood as the new position of the Sun.

et a puncto *D* producantur due linee, una *DK* equidistans linee *CH*, et altera *DI* perpendicularis super lineam *CH*. Quia igitur angulus *HCB* est notus ⟨per⟩ propositum, et triangulus *DIC* est rectangulus, erit linea *DI* sinus rectus arcus noti descripti super centrum *C*, secundum quantitatem semidiametri *CD*; et per consequens nota erit linea *DI*, de partibus 60 parcium linee *CD* note per hypotesim; et similiter nota erit tercia linea *CI*, quia *CI* est sinus rectus residui deficientis[a] a quarta circuli.

Cum igitur abstracta fuerit *IC* nota de tota linea *CH*, nota remanebit *IH*[b] nota. Sed triangulus *HID* est rectangulus, cuius duo latera *DI* et *IH* sunt nota, ergo nota erit *HD* per Dulkarnon. Imaginato itaque circulo super centrum *H* secundum semidiametralitatem *HD*, et posito *HD* 60 parcium de eisdem partibus, nota erit linea *DI*; ergo notus erit eius arcus, qui est quantitas anguli *DHC*, qui eciam est angulus differencie medie elongacionis ad veram. Et per consequens notus erit angulus eidem equalis, scilicet angulus *HDK*.

Si autem per noticiam duorum angulorum *H* et *C* cognoscere volueris quantitatem linee *CD*, converte demonstracionem.

Et nota quod isto modo investigari possunt equaciones cuiuscumque ecentrici, et ecentricitas, equaciones eciam epiciclorum, cum quantitate diametrorum, ut infra[c] patebit.

Invenit autem Ptolomeus per consideraciones lineam *CD* habere 2 par. 30 min. de 60 par. linee *CA*, que est semidiameter deferentis solaris. Secundum autem Albategni, capitulo 28°, habet linea *CD* 2 par. 4 min. et 24[d] sec., per consideraciones longissime virge divise.[e]

CAPITULUM

De modo calculandi tabulas equacionum planetarum.

Modum autem calculandi sic ponit:

Si distancia planete ab auge sua sit minor 90 gradibus, accipe eius distancie cordam rectam et cordam residui de 90. Utrumque per se multiplica in cordam que cadit inter duo centra, si feceris ad equaciones alicuius ecentrici, vel in semidiametrum epicicli, si feceris ad equaciones epicicli; non curo, quia habito modo operandi ad

[a] A deficientes [b] CA *CH* [c] CA intra [d] H 45
[e] H +Secundum Alfonsum 2 par. 16 min. 9 sec.

C and *D* are joined to *H*, and from the point *D* two lines are drawn, one, *DK*, parallel to the line *CH*, and the other, *DI*, perpendicular to the line *CH*. Since, then, the angle *HCB* is known, as presupposed, and the triangle *DIC* is right-angled, the line *DI* will be the sinus rectus of the known arc described with centre *C* and semidiameter *CD*. As a result, the line *DI* will be known in sixtieths of the line *CD*, which is known by hypothesis. Likewise the third line, *CI*, will be known, since *CI* is the sinus rectus of the complementary angle.

When *IC*, which is known, is subtracted from the entire line *CH*—which is also known—*IH* remains. But the triangle *HID* is right-angled, and its two sides *DI* and *IH* are known, therefore *HD* will be found by Dulkarnon. Supposing, then, a circle centre *H* and semidiameter *HD*, and making *HD* of 60 equal parts, the line *DI* will be known, and therefore its arc also—which is the arc of the angle *DHC*. This, moreover, is the angular difference between the mean and true elongation. Consequently an angle equal to this, namely *HDK*, will be known.

If, however, you wish to determine the length of the line *CD* from a knowledge of the two angles *H* and *C*, work backwards through the proof.

Note that in this way equations[1] may be investigated for any eccentricity whatsoever, as may also the eccentricity and the equations of epicycles, with the size of [their] diameters, as will become evident.

Ptolemy found, from his investigations, that the line *CD* is $2^p\ 30'$, where the line *CA*, the semidiameter of the solar deferent, is of 60 parts. According to Albategni's chapter 28, on the other hand, *CD* is $2^p\ 4'\ 24''$, as found from lengthy observations using a divided rod.

On the means of compiling tables of the equations of the planets.

The chapter lays down the method of calculation as follows:

If the (angular) distance of the [centre of the epicycle of the] planet from its aux is less than 90°, find the sinus of its distance and the sinus of the complement.[2] Multiply each, separately, by the chord which falls between the two centres, if you want to work out the equation of an eccentric, or by the semidiameter of the epicycle, if you want to work out the equations of the epicycle.[3] It is of no matter which

[1] The equation ('of the centre') is the difference between the true and mean elongation.

[2] The text has *corda recta* for *sinus rectus* in this sentence, but there is no reason to suppose that Richard was not conscious of the ambiguity, which occurs at a number of points in *Quadripartitum*. The problem is rarely a serious one, since the context is almost always sufficient to give the sense. As Professor Hartner has pointed out to me, it is characteristic of Albategni that he uses the word *watr* (which generally means 'chord') for the sine function, or half-chord, although of course he uses it correctly. For a long discussion of the verbal confusion see Nallino's edition of Albategni, i. 154 seqq.

[3] This bonus solution (for the equation of the argument) can hardly have been understood at this point by the uninitiated.

unum eorum, habetur ad reliquum. Deinde utrumque seorsum servatum divide per 60, et quod erit ex divisione sinus residui de 90 adde semidiametro; hoc est 60 partibus. Et totum illud collectum in se ipsum multiplica. Similiter illud[a] quod exibat ex divisione corde arcus distancie ab auge in se ipsum multiplica. Et ista ultimo duo multiplicata coniunge, et collecti extrahe radicem,[b] et reserva[c] eam.

Deinde illud quod primo habuisti, scilicet cordam distancie ab auge, in 60 multiplica, et collectum istud divide per radicem ultimo reservatam. Et exibit una corda, quam si arcuaveris, ille arcus erit equacio quesita.

Si vero[d] distancia planete ab auge fuerit amplius 90, abice ab eo 90, et residui quere cordam ipsam; eciam arcum tolle de 90 et remanentis cordam similiter quere; et utramque cordam per se multiplica in cordam que cadit inter duo centra, et utrumque multiplicatum seorsum per 60 divide. Et quod exierit per divisionem illius quod fuit ex residuo de 90, tolle a 60; et quod remanet, in se multiplica; et quod surgit ex tali multiplicacione adde illi quod collectum fuit ex multiplicacione corde residui de 90 in[e] se ipsum multiplicate, et tocius agregati quere radicem, et serva. Post hec illam cordam residui de 90 primo habitam in 60 multiplica; et quod surgit per servatam radicem divide. Et quod tunc exit erit corda cuius arcus est equacio argumenti quesita.

Et isto modo calculavit Ptolomeus tabulas equacionum ad singulos gradus 360; et similiter Albategni, ut patet capitulo 28° primi perfecti.

CONCLUSIO 3ª

De centro ecentrici Solis inveniendo

Dato argumento Solis et eius equacione, per alterum duorum centrorum Terre vel deferentis datum alterum reperire.

Fiat circulus deferentis *AB* super punctum *C*, et distancia planete ab auge, quod dicitur *argumentum*, sit[f] arcus *AE*. Et coniunctis *EC*,

[a] A λ [b] C indicem [c] H serva [d] A λ [e] A 18λ (haplog.) [f] A sicut

you do, since if you can work out the one problem you can work out the other. Next, divide each answer separately by 60, and add to the semidiameter the result of dividing [by 60 the eccentricity times] the sinus of the complement; this will be in sixtieths. Next take the square of the result. Similarly, take the square of what results from dividing the sinus[1] of the arc of the distance from the aux. Lastly, add together the two results obtained, extract the square root of the answer, and set the result aside.

Next, take that result which you obtained at the very first, namely the eccentricity times[2] the sinus[3] of the distance from the aux, and divide the result by the answer set aside a moment ago. A sinus[3] results, and if you find the arc corresponding to it, that arc will be the equation sought.

If, in fact, the distance of the planet from the aux turns out to be greater than 90 [degrees], ask what is the sinus[3] of the remainder when 90 is taken from it; also subtract the arc from 90 [degrees], and likewise ask what is the sinus[3] of the remainder. Now multiply each sinus[3] by the chord which lies between the two centres, and divide each product separately by 60. Subtract from 60 that which results from the division of the answer to the product involving the sinus of the complement.[4] Square the result, and add to the answer the square of the result of the [earlier] multiplication involving the sinus.[5] Take the square root of the entire sum, and make a note of it. After this, multiply by 60 that [product involving the] sinus[5] which we had at first; and divide the result by the square root we put aside. The answer will be the sinus[6] whose arc is the equation of the argument which was sought.

In this way Ptolemy worked out tables of equations for every degree up to 360. Likewise Albategni, as is clear from chapter 28 of [his] first completed [book].

PROPOSITION 3

[I. 3]

Given the [mean] argument of the Sun, and its equation, and given one of the two centres—of the Earth or of the deferent—how to discover the other.

On finding the centre of the Sun's eccentric

Let the deferent circle be *AB*, with centre *C*. The distance of the planet[7] from the aux, which is known as the [mean] *argument*, is to

[1] Again the text has the misleading if not erroneous '*corda*', as also on later occasions. Notice, however, that the substitution is not repeated invariably.

[2] The text has 'sixty times', which is a mistake. The author may have written it, but the opening paragraph makes it clear that he did not mean to write it.

[3] The text has *corda*.

[4] That is, a sixtieth of the product of it and the eccentricity.

[5] The text has *corda residui de 90*.

[6] The text has *corda*.

[7] That is to say, the Sun.

super punctum *E* fiat angulus equalis angulo equacionis date, per 23 primi Euclidis, qui sit angulus *CED*, producta linea *CD* ad diametrum *AB*. Dico ergo quod punctus *D* in linea *AB* est orbis signorum centrum, quod sic probatur:

Angulus *ACE* est equalis duobus angulis *E* et *D*, per 32^am^ primi Euclidis, ergo angulus *E* est differencia inter angulum *C* et angulum *D*. Sed ex hypotesi angulus *E* est differencia anguli medie distancie Solis ab auge ad veram, que vocatur *equacio argumenti*. Sed angulus vere distancie est in centro orbis signorum, qui est centrum oculi. Est igitur *D* centrum Terre, quod est pars propositi.

Rursum sit *C* centrum Terre, et angulus medii argumenti *ACG*,[a] et angulus equacionis eiusdem *FCG*; productaque linea a puncto *G* equidistante linee *CF* usque ad punctum *H* in diametro *AB* dico quod *H* est centrum deferentis solaris respectu centri *C*, ut patet per priorem deductionem; quia angulus *H* extrinsecus est equalis duobus angulis *C* et *G*, ergo angulus *G* est differencia inter ipsum et angulum apud *C*; ergo *H* est centrum deferentis solaris.

CONCLUSIO 4^a^

Super omnem punctum linee exeuntis a centro dati circuli circulum ita describere quod erit proporcio semidiametri eius ad id quod cadit[b] *inter ipsa duo centra qualis est proporcio semidiametri deferentis solaris ad distanciam centri eius a centro orbis signorum.*

Fiat circulus *AB* super centrum *C*, producta linea *CB* indifinite quantitatis, et per argumentum et eius equacionem datam per proximam extrahatur triangulus *CHG* ut sit *H* angulus medii argumenti, *C* angulus argumenti equati, et *G* angulus equacionis. Et producta linea *CG* indifinite a puncto assignato in linea *CB* qui sit *D*, producatur linea equidistans linee *HG* usque ad concursum eius cum linea *CG* in puncto *F*. Dico ergo quod proporcio semidiametri *DF* ad lineam *DC* est sicut proporcio semidiametri deferentis solaris ad distanciam centri eius a centro Terre.

Facto circulo super centrum *D* secundum semidiametralitatem *DF*, hec patet quia per 4^am^ sexti Euclidis proporcio *FD* ad *DC* est sicut[c] proporcio *GH* ad *HC*, sed per premissam proporcio *GH* ad *HC* est sicut proporcio semidiametri[d] deferentis ad distanciam centri eius a centro Terre, quod est propositum.

[a] CAHL *ACF* [b] A λ [c] A λ [d] CL diametri

be the arc *AE*. Having joined *EC*, construct (by Euclid I. 23) an angle at the point *E* equal to the angle of the given equation, this being the angle *CED*, the line *DC* being taken on the diameter *AB*. I now maintain that the point *D* on the line *AB* is the centre of the sphere of signs, and this is proved as follows.

Angle *ACE* is equal to the sum of the two angles *E* and *D*, by Euclid I. 32, and therefore the angle *E* is the difference between the angle *C* and the angle *D*. But, by hypothesis, the angle *E* is the difference between the angle from the aux to the mean Sun and that to the true Sun, this difference being called the *equation of the argument*. But the angle equal to the true distance is at the centre of the circle of signs, which is the centre of the eye. *D* is therefore the centre of the Earth. This is part of the proposition.

Conversely, let *C* be the centre of the Earth, and let the angle of mean argument[1] be *ACG*, with *FCG* the angle of its equation. A line being produced from point *G*, parallel to the line *CF* and drawn to the point *H* on the diameter *AB*, I maintain that *H* is the centre of the solar deferent with respect to centre *C*, as is clear from the first deduction. For the exterior angle *H* is equal to the two angles *C* and *G*, wherefore the angle *G* is the difference between [*H*] and the angle at *C*; and therefore *H* is the centre of the solar deferent.

PROPOSITION 4

[I. 4]

How to describe a circle about any point of a line passing through the centre of a given circle, in such a way that the ratio of its semidiameter to the distance between the two centres themselves will be the same as the ratio of the semidiameter of the solar deferent to the distance between its centre and the centre of the sphere of signs.

Let there be a circle *AB* about the centre *C*, the line *CB* being produced indefinitely. The argument and the corresponding equation being known, by the preceding proposition, the triangle *CHG* is to be constructed so that *H* is the angle of the mean argument, *C* the angle of the equated argument, and *G* the angle of the equation. The line *CG* being produced indefinitely from the chosen point *D* on the line *CB*, a line is produced parallel to the line *HG*, as far as *F*, where it meets *CG*. I hold that the ratio of the semidiameter *DF* to the line *DC* is equal to the ratio of the semidiameter of the solar deferent to the distance of its centre from the centre of the Earth.

There being a circle drawn with centre *D* and semidiameter *DF*, [the proposition] is clear from Euclid VI. 4, since the ratio *FD* to *DC* is as the ratio *GH* to *HC*, whereas from the previous proposition the ratio *GH* to *HC* is as the ratio of the semidiameter of the deferent to the distance of its centre from the centre of the Earth; which was proposed.

[1] That is, for argument 90°.

Et nota quod propter istam conclusionem premisse sunt cetere; est enim hec conclusio necessaria ad propositum, quia per eam invenientur omnes ecentrici super quorum centrum[a] est motus equalis—non autem deferentes epiciclorum, sed alio certo modo.

Notandum eciam quod quantitas semidiametri circuli cuiuslibet planete invenitur eodem modo, sicut ecentricitas deferentis ad orbem signorum. Et ideo superflueret recitare demonstracionem nisi omnino non credenti radicibus antiquorum. Invenit quidem Ptolomeus, libro 5° capitulo 5° et libro 4° capitulo 3°, semidiametrum epicicli Lune esse 6 par. et 20 min. de 60 par. sui deferentis, et ecentricitatem deferentis 12 par. et 28 min. Verumptamen hec dicit Albategni, capitulo 28°, semidiameter[b] epicicli cuiuslibet planete est equalis distancie centri ecentrici a centro concentrici. Ideo quelibet equacio motus que potest haberi per unum ecentricum respectu concentrici potest haberi per epiciclum idem eiusdem, et econverso. Et hec est causa quare discordat a sequacibus Ptolomei in equacionibus ecentricorum et epiciclorum, quia discordat ab eis in longitudine ecentricitatum. Sunt autem ecentricitates et semidiametri secundum eum epiciclorum hoc modo:

Lune, 5 par. 15 min. de 60 par. semidiametri deferentis eius;
Saturni, 6 par. 29 min. et 7 sec.
Iovis, 11 par. 30 min. 15 sec.;
Martis, 39 par. 25 min. et 22 sec.;
Veneris, 43 par. 9 min. et 5 sec.;
Mercurii, 22 par. 30 min. et 30 sec., per consideraciones longissimarum regularum. Sed magis puto adherendum experimentis Ptolomei, Arzachelis, et suorum sequacium, qui radices ponunt hoc modo.

Distancia centri deferentis *Lune* a centro orbis signorum est reperta 12 par. et 28 min. de 60 par. semidiametri deferentis; et semidiameter[c] epicicli eius 6 par. et 20 min.;

In *Saturno* distancia centri deferentis a centro Terre 3 par. et 25 min. de 60 partibus semidiametri deferentis; distancia vero centri equantis eius[d] a centro Terre duplum, scilicet 6 par. et 50 min.; semidiameter autem epicicli eius 6 par. et 30 min. de eiusdem[e] partibus;[f]

[a] CH centra [b] A semidiametri [c] A semidiametri [d] A λ
[e] A eiusdem
[f] In MS. A this entire paragraph is misplaced, following the next, on Jupiter

Note that we were leading up to this proposition in placing the others before it. We need the proposition to determine all the eccentrics about whose centres the motion is uniform[1]—not, however, the deferents of the epicycles, which is quite another matter.

It is also to be noted that the size of the semidiameter of the deferent circle of any planet is found in the same way, as is the eccentricity of the deferent circle with respect to the sphere of signs. It would therefore be superfluous to go through the proof unless we were to disbelieve entirely the parameters arrived at by the ancients.[2] Ptolemy, in *Almagest* V. 5 and IV. 3, found the semidiameter of the epicycle of the Moon to be $6^p\ 20'$ (its deferent being 60^p), and the eccentricity of the deferent to be $12^p\ 28'$. Albategni, however, said in chapter 28 that the semidiameter of the epicycle of any planet is equal to the distance between the centre of the eccentric and the centre of the concentric.[3] Any equation of motus, therefore, which may be represented by a circle eccentric with respect to a concentric, may be represented by an epicycle of [semidiameter equal to the eccentricity], and conversely. This is the cause of his disagreement with those who followed Ptolemy over the equations of the eccentrics and of the epicycles; for he disagrees with them over the eccentric distance. The eccentricities and semidiameters[4] according to him are:

Moon—$5^p\ 15'$ (with a deferent semidiameter of 60^p);[5]
Saturn—$6^p\ 29'\ 7''$;
Jupiter—$11^p\ 30'\ 15''$;
Mars—$39^p\ 25'\ 22''$;
Venus—$43^p\ 9'\ 5''$;
Mercury—$22^p\ 30'\ 30''$.

These he found by [? using the instrument known as Ptolemy's] rules, with very long [arms]; but I am more inclined to believe that one should follow the observations of Ptolemy, Arzachel, and those who came after, who use these figures:

The distance of the centre of the deferent of the *Moon* from the centre of the sphere of signs is found to be $12^p\ 28'$ (with a deferent semidiameter of 60^p); and the semidiameter of its epicycle $6^p\ 20'$;

For *Saturn*, the distance of the centre of the deferent from the centre of the Earth is $3^p\ 25'$ (deferent semidiameter 60^p); the distance of the centre of its equant from the centre of the Earth is double, namely $6^p\ 50'$; the semidiameter of its epicycle, however, is $6^p\ 30'$ (with the same units);

[1] The reference is to the equant circles, which are distinguished from the eccentric deferents.

[2] *radices antiquorum.*

[3] That is, assuming that the eccentric is replaced by an epicycle—overlooking any existing epicycle for the time being. If both an eccentric and an epicycle are combined, it is irrelevant whether the eccentricity and the epicycle radius be equal or unequal.

[4] It is not clear that Richard understood Albategni on this point. See our commentary.

[5] The deferent radius given is a mistake. See the note to the table of I. 4 comm.

In *Iove* autem est distancia duorum centrorum Terre et deferentis 2 par. et 45 min.; distancia equantis a centro Terre duplum, scilicet 5 par. et 30 min.; et semidiameter[a] epicicli 11 par. et 30 min.

In *Marte* distancia duorum centrorum Terre et deferentis 6 par. et 30 min., cuius duplum est 13 par., et illa est distancia equantis centri a Terre centro; semidiameter epicicli eius 39[b] par. 30 min.;

In *Venere* distancia centrorum Terre et deferentis est una pars et 15 min., cuius duplum est 2 par. et 30 min., que est distancia centri equantis a centro Solis in modico deferens ab ecentricitate Solis; semidiameter epicicli euis 43 par. et 10 min.;

In *Mercurio* distancia centri equantis a centro Terre est 3 par., et distancia centri deferentis a centro Terre est tripla, scilicet 9 par., quia semidiameter parvi circuli deportantis centrum deferentis est equalis distancie centri equantis a centro Terre—patet hec imaginacio in theorica planetarum; et semidiameter epicicli eius est 22 par. et 30 min.

Super has radices calculantur tabule equacionum Arzachelis, verumptamen quibus non placent hee radices per observanciam rectanguli examinet[c] novas radices, quibus habitis sic demonstrabit equaciones.

CONCLUSIO 5[a]

de equacione argumenti demonstranda

Cuiuslibet argumenti equacionem in omni distancia centri epicicli ab auge deferentis per datam elongacionem planete a longitudine longiori epicicli et per notam qualitatem semidiametri eius ad semidiametrum deferentis ostendere.

Esto primo quod centrum epicicli sit in auge deferentis cuius semidiameter sit *DA*, et centrum orbis signorum *C*, et arcus epicicli datus *EF*; et sit *E* aux epicicli, et *F* locus planete in epiciclo. Productaque linea *EF*[d] loco radii visualis propositum est invenire quantitatem anguli *ECF*, que dicitur *equacio illius argumenti*. Coniungantur itaque *AE*[e] et *AF*, et ab *F* cadat cathetus super lineam *AE* ad punctum *G*. Age igitur de 60 partibus semidiametri *DA* nota est tam linea *CD*[f] quam linea *AE* per ypotesim, ergo nota est tota linea *CE*.

[a] A semidiametri [b] CA 30 [c] H examinent [d] H *CF* [e] A *AC*
[f] A *ED*

For *Jupiter*, the distance between the two centres, those of the Earth and the deferent, is $2^p\,45'$; the distance of the equant from the centre of the Earth is double, namely $5^p\,30'$; and the semidiameter of the epicycle is $11^p\,30'$;

For *Mars*, the distance between the two centres, of the Earth and of the deferent, is $6^p\,30'$,[1] double which is 13^p—and this is the distance of the equant of the centre from the centre of the Earth; the semidiameter of its epicycle is $39^p\,30'$;

For *Venus*, the distance between the centres, of the Earth and of the deferent, is $1^p\,15'$, double which is $2^p\,30'$, which is the distance of the centre of the equant from the centre of the Sun, [the deferent] carrying the planet at an eccentricity somewhat less than that of the Sun; the semidiameter of its epicycle is $43^p\,10'$;

For *Mercury*, the distance of the centre of the equant from the centre of the Earth is 3^p, and the distance[2] of the centre of the deferent from the centre of the Earth is three times as great as this, 9^p; for the semidiameter of the small circle carrying the centre of the deferent is equal to the distance of the centre of the equant from the centre of the Earth—this being evident from planetary theory; and the semidiameter of its epicycle is $22^p\,30^m$.

Arzachel's tables of equations are calculated on the basis of these radices. He, however, who disagrees with these figures can find new ones, having recourse to the rectangulus, after which he will derive equations.

PROPOSITION 5

[I. 5]

To show the equation of any argument, for any distance of the centre of the epicycle from the aux of the deferent, given the elongation of the planet from the far point of the epicycle;[3] *and given the size of its semidiameter in relation to the semidiameter of the deferent.*

On demonstrating the equation of the argument

First, let the centre of the epicycle be at the aux of the deferent, whose semidiameter is *DA*, the centre of the circle of signs[4] *C*, and the given arc of the epicycle *EF*. Furthermore, let *E* be the aux of the epicycle, and *F* the position of the planet on the epicycle. The line *EF* being joined [to *C*] as it were, with a visual ray, it is proposed to find the size of angle *ECF*, which is called the *equation of its argument*. *AE* and *AF* should therefore be joined, and from *F* a line falls at right angles[5] on the line *AE*, meeting it at the point *G*. It is assumed that *CD* and *AE* are of known numbers of parts (of the 60^p which make up the semidiameter *DA*). The whole line *CE* is therefore known.

[1] See I. 4 comm.

[2] The *greatest* distance is that intended.

[3] That is to say, given the true argument.

[4] That is, the centre of the Earth.

[5] *cathetus*.

Item exposito notus est angulus *EAF*, cum igitur triangulus *AGF* sit rectangulus erit notus sinus rectus *GF* de partibus semidiametri *AE* noti; et per consequens nota erit linea *GA*, cum sit residuum deficientis arcus a quarta circuli, ergo nota erit tota linea *CG*. Sed triangulus *CGF* est rectangulus cuius duo latera *CG* et *GF* sunt nota; ergo notum erit tercium latus *CF*[a] de eisdem partibus. Constituatur ergo *CF* 60 parcium et, posito centro in *C*, fiat circulus secundum semidiametralitatem *CF*; et statim erit notus arcus sinus recti *FG* quia cum proporcio *CF* ad *FG* nota sit secundum proporcionem unam numeri certi, nota erit secundum quemcumque alterum numerum quattuor numerorum proporcionalium; ergo notus erit angulus *ECF*, quod est propositum.

Et nota ex hoc quod devenire possimus in noticiam linee *CF* per noticiam linee *CG* ex partibus illis quibus constituitur semidiameter deferentis *DA*[b] 60 parcium. Nota erit nobis distancia linee[c] a centro Terre in omni eius loco deferentis ecentrici epicicli, sicut arguitur in libro 5° *Almagesti* capitulo 7° et capitulo 13°, pro diversitate[d] aspectus habenda, ut infra dicetur capitulo 14°.

Potest autem hec figura variari secundum quod stella est propinquior opposito augis quam augi. Sit igitur arcus distancie ab augis opposito *HI*, et ductis lineis *CI* et *IA* cadat cathetus ab *I* super lineam *HA* in punctum *K*. Et arguatur ut prius: angulus *HAI* est notus, et triangulus *AKI* est rectangulus, ergo de partibus *AI* notis nota erit linea *IK* quia est sinus arcus noti. Et per consequens *KA* latus tercium erit notum quia ipsum est sinus residui deficientis a quarta circuli; ergo residuum erit notum *KC* quia tota *CA* ponitur esse nota. Sed triangulus *CKI* est rectangulus, notis ergo duobus lateribus *KI* et *KC* de eisdem partibus, cognoscitur *CI*; per cuius noticiam devenietur in noticiam anguli *KCI* ut prius. Sed angulus *KCI* est angulus equacionis argumenti *EF* [vel] *HI* quesiti; patet ergo plane propositum.

Modus autem calculandi omnino est similis illi qui ponitur in fine secunde huius, ubi docetur calculare tabulas equacionum que proveniunt ab ecentrico.

Diversitas in figura

Sed pone quod centrum epicicli fuerit alibi in ecentrico quam in auge, ut[e] distet ab auge per arcum *AO*, et distancia planete ab auge epicicli *FH*, productis lineis *DOF* et *COG*, et *CH* et *OH*, et cathetus *HI* super lineam *CG*, et secundus cathetus a puncto *D* quod est centrum deferentis super lineam *CG* que sit *DK*.

[a] AH *EF* [b] A *CA* [c] H Lune [d] H diversa [e] H 4λ (haplog.)

Next in order of composition, the angle *EAF* is known, and since therefore the triangle *AGF* is right-angled, the sinus rectus *GF* will be known as a number of parts of the known semidiameter *AE*. As a consequence, the line *GA* will be known, since it is the arc remaining from a quarter-circle. The whole line *CG* will therefore be known. But the triangle *CGF* is right-angled, and its two sides *CG* and *GF* are known; wherefore the third side *CF* will be known in terms of the same parts. Let *CF* be 60 parts and, with centre *C*, construct a semicircle with *CF* as a semidiameter. The arc of the sinus rectus *FG* will immediately be known because, since the ratio *CF* to *FG* is known as a single numerical ratio (such a term as occurs in a proportionality relating four numbers), therefore angle *ECF* will be known; which is what was proposed.[1]

Note that from this we may arrive at a knowledge of line *CF* from a knowledge of line *CG*, out of sixtieth parts of the semidiameter of the deferent *DA*. The distance of the line from the centre of the Earth at any place on the eccentric deferent of the epicycle will be known to us, as is argued in *Almagest* V. 17 and V. 13, on finding parallax[2] (as is mentioned below in proposition 14).

This figure may, however, be varied in such a way that the planet[3] is nearer the opposite of aux [of the same epicycle] than aux. Therefore let the arc of the distance from the opposite aux be *HI*. The lines *CI* and *IA* being drawn, a perpendicular from *I* should meet the line *HA* in the point *K*. It is to be argued as before that as the angle *HAI* is known, and the triangle *AKI* is right-angled, therefore, since the sinus of the arc [*IH*] is known, the line *IK* will be known in terms of the known parts of *AI*. In consequence, the third side *KA* will be known, since this is the sinus of the complementary angle;[4] and therefore the remaining [side] *KC* will be known, since the whole of *CA* is supposed known. But triangle *CKI* is right-angled, and therefore the two sides *KI* and *KC* being known in the same units, *CI* is known. And from a knowledge of this, a knowledge of angle *KCI* is arrived at, as before. But angle *KCI* is the angle of the equation of the argument required, *EF* [or] *HI*. The proposition is then clearly evident.

The method of calculating is exactly as explained here, and as set out at the end of the second [proposition], where it is explained how to calculate tables of equations arising from the eccentricity.[5]

Place the centre of the epicycle at a point on the eccentric other than at the aux, so that it is distant from the aux by the arc *AO*, and so that the distance of the planet from the aux of the epicycle is *FH*. The lines *DOF*, *COG*, *CH*, and *OH* are produced, as well as a perpendicular *HI* on the line *CG*, and also a second perpendicular *DK*, from the point *D*, the centre of the deferent, on the line *CG*. Proceed then as follows. **A second possible figure**

[1] That is, arc sin(*GF*/*FC*) is determinate. [2] *diversitas aspectus.* [3] *stella.*
[4] Simon Tunsted here has *residuum* rather than *sinus residui.*
[5] That is, equations of the centre.

Age ergo: per noticiam anguli *ADO* et per notam[a] quantitatem linee *OC* ad lineam *DA* ex deductione secunde huius accipe noticiam anguli *COD* cui est equalis angulis *GOF*, et per 15 primi Euclidis. Sed ex hypotesi notus est angulus *FOH*, ergo notus est totus angulus *GOH*. Sed cognito angulo *GOH* per primam partem huius quinte, haberi potest noticia anguli *GCH*, quod est propositum; quia arcus anguli *GCH* est equacio argumenti *FH* equati per arcum *GF*[b] in distancia centri per arcum *AO*. Et hic patet quod equacio centri, que est differencia veri centri ad medium ut arcus anguli *DOC*, est semper equalis illi porcioni qua equatur argumentum planete que dicitur *distancia ab auge epicicli*.

CONCLUSIO 6ª

de ecentrico Lune scribendo

In quocumque puncto linee exeuntis a centro orbis signorum centrum epicicli Lune signetur. Super lineam inter ipsa duo centra cadentem triangulum figurare, cuius unius lateris ad alterum erit proporcio tanquam proporcio[c] *semidiametri deferentis ad distanciam centri eius a centro orbis signorum.*

Fiat circulus orbis signorum *AB* super centrum *C*, producta linea *CE* indifinite quantitatis, sitque *E* punctus communis intersectio linee *CE* et circumferencie *AEB*; sitque angulus *ACE* angulus distancie centri epicicli ab auge deferentis *A* in quavis eius elongacione; productaque diametro *ACB*, signetur centrum epicicli in linea *CE* super punctum *G*. Deinde ab *E* versus *A* accipiatur arcus equacionis centri *AE* suppositi, et sit equacio *EF*, coniunctisque *C* [et] *F*[d] protrahatur eidem equidistans a puncto *G* usque *H* in linea *ACB*. Deinde assignato puncto *D* in linea *CA*, secundum longitudinem *CH* coniungantur *DG*.

Dico ergo quod trianguli *CDG*[e] erecti super lineam *CG* proporcio lateris *DG* ad latus *CD* est sicut proporcio semidiametri deferentis lunaris ad distanciam[f] centri eius a centro Terre.

Probacio. Quia motus Lune in suo epiciclo est ab oriente in occidentem[g] dum est in longitudine longiori epicicli et equacio circuli brevis tollitur ab argumento distancie[h] centri epicicli Lune, Luna

[a] A noticiam [b] AH *GH* [c] H λ [d] CA *EF* [e] CL *EDG*
[f] A instanciam [g] BC occidente [h] CAH distancia

From a knowledge of the angle *ADO*, and by means of the size of the line *OC* in relation to the line *DA*, known from the deduction of the second proposition of this work, find the angle *COD*,[1] equal to angle *GOF* by Euclid I. 15. But, by hypothesis, angle *FOH* is known, and therefore likewise the whole of angle *GOH*. But angle *GOH* being known, a knowledge of angle *GCH* may be had by the method of the first part of this fifth proposition; and this is what was proposed. For the arc of the angle *GCH* is the equation of the argument *FH*, the arc *GF* being the equation of the centre corresponding to the arc *AO*. And here it is clear that the equation of the centre, which is the difference between the true and mean centres, as the arc *DOC*, is invariably equal to that angle by which the argument of the planet is equated. This is called the *distance from the aux of the epicycle*.[2]

PROPOSITION 6

[I. 6]

The centre of the epicycle of the Moon is to be marked at a point of a line drawn from the centre of the sphere of signs. To construct a triangle on the line between the two centres themselves, one side of which triangle shall be to the other as the ratio of the semidiameter of the deferent to the distance of its centre from the centre of the circle of signs. On drawing the eccentric of the Moon

Let *AB* be the circle of the sphere of signs, with centre *C*. The line *CE* being produced indefinitely, the point *E* is to be the point of intersection of the line *CE* and the circumference *AEB*. The angle *ACE* is to be the angular distance of the centre of the epicycle from *A*, the aux of the deferent, at any of its elongations. The diameter *ACB* being produced, the centre of the epicycle on the line *CE* is to be marked *G*. Then from *E* towards *A* is taken *AE*, the arc of the presupposed equation of the centre, and let the equation be *EF*, a parallel to the [line] joining *C* [and] *F* being drawn through the point *G*, as far as *H* on the line *ACB*. Next, with the point *D* on line *CA*, *D* and *G* are to be joined [so that *DG* is equal in] length to *CH*.[3]

I now assert that in the triangle *CDG*, on line *CG*, the ratio of the side *DG* to the side *CD* is equal to the ratio of the semidiameter of the lunar deferent to the distance of its centre from the centre of the Earth.

Proof. The motion of the Moon on its epicycle is from east to west, and therefore while it is at the far point of the epicycle[4] the equation of the small circle[5] is subtracted from the argument of the distance of the centre of the epicycle of the Moon, when the Moon[6] has moved

[1] This is clearly to be found by a procedure equivalent to applying the sine rule. Cf. the method of the *Almagest*.

[2] To which we give the mathematical symbol α.

[3] *secundum longitudinem CH coniungantur DG.*

[4] *in longitudine longiori epicicli.*

[5] That is, the equation of centre.

[6] i.e. its mean motus.

exeunte[a] plus 6 signis, et eadem equacio additur eidem quando centrum, quod vocatur *longitudo duplex*, est minus 6 signis. Necesse est ipsam equacionem centri causari ex concursu duarum linearum in centro epicicli, quarum una exeat a centro Terre, et altera ab opposito puncto centri deferentis. Sed ita se habent due linee *CGI* et *HGK* illud anguli *KGI*, cum sit equalis angulo *FCE* dato mediante angulo *CGH*, per 15^am^ primi Euclidis. Ergo *H* est punctus oppositus centro ecentrici in circulo deferente centrum ecentrici deferentis epiciclum. Ergo *D* est centrum deferentis, cum sint equalis *CH*, *CD*. Erit etiam *DG* semidiameter deferentis quia *G* centrum epicicli nunquam exit circumferenciam deferentis ecentrici. Proporcio ergo *DG* lateris ad latus *DC* est sicut[b] proporcio semidiametri deferentis lunaris ad distanciam centri eius a centro Terre.

Hec deductio nota est satis scientibus theoricam motus Lune. Hec est conclusio necessaria ad operacionem huius instrumenti et sequentes.

CONCLUSIO 7ª

In quocumque puncto linee indifinite exeuntis a centro Terre centrum epicicli Saturni, Iovis, Martis, aut Veneris signetur. Ab eodem puncto unam lineam ad centrum equantis et alteram ad centrum deferentis educere cuius proporcio erit ad id quod cadit inter utrumque centrum qualis est proporcio semidiametri deferentis ad distanciam centri eius a centro Terre.

Fiat circulus *AB* cuius centro *C* exeat una linea indifinita *CF*,[c] et altera ad augem equantis que sit *CA*, que necessario transibit per duo centra equantis et deferentis; sitque arcus medie elongacionis centri epicicli ab auge equantis *AB*; et signato centro epicicli in linea *CF* super punctum *G*, producatur *GH* equidistans linee *CB*. Dividatur *CH* in duo equalia, per 10ªᵐ primi Euclidis, ad punctum *I*, et coniungantur *IG*. Dico ergo quod *H* est centrum equantis quesitum, et *I* centrum deferentis cuiusvis quattuor planetarum; proporcio *GI*

[a] A existente [b] A λ [c] H *EF*

more than six signs [from its aux], whereas the same equation is added to it when the centre, which is known as the *double elongation*,[1] is less than six signs. This same equation of the centre is the angle formed by two lines at the centre of the epicycle, one of which goes out from the centre of the Earth, and the other from the point opposite the centre of the deferent.[2] The two lines *CGI* and *HGK* form an angle equal to angle *KGI*, equal to the given angle *FCE*, which is equal to the angle *CGH*, by Euclid I. 15. *H* is therefore the point opposite the centre of the eccentric on the circle carrying the centre of the eccentric carrying the epicycle. *D* is therefore the centre of the deferent, since *CH* and *CD* are equal. Also *DG* will then be the semidiameter of the deferent, since *G*, the centre of the epicycle, never leaves the circumference of the eccentric deferent. The ratio, therefore, of the side *DG* to the side *DC* is equal to the ratio of the semidiameter of the lunar deferent to the distance of its centre from the centre of the Earth.

This deduction will be sufficiently well known to those who understand the theory of the Moon's motion.[3] The proposition is essential to the working of our instrument, and to those which follow.

PROPOSITION 7

[I. 7]

The centre of the epicycle of Saturn, of Jupiter, of Mars, or of Venus, is to be marked at a point on a line produced indefinitely from the centre of the Earth. To draw from the same point a single line to the centre of the equant, and another to the centre of the deferent, the ratio of the last of which to the line which falls between the two centres shall be as the ratio of the semidiameter of the deferent to the distance of its centre from the centre of the Earth.

Let there be a circle *AB* from whose centre *C* one line *CF* is produced indefinitely, and another, *CA*, going to the aux of the equant, this line necessarily passing through the two centres, of the equant and of the deferent. Let the arc of the mean elongation[4] of the centre of the epicycle from the aux of the equant be *AB*. The centre of the epicycle on the line *CF* being denoted by *G*, *GH* is to be produced parallel to the line *CB*. Let *CH* be divided at the point *I* into two equal parts, by Euclid I. 10, and let *IG* be joined. I hold then that *H* is the required centre of the equant, and *I* the centre of the deferent of any of the four planets. The ratio *GI* to *IH* or *IC*, moreover,

[1] *longitudo duplex*. Cf. I. 7 below, where the word *elongacio* is used.

[2] i.e. from *H*.

[3] That is, it allows a value of e to be determined from a given pair of values of 2η and ξ, a claim which is not as exaggerated as it might sound. (For our symbolism see Appendix 29.)

[4] *medie elongacionis*.

ad *IH* vel *IC* est tanquam proporcio semidiametri deferentis ad distanciam centri eius a centro Terre.

Hec patet quia ex quarta huius convincitur *H* esse centrum equantis. Sed centrum deferentis est in medio inter centrum equantis et centrum Terre, ergo *I* est centrum deferentis. Cum etiam punctus *G* nunquam exit circumferenciam deferentis, erit linea *IG* sicut semidiameter respectu linee *IC* vel *IH*; et hoc est propositum.

Et nota quod omnes iste deductiones dependent a proporcionalitate triangulorum equiangulorum cuiusmodi: est triangulus *CGH* ad triangulum imaginatum ex tribus lineis quarum una imaginarie exit a centro Terre ad centrum epicicli, altera a centro Terre ad centrum equantis, et tercia a centro equantis ad centrum epicicli.

Et nota hic quod auges Solis, Saturni, Iovis, Martis, et Veneris sunt immobiles nisi quatenus moventur motu stellarum fixarum; sed auges Lune et Mercurii sunt mobiles, non habentes locum fixum, et ideo eorum centra et auges inveniuntur[a] alio modo.

De centro et semidiametro deferentis Lune dictum est sexta huius; eius autem aux invenitur quia duplo plus semper distat a centro epicicli quam Sol distat ab eodem. Sed hec alibi habent declarari.

CONCLUSIO 8ª

de centro deferentis Mercurii inveniendo

Data media distancia centri epicicli Mercurii ab auge equantis eius[b] *et equacione eius, augem deferentis et centrum et semidiametrum in debita proporcione ad lineam que est inter centrum Terre et centrum equantis ostendere.*

Esto circulus orbis signorum *AB* super centrum *C* et sit *A* locus augis equantis, qui est immobilis. Deinde per 6am huius, inveniatur centrum[c] equantis per distanciam centri epicicli ab auge equantis et per eius equacionem; et sit arcus elongacionis *AB*, arcus vero equacionis eiusdem arcus[d] *BD*, et centrum equantis *H*. Deinde posito centro in *H*, fiat circulus secundum semidiametrum *HG*, posito *G* esse centrum epicicli.

[a] A inveniantur [b] A λ [c] A centra [d] H λ

is as the ratio of the semidiameter of the deferent to the distance of its centre from the centre of the Earth.

This is evident, since from proposition 4 of this work, *H* is shown to be the centre of the equant. But the centre of the deferent is halfway between the centre of the equant and the centre of the Earth, wherefore *I* is the centre of the deferent. Since, also, the point *G* never leaves the circumference of the deferent, the line *IG* will be exactly as a semidiameter with respect to the line *IC* or *IH*. And this is what was proposed.

Note that all these deductions depend on the proportionality of equiangular triangles: the triangle *CGH* is [similar to] the triangle one of the three sides of which is imagined to go from the centre of the Earth to the centre of the epicycle; another goes from the centre of the Earth to the centre of the equant; and the third goes from the centre of the equant to the centre of the epicycle.

Note also here that the auges of the Sun, of Saturn, of Jupiter, of Mars, and of Venus are immobile, except in so far as they move with the motion of the fixed stars. The auges of the Moon and Mercury, however, are mobile, not having a fixed position, and for this reason their centres and auges are found in a different way.

The centre and semidiameter of the deferent of the Moon are discussed in proposition 6 of this work. Its aux, however, is found, from the fact of its being twice as far from the centre of the epicycle as it is from the Sun. But these things are touched upon elsewhere.

PROPOSITION 8

[I. 8]

Given the mean distance of the centre of the epicycle of Mercury from the aux of its equant, and given its equation, to construct the aux of the deferent, and the centre, and to show the ratio of the semidiameter of the deferent to the distance between the centre of the Earth and the centre of the equant.

On finding the centre of Mercury's deferent

Let *AB* be the circle of the sphere of signs, with centre *C*, and let *A* be the position of the aux of the equant, which is stationary. Next, by proposition I. 6, let the centre of the equant be found from the angular separation of the centre of the epicycle and the aux of the equant, and from the equation of the centre. Also let the arc of the elongation be *AB*, the arc of the equation of the same arc *BD*, and the centre of the equant *H*. Then, with centre *H*, construct a circle having semidiameter *HG*, where *G* is the centre of the epicycle.

Ex altera parte augis *A* accipiatur arcus subduplus ad arcum *AG*, qui sit arcus *DA*, et coniunge *H* cum *D*; et de linea *HA* accipiatur equalis linee[a] *HC* que sit linea *HI*; et super punctum *I*, per 23^am^ primi Euclidis, constituatur angulus extrinsecus duplus ad ipsum angulum intrinsecum *IHD*. Per productionem linee *IL* eritque[b] angulus eius *AIL* equalis angulo[c] *AHG* et angulo *ACB* mediante *AHD*; et ubi linea *IL* intersecat[d] *HD* ponatur *KI*; et coniungantur *KG*[e] producta linea[f] *CKM*.

Dico ergo si centrum epicicli distet ab auge equantis per arcum *AG*, erit *K* centrum deferentis et *M* aux eius; et proporcio *KG* ad *HC* est sicut proporcio semidiametri[g] deferentis ad distanciam centri equantis a centro Terre.

Probacio: Primo angulus[h] *AIL* est equalis angulo *AHG*. Si ergo centrum epicicli sit in aliquo puncto linee indifinite *GH* erit centrum deferentis in aliquo puncto linee *IL*. Hec consequencia est nota scientibus theoricam planetarum. Sed positum est centrum epicicli esse in puncto *G*, ergo necesse est centrum deferentis esse in puncto *K* quia centrum deferentis non potest distare a puncto *I* nisi per quantitatem linee *HI*, cui est equalis *IK*,[i] ex deductione 6^a^ et 32^a^ primi Euclidis.

Item cum centrum epicicli nunquam exit a circumferencia deferentis erit linea *KG* tanquam semidiameter deferentis respectu linee *IH* vel *HC*. Item cum linea exiens a centro Terre per centrum deferentis terminatur in auge eiusdem erit in loco augis deferentis; et hoc est plane propositum.

Per istas conclusiones dividentur ecentrici deferentes 7 planetarum. Quia sicut patet in *Abbreviato Almagesti*, libro 3° capitulo 8°, et libro 4° capitulo 9°, ex 6° Euclidis, conclusione 12^a^,[j] si fuerint centra aliquorum circulorum continue proporcionalium in eadem linea recta constituta, quorum fuerint[k] semidiametri equidistanter protracte, omnis linea priori linee conterminalis easdem semidiametros[l] dividens necesse est arcus resecare proporcionaliter equales.

[a] A λ [b] L 16λ (= haplog.) [c] A angulus [d] C intereat [e] H *K*
[f] A λ [g] CA semidiameter [h] C angulis [i] CAHL *HK* [j] H 7^a^
[k] CA fuerit [l] CA semidiametras

On the other side of the aux A, take an arc equal to half the arc AG,[1] which is to be the arc DA, and join H to D.[2] Let HI be taken on HA, and equal to HC. At the point I, using Euclid I. 23, an exterior angle is to be constructed, double the interior angle IHD_2. By producing line IL, the angle it makes, AIL, will be equal to the angle AHG, and to the angle ACB, where AHD_2 is half of each; and KI is that part of the line IL intercepted by HD_2. The line CKM being produced, K and G are to be joined.

I maintain, then, that if the centre of the epicycle is distant from the aux of the equant by the arc AG,[3] K will be the centre of the deferent and M its aux. And the ratio of the semidiameter of the deferent to the distance of the centre of the equant from the centre of the Earth.

Proof. The angle AIL is equal to the angle AHG. If, therefore, the centre of the epicycle is at a point of the line GH of indefinite length, the centre of the deferent will be at a point of the line IL. This consequence is a standard result of planetary theory. But the centre of the epicycle is taken to be at the point G, and thus it is necessary that the centre of the deferent be at the point K, since the centre of the deferent must be distant from I by none other than the length HI, to which IK is equal, by Euclid I. 6 and I. 32.

Next, since the centre of the epicycle never leaves the circumference of the deferent, the line KG will be equal to the semidiameter of the deferent with respect to the line IH or HC. And since the line going from the centre of the Earth through the centre of the deferent terminates at its aux, it will be at the position of the aux of the deferent. This is clearly what was proposed.

The eccentric deferents of the seven planets will be divided by means of these propositions. For, as is clear in the *Abbreviated Almagest* III. 8 and IV. 9, by Euclid VI. 12, if the centres of any [two] circles on the same straight line are in continued proportion, the semidiameters being drawn parallel,[4] any line through the ends of the radii and meeting the first line will necessarily have intercepts cut off it by the radii in the ratio of the radii themselves.[5]

[1] That is, which is one half of the angle ACG.

[2] This is the second point to be labelled D in the figure. The two are henceforward distinguished in the translation as D_1 and D_2.

[3] That is, more strictly, AD_1.

[4] *equidistanter.*

[5] This is a very free translation of an excessively terse and obscure passage.

CONCLUSIO 9ª

de sinu et arcu eius

Descripto circulo supra semidiametrum alterius circuli a cuius centro due linee exeuntes utriusque circuli arcum includant erit arcus minoris circuli ad arcum maioris circuli proporcionaliter duplus et per applicacionem equalis, et corda duplata minoris ad cordam mediatam maioris equalis.

Hec conclusio hic recitatur quia utilis est ad arcus altitudinis et equaciones epiciclorum sumendas, et est eius declaracio satis plana. Fiat igitur circulus *ABD* supra centrum *C*, cuius semidiametri orthogonales sint[a] *AC*, *DC*. Factoque circulo *AEC* supra semidiametrum *AC*, accipiatur maioris circumferencie arcus notus *AB*, ut puta 60 parcium; et signato centro minoris circuli *G*, producatur linea *CEB*, signato *E* in circumferencia minoris circuli; et iterum coniungamus *AE* et *GE*.

Cum enim angulus *ACB* sit minor angulo recto necesse est *CB* secare circulum *AEC*, per 26am tercii Euclidis. Propositum est ergo probare quod arcus *AE* respectu parcium sui circuli est duplus ad arcum *AB* respectu parcium sui circuli; et tum si *AE* applicaretur ad *AB* non erit eorum alternus[b] excellens. Similiter eciam arcus *CE* applicatus ad arcum *BD*, equantur in quantitate, et in proporcione se excedunt secundum denominacionem dupli et subdupli. Similiter corda *EC* duplata respectu arcus *CE* est equalis corde mediate *BD*; sic eciam corda duplata arcus *AE* est equalis corde mediate *AB*.

Prima pars patet ex 5ª et 23ª primi libri Euclidis, et ex 25ª tercii, et ultima sexti libri. Quia angulus *AGE* est duplus ad angulum *ACB*, ergo arcus arcui.

Secunda pars patet per illam absolutam proposicionem Archimenidis que est: *Proporcio diametri circuli ad diametrum, illa est proporcio circumferencie ad circumferenciam.* Sed medietas diametri *AB* circuli est equalis toti diametro *AC* circuli, ergo medietas circumferencie *AB* erit equalis toti circumferencie alterius; ergo sexta pars maioris, que iam ponitur *AB*, erit equalis tercie parti alterius, scilicet *AE*, et residuum *BD* equale *CE*.

Tercia pars patet quia, cum angulus *AEC* in circumferencia sit rectus, per 20am tercii Euclidis, ergo *AE* est sinus rectus arcus *AB* per difinicionem sinus recti, et idem est sinus duplatus arcus *AE* ut patet. Item *BE*[c] est sinus versus arcus *AB*, ergo residuum semidiametri *BC*, quod est *EC*, est equale corde recte residui de quarta circuli, scilicet

[a] A sunt [b] A alterilis; C (?); H excessus [c] A *BC*

PROPOSITION 9

[I. 9]

On the sinus and its arc

A circle being drawn on the semidiameter of another circle, from whose centre two lines going out include an arc of each circle, the arc of the smaller circle will [in angular measure] be to the arc of the greater circle in the proportion of two to one; and by application [it will in length be] equal; and the whole chord[1] *of the smaller will be equal to the half chord*[2] *of the greater circle.*

This proposition is enunciated here because it is useful for finding the arcs of altitude and the equations of the epicycle,[3] whereas its exposition is straightforward enough. Let there, then, be a circle *ABD*, with centre *C*, orthogonal diameters of which are *AC* and *DC*. The circle *AEC* being drawn on semidiameter *AC*, of 60 parts, take a known arc of the greater circumference *AB*. The centre of the smaller circle being denoted by *G*, let the line *CEB* be produced, *E* being on the circumference of the smaller circle. Next let us join *AE* and *GE*.

Since angle *ACB* is less than a right angle, *CB* will of necessity intersect the circle *AEC*, by Euclid III. 26. It is therefore proposed to prove that the arc *AE*, in terms of the unit degrees of its own circle, is double the arc *AB*, with respect to the unit degrees of its own circle; and that then if *AE* be applied to *AB*, the latter will equal[4] [the former]. Similarly, if the arc *CE* be applied to the arc *BD*, they will be equal in quantity, and in terms of angle be in the ratio of two to one. Likewise, the duplate chord *EC* on the arc *CE* is equal to the mediate chord *BD*; and thus also the duplate chord of the arc *AE* is equal to the mediate chord *AB*.

The first result is evident from Euclid I. 5, and I. 23, and III. 25, and from the last [proposition] of Euclid VI. Since the angle *AGE* is double the angle *ACB*, therefore the arc [is double] the arc.

The second result is evident from that unqualified proposition of Archimedes, namely, *The ratio of the diameters of two circles is as the ratio of their circumferences.* But a half of the diameter of the circle *AB* is equal to the whole diameter of the circle *AC*, and therefore one half of the circumference of *AB* will be equal to the whole of the circumference of the other [circle]; and therefore one sixth part of the greater, which is now taken as *AB*, will be equal to a third part of the other, namely *AE*, and the remainder *BD* equal to *CE*.

The third result is evident since, the angle *AEC* at the circumference [of the lesser circle] being a right angle, by Euclid III. 20, *AE* is the sinus rectus of the arc *AB*, by the definition of sinus rectus, and [*AE*] is also double the sinus of arc *AE*,[5] as is clear. Next, *BE* is the sinus versus of arc *AB*, and therefore, by Euclid I. 33, the remainder of the semidiameter *BC*, namely *EC*, is equal to the *corda*

[1] *corda duplata.*
[2] *corda mediata.*
[3] Namely the equations of the argument.
[4] Literally 'will not surpass'.
[5] But of course with respect to the smaller circle.

arcui *DB*, per 33^{am} primi Euclidis[a] et ipsum est corda duplata arcus *CE*, quod est plane propositum.

CONCLUSIO 10^{a}

De quantitate epiciclorum apparenti

Supra centrum dati circuli epiciclum describere qui oculo existenti in circumferencia apparebit equalis epiciclo cuiuslibet planete imaginaliter viso ab oculo existente in centro orbis signorum.

Fiat circulus ut prius *AB* super centrum *G*, cuius diameter *AB*, et *B* locus oculi in circumferencia positi. Deinde per 5^{am} huius, cognita equacione argumenti maxima possibili in quavis habitudine distancie centri epicicli ab auge deferentis, accipiatur arcus illi equacioni duplus, qui sit *AE*, et protracta linea *BE* cadat super eam cathetus a puncto centri *G*, que sit linea *GDF*. Deinde, posito centro in *G*, fiat circulus secundum semidiametrum *DG* contingens necessario lineam *BE*, per 15^{am} et 17^{am} tercii Euclidis, qui sit circulus *GD*. Dico ergo ubicumque ponatur oculus *B* in circumferencia circuli *ABC*, *GD* circulus apparet tantus eidem quantus apparet epiciclus planete cuius equacio maxima est subduplum ad arcum *AE* in supposita longitudine ab auge ecentrici, oculo exeunti in centro Terre. *Probacio*. Secundum premissam, angulus *AGE* duplus est ad angulum *ABE*, ut patet per 26^{am} tercii Euclidis. Ergo angulus *ABE* est equalis angulo maxime equacionis supposite, et per consequens sub illo angulo precise videtur medietas epicicli ab oculo in centro Terre, et sub eodem angulo precise videtur medietas circuli *GD*[b] ab oculo *B*; ergo iste medietates apparent equales, ergo et eorum tota erunt secundum apparenciam equalia; et ultra ex fine eius quod tangitur in 5 huius, si *B* ponatur centrum Terre, et *BA* que transit per *G* transierit per centrum epicicli, equacio cuiuslibet arcus in longitudine a puncto *G* equalis erit equacioni argumenti eidem equalis ab auge vera epicicli.

Ex hac conclusione secuntur tria corelaria. *Primum* est, argumentum epicicli cui debetur maxima equacio in quacumque distancia ab auge ecentrici est equale composito ex illa equacione et quarta circuli. Nam angulus *GDB* est rectus et constat quod angulus extrinsecus *AGF* excedit angulum *D* rectum per angulum *B*. Et idem verum est de maxima equacione accepta respectu unius ecentrici *D* epicicli; quod argumentum, cui debetur maxima equacio, plus est quarta circuli per ipsam equacionem.

[a] CAλ [b] CAHL *GE*

recta of the remainder from a quarter-circle, namely to the arc *DB*. (This is the whole chord of arc *CE*.) This is the result proposed.

PROPOSITION 10

[I. 10]

To describe an epicycle, about the centre of a given circle, which will appear to an eye at the circumference equal to the epicycle of any planet [as it would be] seen by an eye imagined to be at the centre of the sphere of signs. On the apparent size of the epicycles

As before, let there be a circle *AB*, with centre *G* and diameter *AB*, *B* being the position of the eye at the circumference. Then, by proposition I. 5, the greatest possible equation of the argument being known for any value[1] of the distance of the centre of the epicycle from the aux of the deferent, the double arc of that equation is to be taken, namely *AE*, with the line *BE* intersecting it at right angles. The line *GDF* is perpendicular to *BE*. Now construct a circle with centre *G* and semidiameter *DG*, which circle necessarily touches the line *BE*, by Euclid III. 15 and III. 17. This is the circle *GD*. I maintain, then, that wherever the eye *B* is placed on the circumference of the circle *ABC*, the circle *GD* appears to it to be just as great as does the epicycle of a planet whose greatest equation is one half the arc *AE* at some particular elongation from the aux of the eccentric, to an eye looking out from the centre of the Earth.

Proof. According to the premiss, the angle *AGE* is double the angle *ABE*, by Euclid III. 26. The angle *ABE* is therefore equal to the greatest value of the supposed equation. In consequence, one half of the epicycle seems to subtend to an eye at the centre of the Earth the same angle as one half of the circle *GD* is seen to subtend to the eye at *B*. These half-circles appear equal, and thus the whole circles are likewise equal in appearance. And besides, from what was touched on at the end of I. 5, if *B* be placed at the centre of the Earth, and *BA*, which passes through *G*, pass through the centre of the epicycle, the equation of any arc measured about the point *G* will be equal to the equation of the argument corresponding with the same elongation from the true aux of the epicycle.

Three corollaries follow from this conclusion. *The first* is that the argument of the epicycle giving the greatest equation at any distance from the aux of the eccentric is equal to a right-angle plus the equation itself. For *GDB* is a right-angle, and it is agreed that the exterior angle *AGF* is greater than the right-angle *D* by the angle *B*. The same thing is true of the greatest equation taken with respect to a point *D* of an epicycle on an eccentric. This argument, providing the greatest equation, exceeds a quarter-circle by that same equation.

[1] *habitudo*

Secundum corelarium. Quanto centrum epicicli per eius motum in ecentrico deferente plus appropinquat centro Terre tanto plus apparet planeta removeri a centro epicicli in loco maxime equacionis. Quia quanto *G* ponatur propinquior oculo *B* tanto punctus contingencie *D* plus distat a puncto *A*[a] auge epicicli, alioquin non erit *D* angulus rectus, ergo tanto maior apparebit angulus *ABC*, ergo epiciclus maior; et per consequens maior distancia planete a centro sui epicicli a puncto contingencie.

Tercium corelarium. Oculo exeunti in centro Terre apparet planetam deferri in ecentrico circulo circa centrum epicicli, quia cum planeta per motum in epiciclo equis temporibus equos describat angulos super centrum epicicli, et quamque plus et quamque minus removeri videtur a centro epicicli, istam apparenciam ecentrica deferente[b] oportet salvari. Non quod in celestibus sint ad litteram huius ecentrici aut eciam epicicli, sicut imaginatio mathematica sibi fingit, quod nullus disciplinatus potuit verisimiliter vel putare, sed quia sine huius imaginacionibus mathematicis, vel saltem intellectui plus obscuris, de stellarum motibus regulatis ars certa tradi non potest, que[c] eorum loca ad quodvis momentum certificet, quod a nostris aspectibus non discordat.

CONCLUSIO II[a]

de ecentricis *Ecentricum deferentem planetam circa centrum epicicli describere.*

Super centrum *D* circuli *AB* fiat epiciclus maximus apparens centro epicicli exeunte in auge sui deferentis, per premissam, qui[d] sit *CE*. Item fiat super idem centrum epiciclus maximus apparens eius centro exeunte in opposito eiusdem augis, et sit *FG*. Deinde super communem diametrum, invento centro, fiat circulus contingens utrumque circulum *CE* et *FG* cuius tota diameter erit *EG*[e] qui sit circulus *EHG*. Dico ergo quod circulus *EHG* est ecentricus deferens stellam in epiciclo quesitus, quod sic patet:

[a] H λ [b] CAH deferentem [c] C +sit; H +sint [d] A que
[e] CA *EG*

Second corollary. In so far as the centre of the epicycle through its motion along the eccentric approaches the centre of the Earth, so the planet appears to increase its distance from the centre of the epicycle at the place of greatest equation. For by whatever amount *G* is placed nearer to the eye *B*, to a corresponding extent the point of contact *D* increases its separation from the point *A* of the aux of the epicycle (otherwise *D* will not be a right-angle), and correspondingly does the angle *ABC* appear greater, and therefore also the epicycle. And, as a result, the [angular] distance of the planet from the centre of its epicycle [when at] the point of contact will also be greater.

Third corollary. To an eye looking out from the centre of the Earth, a planet appears to be carried about the centre of an epicycle on an eccentric circle. Since the planet by its motion in the epicycle describes equal angles in equal times about the centre of the epicycle, and seems to be now further and now nearer removed from the centre of the epicycle, that appearance must be saved by means of an eccentric deferent. Not that there are things among the heavenly bodies literally corresponding to this eccentric or epicycle, as the mathematical fancy contrives for its own purposes, a thing which no educated man could even conceive to have a semblance of truth. It is rather that without mathematical conceptions of the state of things, or at least without unfamiliar intellectual ideas, there is no certain way of dealing with the regulated motions of the stars which assures us of their places at any moment, and which provides places differing in nowise from what we see.

PROPOSITION II

[I. 11]

To describe the eccentric carrying the planet about the centre of the epicycle.[1] On eccentrics

On a circle *AB*, with centre *D*, let *CE* be the epicycle[2] which, as explained in the previous proposition, appears greatest when the centre is at the aux of its deferent.[3] Next, draw an epicycle on the same centre, appearing greatest when at the opposite of its aux, and this epicycle is to be *FG*. Then, on a common diameter, the centre having been found, let there be a circle touching each of the circles *CE* and *FG*, the whole diameter of which will be *EG*; and this is to be the circle *EHG*. I say then that the circle *EHG* is the required eccentric, carrying the planet[4] on the epicycle. This is shown as follows.

[1] There is no short and simple way of designating the eccentric except perhaps by the word 'equator' used elsewhere. It is not the ordinary Ptolemaic deferent. See our commentary.

[2] This does not function as an epicycle in the construction which follows, but is simply a circle which from some point *as yet unspecified* is proportional to an apparent size of the epicycle.

[3] Some arbitrary scale must be understood.

[4] *stella*.

Circulus *EHG* est acceptus ex diametri medio circuli *GF* et circuli *CE*; ergo ipse est communis omnibus revolucionibus possibilibus super centrum *D*, inter duos circulos *EC* et *FG*; ergo[a] semper apparet stella in aliquo puncto circuli *GHE* cum centrum eius omnes revoluciones inter *CE* et *FG* super *D* centrum descriptas in solo puncto contingat, alioquin non foret motus epicicli in suo deferente continuus ab auge in eius oppositum. Patet per secundum et tercium corelarium premisse.

Equaciones autem maxime epiciclorum, in auge et in[b] eius opposito deferentis, sunt iste:

Saturni in auge est 5 grad. et 53 min. In opposito eius augis 6 grad. et 37 min.

Iovis in auge 10 grad. 35 min. In opposito 11 grad. 35 min.

Martis in auge 36 grad. 47 min. In opposito 46 grad. et 59 min.

Veneris in auge 44 grad. 49 min. In opposito 47 grad. et 16 min.

Lune in auge 5 grad. 59 sec. In opposito 7 grad. 39[c] min. et 32 sec.

Mercurii in auge 19 grad. 1 min. In opposito 23 grad. et 15 min.

Sed quia Mercurius non habet augem deferentis fixam, non crescunt equaciones eius epicicli per distanciam ab auge deferentis neque per distanciam ab auge equantis simpliciter, sed in respectu maioris approximacionis centri epicicli ad centrum Terre, quod est cum fuerit centrum epicicli in directo 4 aut 8 signorum equantis, quod convincitur ex 15ª primi Euclidis, supposito quod centrum deferentis moveatur uniformiter contra signa in parvo circulo transeunte per centrum equantis, cuius semidiameter est equalis distancie centri equantis a centro Terre, et equivelociter sicut centrum epicicli movetur cum successione signorum super centrum equantis. Sed hec habent alibi declarari.

Equacio autem maxima in hac materia approximacionis[d] centri epicicli ad centrum Terre est 23 grad. et 53 min. Equacio vero ad illam longitudinem ad quam tabule Arzachelis calculantur, scilicet ad 2 signa et 4 grad., est 22[e] grad. et 2 min.

Ex his deducitur unum corelarium, quod deferens planetam circa centrum epicicli habet motum alium propter motum planete in epiciclo. Quia cum omnibus planetis preterquam in Mercurio, quando centrum epicicli est in auge sui deferentis, tunc est centrum

[a] A 18λ (haplog.) [b] CA λ [c] A 59
[d] A approximacione; H approximacionem [e] CAHL 23

The circle *EHG* is determined by one half the sum of the diameters of the circles *GF* and *CE*; and thus this is the same, no matter how many revolutions there are about the centre *D*, between the two circles *EC* and *FG*. The planet therefore always appears at a point of the circle *GHE*, since for any angle turned between *CE* and *FG* about centre *D* there is a unique position for the planet's centre—otherwise the motion of the epicycle on its deferent would not be continuous from the aux to its opposite. This is evident from the second and third corollaries above.

The greatest equations of the epicycles, at the aux and at the opposite of aux on the deferent, are these:

	At aux	At opposite of aux
Saturn	5° 53′ 0″	6° 37′ 0″
Jupiter	10° 35′ 0″	11° 35′ 0″
Mars	36° 47′ 0″	46° 59′ 0″
Venus	44° 49′ 0″	47° 16′ 0″
Moon	5° 0′ 59″	7° 39′ 32″
Mercury	19° 1′ 0″	23° 15′ 0″

Since, however, Mercury has no fixed aux of the deferent, the equations of its epicycle neither increase in accordance with the distance from the aux of the deferent nor in a straightforward manner with the distance from the aux of the equant. They increase the greater the approach of the centre of the epicycle to the centre of the Earth. This approach is closest when the centre of the epicycle is in the direction of the fourth or eighth sign of the equant, which is proved from Euclid I. 15, it being supposed that the centre of the deferent moves in a retrograde sense[1] uniformly along a small circle passing through the centre of the equant. The semidiameter of this small circle is equal to the distance of the centre of the equant from the centre of the Earth. With constant velocity the centre of the deferent, like the centre of the epicycle, moves in the direction of the signs around the centre of the equant. But these matters are explained elsewhere.[2]

At the nearest approach of the centre of the epicycle to the centre of the Earth, the greatest equation [for Mercury] is 23° 53′. The equation for the elongation[3] for which the tables of Arzachel are calculated, namely 2^s 4°, is 22° 2′.[4]

From these results a corollary is deduced, namely, that whatever carries the planet around the centre on the epicycle has another motion on account of the motion of the planet on the epicycle. For, with all the planets apart from Mercury, when the centre of the epicycle is at the aux of its deferent, then the centre of the planet is

[1] *contra signa.* [2] See, for instance, I. 8 above. [3] *longitudo.*

[4] All MSS. have 23° 2′, but 22° 2′ is correct, and the author must have used the correct figure in his calculations. (See, for example, I. 11 comm. (end).)

planete in opposito augis deferentis epicicli. Et[a] cum fuerit centrum epicicli in opposito augis sui deferentis, tunc est centrum planete in auge sui deferentis. Ergo necesse est motum centri epicicli in suo deferente et motum planete in deferente epicicli esse equales. Sed nullius planete motus centri epicicli (qui dicitur medius eius motus) est equalis motui eius circa centrum epicicli (qui dicitur motus medius argumenti); ergo necesse est motum planete in deferente epicicli componi ex duobus motibus, ex motu medii argumenti, et ex motu deferentis contrario motui argumenti quando medius motus est velocior motu argumenti vel motu deferentis stellam in epiciclo, consequente motu argumenti ubi motus medius planete est tardior motu eiusdem circa centrum epicicli.

Secus[b] est de motu Mercurii in suo deferente circa centrum epicicli, quia motus ille non est equalis medio motui centri epicicli circa eius deferentem in comparacione ad augem equantis, que fixa est; sed habet reflexionem, sicut ostenditur in theorica planetarum. Nam cum fuerit centrum epicicli ad 4 signa respectu motus equantis ab auge, erit planeta in longitudine maxima a centro epicicli, et ita in opposito augis deferentis stellam in epiciclo; et iterum cum fuerit centrum epicicli in opposito augis equantis et deferentis, reflectitur planeta in deferente[c] suo proprio ad punctum deferentis equidistantem a centro epicicli, quantum distabat punctus in quo fuerat planeta quando centrum epicicli distabat ab auge equantis per 2 sig. et 27 grad. Et iterum reflectitur ad augis oppositum sui deferentis, et deinde redit ad augem; sed hec actenus quia plane patent *Almagesti* dictione 9ª[d] capitulo 7° et 8° et ultra.

Incipit de eclipsibus

Dicit commentator *Almagesti* libro 3°: In principio communia quedam premittenda sunt, quia hic modus demonstracioni est aptior. Neque hic necesse foret demonstraciones premittere nisi ut

[a] A 10λ (haplog.) [b] A Seccis [c] A +in [d] CAHL 11ª

at the opposite of the aux of the deferent on the epicycle. And when the centre of the epicycle is at the opposite of the aux of its deferent, then the centre of the planet is at the aux of its deferent. The motion of the centre of the epicycle in its deferent, and the motion of the planet on the deferent of the epicycle, are necessarily equal.[1] But for no planet is the motion of the centre of the epicycle (its 'mean motus') equal to its motus about the centre of the epicycle (known as the 'mean motus of the argument'); and thus the motion of the planet on the deferent of the epicycle is necessarily composed of two motions —the motus of the mean argument, and the motus of the deferent in a sense opposed to the motus of the argument. The latter is in the opposite direction to the motus of the argument when the mean motus is faster than the motus of the argument, or the motus with which the planet is carried[2] on the epicycle, and in the same sense as the motus of the argument, where the mean motus of the planet is slower than its motus about the centre of the epicycle.

It is quite different as regards the motion of Mercury around the centre of the epicycle on its deferent, for that motion is not equal to the mean motus of the centre of the epicycle about its deferent with reference to the aux of the equant, which is fixed; but [the centre of the deferent] has a contrary movement,[3] as is shown in planetary theory. For when the centre of the epicycle reaches the fourth sign[4] in motion reckoning from the aux of the equant, the planet will be at its greatest distance from the centre of the epicycle,[5] and so on the epicycle at the opposite of the aux of the deferent of the planet. Again, when the centre of the epicycle reaches the opposite of the aux of the equant and of the deferent, the planet[6] returns on its own deferent to a point of the deferent at a distance from the centre of the epicycle[7] equal to the distance it has when the the centre of the epicycle is $2^s\,27^\circ$ from the aux of the equant. It is there returned to the opposite of the aux of its deferent, whereafter it returns to the aux. But these things are all clearly set down in *Almagest* IX. 7, 8, and following.[8]

Here begins the section on eclipses

The commentator on book III of the *Almagest* says: 'At the outset, certain common assumptions are to be allowed, because the method [of proceeding from such] is more suited to demonstration.' And it should not here be necessary to preface demonstrations, except that by doing so a more subtle way of proceeding may be made available

[1] That is, granted that the two motions around the epicycle and deferent are equal.

[2] *motu deferentis stellam*—'the motion of what carries the planet' is the author's standard construction, as seen elsewhere.

[3] Literally 'a turning back'—*reflexio*.

[4] That is, when $\bar{\gamma} = 120^\circ$ (see comm.).

[5] Presumably this is an error for 'Earth'. Cf. note 7.

[6] That is, the centre of its epicycle.

[7] Again this appears to be a mistake for 'Earth'. See comm.

[8] The sections actually named (XI. 7–8) deal with Saturn. All MSS. have the figure eleven in arabic numerals, but of course the mistake is an easy one with roman numerals.

studiosis quibus radices hic posite micius placent aperiatur via subtilius procedendi, et pocius[a] se opponerent calumpniis detractorum.

CONCLUSIO 12ª

de angulo orizontis et orbis signorum

Si sinus altitudinis gradus medii celi ducatur in semidiametrum, et collectum dividatur per sinum arcus qui est inter ascendens et gradum medii celi, exibit sinus altitudinis gradus nonagesimi ab ascendente.

Hoc satis patet conclusione 12ª libro 2° Gebir, et primo *Almagesti* de figura sectore, commento 30°.

Ille ergo arcus tollatur de 90 et remanebit arcus qui est inter cenith et gradum 90^mu ab ascendente.

CONCLUSIO 13ª

de angulo proveniente ex circulo signorum et circulo altitudinis

Si sinus distancie cenith capitis a gradu nonagesimo ab ascendente multiplicetur in semidiametrum, et productum dividatur per [*sinum*] *distancie*[b] *gradus dati a cenith, exibit sinus arcus anguli constituti super idem punctum ex ecliptica et circulo altitudinis.*

Hic patet per Gebir, in fine secundi et per commentatorem secundo *Almagesti*, commento 36°; et loquimur hic de angulo intrinseco respiciente lineam meridianam, non de angulo orientali extrinseco qui cum eodem facit duos rectos.

Sive autem angulus iste signetur in orbe declini super centrum Lune, sive in ecliptica super verum locum, infra terminos eclipsis solaris non erit differencia sensibilis usquequaque, ut patet 5° *Almagesti* capitulo 19°, et commento 11°, et Gebir, et Albategni de eodem, capitulo 30°.

[a] AC pacius; H paracius [b] CA distanciam

to students who are only moderately pleased with the parameters here set down, and that [the demonstrations] may more effectively forestall the misrepresentations of those who would disparage.

PROPOSITION 12

[I. 12]

If the sinus of the altitude of the mid-heaven[1] *be multiplied by the semidiameter and the result divided by the sinus of the arc between the ascendent and mid-heaven, the final result will be the sinus of the altitude of a point 90°*[2] *from the ascendent.*

On the angle of the horizon and the circle of signs

This is clear enough from Gebir's [*Flores Almagesti*] II. 12, and from *Almagest* I, on the sector-figure, commentary 30.

Let that arc, [namely the altitude which finally results], be subtracted from 90°, and there will remain the arc which lies between the zenith and a point 90° from the ascendent.

PROPOSITION 13

[I. 13]

If the sinus of the distance from the zenith overhead of the point 90° from the ascendent be multiplied by the semidiameter, and the product be divided by the [sinus of] the zenith distance of a given point [on the ecliptic],[3] *the result will be the sinus of the arc of the angle formed at the same point by the ecliptic and the circle of altitude.*

On the angle formed by the circle of signs and the circle of altitude

This is evident from Gebir, at the end of the second book, and from the commentator on the second book of *Almagest*, commentary 36. We are speaking here of the interior angle on the side of the meridian line, not of the exterior angle to the east, which makes with it two right angles.

The angle in question can be taken as that at the centre of the Moon formed by its inclined orbit [and the circle of altitude], or [that] at the true place of the Moon on the ecliptic. Within the solar eclipse limits there is no appreciable difference, as is clear from *Almagest* V. 19, and commentary 11, and from Gebir, and Albategni on the same, chapter 30.

[1] The upper point where the ecliptic crossed the meridian. The phrase was encountered in *Exafrenon*.

[2] Literally 'of the ninetieth degree'.

[3] This is perhaps implied by the word *gradus*.

CONCLUSIO 14ª

de diversitate aspectus maxima que contingit apud orizontem

Maxima diversitas aspectus in circulo altitudinis apud orizontem cum fuerit Luna in auge ecentrici et epicicli est 54 minuta; cum autem fuerit Luna in auge ecentrici et opposito augis epicicli, 64 minuta.

Hic recitatur ab Alfragano in *Floribus Almagesti*, capitulo 27°, et potest probari per distanciam Lune a Terra ad predicta loca investigatam[a] per conclusionem 5ᵃᵐ huius primi, et per modum arguendi Albategni, capitulo 30°.

Distancia quidem[b] Lune a Terra cum fuerit in utraque auge secundum posicionem Ptolomei, libro 5° *Almagesti*, capitulo 17°, est 64 par. 10 min. de partibus quarum semidiameter Terre est pars una. Cum vero fuerit Luna in auge ecentrici, et in opposito augis epicicli, est distancia Lune a Terra de eisdem partibus 53 par. et 50 min. De ceteris locis hic taceo, quia in his solis duobus locis possibiles[c] sunt eclipses.

CONCLUSIO 15ª

Proporcio diversitatis aspectus in circulo altitudinis ad distanciam stelle a cenith capitis datam ad maximam diversitatem aspectus apud orizontem, imitatur proporcionem corde elongacionis stelle a cenith date ad cordam totam, absque differencia perceptibili in fracturis.

Hic satis patet ex modo operandi Albategni, et multum appropinquat veritati in comparacione eius quod ponit commentator in *Almagesti* libro 5°, capitulo 5°, et Ptolomeus, opere tabulari.

CONCLUSIO 16ª

Cum in superficie spere signatum fuerit quadrilaterum orthogonium, cuius duo latera sese et prima duo latera ad quantitatem unius partis de 360 partibus circulorum magnorum vel circiter orthogonaliter secaverint, si depressa fuerit spera super angulum rectum a duobus circulis magnis contentum, ipsa figura quadrilatera a quadrato rectilineo insensibiliter discrepabit.

[a] A investigandam [b] CA quidam [c] C possibilem; C′ possibiles

PROPOSITION 14

[I. 14]

The greatest parallax in the circle of altitude at the horizon,[1] *when the Moon is at the aux of the eccentric and of the epicycle, is 54′. It is, however, 64′ when the Moon is at the aux of the eccentric and the opposite of the aux of the epicycle.*

On the greatest parallax occurring at the horizon

This is stated by Alfraganus, in *Flowers of the Almagest*, chapter 27, and may be proved from the distance from the Earth of the Moon when it is in the aforesaid positions, [the distance] being investigated by proposition 5 of the first book of this work, and by the method of argument of Albategni, chapter 30.

The distance, indeed, of the Moon from the Earth when it is at either aux, according to Ptolemy, *Almagest* V. 17, is $64^p\ 10^m$ (where 1^p is the semidiameter of the Earth). When the Moon is at the aux of the eccentric, and at the opposite of the aux of the epicycle, the distance of the Moon from the Earth in the same units is $53^p\ 50'$. I shall say nothing here of the other positions [of the Moon], since only [when it is] in[2] these two positions are eclipses possible.

PROPOSITION 15

[I. 15]

The ratio of the parallax, in the circle of altitude, of a star at a given zenith distance, to the maximum horizontal parallax[3] *is approximately equal to*[4] *the ratio of the chord*[5] *of the zenith distance of the star to the whole chord, without any appreciable error.*

This is quite clear from Albategni's working, and is in close agreement with the truth as revealed in the remarks made by the commentator on *Almagest* V. 5, and with Ptolemy's [parallactic] table.[6]

PROPOSITION 16

[I. 16]

When a right-angled quadrilateral is drawn on the surface of a sphere, two sides being [parts of] great circles, and the remaining two sides cutting themselves and the first two sides orthogonally, [so that the intercepts are] of order of magnitude one degree,[7] *if the sphere be depressed at the right-angle contained by the two great circles, the actual quadrilateral will differ insensibly from a four-sided rectilinear figure.*

[1] That is, in modern parlance, the horizontal parallax in altitude.

[2] This is expressed too loosely, for eclipses may take place when the Moon is in the neighbourhood of these positions, a fact which is acknowledged in the discussion of eclipse limits and conjunction (III. 15 to III. 20 below).

[3] Literally, 'at the greatest diversity of aspect at the horizon'.

[4] *imitatur.* [5] See our commentary. [6] The table occupies *Almagest*, V. 18.

[7] Literally, 'one part out of 360 parts of the great circles, or thereabouts'.

Hec patet in spera modica, quanto plus in spera maiori; et est hec communis supposicio Ptolomei, Gebir, et commentatoris, ut patet in libro 5°, capitulo 19°, et commento 8°.

CONCLUSIO 17ª

de figura diversitatis aspectum

Cum signate fuerint in spera diversitates aspectus in longitudine et latitudine et altitudine, erunt diversitates signate in longitudine et latitudine tanquam latera quadranguli orthogonalis, et diversitas aspectus in altitudine tanquam diagonalis eiusdem.

Hec satis patet libro 5° *Almagesti* capitulo 19°, et per commentatorem ibidem.

CONCLUSIO 18ª

de quantitate diametri Lune

Diameter Lune cum fuerit in auge ecentrici et epicicli cordat arcum circuli magni 29 minuta et 20[a] *secunda. Cum vero fuerit in opposito augis epicicli, cordat arcum 35 minuta et 20 secunda.*

Hec patet per Albategni, Gebir, et commentatores[b] libro 5° capitulo 4°, qui hic corrigunt Ptolomeum libro 5° capitulo 15°.

CONCLUSIO 19ª

de semidiametro umbre

Semidiameter umbre in loco transitus Lune eclipsate continet semidiametrum Lune bis et eius tres quintas.

Hic proporcio patet per Albategni, capitulo 30°, et per commentatorem *Almagesti,* libro 6°, commento 4°.

Cum igitur fuerit in utraque auge, erit semidiameter umbre 38 min. et 10[c] sec., cui cum addiderimus 14 min. et 40 sec. semidiametri Lune erit maxima eclipsis in illo statu possibilis 52 min. et 50 sec.

Cum autem fuerit Luna in opposito augis epicicli, erit semidiameter umbre 45 min. et 56 sec., et aggregatum ex illo et semidiametro Lune erit 63 min. et 36 sec.; a qua summa si extraxerimus totam diametrum Lune in illo situ remanebit dimidium more, quod est 28[a] min. 16 sec. In auge vero epicicli erit dimidium more 23 min. et 30 sec. Hec Albategni, et commentatoris libro 6°, commento 4°:

[a] A 29 [b] H commentatorem [c] A 8 [d] H 38

This is clear on a sphere of moderate size, and even more so on a larger sphere. It is a common supposition of Ptolemy, Gebir, and the commentator, as is evident from book V, chapter 19, and commentary 8.

PROPOSITION 17

[I. 17]

When the parallaxes have been marked on a sphere, in longitude, latitude, and altitude, the differences in longitude and latitude will be marked like the sides of an orthogonal quadrangle, while the parallax in altitude will be as the diagonal of the same. On the figure [illustrating] parallax

This is quite clear from *Almagest* V. 19, and from the commentator on the same.

PROPOSITION 18

[I. 18]

When the Moon is at the aux of the eccentric and epicycle, its diameter spans a chord of arc 29′ 20″ of a great circle. When, however, it is at the opposite of the aux of the epicycle, it spans a chord of arc 35′ 20″. On the size of the Moon's diameter

This is made clear by Albategni, Gebir, and the commentator, book V, chapter 4, who here correct Ptolemy, [*Almagest*] V. 15.

PROPOSITION 19

[I. 19]

The semidiameter of the shadow, at the place where it is crossed by the eclipsed Moon, contains $2\frac{3}{5}$ semidiameters of the Moon. On the semidiameter of the shadow

Here the ratio is evident from Albategni, chapter 30, and the commentator on the *Almagest*, book VI, commentary 4.

When, therefore, [the Moon] is at each aux,[1] the semidiameter of the shadow will be 38′ 10″. When we have added 14′ 40″ to this for the semidiameter of the Moon, the greatest eclipse possible at that position will be 52′ 50″.

When, however, the Moon is at the opposite of the aux of the epicycle, the semidiameter of the shadow will be 45′ 56″, and the sum of this and the semidiameter of the Moon will be 63′ 36″. If we subtract from this sum the total diameter of the Moon when it is in that position, the half-measure[2] remains, namely 28′ 16″. At the aux of the epicycle, indeed, the half-measure will be 23′ 30″. These data are from Albategni, and the commentator [on *Almagest*], book VI, commentary 4.

[1] That is, of epicycle and deferent.

[2] *dimidium more*. The qualification 'for a central eclipse' should have been added here and in the next sentence, if the *dimidium more* is here being used in the usual technical sense.

Et nota quid dicit hic commentator, quod omnes indifferenter utuntur hic[a] lineis rectis pro arcubus, propter hoc quod insensibilis est eorum differencia in tam modica quantitate. Ponit igitur semidiametrum umbre cum semidiametro Lune adequari declinacioni circuli declivi ab ecliptica in termino eclipsis lunaris a nodo.

CONCLUSIO 20ª

Vestigium transitus Lune per umbram apparet linea recta equidistans ab ecliptica.

Hec patet commento 4° libro 6° *Almagesti*; et motus Lune in ipso circulo declivi insensibiliter differt ab eius motu in ecliptica in tam modica quantitate, quia in quarta circuli differre non possunt ultra 5 min. Hec commentator libro 6°, commento 12°.

CONCLUSIO 21ª

de diametro Solis

Diameter Solis cordat arcum 32 minuta et 30 secunda in longitudine sua media.

Hoc patet libro 5°, commento 4°, et capitulo 15°, ubi[b] corrigitur Ptolomeus per Albategni et Gebir. Nec causatur sensibilis differencia ex diametro Solis exeuntis[c] in auge vel in eius opposito, in eclipsi solari vel lunari, propter modicam ecentricitatem orbis sui, ut dicit hic commentator.

CONCLUSIO 22ª

de quantitate Solis et Lune demonstranda

Si de quadrato linee aggregate ex semidiametro umbre et Lune in eclipsi lunari, vel semidiametro Solis et Lune in eclipsi solari, extraxerimus quadratum visibilis latitudinis Lune ab[d] ecliptica, radix residui erit medietas eclipsis eius. Et si de medietate eclipsis extraxerimus totam diametrum Lune, et si de medietate umbre

[a] H hiis [b] C vel [c] A existentis [d] CA ad

And note something that the commentator says here, that is, that everyone is indifferent here about the use of straight lines for arcs, since the difference between them is so slight and imperceptible. He therefore takes the sum of the semidiameter of the shadow and the semidiameter of the Moon as equal to the declination[1] of a point on the Moon's path[2] at the [distance of the] lunar eclipse limit from the node.

PROPOSITION 20

[I. 20]

The track of the passage of the Moon through the shadow appears as a straight line parallel to the ecliptic.

This is clear from commentary 4 to *Almagest* VI. The motion of the Moon in its path[2] differs imperceptibly from its motion at the ecliptic, and the motions differ to so small an extent since in one quadrant there can be a discrepancy of no more than 5′.[3] And this according to the commentator, book VI, commentary 12.

PROPOSITION 21

[I. 21]

The diameter of the Sun at its mean distance covers the chord of an arc of 32′ 30″.

On the diameter of the Sun

This is evident from book V, commentary 4, chapter 15, where Ptolemy is corrected by Albategni and Gebir. No perceptible difference is caused in the passage of the Sun from its aux to the opposite of its aux, whether in a solar or lunar eclipse. This is on account of the small eccentricity of its path,[4] as the commentator says.

PROPOSITION 22

[I. 22]

On proving the extent of an eclipse of the Sun and Moon

If from the square[5] of the sum of the semidiameters of the Moon and the shadow in a lunar eclipse, or the semidiameters of Sun and Moon in a solar eclipse, we subtract the square of the visible latitude of the Moon from the ecliptic, the [square] root of the remainder will be the half-eclipse.[6] (If from the half-eclipse we subtract the whole diameter of the Moon),[7] and if from the half-shadow we

[1] That is, ecliptic latitude.
[2] 'the inclined circle'.
[3] The meaning is obscure. See I. 20, comm.
[4] *orbis*.
[5] Literally, 'the square of the line of . . .'.
[6] *medietas eclipsis*.
[7] The clause in parentheses should be omitted as unnecessary and as suggesting something which is in fact untrue. It might have been a false start to a sentence which the author forgot to strike out.

extraxerimus medietatem diametri Lune, et[a] *de quadrato residui extraxerimus quadratum declinacionis vere Lune, radix residui erit dimidium more.*

Hec patent 6° *Almagesti*, capitulo 8°, et commento 14°; nec opus est circa eius demonstracionem hic morari.

CONCLUSIO 23ª

de multiplicacione et divisione

Datis tribus sinibus, quorum unus est sinus totus, quartum eis proporcionalem subiungere, circa circulum cuius diameter est divisa in 60 partes equales et semicirculus eius in 90 partes equales.

Ex quo patet, absque calculacione haberi poterit tota operacio multiplicacionis et divisionis pertinens ad sinus rectos in figura sectore per 6 quantitates proporcionales, modo quo procedit Ptolomeus, vel ex 4 quantitatibus proporcionalibus, eo modo quo Gebir procedit.

Sit ergo circulus *ABC* super centrum *D*, cuius diameter *AC* sit divisa in 60 partes equales; et uterque semicirculus *ABC* et *AEC* in 90 partes equales; et capiatur arcus unus notus *CB*, producta linea *CB*. Item accipiatur alter arcus notus *FC* producta linea *CF*. Quia ergo, per 9am primi huius, linea *CB* est corda recta arcus *CB* noti, et *CF* linea est corda recta arcus *CF*, et utraque est nota respectu parcium tocius sinus *CA*, propositum est[b] invenire sinum quartum qui se habet ad sinum *CF* sicut *CA* se habet ad *CB*;[c] ad quem inveniendum producatur linea perpendicularis super punctum *G* in linea *CB* cum *CG* facta fuerit equalis linee *CF*, quod facile est dare si arcus *HB* et *AE*[d] capiantur equales, et ubi hec linea abscindit sinum *CA* nota punctum *I*. Dico ergo quod sinus *CI* est sinus quesitus, quod patet per 2am libri 6i Euclidis, cum linea *GI* sit equidistans basi *BA* in triangulo *CBA*; et hoc est propositum.

CONCLUSIO 24ª

de radicum extractione per figuram

Si fuerint duo semidiametri circuli orthogonales in 60 partes equales divise, duobus lateribus trigoni orthogonii cognitis, absque radicum extractione tercium latus cognoscere.

Ex quo patet quod absque calculacione proveniet illud quod

[a] CA vel [b] A λ [c] H *CH* [d] A *AC*

subtract one half the diameter of the Moon, and from the square of the difference we subtract the square of the declination[1] *of the true Moon, the root of the remainder will be the half-measure.*

These things are clear from *Almagest* VI. 8,[2] commentary 14, and there is no need to delay by proving them here.

PROPOSITION 23

[I. 23]

Given three sinus, one of which is the sinus of a quarter-circle,[3] *to find a fourth proportional to them about a circle whose diameter is divided into 60 equal parts, and the semicircle of which is divided into 90 equal parts.*

On multiplication and division

[Given a solution], it is clear that without calculation the entire operation of multiplication and division relating to the sinus rectus can be done—with the sector-figure by six proportional quantities, in the way Ptolemy works, or from four proportional quantities, in the way Gebir works.

Let there be, then, a circle *ABC* with centre *D*, whose diameter *AC* is divided into 60 equal parts. Each of the semicircles *ABC* and *AEC* is to be divided into 90 equal parts. One of the known arcs *CB* is to be taken, and the line *CB* produced. Next, another known arc is taken, *FC*, and the line *FC* is produced. Since, by the ninth proposition of this book, the line *CB* is the corda recta of the known arc *CB*, and *CF* is the corda recta of the arc *CF*, and each is known as a certain number of parts of the sinus *CA*, it is proposed to find a fourth sinus which bears to sinus *CF* the same [ratio] as sinus *CA* has to [sinus] *CB*. To find this, a line is drawn at right angles to the line *CB* through the point *G*, with *CG* equal to the line *CF*—and it is easy to tell when this is done properly, for the arcs *HB* and *AE* will be equal. *I* is the point at which [the perpendicular] cuts the sinus *CA*. I hold, then, that the sinus *CI*[4] is the required sinus, which is evident from Euclid VI[. 4], since the line *GI* is parallel to the base *BA* in the triangle *CBA*. And this is what was proposed.

PROPOSITION 24

[I. 24]

If there are two orthogonal semidiameters of a circle, these being divided into 60 equal parts, and two sides of a right-angled triangle are known, to discover the third side without extracting roots.

On extracting roots by means of a figure

From [this proposition] what emerges without calculation is exactly

[1] This must be understood as 'latitude', in modern terms.

[2] That chapter comprises merely a table of solar and lunar eclipses. Perhaps the *et* of the text means that VI. 7 (an explanation of the tables) was intended in addition to the commentary.

[3] *sinus totus*, a whole sinus, that is.

[4] Measured on the diametral scale of sixty parts.

provenire solet per radicum extractionem ex sinibus datis in figuris in quibus demonstrant philosophi quantitates angulorum, et distancias centrorum, et equaciones per triangulos rectangulos assignatas.

Fiat ergo circulus *ABCD*, ductis diametris *AC BD*, assignatoque uno latere trianguli rectanguli ut *OF* et altera ut *OT*. Propositum est invenire quantitatem tercii lateris *TF*.

Age:[a] A puncto *T* ducatur perpendicularis super diametrum *DB* ad punctum *P*. Cum[b] linea *OP* sit equalis linee *FT*, et *OP* ex supposito est notarum parcium, per consequens nota erit linea *FT*, quod est propositum.

Hec conclusio hic posita est gracia eclipsium Solis et Lune, sicut premissa ne in aliquo hesitent operantes.

Explicit pars prima Albionis

II

Incipit secunda pars.

ALBION ad instar planisperii tocius spere celestis concentricos, ecentricos, et eciam epiciclos, pro variis stellarum motibus, machinatos errancium seu[c] fixarum mira continet brevitate. Cuius composicio ad modum facilis est, eciam rudibus, prehabita[d] proporcionalium noticia circulorum ad tam dispares circuitus[e] planetarum. Huius igitur gracia opere precium fuerat premonstrare qualiter huiusmodi[f] circuli in apta materia poterunt inveniri; et deinde subdere consequenter quo artificio idem circuli dividi debeant et inscribi; quia in horum divisione proporcionalium circulorum verifica et subtili plurimum consistit huius bonitas instrumenti.

CAPITULUM PRIMUM

De arte componendi materiam tocius instrumenti.

Aptare materiam operi est accipere unam laminam auricalci pollitam et planam, et in ea describere circulum super *G* centrum, cuius diameter 12 ad minus contineat pollices; et eidem circulo maximum quadratum inscribere quod sit *ABCD*; et super summitatem

[a] L *AGE* (as though a triangle) [b] A +ergo [c] CAHL scilicet
[d] CAL prehabitatum [e] A λ [f] H huius

what usually results from the extraction of roots involving given sinus in the diagrams which philosophers use to establish the sizes of angles, the distances of centres, and equations worked out on the basis of right-angled triangles.

Let there be, therefore, a circle *ABCD*, the diameters being *AC* and *BD*, and one side of the right-angled triangle being denoted by *OF*, the other *OT*. It is proposed to find the size of the third side, *TF*. *Procedure*. From point *T*, a perpendicular is dropped on the diameter *DB*, [meeting it] at the point *P*. Since the line *OP* is equal to the line *FT*, and *OP* is—by hypothesis—of a known number of parts, the line *FT* will in consequence be known, which is what was proposed.

This proposition is here set down for the sake of [calculating] eclipses of the Sun and Moon, just as the previous [propositions are included] lest, in other matters, those using[1] [the instrument] should be uncertain.

Here ends the first part of Albion, and begins the second

II

Albion, like a planisphere of the whole celestial sphere, reproduces mechanically, with remarkable conciseness, the concentrics, eccentrics, and epicycles, for the various movements of the stars, whether wandering or fixed. Its construction is quite simple even for the novice, provided you know the proportions of the circles for the different revolutions of the planets. For this reason, therefore, it was worth giving in advance instructions for constructing circles of this sort out of suitable material, and then going on to give the method whereby these same circles should be divided and inscribed; for the excellence of this instrument lies chiefly in the accurate and minute division of these proportional circles.

CHAPTER I

[II. 1]

On the art of preparing the material of the entire instrument.

To prepare the material for the work, take one flat polished sheet of brass, and describe upon it a circle with centre *G*, whose diameter should be at least 12 inches;[2] and inscribe in the same circle the greatest square possible, *ABCD*; and at the top of the line of the

[1] *operantes*.

[2] Originally meaning 'thumb-[breadth]', by the fourteenth century the word *pollex* had become a technical term for the inch.

linee *GF* semidiametri orthogonaliter secantis costam *CB* aliquid superfluum extra circulum designare pro armilla suspensoria instrumenti; et duas superfluitates in summitatibus duarum costarum *DC* et *AB* pro pinnulis aspectuum retinendis; et deinde cum circino peracuto quicquid de lamina *ABCD* extra circumferenciam superfluit resecare, preter has tres superfluitates que sunt apud *F*, et *C*, et *B*.

Deinde huic lamine sic formate, inquirantur alie due lamine per omnia similes et equales, et sit[a] una earum *EHI*, et eius centrum *K*, altera vero sit *LMN*, et eius centrum *O*. Et super laminam *EHI* fiat circulus equalis maximo circulo inscriptibili quadrato *ABCD* et sit iste circulus *PGQ*, supra quam circumferenciam cum circino peracuto dividamus laminam ipsam *HGI* in duas partes equales, cuius pars exterior erit armilla *EHI*; et similiter lamina circularis *PGQ*. Eodem modo de tercia lamina *LMN* abradamus subtiliter cum circino peracuto unam armillam, cuius latitudo sit subtripla ad latitudinem armille *HI*, et sit hec armilla *LMN*;[b] et residuum interius circulare sit lamina *ROS*.

Post hec conclaventur iste due armille *HI* et armilla *LMN* altrinsecus limbo lamine prime *AB*, et cum solidatura competenti firmissime conglutinentur. Et tunc aptari faciamus duas residuas laminas *PGQ* et *ROS*, ut recipi possint inter concavitates limborum aptissime, ita quod massa trium laminarum coherencium appareat tanquam corpus unius superficiei ex utraque parte planissimum.

Ut[c] posset autem lamina media[d] *ABCD* cum suis limbis altrinsecis opere tornatili de eadem massa componi, et due lamine breviores *PGQ*[e] et *ROS* cum concavitatibus replendis aptari. Quibus meliori modo compositis, in duabus superfluitatibus *C* et *B* erigantur due pinnule perforate equidistantes linee *GF* ex parte limbi *LMON*, et superfluitati *F* inponatur armilla suspensoria; et completum est.

CAPITULUM 2^m

De modo inscribendi circulos limbi primi.

Et cum preparata fuerit materia instrumenti, consolidemus utramque laminam *PGQ* et *ROS* in locis suis, ita quod ante completam instrumenti facturam nullatenus moveantur e loco. Et incipientes a divisione limbi maioris, protrahamus occulte diametrum *AGC*, et partem diametri que cadit in limbo *HI* dividamus in decem partes

[a] CA fit; L fiat [b] CAH *LM* et *N* [c] CL λ [d] A λ
[e] A *QGQ*

semidiameter *GF* cutting the side *CB*, reserve any spare [metal] outside the circle [to hold] the suspension-ring of the instrument. Two spare [pieces of metal] at the tops of the two sides *DC* and *AB* are to be retained for sighting pinnules. Next, with a pair of sharp dividers, cut away any of the plate *ABCD* which is superfluous, beyond the circumference, apart from the three pieces left at *F*, *C*, and *B*.

After this, two more plates are to be found, in all respects similar and equal to the plate so formed. Let the first of them be *EHI*, with its centre *K*, and the other *LMN*, with its centre *O*. On the plate *EHI* draw a circle *PGQ*, equal to the greatest circle which it is possible to inscribe in the square *ABCD*. With a pair of sharp dividers we are to divide this plate *HGI* at the [inner] circumference into two equal parts, the outer of which will be the ring[1] *EHI*, and the plate [remaining] *PGQ*. In the same way, with the third plate *LMN*, we accurately cut away a ring with a pair of sharp dividers, the ring being *LMN*, and having a thickness one third that of the ring *HI*. The inner circular part which remains is to be [called] the plate *ROS*.

The two rings *HI* and *LMN* are now to be riveted, one on each side of the limb of the first plate *AB*, and cemented firmly with solder. We are now to prepare the two remaining plates *PGQ* and *ROS*, which may be fitted precisely into the concavities [formed by the rings at] the limbs, and in such a way that from both sides the block of the three plates combined will appear exactly as though it was a body with a single surface absolutely flat on both sides.

It is possible to make the plate *ABCD*, with its [raised] edges on both sides, out of a single piece [of metal] by the use of a lathe. The two smaller plates, *PGQ* and *ROS*, may then be made to fit into the concavities. Having made it in this better way,[2] two perforated pinnules are to be erected on the side *LMON* at the two spare pieces *C* and *B*, equidistant from the line *GF*; and at the spare piece *F* a suspension ring is to be fitted. This completes [the first stage].

CHAPTER 2

[II. 2]

On the method of inscribing the circles of the first limb.

When the material[3] of the instrument has been prepared, we are to make fast each of the plates *PGQ* and *ROS* in its appropriate place, so that neither moves from its place in any way until the instrument is finished. Beginning with the division of the greater limb, let us lightly[4] draw the diameter *AGC*, and divide that part of the diameter

[1] *armilla*.

[2] The author's intention was perhaps to say 'whether or not it be made in this (preferred) way . . .'.

[3] That is, the body.

[4] *occulte*. (The line is to be a construction line only.)

equales; et faciamus per quamlibet divisionem transire circulum unum super *G* centrum descriptum. Et habeat quodlibet spacium inter duos circulos extra se unam brevem marginem pro divisionibus graduum capiendis, excepto secundo spacio quod habebit marginem intra se. Erunt itaque decem spacia pro litteris numerorum et totidem margines pro gradibus circulorum.

Deinde tercium marginem, numerando ab extra, dividamus fideliter in 360 partes equales, et spacium eius sequens per senas et senas partes equales pro litteris graduum inscribendis; et describamus gradus[a] marginis in 12 sig., et in spacio primorum 6 gra. cuiusvis signi intituletur nomen signi illius, vel in spacio ultimorum 6 gra. distincte sculpetur quocum fuerit illud signum. In residuis autem spaciis 6 gra. eorum numeri insculpentur, et incipiat numerus signorum ex opposito *C* ad punctum *A*. Hoc nota.

Post hoc dividemus extimum spacium cum suo margine tali arte: a puncto eius in margine quod est in directo *A*, separabimus contra naturalem successionem signorum unum arcum 5 gra. 10 min. 27 sec., quam precise fieri poterit, qui sit in extimo margine arcus *AX*; et totam residui[b] marginis, scilicet arcum *ACX*, dividamus in 360 partes equales et, si fideliter sit divisum, arcus *XA* de eiusdem partibus continebit 5 par. et quartam unius.

Post hec, incipientes a puncto *A*, distribuamus easdem divisiones per menses, dando cuilibet mensi tot divisiones quot habuerit ipse dies; ponentesque principium mensis Marcii apud *A*, dividamus spacium eiusdem marginis per 6 et 6 partes, inscribendo nomen mensis in primo spacio 6 dierum mensis eiusdem; et habebit ultimus mensis superfluum unius quarte diei.

Medium spacium, cum margine nondum diviso, dividemus hoc modo: semicirculum eius *BAD* dividemus in 90 partes equales, et spacium marginis[c] proximi in 18, et inscribantur numeri[d] graduum incipientes apud *D* versus *A*. Reliquam vero medietatem marginis in 12 partes equales secemus, quas *12 horas* in spacio margini proximo[e] titulabis. Quamlibet illarum 12 parcium in 60 partes dividemus si latitudo marginis hoc permittit, et erunt minuta horarum; incipietque earum[f] computacio a puncto *D* versus *C*; vocabiturque circulus

[a] A graduum [b] C residuum [c] CAH margini [d] CA numerum
[e] C proximum; AL proximi [f] A λ

which falls on the limb *HI* into ten equal parts. We draw a circle about centre *G*, through each [of the ten points of] division [of the diameter]. Every space between two circles is to have, towards its outside, a narrow margin [set apart] to take the degree-divisions, the only exception being the second space,[1] which will have the margin to the inside. There will thus be ten spaces for the numerals and as many margins for the degree-divisions of the circles.

We must next accurately divide the third margin from the outside into 360 equal parts, the space next following being divided at intervals of six degrees,[2] in which to inscribe the figures. The degrees of the [third] margin are to be grouped into twelve signs, the name of each sign being incised distinctly, no matter what be the sign, either in the first or the last six degrees of that sign. In the remaining spaces of six degrees their numbers are to be engraved, the number[ing] of the signs beginning opposite *C* and going towards the point *A*. Take note of this.

After this we are to divide the outermost space with its margin as follows. From the point on the margin in question in the direction *A*, we shall mark off, as precisely as may be done, an arc of 5° 10′ 27″,[3] reckoning in a direction opposite that of the natural succession of the signs. What is thus deducted from the margin is to be called the arc *AX*. We must divide the entire remaining margin into 360 equal parts and, if the division be done accurately, the arc *XA* will contain 5¼ of these same parts.

We are now to illustrate the same divisions by months, beginning from the point *A*, and giving to each month just as many divisions as it has days. Putting the beginning of March at *A*, we should divide up the space [adjacent to] this same margin at intervals of six parts, the name of the month being inscribed in the first six-day space of the corresponding month. The last month will have a superfluous quarter of a day.

The intervening space, with its margin not yet divided, we shall divide in the following manner. We shall divide its semicircle *BAD* into 90 equal parts, and the space of the adjacent margin into 18.[4] The numbers of degrees should be inscribed, beginning at *D* [and counting] towards *A*. We must mark out the remaining half of the margin in 12 equal parts, which you shall entitle '12 hours' in the space next to the margin. We are to divide into 60 parts, if the breadth of the margin permits it, each of the twelve parts; and the result will be minutes of an hour, these being counted from *D* towards *C*.

The first circle will be called that of the *solar year*, and the semi-

[1] Counting inwards. [2] *per senas et senas partes equales.*

[3] In order to give 365¼ equal divisions, it will be seen that the quantity *AX* must be (5¼/365¼) × 360°, or 5° 10′ 28″.

[4] That is, the scale is unlike the others in that the degrees are grouped in fives.

primus *anni Solis*; et semicirculus *BCD* circuli secundi vocabitur *arcus horarum*; et alter semicirculus *DAB* vocabitur *altitudinis*. Tercius autem circulus est vice *zodiaci spere octave*; et completum est.

Verumptamen antequam fiunt septem interiores circuli cum suis marginibus, super diametrum limbi apud *A* et apud *C* producantur due linee altrinsecus equidistantes diametro, pro margine divisionum et spacio litterarum inscribendo, sicut interius dicetur capitulo 34°,[a] sicut in presenti patet in figura.

CAPITULUM 3ᵐ

De[b] *divisione orbis Saturni.*

Quartus circulus cum suo margine vocabitur *orbis Saturni*, cuius marginis divisio fiat ita: in vacuo spacio lamine *PGQ* super semidiametrum *GA*, per quartam et septimam prime huius, inveniantur centrum equantis et deferentis Saturni cum proporcione semidiametri eorundem ad id quod cadit inter centrum deferentis et *G* centrum Terre. Quibus compertis, protrahantur occulte et delebiliter deferens atque equans; et dividatur equans in 360 partes equales occulte, incipiendo in directo *A* puncti ubi ponitur aux equantis, et regula exiens a centro equantis per divisiones has singulas signet easdem in circulo deferentis. Et iterum alia regula fixa in centro Terre *G*, transiens super omnes divisiones iam in deferente signatas, transferat omnes ad marginem orbis Saturni.

Post hec ille 360 partes in orbe Saturni signate distribuantur per 12 signa, incipiendo in directo *A*, et spacium margini adiunctum per 6 et 6 partes dividatur omnino sicut dictum est in divisione zodiaci octave spere; et in ultimo spacio 6 graduum signi sculpentur distincte quocum[c] fuerit illud signum.

Isto modo autem omnino dividemus quintum circulum cum suo margine, qui est[d] vice *orbis Iovis*; et sextum circulum, qui erit *orbis vice Martis*. Nullam tamen harum divisionum signemus infra spacium transversum apud *A* et apud *C*, sed omnino intacte[e] permaneant usque ad capitulum 24ᵐ, ubi docebitur quomodo dividentur; et completum est.

[a] H 24° [b] C λ [c] AH quorum [d] CL λ [e] C intacta

circle *BCD*, of the second circle, the *arc of hours*, the other semicircle, *DAB*, being the [*arc of*] *altitude*. The third stands for the *zodiac of the eighth sphere*.

Nevertheless, before the seven inner circles are made with their margins, two lines are to be drawn at *A* and *C*, on both sides of, and parallel to, the diameter of the limb. This is for inscribing a margin which may be divided, and a space for labelling, in a way which is explained in the middle of chapter 34, and such as is evident from the present figure.

CHAPTER 3

[II. 3]

On the division of the circle[1] *of Saturn.*

The fourth circle with its margin is called the *circle of Saturn*, and the division of its margin is to be made as follows. In the empty space on the plate *PGQ*, on the semidiameter *GA*, using I. 4 and I. 7 of this work, the centres of the equant and deferent of Saturn should be found, together with the ratio of their semidiameter to what is intercepted between the centre of the deferent and *G*, the centre of the Earth. Having found these, draw in lightly, so that they may be erased, the deferent and equant [circle], the equant [circle] being lightly divided into 360 equal parts. Starting from the direction of the point *A*, where the aux of the equant is taken to be, a rule passing through the centre of the equant marks divisions on the deferent circle corresponding to the separate divisions [on the equant circle]. Next, another rule, fixed at the centre of the Earth *G*, and passing through each of the divisions now marked on the deferent, is to transfer them all to the margin of the circle of Saturn.

The 360 parts in the circle of Saturn are subsequently to be distributed into 12 signs, starting from the direction *A*, and the space adjoining the margin should be divided up entirely in groups of six parts, as explained of the division of the zodiac of the eighth sphere. In [each] last space of six degrees the [name of the] appropriate sign is to be clearly engraved.

In this way we are to divide entirely the fifth circle with its margin, which stands for the *circle of Jupiter*; and the sixth circle, which stands for the *circle of Mars*. No divisions are, however, to be marked on the strip running from *A* across to *C*, but they are to be left entirely untouched until chapter 24, where the division will be explained.

And [the division] is complete.

[1] *orbis*. This is here generally translated as 'circle' but occasionally as 'scale'. (The former is still in astronomical use, and *circulus* is used later in *Albion* as an alternative to *orbis*.)

CAPITULUM 4^m

De modo communi operandi in divisione orbis.

Ad maiorem huius operis declaracionem, sit circulus orbis Saturni *ABCD* super centrum *G*. Et quia quarta et septima conclusio prime huius sunt generales ad inveniendum centrum equantis et deferentis in quovis puncto linee exeuntis a centro *G*[a] Terre, et circulos proporcionales quantumlibet magnos vel breves, per quantamcumque[b] distanciam centri epicicli ab auge deferentis suppositam et per notam eiusdem distancie equacionem, que vocatur equacio centri, volumus[c] deinceps uti distancia centri ab auge *A* trium signorum, qui[d] sit arcus *AB*, et separemus a puncto *B* versus *A* arcum equacionis talis centri, qui[e] sit arcus *BF*. Nam per calculum[f] Ptolomei in radicibus *Almagesti* fiunt equaciones centri ad tercia signa sic:

Saturni, equacio centri trium signorum 6 gra. 31 min.;
Iovis, 5 gra. et 15 min.;
Martis, 11 gra. et 23 min.;
Solis et *Veneris*, 1 gra. 59[g] min.;
Mercurii vero 3 gra. et 1 min.

Sit igitur arcus *BF* 6 gra. et 31 min. et, producta linea *GF* in eadem, signemus punctum *H*[h] vice centri epicicli, caveaturque in operando ne linea *GH* sit maior vel equalis linee *GB*; nam si ita fuerit, non cadent isti ecentrici infra spacium *PGQ*. Deinde a puncto *H* producatur [linea] equidistans linee *GB* super lineam *GA* ad punctum *I*. Patet ergo, per 4^am prime, quod *I* est centrum proporcionale equantis.

Posito ergo centro in medio inter *G* et *I*, circumducatur circulus per *H* punctum, cui fiat equalis super centrum *I*, quia equans et deferens imaginantur a disciplinatis esse equales. Deinde dividatur circulus equans, qui est super centrum *I*, in 360 partes equales. Et cum fixa regula in puncto *I* transeunte per singulos illos gradus, transferantur singule divisiones ille ad alterum circulum prius factum, scilicet ad deferentem. Post hec regula fixa in *G* centro Terre, circumvolvatur regula super illas 360 divisiones in deferentem translatas, et signentur eedem divisiones in circulo *ABCD* ad contactum regule cum circumferencia *ABCD*, que est loco[i] orbis Saturni. Et sic patet eius artificialis divisio.[j]

[a] A λ [b] CAH quantumque; L quantamlibet
[c] C volimus (*vel* velimus?); L nihilominus [d] A que [e] A que
[f] L calculacionem [g] H 29 [h] A λ [i] A locus; H λ
[j] H +Ultimo dividatur circulus decimus qui est dictus *circulus verus Lune* omnino sicut dividitur circulus orbis Solis per maximam equacionem argumenti ad 3 signa, que est 5 gra. et 2 min. Et distribuantur divisiones in signa et gradus ut pretactum est. (Cf. cap. 7.)

CHAPTER 4
[II. 4]

On the common method of dividing the circle.

To make this operation clearer, let the circle of the sphere[1] of Saturn be *ABCD*, with centre *G*. Propositions I. 4 and I. 7 above are designed to find the centre of the equant and deferent on any point of a line drawn out from *G*, the centre of the Earth, and to find the proportional circles,[2] however great or small, whatever the distance supposed between the centre of the epicycle and the aux of the deferent, and to do so by means of the known equation of this same distance, which is called the 'equation of the centre'. We now want to make use of a distance of the centre of three signs from the aux *A*, which is to be arc *AB*, and we should mark off the arc of the equation of such a centre from *B* towards *A*, this being the arc *BF*. For, by Ptolemy's calculation based on the parameters of *Almagest*, the equations of the centre for three signs turn out to be:

of *Saturn*, the equation of the centre for [an argument of] three signs, 6° 31′;
of *Jupiter*, 5° 15′;
of *Mars*, 11° 23′;
of the *Sun* and *Venus*, 1° 59′;
of *Mercury*, 3° 1′.

Let the arc *BF*, therefore, be 6° 31′ and the line *GF* being drawn to it, let the point *H* stand for the centre of the epicycle; and in the working beware lest the line *GH* be greater than or equal to the line *GB*, for in this case these eccentrics will not fall within the space *PGQ*.[3] Next, let a line be produced from the point *H* to the point *I*, parallel to the line *GB* and meeting line *GA* at *I*. It is plain, then, by I. 4 above, that *I* is the centre of the equant in correct proportion.

With centre midway between *G* and *I*, therefore, a circle is to be drawn through the point *H*, and an equal circle drawn with centre *I*, since equant and deferent are by trained astronomers[4] imagined equal. The equant circle about centre *I* is now to be divided into 360 equal parts. With a rule passing invariably from the point *I* to the individual degree marks, divisions are to be transferred one by one to the circle first drawn, that is, to the deferent. After this, a rule being fixed at *G*, the centre of the Earth, the rule is to be moved round over the 360 divisions transferred to the deferent and the same[5] divisions marked on the circle *ABCD* at the point where the rule meets the circumference *ABCD*, which is at the circle[6] of Saturn. And this is the art of dividing it.

[1] *orbis*.

[2] Presumably meaning equant and deferent circles of the appropriate proportions.

[3] Making *GH* less than *GB* is, on the other hand, no guarantee that the eccentric will fall within the space.

[4] *disciplinati*.

[5] i.e. corresponding.

[6] *orbis*; the usage is not noted beyond this point.

CAPITULUM 5^{m}

De Solis orbe.

Post divisionem orbis Saturni et Iovis et Martis restat dividere orbem Solis. Invento ergo ecentrico deferente Solis, per 4^{am} prime huius, et ipso diviso in 360 partes equales, statim posita regula in centro *G*, revolvatur ipsa super singulas has divisiones et transferat eas ut prius ad marginem orbis Solis. De quibus gradibus fiant 12 signa, et intitulentur in spacio margini annexo omnino sicut predictum est de ceteris orbibus. Et completum est.

CAPITULUM 6^{m}

De divisione orbis Mercurii.

Nunc superest dividere orbem Mercurii, quia orbes Solis et Veneris sunt unius differencie in movendo. Per 4^{am} prime huius inveniatur ut prius centrum equantis Mercurii in semidiametro *GA*, per centrum trium signorum et per equacionem eiusdem centri. Et per 8^{am} prime huius inveniatur quantitas semidiametri deferentis in debita proporcione ad distanciam centri equantis a centro Terre; ut verbi gracia:

Sit circulus orbis Mercurii *ABCD* super centrum *G*, et centrum equantis *F*, et circulus equantis *AS* secundum mensuram semidiametri *FK*. Item circulus brevis transiens per centrum equantis et centrum deferentis, cuius semidiameter est equalis linee *FG*, sit *LFM*. Deinde dividatur equans in 360 partes equales, et circulus brevis in totidem, vel in quot eorum fuerit possibile. Item a centro equantis producantur linee occulte indifinite per singulas divisiones equantis, quarum 30, que sunt altrinsecus iuxta augem equantis *A*,[a] excedant semidiametrum equantis. Deinde circino quo fiebat circulus equans non mutato, ponatur unus eius pes in circumferencia parvi circuli in prima divisione, et signet unam divisionem in prima linea que exit a centro *G* per augem equantis ad punctum *N*. Item posito pede circini in secunda divisione parvi circuli, computando contra

[a] C *A* (but resembling 'et'); H *A*; L et

CHAPTER 5

[II. 5]

On the Sun's circle.

After the division of the circles of Saturn, Jupiter, and Mars, it remains to divide the circle of the Sun.[1] The eccentric deferent of the Sun having been found by I. 4 above, and having been divided into 360 equal parts, a rule at the centre is to be moved round over the individual divisions, which are to be transferred as before to the margin of the Sun's circle. Let there be 12 signs made up out of these degrees, and let them be labelled in the space next to the margin, exactly as was previously explained for the other circles. And [the division] is complete.

CHAPTER 6

[II. 6]

On the division of the circle of Mercury.

It now remains to divide up the circle of Mercury, since the spheres of the Sun and of Venus are one in their movement. By I. 4 above, as before, the centre of the equant of Mercury on the semidiameter *GA* is to be found, working from a centre at three signs [elongation from aux], and the equation of the centre of the same. By I. 8 above, the size of the semidiameter of the deferent is to be found, in correct ratio to the distance of the centre of the equant from the centre of the Earth.

Example. Let the circle of the sphere of Mercury be *ABCD*, with centre *G*, and let the centre of the equant be *F*, the equant circle being *AS* with *FK* as semidiameter. Next, let there be a small circle *LFM* through the centre of the equant and the centre of the deferent, and having semidiameter equal to the line *FG*. Let the equant now be divided into 360 equal parts, and the small circle into the same number, or into as many as possible. Next, lines of arbitrary length are to be lightly drawn from the centre of the equant and through the individual divisions of the equant; and 30 [60?] of these, adjacent to and on both sides of the aux of the equant *A*, are to pass beyond the equant circle.[2] Then with dividers set for describing the equant circle, place one foot at the circumference of the small circle, at the first division, and let a division be marked on the first line, which passes from the centre *G*, through the aux of the equant, to the point *N*. Next, with one foot of the dividers on the second division

[1] This serves also for Venus.

[2] Literally 'beyond the semidiameter of the circle'. The reason for this stipulation is clear, although the number 30 seems to be a mistake. The construction which follows is such that lines through *F* are to intersect certain arcs of circles, and within about 60° on either side of *FN* they intersect outside the equant circle (see the scale diagram, vol. iii, p. 30). The figure of 30° would certainly be an easy mistake for 60°, its complement.

naturalem successionem signorum, extendatur alter pes circini non variati ad secundam lineam exeuntem a centro equantis *F*, computando lineas ab auge eius secundum successionem signorum, signando divisionem ad punctum *O*;[a] et sic[b] deinceps currat alter pes circini in parvo circulo trecenties sexagesies, centro mutato quousque alter pes signaverit divisionem unam in singulis lineis a centro equantis eductis.

Deinde fixa regula in centro *G*, volvatur super singulas has divisiones et transferat[c] eas ad marginem orbis Mercurii *ABCD*, sicut pretactum est de ceteris; et distribuantur divisiones per 12 signa, et gradus in spacio margini coherenti, et intitulentur ut cetere. Et examinetur opus vel[d] similiter[e] perficiatur per tabulam centri equati Mercurii, et redit in idem.

Et nota quod si equans et circulus ille parvus non possent commode signari in spacio lamine *PGQ*, fixo instrumento in aliqua plana superficie invenire poterimus centrum equantis extra superficiem[f] limbi, et quantitatem semidiametri deferentis, et circulum illum brevem in quantalibet longitudine et magnitudine, nec omnino variabit divisiones recipiendas in margine orbis Mercurii; et causa patet eo quod conclusio 7ª prime huius generalis est et causa tangitur paragraffo finali. Et completum est.

CAPITULUM 7ᵐ

De orbe Lune.

Adhuc restat nonus circulus dividendus, qui erit circulus orbis Lune, cuius divisio fiet ita:

Per 6ᵃᵐ prime huius inveniamus ecentricum deferentis lunaris et, posita regula in centro *G* Terre, circumducatur regula super singulas divisiones marginis zodiaci spere octave et transferantur eedem divisiones ad deferentem lunarem. Deinde regula fixa in puncto opposito[g] centro[h] deferentis in diametro *AC*, transferantur omnes divisiones deferentis ad marginem orbis lunaris; et tunc spacium eius omnino intituletur sicut spacia[i] orbium aliorum.

[a] A *D* [b] A λ [c] A transferas [d] A ut [e] CAH similibet
[f] CA superficie [g] CA opposite [h] A centrum [i] A spacium

of the small circle, counting against the natural succession of signs, the other foot is to be moved (without altering the opening of the dividers further) to the second of the lines through the centre of the equant *F*, counting the lines from its aux in the order of the signs, and marking the point of intersection *O*. And thus in succession the other foot of the dividers should intersect the small circle 360 times. Moving the centre, the other foot will invariably mark the point of division on the individual lines through the equant centre [*F*].

The rule now being placed at centre *G*, let it be turned so that it passes through these divisions, and let it [be used to] transfer them to the margin of the circle of Mercury, *ABCD*, as was touched on earlier in connection with the other [circles]. The divisions should be distributed into twelve signs, with the degrees in the space of the neighbouring margin, the labelling as with the others. The work should be tested or, as before,[1] perfected by a table of the equation of the centre of Mercury—which leads to the same results [as the geometrical construction].

Note that if the equant and the small circle may not be conveniently marked in the space [available] on the plate *PGQ*, having fixed the instrument on any plane surface we may find the centre of the equant outside the limb, and also a semidiameter for the deferent, and the small circle of whatever dimensions, and this will not make any difference to the final divisions on the margin of the circle of Mercury. The justification for this is that proposition I. 7 above is of general application—the reason is touched on in the final paragraph.

And [the division] is complete.

CHAPTER 7

[II. 7]

On the circle of the Moon.

The ninth circle still remains to be divided, and this will be the circle of the sphere of the Moon, the division of which will be as follows.

By I. 6 above, let us find the eccentric of the lunar deferent. A rule having been placed at *G*, the centre of the Earth, let it be moved round over the individual divisions of the margin of the zodiac of the eighth sphere, these divisions being transferred to the lunar deferent. The rule being now placed permanently on the point opposite the centre of the deferent on the diameter *AC*,[2] all the divisions of the deferent are to be transferred to the margin of the lunar circle. The space belonging to it is then to be everywhere labelled as the spaces of the other circles.

[1] *similibet* in all MSS. I find no such word in any dictionary. S reads this *vel si libet*, but the MSS. are explicit. Perhaps '*sibimet*' was originally written.

[2] That is, at the same distance from *G* as the centre, but on the opposite side.

Potest eciam hoc fieri vel examinari per tabulam centri equati Lune, et redit[a] opus in idem. Et completa est divisio primi[b] limbi.

Ultimo dividatur circulus decimus, qui[c] dicitur *circulus verus Lune*, omnino sicut dividitur circulus orbis Solis per maximam equacionem argumenti ad tria signa, que est 5 gra. et 2[d] min. Et distribuantur divisiones in signa et gradus ut pretactum est. Et completum est.

CAPITULUM 8m

De ecentricis epiciclorum

Post divisionem orbium planetarum in spacio limbi primi *HI*,[e] restat describere deferentes planetarum circa centra[f] epiciclorum suorum in spacio lamine *PGQ*; et inscribantur sensibiliter per doctrinam undecime prime huius, et eorum centra accipiantur in semidiametro *GC*, eruntque auges earum in directo pinnule *C*. Et capi potest eorum mensura a gradibus marginis circuli altitudinis indifferenter vel a gradibus marginis zodiaci orbis octavi, sed tunc oportet duplare equacionem, quia respectu centri *C* quilibet gradus est duplus in comparacione ad centrum *G*, sicut patet per 9am conclusionem prime huius. Regula autem per quam capientur huiusmodi[g] arcus fixa erit in *C* puncto in circulo medio communi duabus marginibus, circuli altitudinis et zodiaci; et erit ordo istorum ecentricorum primo Veneris, secundo Martis, tercio Mercurii, quarto Iovis, quinto Lune, ultimo Saturni; et intitulentur quorum sint ecentrici. Et completum est.[h]

[a] C rediddit [b] H 2λ
[c] H 38λ; H +et sit in fine capituli quarti precedentis [d] L 20 [e] A λ
[f] CA centrum [g] H huius
[h] H +Si velis ecentricum Saturni prout habere cum eius equacio sit 5 gra. et 53 min. pone regulam supra centrum C et protradatur [*sic* MS.] usque ad 5 gra. circuli latitudinis et 53 min. vel usque ad 11 gra. et 6 min. zodiaci quod idem est, et fiat occulta linea. Deinde accipiatur punctus medietatis illius linee et ponatur pes circini super centrum *G* et ducatur iste punctus in diametrum versus *C*. Deinde ponatur pes circini in *C* et ducatur linea in circulo altitudinis 6 gra. et 37 min. et capiatur punctus medietatis illius et ponatur pes circini in *G* et ducatur punctus in diametro versus *A*. Deinde ponatur pes circini in medio istarum sectionum et fiat circulus; et habetur ecentricus Saturni.

It is also possible that this should be made or tested by means of a table of the equated centre of the Moon, and the result will be exactly the same. This completes the division of the first limb.

Finally the tenth circle is to be divided, this being known as the *true circle of the Moon*. It is divided exactly as the circle of the sphere of the Sun, by means of the greatest equation of the argument at three signs,[1] this being 5° 2′. The divisions are distributed in degrees and signs, as was touched on earlier. And [the circles of the Moon are now] completed.

CHAPTER 8

[II. 8]

On the eccentrics of the epicycles.

After the division of the circles of the planets in the space of the first limb *HI*, it remains to describe the circles carrying the planets about the centres of their epicycles in the space on the plate *PGQ*.[2] They are to be inscribed permanently,[3] using the ideas of I. 11 above,[4] and their centres are to be taken on the semidiameter *GC*, the auges [of the planets] being in the direction of the pinnule *C*. They may be measured out with the help of the degrees along the margin of the circle of altitude or, equally well, along the margin of the zodiac of the eighth sphere, although in this case it will be necessary to double the equation, since with respect to the centre *C* any angle[5] is twice as large as with respect to centre *G*,[6] as is evident from proposition I. 9 above. The rule used to find these arcs is placed at the point [in direction] *C* on the two adjacent margins, those of the circle of altitude and the circle of the zodiac. The order of these eccentrics will be: first that of Venus, second that of Mars, third that of Mercury, fourth that of Jupiter, the fifth of the Moon, and the last that of Saturn; and they are to be labelled appropriately. And [these eccentrics are] complete.

1 See our commentary.

2 The circles are subsequently (III. 1, etc.) called the 'equators' of the planets, and this name conveys their purpose well, for they will be used subsequently (III. 9, etc.) to determine the equations of the argument.

3 *sensibiliter*—presumed to be in contrast with lines drawn *occulte*, which were to be erased.

4 The purpose of this section cannot be properly understood unless reference be made to I. 11 and III. 9.

5 Literally, 'any degree'.

6 This is to say that a line through *C*, at angle θ to *CG*, will cross the scale at a point *R* with graduation 2θ (that is, the angle *RGA* which is twice the angle at the circumference). The circle of altitude (67), having 90 divisions distributed over a semicircle, may be used to give a direct reading for θ.

CAPITULUM 9m

De ecentrico Mercurii.

Ecentricum epicicli Mercurii oportet alio modo depingere, quia non habet augem deferentis fixam, et quia prius est planeta in maxima elongacione eius apparenti a centro epicicli quam fuerit centrum epicicli in augis opposito deferentis; et ideo ad removendum difficultatem inveniendi locum stelle in ecentrico epicicli, [et] proporcionem motus centri epicicli in suo ecentrico deferente, et ad abbreviandum multitudinem circulorum, maxime autem ut hoc instrumentum in nullo videatur a tabulis discrepare, fiat ecentricus epicicli Mercurii isto modo:

In spacio lamine *PGQ*, posita regula in centro *G*, separemus alterutra[a] parte *A* duos arcus 64 graduum, qui sunt arcus *AF* et *AH*, protractis occulte lineis *GF* et *GH*. Item separabimus altrinsecus duos arcus 4 signorum, qui sint *AI* et *AK*, productis lineis occultis *GI* et *GK*. Et iterum tercio separemus duos arcus utrumque 5 signorum, et sint *AL* et *AM*, protractis lineis *GL* et *GM*.

Deinde per 10am prime huius[b] inveniatur epiciclus occultus proporcionatus equacioni maxime 23 gra. et 53 min., et ubi iste circulus abscindet[c] duas lineas *GI* et *GK* ponantur note *I* et *K*. Secundo, per eandem decimam prime huius, inveniatur epiciclus proporcionatus maxime equacioni 23 gra. et 32 min., et ubi iste epiciclus occultus abscindet duas lineas *GL* et *GM* ponantur due note *L*,[d] *M*. Tercio, inveniatur epiciclus[e] proporcionalis equacioni 23 gra. et 15 min., et in[f] abscisione linee *GC* ponatur nota *R*. Quarto, inveniatur epiciclus proporcionalis equacioni maxime 21 gra. et 44 min., et ubi abscindit duas lineas *GF* et *GH* ponantur note *F*, *H*. Quinto, inveniatur epiciclus proporcionalis equacioni maxime 19 gra. et 1 min., et ubi abscindit lineam *GA* ponatur nota *S*.

Deinde posito centro in linea *GCA*, inveniatur arcus circuli qui transeat per tria puncta *L*, *R*, *M*; et fiat arcus *LRM* tantum satis patens. Item[g] posito centro in eadem linea *CGA*, describatur arcus manifestus transiens per tria puncta *F*, *S*, *H*.[h] Item posito centro in linea *GK*, fiat arcus patens transiens per tria puncta *F*, *I*, *L*. Quarto, posito centro in linea *GI*, fiat arcus patens super tria puncta *H*, *K*, *M*, et complebitur circumferencia *RLIFSHKMR*, fractis in quattuor punctis *F*, *H*, *L*, *M*, habens staciones ad quattuor puncta *S*, *R*, *I*, *K*,[i]

[a] CA alterencia; C′ alterutra; H ex utraque
[b] CA λ [c] A abscinderet [d] A *I* [e] H ciclus [f] CAH λ
[g] CL Iterum
[h] C middle letter erased; A leaves blank space; HS *FSH*; L *F, O, H*
[i] CA super *IK*; S *SCIK*

CHAPTER 9

[II. 9]

On the eccentric of Mercury.

The eccentric of the epicycle of Mercury[1] must be depicted in a different way, since Mercury does not have a fixed aux of the deferent, and since the planet reaches its greatest apparent elongation from the centre of the epicycle before the centre of the epicycle reaches the opposite of the aux of the deferent. Thus, to remove the difficulty of finding the place of the planet on the eccentric of the epicycle, and that of finding the ratio of the motus of the centre of the epicycle on its eccentric deferent, and to cut down the number of circles, but above all in order that there should be no apparent discrepancy between this instrument and the tables, let the eccentric of the epicycle of Mercury be made in the following manner.

In the space on the plate *PGQ*, a rule being placed at the centre *G*, we are to mark out two arcs of 64°, namely *AF* and *AH*, one on each side of *A*, the lines *GF* and *GH* being drawn in lightly. Next we shall mark out on each side [of *A*] two arcs of 4 signs, namely *AI* and *AK*, the lines *GI* and *GK* being lightly drawn. Then, thirdly, we shall mark out two arcs of 5 signs each, namely *AL* and *AM*, drawing in the lines *GL* and *GM*.

By I. 10 above, the epicycle proportional to the greatest equation of 23° 53′ is now to be lightly drawn in, the points in which this circle cuts the two lines *GI* and *GK* being *I* and *K*. Secondly, again by I. 10, the epicycle proportional to a greatest equation of 23° 32′ is to be found, the points in which this lightly drawn epicycle cuts the lines *GL* and *GM* being denoted *L* and *M*. Thirdly, the epicycle proportional to an equation of 23° 15′ should be determined, *R* denoting the point at which it cuts the line *GC*. Fourth, the epicycle proportional to a greatest equation of 21° 44′[2] is to be determined, *F* and *H* denoting the points at which it cuts the two lines *GF* and *GH*. Fifth, the epicycle proportional to a greatest equation of 19° 1′ should be determined, *S* denoting the point in which it cuts the line *GA*.

With centre on line *CGA*, the arc of the circle is now found which passes through the three points *L*, *R*, and *M*; and let the arc be drawn sufficiently clearly. With centre of the same line *CGA*, the arc through the three points *F*, *S*, and *H* is next clearly marked in.[3] With centre on the line *GK*, an arc is to be clearly drawn through the three points *F*, *I*, and *L*. Fourth, with centre on the line *GI*, let an arc be plainly drawn through the three points *L*, *K*, and *M*, and the circumference *RLIFSHKMR* will be completed, broken at four points *F*, *H*, *L*, and *M*, and having stations[4] at the four points *S*, *R*, *I*, and *K*. This

[1] Subsequently called the 'equator of Mercury' (III. 1 and subsequently).

[2] This is an error, and should have been 22° 2′. See II. 9 comm.

[3] *describatur . . . manifestus.*

[4] That is, maxima or minima of the radial coordinate.

in quo innuitur causa quare necesse est eam fieri super quattuor centra et super tres diametros.[a] Et iste circulus ecentricus non habet errorem sensibilem in instrumento cuius diameter est 60 cubitorum, et est abrevians multos circulos ab instrumento in quo est ecentricus epicicli Mercurii circulus vere unus, sicut contingit in ceteris quinque planetis; et fiant isti sex ecentrici patentes valde. Et completum est.

CAPITULUM 10^m

De inscripcione circulorum continencium circulum anni solaris et iomyn.

Perfectis sex ecentricis planetarum incipiamus novam operacionem in lamina *PGQ* ac si nichil adhuc inibi sculperetur; et primo separemus unum spacium cum margine, et perficiamus circulum anni Solis omnino sicut dictum est de primo limbo, capitulo 2°. Deinde separemus[b] duo spacia competencia pro litteris numerorum, et marginem interceptum pro gradibus protrahendis, quem marginem dividamus in 360 partes equales; et compleamus orbem signorum ut sepe dictum est scribentes in exteriori spacio numerum signorum et graduum, sicut exigit signorum successio naturalis, in interiori vero spacio scribantur idem[c] numeri signorum et graduum, sed contra successionem signorum. Et erit incepcio huius orbis[d] in directo *A*.

Deinde separemus *ABD*[e] spacium cum margine, et marginem eius dividemus per unam tabulam ad[f] hoc factam que intitulatur *Tabula motus iomyn equati*, cuius composicio patebit infra parte tercia[g] huius, ubi traditur ars componendi tabulas[h] quibus alio modo operandi possent dividi circuli antedicti, et distribuantur illi gradus in 12 signa et intituletur spacium marginis consimiliter ceteris. Et completum est.

[a] H diametras [b] CA separamus [c] H hijd (?) [d] H +signorum
[e] A *ABC* [f] C 2λ+adhuc [g] L quarta [h] CAH tabulis

explains why it is necessary that the circumference be drawn about four centres, and three diameters. And this eccentric circle is not appreciably in error on an instrument of 60 cubits diameter, and it eliminates the need for many of the circles on an instrument with only a single circular eccentric for Mercury's epicycle, such as was [all that we needed] with the other five planets. These six eccentrics are to be [inscribed] extremely clearly. And [the six are now] complete.

CHAPTER 10

[II. 10]

On inscribing the circles containing the circle of the solar year and iomyn.[1]

Having completed the six eccentrics of the planets, let us now set to work on the plate *PGQ*, [in the outer region] where nothing has so far been inscribed.[2] First let us mark out a space with margin, and fill in the circle of the solar year entirely as explained of the first limb, in chapter 2 above.[3] Then let us mark out two adjacent spaces for the numerals, and a margin between them for the degrees. This margin we are to divide into 360 equal parts. We must complete the circle of signs as often explained: in the outer space there is to be the number of signs and degrees, in the natural order, while in the inner space are to be written the same numbers of signs and degrees, but in the reverse order.[4] The beginning of this circle will be in the direction of *A*.

We must now mark out a space *ABD*, with margin, dividing up the latter by means of a table compiled for this purpose, entitled *Table of the equated motus of iomyn.*[5] Its method of construction will be evident after part III, below, where the rules for drawing up [all] the tables are given.[6] With these tables it is possible to divide, in an alternative way, [all] the circles discussed earlier. The degrees are to be distributed into twelve signs and the margin labelled, exactly as with the others.

[1] The 'equation of days'—see glossary and commentary.

[2] There are eleven scales still to come in the space between the outermost eccentric (that of Venus) and the limb of the moving plate.

[3] In other words, two scales (this scale, which we later number (17), and scale (7)) are alike, but one is fixed while the other is movable.

[4] The scale, numbered twice over as explained, is here later designated (18).

[5] The resulting scale is later designated (19).

[6] See IV. 8 below.

CAPITULUM 11^{m}

De inscripcione circuli medii motus Saturni.

Post circulum equacionis iomyn immediate[a] separabimus unam brevem marginem cum duobus spaciis pro litteris numerorum; eritque circulus iste medii motus Saturni, cuius marginem taliter dividemus:

In directo *A* puncti separabimus de margine contra successionem signorum arcum *AX* secundum quantitatem 6 gra. marginis et 29 min. et 19 sec., quam precise estimari poterit a subtili. Deinde totum residuum arcum marginis ab *A* in *X* secundum successionem signorum dividamus in 12 partes equales, que divisiones transeant per ambo spacia margini coadiuncta; et quamlibet partem[b] duodecimam in margine dividamus per 60 minuta, de quibus partibus continebit arcus residuus *XA* 13 min. et 12 sec. si fideliter fuerimus operati. Et continue quinta vel sexta divisio marginis transeat per spacium margini proximum;[c] in spacio sex vel quinque minutorum scribi poterit numerus eorum; et 12 partes in spacio secundo intitulentur gradus 12, ut patet in presenti figura. Verumptamen summo opere caveatur quod ita discrete fiant isti margines cum spaciis suis adiunctis, quod non confundant circulos sex ecentricos prius factos. Et completum est.

CAPITULUM 12^{m}

De circulo anni Iovis.

Circulum anni Iovis restat dividere isto modo:

Faciamus ut prius super centrum *G*, immediate post circulum anni Saturni, brevem marginem cum duobus spaciis, et a principio marginis separemus contra signa arcum 4 gra. 8 min. et 51 sec., quam precise poterimus, et sit arcus *AX*. Deinde totum arcum marginis residuum ab *A* in *X* dividamus in 30 partes equales, et transeat quelibet[d] divisio per ambo spacia marginis. Post hoc quamlibet illarum 30 parcium dividamus in 60 minuta, et istarum divisionum sexta et sexta transeat per spacium margini proximum,[e] et in primo spacio margini proximo scribantur littere numeri minutorum. In secundo vero spacio scribantur littere numeri graduum.

[a] A in mediate [b] A parte [c] CAH proximi [d] CA qualibet
[e] CAL proximi

CHAPTER 11

[II. 11]

On inscribing the circle of mean motus of Saturn.[1]

After the circle of the equation of iomyn, we shall immediately mark out one narrow margin with two spaces for numerals. This will be the circle of the mean motus of Saturn, whose margin we shall divide as follows.

From the point *A*, we shall mark off from the margin, against the [natural] order of the signs, an arc *AX* equal to 6° 29′ 19″, as precisely as may be estimated by one who is careful. We are then to divide into 12 equal parts the remaining arc of the margin, from *A* to *X* working in the order of the signs, the divisions crossing both of the spaces adjacent to the margin. Each twelfth part of the margin should be divided into 60 minutes, of which parts the remaining arc *XA* will contain 13′ 12″ [of the new parts], if the job be done faithfully.[2] Every fifth or sixth division of the margin should cross into the space next to the margin, the number, [a multiple of] 6 or 5, being written [there]. And the twelve parts in the second space should be numbered up to 12, as is clear from the present figure. Take care, however, in making the instrument as a whole, that the margins with their adjoining spaces are so separated that they are not mixed up with the six eccentric circles made beforehand. And [the circle] is complete.

CHAPTER 12

[II. 12]

On the circle of the year of Jupiter.[3]

It remains to divide the circle of the year of Jupiter, and in this manner:

As before, with centre *G* and immediately within the circle of the year of Saturn, let us draw a narrow margin with two spaces, and from the beginning of the margin let us mark off as accurately as we can, against the [order of the] signs, an arc *AX* of 4° 8′ 51″. We are then to divide the whole remaining arc of the margin, from *A* to *X*, into 30 equal parts, letting each division cross both spaces of the margin. After this, let us divide each of the 30 parts into 60 minutes, with every sixth[4] division crossing into the space next to the margin, and in the first space nearest to the margin let the numbers of the minutes be written. In the second space are to be written the numbers of degrees.

[1] Which we number (20).

[2] Obtained (and a little over 1″ arc too great) by dividing the product of 6° 29′ 19″ and 720 by 354° 30′ 41″.

[3] Numbered (21) below.

[4] Notice the apparent preference for division into sixes, rather than into fives.

Si autem bene diviseris arcum magnum *AX*, continebit arcus residuus *XA* 20 par. et 58 sec., hoc est 21 min. fere. Et completum est.

CAPITULUM 13^m

Pro Marte.

Deinde perficiemus circulum anni Martis hoc modo:

Super *G* centrum faciamus marginem cum unico spacio vel duobus, et a principio marginis, contra signa, separabimus arcum *AX* ut prius 21 gra. et 27 min. et 2 sec.; et arcum residuum dividemus secundum successionem signorum in 6 partes equales, que divisiones transeant per spacium et per marginem. Erunt quidem ille partes sex signa. Post hec dividamus quamlibet illarum 6 par. in 30 gra., et de illis partibus graduum continebit arcus residuus *XA* 11 gra. et 24 min. fere, quia ultra hoc continebit 18 sec. Et cum diviserimus similiter arcum *XA*, intitulabimus spacium cum numeris signorum et graduum satis distincte. Et completum est.

CAPITULUM 14^m

Pro circulo anni Veneris.

Adhuc dividamus circulum anni Veneris isto modo:

Separabimus ut prius unum vel duo spacia cum stricto margine super commune[a] centrum *G*, et a principio marginis resecemus arcum *AX* contra signa, qui sit 17 min. et 52 sec. fere. Deinde totum arcum residuum marginis dividamus in 15 partes equales, et quamlibet illarum parcium iterum in 15 dividamus. Deinde de[b] his partibus faciamus signa, dantes cuilibet signo 30 partes; et erunt 7 sig. et 15 gra.; et cum fuerit quilibet illorum graduum divisus[c] in 60 min., vel quot fuerit possibile de illis fractionibus, continebit residuus arcus *XA* 11 min. et 9 sec. Deinde intituletur spacium litteris numerorum signorum et graduum, sicut prius. Et completum est.

[a] CAH communem [b] A λ [c] CAL divisum

If you have accurately divided the large arc AX, the remaining arc XA will contain 20 parts 58″, that is, almost 21′. And [the circle is] complete.

CHAPTER 13

[II. 13]

On Mars.

Let us complete the circle of the year of Mars[1] in the following manner:

With centre G, let us make a margin with one space or two. From the beginning of the margin, against [the order of] the signs, we shall mark off, as before, an arc AX of 21° 27′ 2″. The remaining arc we shall divide, according to the order of the signs, into 6 equal parts, the divisions crossing the space and the margin. Those parts will comprise six signs. After this, we are to divide each of the six parts into 30 degrees. The remaining arc XA will contain, in terms of these reduced degrees,[2] approximately 11° 24′. (It will actually contain 18″ more than this.) When we have similarly divided the arc XA, we shall inscribe the space sufficiently clearly with the numbers of signs and degrees. And [the circle] is complete.

CHAPTER 14

[II. 14]

On the circle of the year of Venus.[3]

Let us now divide the circle of the year of Venus as follows:

As before, we shall draw, about a common centre G, one or two spaces with a narrow margin, and from the beginning of the margin let us cut off arc AX, against [the order of] the signs, the arc being about 17′ 52″. Let us then divide the whole of the remaining arc of the margin into 15 equal parts, and each of these again into 15. From these parts we are then to make signs, giving to any one sign 30 parts. There will, [in these 225 degrees], be seven signs and fifteen degrees. When any of these degrees has been divided into 60 minutes, or into whatever is possible with these small intervals, the remaining arc XA will comprise 11′ 9″. The space is then to be inscribed with the numbers of signs and degrees, as before. And [the circle] is complete.

[1] Numbered (22) below.

[2] *de illis partibus graduum.*

[3] Numbered (23) below.

CAPITULUM 15^m

De divisione circuli Geuzahar.

Superest circulum anni Capitis Draconis Lune dividere, consequenter separemus duo spacia cum margine; et accipiamus a capite marginis arcum *AX* contra signa 6 gra. 37 min. 56 sec., et residuum dividamus in 19 partes equales que erunt gradus, et quemlibet gradum dividemus in 60 min.; et intitulabimus spacia gradibus et minutis ut prius; et arcus residuus *XA* de illis partibus continebit 21 min. et 22 sec.

Et si videatur divisio arcus[a] *AX* in 19 partes equales aliqualiter tediosa, faciamus arcum *AX* contra signa 25 gra. 13 min. 31 sec. quam precise poterimus; et residuum secundum successionem signorum dividamus in 18 partes equales, et tunc continebit arcus *XA* de illis partibus 1 gra. et 21 min. et 22 sec.; et redibit utraque operacio in id ipsum. Hec autem operacio presupponit certum motum medium omnium predictorum in una hora.

CAPITULUM 16^m

De circulo coniunctionis.

His succedat circulus coniunctionis et prevencionis equate Solis et Lune, pro quo separabimus circulum habentem marginem et spacium unum pro litteris numerorum; et a principio marginis separabimus arcum *AX* contra signa 2 gra. et 30 min.; et residuum dividemus in 11 partes equales, et erunt 11 gra. Quantitas autem unius illarum 11 parcium erit 32 par. et 30 min. de gradibus zodiaci limbi eius. Et quemlibet gradum dividemus in 60 minuta, quo facto, dividemus spacium marginis per 10 et 10 minuta, et in illis spaciis scribentur littere numerorum pro minutis, preterquam in spacio ultimorum 10, et ibi scribentur distincte numerus graduum.

Post hoc, quam remoti poterimus commode prope centrum *G* ut puta inter ecentricum Saturni et Iovis, separabimus circulum unum

[a] H λ

CHAPTER 15

[II. 15]

On the division of the circle of geusahar.[1]

It remains to divide the circle of the year of the head of the dragon of the Moon,[2] and therefore let us mark out two spaces with margin. From the beginning of the margin let us take an arc *AX* of 6° 37′ 56″, against the signs, dividing the remainder into 19 equal parts (degrees), and each degree into 60 minutes. We shall put in the numbers of degrees and minutes as before. The arc *XA* of the remainder, in terms of the [newly defined] parts, will be 21′ 22″.

If the division of the arc *AX* into 19 equal parts should seem somewhat tedious, let us make the arc *AX* 25° 13′ 31″, against the signs, as precisely as we can. Let us divide the remainder, in the order of the signs, into 18 equal parts, when the arc *XA* will comprise 1° 21′ 22″ in terms of these parts. Both methods of graduation will give the same result.[3] This method of working, however, presupposes [a knowledge of] a definite mean motus of all the aforesaid in one hour.

CHAPTER 16

[II. 16]

On the circle[s] of conjunction.[4]

After these comes the circle of conjunction and prevention[5] of the equated Sun and Moon, for which we shall mark out a circle having a margin and a space for the figures. From the beginning of the margin, and against the signs, we shall mark off an arc *AX* of 2° 30′, dividing the remainder into 11 equal parts, which will be 11 degrees. Each of these 11 parts will be of 32 parts and 30 minutes, in terms of the degrees of the zodiac of the limb. Each degree we shall divide into 60 minutes, having done which we shall divide the space [next to] the margin at intervals of 10′, writing the number of minutes in each space, with the exception of the last, in which the number of degrees is to be written distinctly.

After this, as far away [from the previous scale] as we conveniently can and as near as possible to the centre *G* (as, for instance, between the eccentric of Saturn and that of Jupiter), we draw a circle having

[1] Numbered (24) below.

[2] That is, the ascending node of the Moon's path at its intersection with the ecliptic.

[3] Literally, 'each operation will return to that very point'.

[4] Numbered (25) below, but comprising two scales: (26) the 'circle of longitude between the Sun and Moon'; and (27) the 'circle of lunations'.

[5] This turn of phrase is found in the Latin tradition of the canons to the Toledan tables. See MS. Laud misc. 644 at f. 96rb, and (in the version associated with the town of Cremona) Bodleian MS. Canonici misc. 51, at f. 81vb.

habentem marginem et spacium; et a capite marginis contra signa resecemus arcum *AX* ut prius, cuius quantitas sit 9 gra. 34 min. et 34 sec., et residuum, secundum successionem signorum, dividemus in 12 partes equales. Et si pars terciadecima extenderit se ultra *A* usque 19 gra. 31 min. et 45 sec., fideliter est divisum.

Et erunt hee[a] partes 12 lunaciones, quarum quamlibet dividamus in 4 partes equales, terciadecima lunacione que incompleta est[b] totaliter indivisa; cuius tamen partis[c] terciedecime finem in margine sensibiliter signemus, intitulaturque pars hec incompleta ultra 12 lunaciones *Epacta*. Et nota quod quanto[d] fuerit maius spacium inter istum circulum lunacionis et priorem, qui est circulus longitudinis, cum sua parte duodecima, tanto congruencius, dum tamen iste circulus lunacionum non fuerit nimis prope centrum pro suis divisionibus, quia in spacio intercepto describuntur hore Lune inequales hoc modo.

CAPITULUM 17m

De horis lunacionis.

Inter intimam circumferenciam circuli exterioris, que dicitur *circulus longitudinis*, et extimam circumferenciam huius interioris, que[e] est *circulus lunacionum*, productis diametris orthogonalibus sed occulte *AC* [et] *BD*,[f] inveniatur unus circulus contingens ipsas circumferencias ad duo puncta opposita in linea *AC*, ita ut[g] contingat exteriorem versus *A* et interiorem versus *C*; et fiat iste[h] ecentricus sensibilis. Et ubi iste ecentricus dividit lineam diametrabilem *DB* ponantur puncta[i] *D*, *B*, et posito centro in *G*, fiat circulus unus occultus transiens per *D* et *B*, ut patet in presenti figura.

Deinde extimam[j] circumferenciam circuli coniunctionis dividemus sic: a principio[k] secundum successionem signorum, separabimus unum arcum 18 gra. et 2 min., et secundum quantitatem huius partis prime dividemus totam circumferenciam ad sensibilia puncta ut sint *A*, *F*, *G*, *H*, *I*, *K*, *L*, et cetera. Nam ista circumferencia sic divisa continebit 24 divisiones, excepto quod ultima divisio erit incompleta per 2 gra., si bene fuerimus operati.

[a] A λ [b] H λ [c] A partes [d] H quantum [e] CAHL qui
[f] H *ABCD* [g] CAL vel; H vel ut [h] A λ [i] H λ [j] H intimam
[k] A +eius *A*; H +eius sic ad *A*

a margin and space. From the head of the margin, against the signs, we shall cut an arc *AX*, as before, whose size is to be 9° 34′ 34″, and we shall divide the remainder, in the order of the signs, into 12 equal parts. And if the thirteenth part extends beyond *A* by 19° 31′ 45″,[1] then [the scale] has been accurately divided.

These parts will be 12 lunations, each one of which we are to divide into four equal parts, with the thirteenth lunation, which is incomplete, wholly undivided. But let us mark distinctly the beginning[2] of the thirteenth part in the margin, and let this incomplete part beyond 12 lunations be inscribed 'Epact'. And note that the greater the space between that circle of lunation, with its twelfth part, and the former, the circle of longitude,[3] the better the agreement,[4] provided that the circle of lunations is not too near the centre for its own divisions, for in the intervening space are drawn the unequal hours of the Moon, [this being done] as follows.

CHAPTER 17

[II. 17]

On the hours of the lunation.[5]

Between the inmost circumference of the outer circle, which is called the *circle of longitude*, and the outermost circumference of the inner one, which is the *circle of lunations*, orthogonal diameters *AC* and *BD* having been drawn in lightly, a circle is to be found touching the circumferences themselves at two points at opposite ends of the line *AC*, in such a way that it touches the outer one at *A* and the inner one at *C*. And let that eccentric be [inscribed] plainly. Where that eccentric intersects the diametral line *DB*, the points *D* and *B* are to be placed. With centre *G*, let there be a circle drawn in lightly which passes through *D* and *B*, as is clear in the present figure.

We shall then divide the outermost circumference of the circle of conjunction in this way: from its beginning, in the order of the signs, we shall mark off an arc of 18° 2′. We shall divide up the entire circumference at the clearly marked points *A*, *F*, *G*, *H*, *I*, *L*, etc., all separated by this same arc-length. The circumference so divided will contain 20[6] divisions, except that the last will be incomplete by an amount of 2°,[7] if we have worked carefully.

[1] The correct figure is 19° 37′ 45″, assuming no error in the preceding figures.

[2] *finis* (!)

[3] That is, between (26) and (27).

[4] The passage is obscure.

[5] The lines are in due course here numbered (28).

[6] Corrected from 24—see the text.

[7] This is almost certainly a mistake for 40′. Although the earlier figure of 18° 2′ could be a mistake for 18° 6′, reasons will be found in II. 17 for thinking that this is not so. There is an intelligent gloss on MS. C, f. 59r, at this point, altering the text in the way suggested here, and saying that the instructions ('canon') do not agree perfectly with the use explained in III. 19 below unless amended. This was the glossator who had clearly seen Richard's own albion (see II. 20 comm. below).

Item[a] circulus occultus *DB* dividatur hoc modo: ab eius principio, secundum signa separemus arcum *AM* 16 gra. et 21 min., et secundum illam quantitatem dividamus totam circumferenciam *AB* occultam ad notas *A*, *M*, *N*, *O*, *P*, *Q*, et cetera; et habebit ista circumferencia tales partes 22 fere.

Tercio, de extima circumferencia circuli lunacionum ab eius principio consimiliter separemus arcum unum 15 gra. 9 min. et secundum quantitatem illius arcus dividamus totam circumferenciam ad notas *A*, *R*, *S*, *T*, *U*, *X*, *Y*, et cetera; et habebit ista circumferencia tales partes 24 fere.

Ut autem absque tedio hos tres circulos dividamus per motus Lune in una hora ad hos tres status, scilicet ad longitudinem maximam, mediam, et minimam, ad quos est motus horarius vel 36 min. et 4 sec. in proxima, vel est 32 min. et 42 sec. in media, vel est 30 min. et 18 sec. in maxima, composite sunt tabule sicut infra dicetur parte tercia; sed tunc de gradibus limbi oportet computare unum signum pro uno gradu et unum gradum pro duobus minutis, et ista divisio redit in idem cum prima.

Divisis autem his tribus circulis, ducatur primo linea una[b] recta patens per eorum principia, *A*, *A*, et *A*. Item invento centro, ducatur linea arcualis[c] per[d] tria puncta *F*, *M*, *R*; et tercia[e] per tria puncta *G*, *N*, *S*; et quarta per puncta *H*, *O*, *T*,[f] et sic deinceps. Caveatur tamen in fine quod nulla linea horaria non transgrediatur[g] primam lineam horarum, que est *AAA* recta.

Post hec abraso occulto circulo *DB*, scribantur in spaciis horarum littere numerorum horarum sicut in astrolabio solet. Novissimo, infra ecentricum Saturni vel ubi commodius poterit, fiant duo parvi circuli in quibus signari poterunt auges et geusahar planetarum, qui habent auges[h] fixas et geusahar.

Et completa est prima facies instrumenti.

CAPITULUM 18^m

De inscripcione circulorum in limbo primo secunde faciei instrumenti.

Alteram faciem instrumenti iam restat dividere, et erit inchoacio operis a latitudine limbi *LM*. Dividamus igitur limbi latitudinem in tres partes equales cum totidem strictis marginibus, et primum marginem dividamus in 360 partes equales, et distribuamus eas in

[a] H iste [b] H λ [c] H +pro [d] CAHL proxima [e] A secunda
[f] L hoc [g] CAL transgrediantur [h] H causas

Next, the lightly drawn circle *DB* is to be divided thus: from its beginning, and in the order of the signs, we are to mark off an arc *AM* of 16° 21′, and to divide the whole of the lightly drawn circumference *AB* at [points] denoted by the letters *A, M, N, O, P, Q*, etc., [all separated by] this same quantity. The circumference will have approximately 22 such parts.

Thirdly, from the beginning of the outermost circumference of the circle of lunations let us similarly mark off an arc of 15° 9′, and divide the entire circumference of the circle at *A, R, S, T, U, X, Y*, etc., [all separated by] this same quantity. The circumference will have about 24 such parts.

In order to divide these three circles without tedium, by means of the motion of the Moon in one hour at its three positions, namely, at maximum, mean, and minimum distance, at which the hourly motion is 36′ 4″ (at the Moon's closest), 32′ 42″ (at the mean position), or 30′ 18″ (at its greatest distance), tables are drawn up as is explained below, in part III. In terms of degrees of the limb, one sign should be taken as one degree, and one degree for two minutes. This [method of] division gives the same results as the first.

These three circles having been divided up, first let a straight line be drawn plainly through their starting-points *A*, *A*, and *A*. Next, its centre having been found, an arc-line is to be drawn through the three points *F*, *M*, and *R*; a third [arc] through the three points *G*, *N*, and *S*; a fourth through the points *H*, *O*, and *T*; and so on. Towards the end, however, beware that no hour-line cross the first line of hours, which is the straight line *AAA*.

After this, the lightly marked circle *DB* having been rubbed away, the hour-numbers are to be inscribed in the appropriate spaces, as is customary on an astrolabe. Lastly, below the eccentric of Saturn, or wherever convenient, let two small circles be drawn on which the auges and nodes of the planets may be marked. (The planets have fixed auges and nodes.)[1]

The first face of the instrument is now complete.

CHAPTER 18

[II. 18]

On the inscription of the circles on the first limb of the second face of the instrument.

It now remains to divide up the other face of the instrument, and the task will begin with the border of the limb *LM*. Let us, then, divide the breadth of the limb into three equal parts, with as many narrow margins, and let us divide the first margin[2] into 360 equal

[1] This is wrong, but was widely believed, following Ptolemy. See the commentary to this chapter, and also I. 7 comm.

[2] Later numbered (54).

12 signa[a] intitulantes, ut prius, nomina vel numerum signorum cum numero graduum in marginis spacio, secundum successionem signorum,[b] incepcio autem signorum in hoc limbo fiat apud *D*.[c]

Item ab eodem puncto incipiamus dividere secundam marginem per tabulam ascensionum in circulo directo, sed contra naturalem successionem signorum ab oriente in occidente,[d] a puncto *D* per *C* armillam; et dividamus in signa et scribamus ut prius numeros in spacio marginis.

Item ab eodem puncto *D* contra naturalem successionem [signorum] dividemus terciam[e] marginem cum suo spacio, per tabulam ascensionum[f] ville vel climatis ubi magis intendimus commorari et consideracionibus magis uti; et spacium huius marginis intitulabimus nominibus et numeris signorum et graduum, sicut sepe dictum est. Et completum est.

CAPITULUM 19m

De composicione circuli involuti.

Laminam interiorem *ROS* postea dividemus hoc modo:

Primo separabimus brevem marginem cum spacio. Deinde ipsa lamina immobiliter loco fixa, producemus occulte diametros orthogonales *AC* et *DB* super centrum *O*. Super quattuor centra in diametris accepta prope faciemus unum circulum involutum 32 revolucionum hoc modo:

Prope *O* in diametro *AC* signentur duo centra altrinsecus *E* [et] *F*, in quorum medio equidistanter sit *O* centrum. Item in linea *DB* signentur alia duo centra equidistanter ipsi *O*, qui sint *G* [et] *I*, ita quod in eadem linea[g] sint consequenter puncta *D*, *G*, *O*, *I*, *B*, et in altera *A*, *E*, *O*, *F*, *C*, hoc modo: apud *B* accipiatur una linea, que sit *BH*, quantum sufficit ad latitudinem unius spacii circularis cum suo margine, et dividamus ipsam lineam in quattuor partes equales; et a prima divisione prope *B* accipiatur linea versus *O* equalis linee *BO* cuius terminus sit *G*. Erit[h] distancia quattuor centrorum *F*, *G*, *E*, *I* a centro *O* equalis distancie linee *GO*.

Deinde posito pede circini in centro *I*, extendatur alius pes ad[i] *B* et circumducatur usque ad semidiametrum *OC* ad punctum *K*. Item

[a] H supra

[b] H +vel potest accipi una margo inter duo spacia quorum unum spacium erit intitulatum secundum successionem signorum

[c] H +et erit *D* in principio quarti circuli ab armilla versus sinistrum et *A* directe ex opposito armille.

[d] CA oriens [e] CAL tercium [f] A ascensionem [g] A λ

[h] CAHL +*G* (a relic of 'ergo'?) [i] CAHL *AD*

parts, distributing them into 12 signs whose name, or number, is filled in as before, following [the natural] order of the signs, in the space [next to] the margin, together with the number of degrees. The point from which the signs are to commence on this limb, however, is at *D*.

Let us next start from the same point, but against the natural order of the signs, from east to west, to divide the second margin, working from the point *D* through *C* [and] the ring,[1] by means of a table of ascensions on the direct circle. Let us divide into signs, and as before inscribe the numbers in the space [next to] the margin.

Working against the natural order of signs, from the same point *D*, let us now divide up the third margin with its (adjacent) space, by means of a table of ascensions [calculated] for a town or climate in which we intend to stay for a long time and make many observations.[2] In the space [next to] this margin we shall fill in the names and numbers of the signs and degrees, as frequently explained [already].

CHAPTER 19

[II. 19]

On the construction of the spiral.[3]

Let us next divide up the inner plate *ROS* in the following manner: We shall first mark out a narrow margin with a space. Then, the plate itself being firmly fixed in position, let us lightly draw orthogonal diameters *AC* and *DB* through centre *O*, and let us construct, as explained below, a spiral of 32 turns about four centres taken on the diameter near [*O*].

On the diameter *AC*, near and equidistant from *O*, two centres *E* and *F* are to be marked, one on each side of *O*. Next, on the line *DB* two other centres are to be marked, equidistant from *O*, these being *G* and *I*; so that there are the points *D*, *G*, *O*, *I*, and *B* in that order on the one line, and *A*, *E*, *O*, *F*, and *C*, on the other. [The points are to be chosen] in this way: a line *BH* is to be taken at *B* of length equal to the width of one circular space with its margin, and we are to divide the line [*BH*] into four equal parts. From the division nearest *B* a line is to be taken in the direction of *O* equal to the line *BO*, and terminating at [a point to be known as] *G*. The distance of the four centres *F*, *G*, *E*, and *I* from the centre *O* will be equal to the length *GO*.

One foot of the dividers being placed at centre *I*, the other foot is to be extended to *B*, and drawn round to the semidiameter *OC*,

[1] The lettering of the figure has not changed from that of II. 1. Although at first sight it might appear that the point *C* is now at the ring, this cannot be (see II. 30 below).

[2] *consideracio*. The word perhaps assumes what some have suggested to be its original Latin meaning (examination of the stars, from *con*+*siderare*).

[3] *circulus involutus*.

posito centro in *F*, et alter pes circini ponatur in *K*, et circumducatur usque ad semidiametrum *DO*, terminans in punctum *L* prope *D*. Item posito centro in *G* et altero pede in *L*, circumducatur usque ad semidiametrum *OA*[a] in punctum *M*. Item posito centro in *E* ⟨et altero pede in *M*⟩, circumferatur alter pes usque ad semidiametrum *OB*; et iste vocatur unus circulus involutus. Tales revoluciones involutas faciamus 32, et super eisdem centris, constricto modicum[b] circino, marginabimus eas. Et completum est.

CAPITULUM 20m

De arte componendi circulum involutum.

Circulus involutus continebit medium motum anni Mercurii, et medium motum anni Lune, et medium motum argumenti Lune; ideoque primo dicamus de[c] descripcione[d] medii motus Mercurii.

Inchoantes igitur in prima revolucione apud *D* dividamus illam continue per tabulam medii motus Mercurii expansam ad dies singulos anni solaris, scilicet ad 365 dies et ultra quantum est eius motus ultra in residuo anni, scilicet 6 hore fere, que tabula ponetur in parte huius [quarta], inter ceteras tabulas. Ponemus autem omnes divisiones in margine, et sextam et sextam in spacio marginis, et tunc distribuemus illos 365 dies in menses dando cuilibet mensi dies suos, inchoando a mense Ianuario vel Marcio, sicut in aliis. Et continebit ultimus mensis ultra dies illud superfluum quarte diei fere. Et completum est.

CAPITULUM 21m

De inscribendo medio motu Lune.

Ad inscribendum medium motum Lune in residuo circuli involuti, primo amoveamus laminam *ROS* de loco suo fixo, et ponamus ultimam divisionem circuli anni Mercurii in directo contra punctum *D*, et tunc consolidemus laminam *ROS* in loco illo; et incipientes a puncto quod est in directo *D* dividamus circulum involutum continue per motum Lune in singulis diebus anni, et ultra quantum superfluit annus ultra 365 dies; et ponemus has divisiones in margine

[a] CA *OS* [b] A modico [c] CAH λ [d] H descripcionem

[which it meets] at the point *K*.[1] Next, with centre *F*, and the other foot of the dividers at *K*, this is to be drawn round to the semidiameter *DO*, ending at the point *L* nearer *D*. Next, with centre *G* and the other foot at *L*, the latter is to be drawn round as far as the semidiameter *OA* at the point *M*. Next, with centre *E*, [and the other foot at *M*], let this other foot be carried round to the semidiameter *OB*; and this [completes] what is called a single involuted circle, [or turn of the spiral]. We are to mark 32 such turns, and with the same centres, closing up the dividers somewhat, we are to provide them with a margin.

CHAPTER 20

[II. 20]

On the method of arranging the spiral.

The spiral will contain the mean motus[2] of the year of Mercury, the mean motus of the year of the Moon, and the mean motus of the argument of the Moon; and first, therefore, let us explain how to inscribe the mean motus of Mercury.

Beginning on the first turn, at *D*, let us divide it continuously by means on the table of mean motus of Mercury, compiled for intervals of a single day of the solar year, that is, up to 365 days, and beyond that as much as is its additional motus for the rest of the year, that is, approximately 6 hours. This table is placed amongst the other tables, in part [IV][3] below. We shall, however, put all the divisions in the margin, with every sixth continued into the adjoining space, and then we shall distribute the 365 days into months by giving to each month its appropriate number of days, beginning with the month of January or March, as elsewhere. And the last month will contain beyond the last day the remainder of approximately a quarter of a day.

CHAPTER 21

[II. 21]

On inscribing the mean motus of the Moon.

In order to inscribe the mean motus of the Moon on the remaining part of the spiral let us first move the plate *ROS* from its fixed position, and put the last division of the circle of the year of Mercury in the direction of *D*, fixing the plate *ROS* in that position. Beginning from the point in the direction of *D*, let us divide the spiral continuously for the motus of the Moon at daily intervals over a year, and as far beyond as the year exceeds 365 days. We shall put these divisions in

[1] The direction of the arc must be such that arc $BC = 90°$, and so on.

[2] Strictly, its mean movement in *argument*.

[3] Numbered here IV. 13.

involuto, et similiter in margine illo facto in limbo lamine *SOR* nondum diviso. Deinde quamlibet divisionem factam in illo margine extimo dividemus in 24 partes equales, et erunt hore motus Lune.

Et spacium marginis involuti dividemus per binos et binos dies, et in directo dierum scribemus in spaciis numerum dierum, et litteras feriarum, et nomina festorum secundum commoditatem instrumenti, ne confusio set pocius[a] distinctio[b] appareat manifesta.

Et cum hoc perfectum fuerit, iterum amovebimus laminam *ROS*, et ponemus finem anni lunaris contra *D*, et ibi incipiemus dividere residuum circuli involuti per tabulam[c] argumenti Lune, omnino sicut dictum est de medio motu Lune; fiatque apparens signum ubi incipit primus mensis motus Lune et argumenti Lune. Et completum est.

CAPITULUM 22^m

De climate et almucantarath.

Vertamus laminam *ROS* et consolidemus eam in loco suo. Deinde separemus de limbo eius unum marginem cum unico spacio, et dividemus marginem eius in 12 signa et gradus equales; et intitulabimus eius[d] spacium ita quod caput Capricorni sit apud *A*, et caput Arietis sit apud *D*, cum successu et contra naturalem successum signorum. Deinde patenter productis diametris *AC* et *BD* extrahatur ab intimo circulo[e] arcus 30 graduum computando ab *A* versus *D*. Deinde posita regula in[f] *B*, linea a puncto *B* usque ad finem arcus 30 graduum[g] ibi ponatur *E*. Deinde ubi linea *BE* abscindit semidiametrum *OA*, ponatur nota *F*; fiatque circulus secundum quantitatem semidiametri *OF*, qui erit vice circuli equinoxialis. Deinde, per artem traditam in operacione astrolabi, extrahatur linea ecliptica, cuius arcus meridionalis sit versus punctum *A* et arcus septentrionalis versus *C*; et fiat ecliptica sensibilis atque patens.

[a] H λ [b] CAL distincte [c] A tabulas
[d] H 2λ+et erit interior circulus dicti spacii circulus Capricorni
[e] H +rote
[f] H +intersectione ubi diameter *OB* intersecat circulum Capricorni in
[g] H +in Capricorno in parte interiori

the spiral margin, and similarly in that margin made on the limb of the plate *SOR* which is not yet divided. Let us now divide into 24 equal parts each division made on the outer margin, and these will be hours of the Moon's motus.[1]

We are to divide the space of the margin of the spiral into groups of two days, and opposite the days write in the numbers of the days, the day letters,[2] and the names of the feasts so far as the instrument permits, preferring distinct to confused lettering.

When this is done, again we shall move the plate *ROS*, and put the end of the lunar year against *D*; and there we shall begin to divide the remainder of the spiral by means of a table of the argument of the Moon, entirely as explained of the mean motus of the Moon. And let there be a sign marking the point where begins the first month of the motus of the Moon and of the argument of the Moon.

CHAPTER 22

[II. 22]

On the climate, and on the almucantars.

Let us turn the plate *ROS* over, and fix it in position. Next, let us mark out on its limb a margin with a single space, and divide the margin into 12 signs and into equal degrees. We shall label the corresponding space in such a way that the beginning of Capricorn is at *A*, and the beginning of Aries is at *D*, with the signs in the right sequence, but running counter to the natural order. The diameters *AC* and *BD* having then been clearly produced, let an arc of 30° counting from *A* towards *D* be taken off the inner circle. The rule being placed at *B*, let the point *E* be [one extremity of] the line from the point *B* to the end of the arc of 30°. *F* denotes the point at which *BE* intersects the semidiameter *OA*, and let there be a circle with semidiameter *OF*, which will represent the equinoctial circle.[3] By means, then, of the theory taught in connection with the working of an astrolabe, an ecliptic is to be drawn, whose southerly arc[4] is towards the point *A*, and whose northerly arc[5] is towards *C*. Let this ecliptic be clearly perceptible.

[1] This last scale is not well explained. See the commentary.

[2] *littere feriarum*. *Feria* generally means 'holiday' or 'weekday'. *Feria prima* is 'Sunday'. To determine Easter and those feasts with a related position in the calendar it is necessary to know the day of the week of the paschal full moon. To assist in this and other calculations of the day of a given date, the seven letters A to G are given cyclically to the days of the year, beginning at 1 January. The *dominical letter* of a year is the letter assigned to the first Sunday in the year. Matters are slightly complicated, of course, by the existence of leap years.

[3] The celestial equator.

[4] That is, the half below the equator, with the head of Capricorn as its lowest extreme.

[5] That is, the half above the equator, whose most northerly point is the first point of Cancer.

Deinde extra eclipticam fiant sex circuli subtiles ostendentes latitudinem sex graduum meridionalium;[a] et totidem circuli fiant intra eclipticam ostendentes latitudinem sex graduum septentrionalium. Deinde dividatur sola ecliptica per ascensiones in circulo directo, et invento polo ecliptice per totam declinacionem a polo equinoxiali *O*, qui polus ecliptice sit *G*,[b] inveniantur circuli qui possunt transire per polum *G* et per singulam divisionem eliptice et eius nadir; et sensibiliter divident isti circuli totam latitudinem zodiaci, et non appareat quicquam amplius de illis circulis.

Deinde fiat orizon et clima illius latitudinis cuius ascensiones scribuntur in limbo armille *LM*; et fiant perfecte azimuth et almucantarath ita tamen quod non confundant nimis gradus zodiaci in longitudine et latitudine, nec appareat quicquam de circulo equinoxiali aut de tropicis nisi sub almucantarath inter lineas horarum. Et completum est.

CAPITULUM 23^m^

De eodem.

Si autem placuerit habere plures tabulas climatum, faciamus illas omnino sicut dictum est uno addito, quod in limbo eiusdem lamine oportet in loco circuli zodiaci eius, vel cum[c] illo, inscribere eciam ascensiones eiusdem climatis ad quod illa tabula perficitur. Et non est opus habere rethe stellarum in isto opere, sed imprimantur stelle et eorum nomina in singulis tabulis climatum. Sed excedit hoc opus astrolabium in facilitate operandi etprec isione; et superaddit quedam quia cum hoc operari possumus cum omnibus stellis erraticis de nocte equifaciliter sicut cum stellis fixis, quod non contingit in astrolabio, sicut patebit.

CAPITULUM 24^m^

De saphea.

In fundo etiam medie lamine perficere possumus instrumentum saphee, quod est quasi astrolabium universale, vel saltem ad abbreviandum difficultatem operis.

[a] L 9λ (haplog.) [b] L 8λ (haplog.) [c] A tamen

Outside the ecliptic let there be six circles, accurately showing a latitude up to 6 degrees south, and let there be as many circles within the ecliptic showing a latitude of 6 degrees north. The ecliptic alone is then to be divided by means of the ascensions on the direct circle; and the pole of the ecliptic, *G*, having been found through its declination[1] from the equinoctial pole *O*, let those circles be found which pass through the pole *G* and through each separate [point of graduation] of the ecliptic and its nadir.[2] These circles will clearly divide up the whole width of the zodiac; and let these circles be drawn only so far as is necessary to cover the band of the zodiac.[3]

The horizon and climate are next to be added, for that latitude appropriate to the ascensions inscribed on the limb of the ring *LM*;[4] but the azimuths and almucantars are to be drawn in such a way that they are not seriously mixed up with the degrees of the zodiac in longitude and latitude. And no hour line is to be drawn from the equinoctial circle or the tropics, unless it be below the almucantars.

CHAPTER 23

[II. 23]

On the same subject.

If it be desired to have several tablets of climates, let us make them exactly as explained already, but with one additional [plate], on the limb of which it is necessary to inscribe, either in place of, or together with, its zodiac circle, the ascensions of the climate at which that table is drawn up.[5] There is no need to have a rete of stars on that instrument, but let the stars and their names be inscribed on the individual tablets [for the different] climates. This instrument, even so, surpasses the astrolabe in simplicity of operation and precision; and it adds something more, for with it we can work by night as easily with the planets as with the fixed stars (which is not true of the astrolabe), as will become clear.

CHAPTER 24

[II. 24]

On the saphea.[6]

In the recess[7] in the mater we can make the saphea instrument, which is a sort of universal astrolabe, or one which at least enables us to reduce the difficulty of working.

[1] The obliquity of the ecliptic, whose value is not stated here.

[2] That is, each point 180° removed in longitude. See the glossary. The centres of the circles described (loci of constant ecliptic longitude) lie on a straight line. See our commentary.

[3] This is a free translation of an obscure passage.

[4] That is to say, on scale (56) of the mater. See II. 18 above.

[5] Necessary, since scale (56) on the mater serves only the principal climate chosen.

[6] More specifically *saphea Arzachelis*. See glossary and commentary.

[7] *In fundo.*

In saphea, factis circulis ascensionum et revolucionum saphee, extrahi poterit orizon certi climatis et cenith capitis; et per equidistantes lineas orizonti usque ad cenith capitis, quod est polus[a] omnium illorum, et per lineas quotlibet[b] exeuntes a cenith cadentes orthogonaliter super orizontem, perficere poterimus almucantarath et azimuth eiusdem climatis; et erit eius operacio expedita ad horas diei et noctis, et mirabiliter manifesta.

Et si placuerit, sub illo orizonte obliquo signari poterit orizon non multum obliquus a priori, et eius polus[c] extrahi poterit in parte opposita quasi in angulo Terre; et perficere poterimus almucantarath et azimuth secundum exigenciam huius orizontis et cenith capitis, et sic in eadem superficie satis commode habebimus quasi duo climata; et marginentur eorum orizonta et circulus zodiaci distincte et aperte, et non[d] erit confusio quamvis multi circuli condense sese intersecent.

Qui autem conceperit artem componendi sapheam de facili potest ista concipere, et ideo non est opus hic diucius immorari circa exemplarum descripcionem istorum; ita igitur fiat. Et completum est.

CAPITULUM 25^m

De linea corde recte et linea minutorum dividenda.

Vertamus insuper laminam *PGQ*, et consolidemus eam in loco suo, quia ex parte hac describimus circulos eclipsium Solis et Lune; et primo perficiamus in ea lineam corde recte et verse hoc modo:

Productis diametris orthogonalibus *AGC* et *BGD*, perficiamus marginem cum spacio, equidistanti diametro *AC*; iam capitulo secundo inceptum. Deinde totam diametrum medii circuli communis duobus marginibus limbi primi, scilicet circuli altitudinis et zodiaci octave spere, dividemus in 60 partes equales, et eius spacium per quinos et quinos gradus, et quamlibet partem illarum 60 dividamus ad minus in tres partes, vel quot minuta fuerit possibile; et intitulabimus[e] in spacio illos gradus duplici numero quorum unus incipiet a[f] *C* versus *A*, et alter ab *A* versus *C*, et erit uterque numerus 60; nec

[a] H plus [b] A quodlibet [c] A λ [d] A nota [e] A intitulamus
[f] H *A*

Circles of ascensions and revolutions[1] having been made for the saphea, it will be possible to mark out the horizon of a definite climate and the zenith;[2] and by means of lines equidistant from the horizon[3] going right up to the zenith, which is the pole of all of them, and by means of as many lines as you like going out from the zenith and falling orthogonally on the horizon,[4] we shall be able to complete the [system of] almucantars and azimuth-lines for the same climate. And working with this is rapid, whether by day or night, and is wonderfully clear.

If desired, another oblique horizon[5] may be marked in under the former, but not greatly inclined to it, and its pole may be joined to the opposite pole, with the angle the same as in the case of the Earth. And we can put in the almucantars and azimuth-lines according to what is needed with this horizon and zenith, and thus we shall have quite conveniently on the same surface, as it were, two climates. Their horizons and the circle of the zodiac are to be provided with a margin, clearly and distinctly, and there will be no confusion, even though many circles crowded together are mutually intersecting.

Anyone who wishes to learn the art of constructing the saphea may easily do so, and therefore there is no need to spend longer here on drawing an example. Let [the saphea] be made, then, in this way.

CHAPTER 25

[II. 25]

On dividing the scale[6] of corda recta and the scales of minutes.

Let us turn over the plate *PGQ*, and fix it in position.[7] It is on this side that we draw the circles of the eclipses of the Sun and Moon. Let us first execute on it the scale of corda recta and corda versa in the following way:

Having drawn orthogonal diameters *AGC* and *BGD*, let us make a margin with corresponding space, parallel to the diameter *AC*, a matter which was introduced [at the end of] chapter 2 above. Let us then divide up into sixty equal parts the entire diameter of the middle circle common to the margins of the scale of altitude and that of the zodiac of the eighth sphere. Let us divide the corresponding space at intervals of five units, and each of the sixtieth parts [of the whole] let us divide into three, or as many minutes as possible. We shall label those units twice over in the space, one set of numbers running from *C* to *A*, and the other from *A* to *C*; and both [*C* and *A*]

[1] That is, lines of constant ascension and declination.

[2] *cenith capitis.*

[3] Not, that is to say, parallel straight lines on the plate, but parallels of altitude, i.e. almucantars.

[4] These are not, of course, straight lines, but the arcs of circles.

[5] *orizon obliquus.* See glossary.

[6] *linea.*

[7] For the inscription of the first side see II. 8 sqq.

omnino transeat aliquis circulus per hanc lineam cordarum, ne confundat eius divisiones omnino.

Deinde posito circino in centro communi *G*, de medietate diametri iam divise,[a] accipiamus lineam 21 gra. et 20 min. et secundum illam quantitatem resecabimus de utraque parte alterius diametri orthogonalis; et marginabimus illud altrinsecus cum spacio, ad instar posterioris, et utramque illarum parcium *GD* et *GB* dividemus in 64 partes equales: et spacium eius per 5 et 5, inscribendo numeros incipiendo a centro *G* versus *B* et versus *D*. Et completum est.

CAPITULUM 26m

De figura eclipsis Lune scribenda.

Circulum eclipsis Lune perficiamus hoc modo: Posito centro in *G*, extendatur alter pes super alteram linearum divisarum per 64[b] partes equales usque ad 63 par. et 36 min. et secundum illam[c] quantitatem fiat semicirculus *DEB* super centrum *G*, nullo tamen modo confundat eius transitus divisiones signatas[d] in linea *CA* divisa in 60. Item super idem centrum fiat alter semicirculus *FHI*, ita quod tam *GF* semidiameter quam *GI* sit 52 par. et 50 min. de eisdem partibus. Deinde divisa linea *DGI* in duas partes equales, et in eius medio facto centro, ducatur semicirculus *DKI*, contingens utrumque duorum semicirculorum priorum.

Post hec extra circulum *DEB*[e] perficiamus primo marginem prope *DEB* et deinde spacium; et eius marginem dividamus hoc modo: super semidiametrum *GA* divisam[f] in 30 partes equales, quarum quelibet[g] similiter est divisa in tres partes que sunt minuta 20, scilicet pars tercia unius gradus, extendatur regula orthogonaliter super[h] quamlibet partem 20 minutorum; et ubi hec regula contingit semicirculum *DEB* signetur punctus ex utraque parte, quinta tamen divisio[i] transeat super totum spacium. Quod est inter duos semicirculos *DEB* et *FHI* usquequo totus semicirculus *DEB* fuerit per talia puncta divisus, et spacium quod est inter duos semicirculos predictos per lineas rectas, que non extendantur ultra spacium inter dictos semicirculos interceptum.

Post hec posita regula in centro communi *G*, transeat per quemlibet[j] punctum signatum in semicirculo antedicto *DEB*, et dividamus marginem antedictam operacioni communi quarum quinta et quinta[k] divisio ultra extendatur per spacium pro litteris assignatum.

Deinde de semidiametro *GD* capiatur linea 45 par. et 56 min., que sit linea *GL*. Item de linea *GB* capiatur altera linea 38 par. et 10 min.,

[a] CAHLS divisi [b] CAHL 60 [c] H λ [d] A asignatas [e] L 5λ (haplog.) [f] CAHLS divisum [g] CA quilibet [h] H secundum [i] H λ [j] CA quamlibet [k] CALS quarta

will be numbered '60'. And do not let any circle pass through this scale entirely, lest the divisions on it be completely confused.

Having then placed the dividers on the common centre *G*, from the middle of the diameter now divided let us take a length of 21° 20′, which we shall mark off along the other orthogonal diameter on both sides. We shall provide it with a margin with an appropriate space, on both sides, like the latter, and we shall divide both *GD* and *GB* into 64 equal parts, dividing the space adjoining at intervals of 5 units, and numbering from the centre *G* towards *B* and towards *D*.

CHAPTER 26

[II. 26]

On drawing the figure for a lunar eclipse.

Let us make the circle for an eclipse of the moon in this way. With centre *G*, let the other foot of the dividers be extended to 63^p 36′ (with regard to the scale divided into 64^p),[1] and describe a semicircle *DEB* about this centre and with this as semidiameter, but do not allow it, in crossing the scale *CA* divided into 60, to confuse the scale-divisions. Next, about the same centre, draw another semicircle *FHI*, so that (in the same units as before) both semidiameter *GF* and *GI* are of 52^p 50′. The line *DGI* having then been divided into two equal parts, a semicircle *DKI* is to be drawn with centre in the middle of *DGI*, the semicircle touching each of the two former semicircles.

Let us subsequently draw, outside the circle *DEB* and near it, first a margin and then a space. We are to divide the margin in this way: the rule is to be placed at right-angles to the semidiameter *GA* (divided into 30 equal parts, each of which is further divided into 3 parts, these being of 20′, namely $\frac{1}{3}$°), at each 20′-division. Where the rule meets the semicircle *DEB* on each side, let a short stroke be marked, letting every fifth division, however, cross the entire space. What lies between the two semicircles *DEB* and *FHI*, up to the whole semicircle *DEB*, will be divided by such strokes, and the space between the two aforesaid semicircles will be divided up by straight lines, which are not to extend beyond this enclosed space.

The rule at the common centre *G* having been placed, let it pass [in turn] through each point marked on the aforesaid semicircle *DEB*, and let us divide the previously mentioned margin in the usual way, whereby every fifth division crosses through the space reserved for the numerals.[2]

A length of 45^p 56′ is then to be taken from the semidiameter *GD*, and this is to be [the length of] *GL*. Next, from the line *GB* is to be

[1] The MSS. have '60^p', but the plate would not accommodate a semicircle of 63^p 36′ on this reading.

[2] *littere*.

que sit linea *GM*. Et super lineam *LM* perficiamus semicirculum. Item de linea *GD* capiamus lineam 28 gra. 16 min. que sit linea *GN*, alteram et[a] lineam *GO* de linea *GB*, que sit 23 par. et 30 min.; et perficiamus semicirculum super lineam *ON*. Et completum est.

CAPITULUM 27m

De arte componendi circulum latitudinis Lune.

Post hec super[b] spacium alterius semicirculi *DEB*, perficiamus circulum latitudinis Lune hoc modo:

Prope limbum eiusdem separemus unicam marginem cum unico spacio, et dividamus eius marginem in 10 partes equales, que erunt gradus, et quamlibet illarum parcium in 60 min.; et posita regula super commune centrum *G*, dividamus marginem per singulas divisiones, et spacium per 5 et 5 vel per 6 et 6, modo communi; et intitulabimus ipsas divisiones graduum a puncto *D* versus *C*, similiter[c] a puncto *B* versus *C*, et habebit utraque quarta circuli 5 gra., et quilibet gradus 60 min. Deinde separabimus consequenter unam marginem cum duobus spaciis, et ipsam marginem modo communi dividemus, per tabulam latitudinis Lune. Postea incipientes a puncto *B* versus *D*, distribuamus priori spacio gradus per 6 signa, et iterum redeundo a puncto *D* versus *B* intitulabimus residua 6 signa, videlicet a septem usque 12. Nullo tamen modo transeant alique[d] harum divisionum aut circuli super divisiones diametrorum, ne forte confundentur, quia necesse est eas esse distinctas. Et completum est.

CAPITULUM 28m

De complemento eclipsis Solaris.

Novissimo compleamus figuram eclipsis solaris hoc modo:

Primo, super centrum *G* fiat unus circulus intra semicirculum latitudinis Lune, cuius semidiameter *GP* contineat 34 par. de partibus linee *GD*, et alterum semicirculum cuius semidiameter *GQ* contineat 31 par.; et divisa linea *PQ* in duo equalia, fiat super ipsam semicirculus *PRQ* contingens utrumque semicirculum antedictum. Deinde extra semicirculum *P* concentricum faciamus marginem et spacium, et dividamus eas per regulam iacentem orthogonaliter super semidiametrum *GO*[e] et transeuntem super singulas tercias unius gradus omnino sicut dictum est de figura eclipsis lunaris. Postea posito centro

[a] A 2λ+eciam [b] C +*P* [c] A 6λ (haplog.) [d] H aliam [e] AHS *GC*

taken another line, of 38^p 10′, and this is to be the line *GM*. A semicircle is drawn on *LM*. Next, from the line *GD* we are to take a length of 28° 16′, which is to be line *GN*, and from the line *GB* another line, *GO*, of 23^p 30′; and let us draw a semicircle on the line *ON*.

CHAPTER 27

[II. 27]

On the method of constructing a circle for the latitude of the Moon.

After this, let us draw the circle of lunar latitude above the space of the semicircle *DEB*, in the following manner:

Near the limb [of the plate *PGQ*] let us mark out a single margin with a single space, and divide the margin into 10 equal parts, which will be degrees, and each of these into 60 minutes. Placing the rule at the common centre *G*, let us divide the margin with single divisions, and the space at intervals of 5 or 6, in the usual way. We shall label the degree-divisions from *D* towards *C*, and similarly from *B* towards *C*; and each quadrant will contain 5 degrees, and each degree 60 minutes. We shall then subsequently draw a margin with two corresponding spaces, dividing the margin itself in the usual way, by means of a table of the latitude of the Moon. Working from point *B* towards *D*, we are then to distribute the degrees of six signs in the first space; and then, returning from *D* towards *B*, we shall give the numbers of the remaining 6 signs, namely, from 7 to 12. In no way should any of these divisions, or the circle, cross the divisions of the diameters, lest they be confused; for it is essential that they be kept distinct.

CHAPTER 28

[II. 28]

On the completion of the [figure for the] eclipse of the Sun.

Lastly, we are to complete the figure for the eclipse of the Sun, in the following way:

First, with centre *G*, there is to be a circle below the semicircle of the latitude of the Moon, with a semidiameter *GP* of 34^p (the units being parts of the scale *GD*); and there is to be another semicircle with semidiameter *GQ* of 31^p. The line *PQ* having been divided into two equal parts, there is to be a semicircle *PRQ* drawn on *PQ*, touching each of the semicircles already mentioned. Outside the concentric semicircle *P*, let us then make a margin and a corresponding space, and divide them by means of a rule at right-angles to the semidiameter *GO*, and crossing the individual $\frac{1}{3}$°-divisions, entirely as explained of the figure for a lunar eclipse. With centre *G*, another

in *G* fiat alter semicirculus cuius semidiameter contineat 16 par. et 20[a] min., ut *GS* omnino sicut patet in presenti figura. Et completum est.

CAPITULUM 29^{m}

De paxillo.

Tandem perficiamus paxillum qui possit transire super centrum *G* per medium tocius instrumenti ad constringendum utrasque tabulas instrumenti cum rethi, sicut in astrolabio, excepto quod non erit illi equus nec aliqua pars multum supereminens super superficiem laminarum. Sed erit illi firmamentum hoc modo: ex una parte instrumenti habeat paxillus caput subtile non multum supereminens, sed sufficiens ad constringendum laminam cum matre. Ex parte altera fiat una concavitas in paxillo, prope superficiem alterius faciei instrumenti in circuitu eius, ut possit recipere tenuem laminam hoc modo: coniungantur due breves lamine circulares, in medio perforate ita quod possent recipere ipsum paxillum inter illas breves laminas. Comprehendatur alia tenuis lamina semicircularis firmata tantum cum una clavi, ita quod possit extrahi per eius motum tanquam in cardine extra locum suum inter illas duas laminas breves; et habeat in eius[b] diametro concavitatem qua possint applicari ad paxillum et constringet alteram laminam ac si foret alterum caput paxilli; et hec erit forma istius capitis, sive firmamenti, ex his tribus partibus compositi.[c]

Habeatque ipsa media lamina semicircularis unam superfluitatem prominentem qua possit faciliter extrahi de loco suo, cum voluerimus separare laminas instrumenti ab invicem.

Et perforemus paxillum veraciter super centrum eius *G*, ita quod filum possit pertransire per medium eius ex utraque facie instrumenti. Et completum est.

CAPITULUM 30^{m}

et ultimum secunde parte

De[d] armilla.

Cum ista perfecerimus, aptemus armillam unam suspensoriam sive portatilem. Et in circulo illo tercio communi duabus marginibus circuli altitudinis et zodiaci primi limbi, apud *C*, faciamus foramen.

[a] H 30 [b] CAL eo [c] CAH compositum [d] CA 2λ

semicircle is to be made with semidiameter 16p 20′, exactly as *GS* appears in the present figure.

CHAPTER 29

[II. 29]

On the pin.[1]

Finally, let us make a pin which can, at centre *G*, pass through the middle of the whole instrument in order to hold together the two tablets of the instrument and the rete, just as with an astrolabe, except that here there will be no horse,[2] nor anything else raised up appreciably above the area of the plate. Rather will the pin be made fast as follows: on one side of the instrument the pin should have a head carefully made so as not to be greatly raised, but sufficient to hold the plate to the mater. On the other side, let there be a peripheral groove[3] in the pin, near the surface of the other face of the instrument, such as will take a thin plate in this way: two small circular plates, perforated in the middle so as to be able to take the pin itself through them, are to be joined together. The other thin semicircular plate is to be held between them with a rivet, just so firmly that it may be drawn out of position as though on a hinge. And it is to have an aperture[4] on its diameter by which [the parts of the clip] may be fitted into the pin at its groove. This will retain the pin and the other plate, as if it were another head on the pin. This then will be the form[5] of the head, or securing device,[6] made out of the three parts.

The central semicircular plate should also have a small protruding part by which it may be easily pulled out of position, when we wish to take apart the plates of the instrument.

We are also to drill the pin exactly through its centre *G*, so that a thread may be drawn through its centre to both faces of the instrument.

CHAPTER 30, and the last of Part II

[II. 30]

On the ring [and plumb-line].

When we have done that, let us fit a suspension- or carrying-ring. At *C* on that third circle common to the two margins of the circle of altitude and the zodiac of the first limb, a hole is to be made.[7] We

[1] *paxillus—clavus* would perhaps have been more usual.

[2] *equus*—the term used for the wedge through the pin of an astrolabe, this wedge having been often formed in the shape of a horse's head.

[3] *concavitas in paxillo . . . in circuitu eius.*

[4] *concavitas.*

[5] I have found no illustration of the sort suggested, if not exactly demanded, by this sentence.

[6] *firmamentum.*

[7] That is, on the circle common to scales (8) and (9), for the division of which see II. 2 above.

Et firmabimus ibi filum cum perpendiculo ad capiendum altitudines (et per utramque pinnulam perforatam), et ad perficiendum alia opera ad quod institutum est, sicut loco suo docetur. Et completum est hoc opus cum adiutorio Dei, Anno Christi[a] 1327 inchoante.

III

Incipit pars tercia de modo operandi per instrumentum.

ALBION ad singulos motus novem orbium se extendit eo quod stellas erraticas omni loco et tempore pariter equat[b] aspectui, et per eam fixaque sidera proprietates omnes ostendit motus diurni. Armille quidem,[c] rectangulus, et turketus omnium[d] radices stellarum verificant. Almanac pluresque tabule[e] calculate motus varios planetarum verificant. Astrolabium, quadrans, horologium, et saphea temporum et horarum momenta certificant. Hoc autem unicum instrumentum, si omnium et singulorum commoditates corpore tam brevi contineat, et quedam alia forsitan superaddat, inter cetera non erit abiectum, maxime cum eius invencio multorum ingenia excitare poterit ad maiora.[f]

CAPITULUM sive utilitas prima tercie partis huius

de nominibus circulorum huius instrumenti, et ordine

Omnes circulos in prima facie instrumenti nominate discernere.

Mater[g] (1) tocius instrumenti dicitur lamina media concavata que continet ex utraque sui parte laminam circularem.

Prima facies (2) instrumenti est illa in qua lacior limbus matris est.

Rota prima (3) dicitur illa lamina que volvitur in prima facie instrumenti, cuius *facies* (4) est illa que continet ecentricos planetarum. *Facies*[h] eius *secunda* (5) est illa[i] que continet *lineas corde recte et verse.*

Secunda rota[j] (6) est illa quam continet facies matris secunda.

Primus circulus cum sua margine in limbo matris in prima facie instrumenti dicitur *circulus anni solaris* (7).

Secundus circulus cum sua margine dicitur *circulus altitudinis et horarum* (67, & 8).

Tercius circulus cum sua margine dicitur *zodiacus octave spere,* sive *zodiacus primi limbi* (9).

[a] H Domini Nostri [b] H equet [c] CA quidam [d] CA omnique
[e] H tabulate [f] A memoria [g] CA materia [h] H 2λ
[i] A λ [j] L (and MSS. Joh. Gmunden) 9λ

shall there fix a thread, with a plummet for taking altitudes (with the help of both perforated pinnules), and for performing the other tasks for which it was designed; as is shown in its proper place.

The instrument is now complete, with the help of God, in the year of Christ beginning 1327.

III

ON THE METHOD OF WORKING WITH THE INSTRUMENT

ALBION applies to the individual motions of the nine spheres, since it corresponds exactly to the appearance of the planets in every position and at all times. By its means and by means of the fixed stars it shows all the properties of the diurnal motion. The *armillae*, the *rectangulus*, and the *turketum* certainly confirm the parameters of all the stars. The *almanac* and several calculated *tables* confirm the various motions of the planets. The *astrolabe*, *quadrant*, *horologe*, and *saphea* make known the movements of times and hours. Now if this unique instrument contains in so small a frame easy access to all this information, and perhaps furnishes other information in addition, then its place amongst the others will not be a humble one, especially as its design will be capable of exciting the minds of many to greater things.

[III. 1]

FIRST CHAPTER, or 'USE', in this third part.

To distinguish by name all the circles on the first face of the instrument.

On the names of the circles on this instrument, and their arrangement

The hollowed-out middle plate, containing a circular plate on either side of it, is called the *mater* (1) of the whole instrument.

The *first face* (2) of the instrument is that on which is the broader limb of the mater.

That plate which revolves on the first face of the instrument is called the *first wheel* (3). Its *face* (4) is that which contains the eccentrics of the planets. Its *second face* (5) is that which contains the *scales of corda recta and versa.*

The *second wheel* (6) is that on the second face (6a) of the mater.

The first circle with its margin, on the limb of the mater on the first face of the instrument, is called the *circle of the solar year* (7).

The second circle with margin is called the *circle of altitude* (67) *and of hours* (8).

The third circle with margin is called the *zodiac of the eighth sphere*, or *zodiac of the first limb* (9).

Quartus circulus dicitur *circulus orbis Saturni* (10).
Quintus circulus dicitur *circulus orbis Iovis* (11).
Sextus dicitur *circulus orbis Martis* (12).
Septimus dicitur *circulus orbis Solis et Veneris* (13).
Octavus dicitur *circulus orbis Mercurii* (14).
Nonus dicitur *circulus orbis Lune* (15).
Decimus circulus et intimus dicitur *circulus verus Lune* (16).

Primus circulus in prima facie rote prime dicitur, ut prius, *circulus anni Solis* (17).

Secundus et tercius circulus cum margine intercepto dicitur *zodiacus rote prime* (18).

Tercius circulus dicitur *circulus iomyn equati*, sive *equacionis dierum* (19).

Quartus circulus cum suis marginibus dicitur *anni Saturni* (20); quintus circulus cum suis marginibus *Iovis* (21); sextus *anni Martis* (22); septimus *anni Veneris* (23); octavus *anni capitis atque caude* (24).

Nonus dicitur *circulus coniunctionis equalis* (25), qui habet plures partes quarum prima dicitur *circulus longitudinis inter Solem et Lunam* (26).

Intimus circulus dicitur *circulus lunacionum* (27).

Ecentricus interceptus dicitur *equator velocitatis lunaris* (28).

Linee transversales dicuntur *hore motus lunaris* (29).

Intimus circulus dicitur *circulus augium planetarum* (30).

Primus ecentricus in facie prime rote est *equator Veneris* (31); secundus *equator Martis* (32); tercius *equator Mercurii* (33); quartus *Iovis* (34); quintus *equator Lune* (35); brevissimus *equator Saturni* (36).

In secunda facie rote prime linea[a] transversalis que erit super primum limbum matris est *linea corde recte et verse* (37).

Secunda linea, orthogonaliter ipsam secans, dicitur *linea minutorum casus et more* (38).

Semicirculus in limbo rote, cuius divisio est in 10 partes equales cum suis fractionibus, dicitur *circulus latitudinis Lune* (39).

Duo circuli sequentes cum margine dicuntur *geusahar Lune* (40).

Semicirculus interior cuius[b] semidiameter est 34 min. de linea minutorum casus et more dicitur *tropicus eclipsis solaris ad oppositum augis epicicli*[c] *lunaris* (41).

[a] C Luna [b] CA eius [c] A eclipsi

The fourth circle is called *circle of the sphere of Saturn* (10).
The fifth circle is called the *circle of the sphere of Jupiter* (11).
The sixth is the *circle of the sphere of Mars* (12).
The seventh is the *circle of the sphere of the Sun and of Venus* (13).
The eighth is called the *circle of the sphere of Mercury* (14).
The ninth is called the *circle of the sphere of the Moon* (15).
The tenth and innermost circle is the *true circle of the Moon* (16).

The first circle on the first face of the first wheel is called, as before, the *circle of the solar year* (17).[1]

The second and third circle, with the contained margin, is called the *zodiac of the first wheel* (18).[2]

The third circle is called the *circle of the equation of iomyn*, or *of the equation of days* (19).

The fourth circle with its margins is called the [*circle of the*] *year of Saturn* (20); the fifth with its margins [is *of the year of*] *Jupiter* (21); the sixth is of *the year of Mars* (22); the seventh is of *the year of Venus* (23); the eighth of *the year of Caput and Cauda* [*Draconis*] (24).

The ninth is the *circle of equal conjunction* (25), which has several parts, of which the first is called the *circle of longitude between Sun and Moon* (26).

The innermost circle is called *circle of lunations* (27).

The eccentric inside is called the *equator of lunar velocity* (28).

The transverse lines are called the *hours of lunar motion* (29).

The innermost circle is called the *circle of the auges of the planets* (30).[3]

The first eccentric on the face of the first wheel is the *equator of Venus* (31); the second is the *equator of Mars* (32); and the third is the *equator of Mercury* (33); the fourth *of Jupiter* (34); the fifth the *equator of the Moon* (35); the smallest the *equator of Saturn* (36).

On the second face of the first wheel the transverse line reaching to the first limb of the mater is the *scale of corda recta and versa* (37).

The second line, cutting (37) at right angles, is called the *scale of minutes of partial and total phase* (38).[4]

The semicircle on the limb of the wheel, which is divided up into 10 equal parts, with appropriate subdivisions, is called the *circle of the latitude of the Moon* (39).

The two circles following, with margin, are called *geusahar of the Moon* (40).

The inner semicircle whose semidiameter is 34 minutes of the scale of minutes of partial and total phase is called the *tropic of a solar eclipse at the opposite of aux of a lunar epicycle* (41).

[1] This duplicates scale (7).

[2] A simple scale (cf. scale (9)) but labelled twice over, once in a direct sense, and once in the contrary sense. See II. 10.

[3] In II. 17 above, this scale is said to have the planetary nodes marked on it, as well as the auges.

[4] *casus et mora*. See the note to the translation of III. 24 below.

Sequens semicirculus cuius[a] semidiameter est 31 min. est *tropicus eclipsis solaris in auge epicicli lunaris* (42).[b]

Ecentricus interceptus dicitur *equator eclipsis solaris in auge* (43).

Divisiones marginis maioris tropici et minoris sunt *minuta latitudinis Lune* (44).

Intimus semicirculus, cuius semidiameter[c] est 16 min. et 20 sec., est *orbis solaris in eius longitudine media* (45).

Semicirculus in altera medietate rote, cuius semidiameter est 63 min. et 36 sec., est *tropicus eclipsis lunaris ad longitudinem propiorem* (46).

Interior semicirculus, cuius semidiameter[d] est 52 min. et 50 sec. est *tropicus eclipsis lunaris ad oppositum augis* (47).

Ecentricus interceptus est *equator eclipsis lunaris* (48).

Tercius semicirculus ecentricus dicitur *equator umbre* (49).

Quartus semicirculus, qui similiter est ecentricus, dicitur *equator casus et more* (50).

Divisiones vero in margine tropici eclipsis lunaris sunt *minuta latitudinis Lune ab orbe signorum* (51), (52), et *linea minutorum casus* (38) habet hic vicem ecliptice.[e]

UTILITAS 2ª huius tercie partis

de nominibus circulorum secunda faciei instrumenti

Omnes circulos in secunda facie instrumenti nominatim discernere.

Secunda facies (53) instrumenti est illa in qua est strictior limbus matris, cuius primus circulus cum sua margine dicitur *zodiacus primi limbi* (54).

Secundus circulus dicitur *circulus ascensionum in circulo directo* (55).

Tercius circulus dicitur *circulus ascensionum in circulo obliquo* (56).

Prima facies rote secunde (57) dicitur illa que continet circulum involutum, cuius primus circulus cum margine est *circulus horarum medii motus Lune* (58). Deinde a principio circuli involuti continetur *medius motus anni Mercurii* (59) in tres eius involuciones et circiter. Deinde sequitur *motus anni lunaris* (60) citra 13 involuciones et ultimo sequitur *motus argumenti Lune* (61).

Secunda facies rote secunda (62) est illa[f] in qua describitur unum clima opere astrolabii, nichil superaddens preter *gradus latitudinis* (63) septentrionalis et meridionalis ab ecliptica.

[a] CA eius [b] Sentence in C only as a footnote [c] H diameter [d] H diameter

[e] H +*Disposicio parcium et circulorum albyon secundum correctionem W. S.* Mater tocius instrumenti dicitur lamina . . . et intimus semicirculus omnium semicirculorum dicitur orbis lunaris. Hec in prima rota continentur. (See III. 1 seq. in the commentary below.)

[f] AH λ

The following semicircle of semidiameter 31 minutes is the *tropic of a solar eclipse at the aux of a lunar epicycle* (42).

The eccentric contained [within (41) and (42)] is called the *equator of a solar eclipse at the aux* (43).

The divisions of the margin of the greater and lesser tropic are the *minutes of the latitude of the Moon* (44).

The innermost semicircle, whose semidiameter is 16′ 20″, is *of the solar sphere at its mean distance* (45).

The semicircle on the other half of the wheel, having semidiameter 63′ 36″, is the *tropic*[1] *of a lunar eclipse at its nearest distance*[2] [*on the epicycle*] (46).

The inner semicircle, whose semidiameter is 52′ 50″, is the *tropic of a lunar eclipse at far aux* (47).

The eccentric contained [within it] is the *equator of a lunar eclipse* (48).

The third eccentric semicircle is called the *equator of the shadow* (49).

The fourth semicircle, which is similarly eccentric, is called the *equator of partial and total phase* (50).[3]

The divisions of the margin of the tropic of a lunar eclipse are *minutes of the latitude of the Moon from the sphere of signs* (51) and (52),[4] and here the *scale of minutes of the partial phase* (38) plays the part of the ecliptic.

USE 2

[III. 2]

To distinguish all the circles of the second face of the instrument.

On the names of the circles of the second face of the instrument

The *second face* (53) of the instrument is that on which lies the narrow limb of the mater, and whose first circle, with its margin, is called the *zodiac of the first limb* (54).

The second circle is called the *circle of ascensions on the direct circle* (55).

The third circle is called the *circle of ascensions on the oblique circle* (56).

The *first face of the second wheel* (57) is that which contains the spiral, but its first circle with margin is the *circle of hours of mean motus of the Moon* (58). Then, from the beginning of the spiral, for approximately three turns, there comes the *mean motus of the year of Mercury* (59). Next follows the *motus of the lunar year* (60) of about thirteen turns, and finally the *motus of the argument of the Moon* (61).

The *second face of the second wheel* (62) is that on which a single climate is described, as when making an astrolabe, adding nothing beyond *degrees of latitude* (63) measured north and south of the ecliptic.[5]

[1] 'Tropic' is here contrasted with 'eccentric' or 'equator'.

[2] *longitudo propior* (the epicycle being understood nevertheless to be at the aux of the deferent).

[3] Cf. the note to (38).

[4] Reserving (51) for the parallel graduations, and (52) for their continuation in the margin.

[5] Actually degrees of ecliptic longitude and stars were also to be added, and in a somewhat unusual way. See II. 22–3.

Paxillus[a] (64) dicitur ille axis super quem rotantur due rote predicte.

Filum centri (65) dicitur filum qui erigitur a centro paxilli.

Filum limbi (66) dicitur illud filum cuius fixio est in limbo matris in prima facie instrumenti, in puncto 6 signorum circuli altitudinis et zodiaci limbi primi.

Principium circuli altitudinis (67) est in directo puncti 9 signorum zodiaci primi limbi.

Principium circuli horarum (68) est in directo puncti trium signorum zodiaci primi limbi.

UTILITAS 3ª tercie partis huius

Medios motus omnium planetarum per addicionem et subtractionem graduum calculare.

Quod si placuerit, pone principium limbi secundi in directo contra gradus in zodiaco limbi primi, quibus aliquid fuerit addendum vel subtrahendum. Quod si addere aliquid volueris, pone filum centri in zodiaco limbi secundi super summam signorum, graduum, et minutorum addendorum, et collectum[b] reperietur sub filo in zodiaco limbi primi. Si autem subtrahere volueris, posito principio limbi secundi ut prius citra summam in qua debet fieri subtraxio, transfer filum super zodiacum limbi secundi contra signa usque ad summam subtrahendam, et illud quod remanet necessario reperietur sub filo in zodiaco limbi primi.

Et eodem modo fieri poterit in fractionibus si 6 grad. usus fueris pro uno minuto.

UTILITAS 4ª huius tercie partis

Medios[c] *motus omnium planetarum et capitis geusahar, argumenta eciam Veneris et Mercurii atque Lune, pro quolibet tempore ad quamvis*[d] *meridiem invenire.*

Quod si placuerit, scias semel in anno loca omnium eorum ad principium Marcii, necnon[e] et locum vel tempus prime coniunctionis ante principium Marcii vel post; et redigas in tabulam et reponas

[a] CAH paxillum [b] C collectui [c] A Medius [d] A quavis [e] H ιολ (haplog.)

The axle about which the two aforementioned wheels rotate is called the *pin* (64).

The thread stretching from the centre of the pin is called the *thread of the centre* (65).

The thread whose attachment is to the limb of the mater on the first face of the instrument, at the point of 6 signs of the circle of altitude and of the zodiac of the first limb, is called the *thread of the limb* (66).

The *beginning of the circle of altitude* (67) is in the direction of the point of 9 signs of the zodiac of the first limb.

The *beginning of the circle of hours* (68) is in the direction of the point of 3 signs of the zodiac of the first limb.[1]

USE 3
[III. 3]

To calculate the mean motus of all the planets by means of the addition and subtraction of degrees.

Place, if you please, the beginning of the second limb against the degree on the zodiac of the first limb to which something is to be added or subtracted. If you wish to add something, place the thread of the centre on the zodiac of the second limb over the number[2] of signs, degrees, and minutes, to be added. The result of the addition will be found under the thread at the zodiac of the first limb. If, however, you wish to subtract, having placed the beginning of the second limb, as before, against the number from which the subtraction is to be made, move the thread over the zodiac of the second limb, against the signs, as far as the number to be subtracted. The remainder will necessarily be found under the thread on the zodiac of the first limb.

The same procedure can be adopted with fractions, if you use 6° for one minute.

USE 4
[III. 4]

To find the mean motus of all the planets, and of the head of geusahar, and also the arguments of Venus, Mercury, and the Moon, for any time and at any meridian.

If it please you,[3] you are to find out for one time in the year—the first of March—the places of all of them, and also the place or the time of the conjunction next before or next after the first of March;[4]

[1] The list ends, and Richard appears to have overlooked (69) the *saphea* in the recess on the second face (52) of the mater. See II. 24 above. For minor omissions see the commentary.

[2] *super summam*. This refers only to the number to be added, and not the final sum.

[3] 'Quod si placuerit'. Henceforth this formal beginning will not be translated.

[4] Of the Sun and Moon? This last remark reflects the practice of those who drew up astronomical tables.

in fundo instrumenti quod per totum annum habebitis[a] pro radice. Deinde si fuerit circulus anni Solis descriptus in prima facie limbi primi, pone radicem planete signatum in zodiaco rote prime in directo capitis limbi primi. Deinde transfer filum centri super circulum anni Solis in limbo primo descriptum usque ad diem et momentum in quo vis certificari de illo quod queris. Et si quesieris locum Solis, ipse erit quem sub filo inveneris in zodiaco rote[b] prime. Si autem quesieris locum Saturni, Iovis, Martis, capitis geusahar, aut argumenti Veneris, vide in circulo anni eiusdem planete super quot gradus cecidit filum centri, et super totidem gradus pone idem filum super zodiacum limbi primi;[c] et cadet filum in zodiaco rote[d] prime super locum quesitum verissime.

Si autem volueris operari per circulum anni Solis in zodiaco rote[f] prime repertum, operaberis econverso, et idem proveniet sine dolo.

Quod si quesieris medium motum Lune, et argumenti eius et Mercurii, pone principium anni planete in rota secunda directe contra radicem planete signatum in zodiaco limbi primi in altera facie instrumenti. Deinde transfer filum centri super circulum involutum usque ad momentum temporis de quo queris, et cadet filum centri in zodiaco limbi primi super locum planete quesitum.

Huius tercie partis UTILITAS 5ª

Verum locum Solis pro quolibet tempore una cum argumento et equacione debita reperire.

Quod si placuerit, pone augem Solis in secundo limbo signatam in directo capitis primi limbi. Deinde transfer filum centri super zodiacum secundi limbi ad medium motum Solis, per precedentem inventum; et quod in primo limbo de gradibus zodiaci octave spere sub filo ceciderit, ipsum est argumentum Solis. Considera igitur et

[a] A habebis; H habebitur [b] H limbi [c] H secundi in rota
[d] H limbi primi [e] H λ [f] H 2λ

and return to the table and set at the bottom of the instrument[1] what you have as the radix for the year.[2] If the circle of the year of the Sun is [taken as that][3] described on the first face of the first limb, set then the planet's radix, marked on the zodiac of the first wheel, in the direction of the beginning [of the scale][4] of the first limb. Then move the thread of the centre over the circle of the year of the Sun (described on the first limb) to the day and moment for which you would like certifying whatever it is you seek; and if you want the place of the Sun, it will be what you find under the thread on the zodiac of the first wheel. If, however, you want the place[5] of Saturn, Jupiter, Mars, the head of geusahar, or the [mean] argument of Venus, look on the circle of the year of the planet in question for the number of degrees indicated by the thread of the centre; and set this number of degrees on the zodiac of the first limb. The thread will certainly fall on the zodiac of the first wheel at the required position.

If, however, you wish to work with the circle of the year of the Sun in the zodiac which is found on the first wheel,[6] you will work in the converse way, achieving the same end without any error.

If you want the mean motus of the Moon, and its argument, and that of Mercury, set the beginning of the [spiral] of the year of the planet on the second wheel exactly against the radix of the planet marked on the zodiac of the first limb on the other face of the instrument. Then move the thread of the centre across the spiral as far as the moment of time with which your inquiry is concerned. The thread of the centre will fall on the zodiac of the first limb over the required place[7] of the planet.

USE 5

[III. 5]

To find the true place of the Sun for any time, together with the argument and corresponding[8] *equation.*

Place the aux of the Sun, marked on the second limb,[9] at the origin of the first limb. Move the thread of the centre, next, over the zodiac of the second limb up to the mean motus of the Sun, found from the preceding [Use 4]. The number of degrees of the zodiac of the eighth sphere indicated by the thread is the argument of the Sun. Find out

[1] 1 March is '4.30 o'clock' of the instrument, and thus near the bottom, but 1 January is near '6 o'clock'. The former year-beginning seems to be intended, nevertheless.

[2] 'For the year as a whole', that is, up to the end of the complete year.

[3] There are two such circles, one fixed (7) on the limb, the other movable (17) on the first face of the first wheel. The former is intended.

[4] That is, 1 March, or whatever date is chosen for the radix.

[5] Rather the mean motus.

[6] That is, with scale (17) rather than (7).

[7] The mean motus, strictly speaking.

[8] *debita.*

[9] The phrase is often used in Part III to denote the movable disc under discussion, as opposed to the outer limb.

pone filum super consimilem numerum argumento in circulo orbis Solis, quo facto, vide locum qui est sub filo in zodiaco limbi secundi, et ipse est verus locus Solis quem queris. Differencia vero que est inter locum primi limbi sub quo cecidit prius filum et locum super quem cadit in eodem limbo post hanc translacionem est equacio argumenti. Est autem aux Solis prope 3 signa.

Si autem locum[a] Solis in spera nona quesieris, per terciam huius adde super locum equatum motum octave spere, puta circiter 10 gra.

UTILITAS 6ª

Argumenta media Saturni, Iovis, et Martis similiter invenire.

Quod si placuerit, pone filum centri super medium motum Solis in zodiaco primi limbi, et medium motum planete in secundi limbo signatum pone sub eodem filo. Postea transfer filum ad caput limbi secundi, et ubi ceciderit[b] filum in primo limbo ibi est argumentum medium planete quesitum.

UTILITAS 7ª

De distancia planetarum ab augibus[c] deferentium que vocatur centrum

Centra media Saturni, Iovis, Martis, Veneris, et Mercurii similiter invenire.

Quod si placuerit, pone augem planete, in rota prima signatam, in directo capitis primi limbi. Deinde pone filum super medium motum planete in secundo limbo, et locus super quem cecidit filum in primo limbo ipse est centrum planete quesitum.

UTILITAS 8ª

De centris planetarum equandis

Centra et argumenta Saturni, Iovis, Martis, Veneris, et Mercurii equare; et equaciones centri similiter reperire.

Quod si placuerit, pone filum centri super centrum[d] in zodiaco octave spere primi limbi, et pone argumentum medium repertum in zodiaco secundi limbi, qui vadit contra naturalem successionem

[a] A locus [b] A cecidit [c] C +suo; C'H +suorum [d] H +medium

what this is, therefore, and place the thread at the same argument in the circle of the sphere of the Sun. Having done so, see where the thread comes on the zodiac of the second limb, for this is the true place of the Sun which you require. The difference between the positions of the thread on the first limb before and after it was moved is in fact the equation of the argument. (The aux of the Sun is near 3 signs.)

If, however, you require the place of the Sun in the ninth sphere, add the motion of the eighth sphere, with the help of Use 3, to the equated place.[1] [The motion of the eighth sphere] is, say, about 10°.[2]

USE 6

[III. 6]

To find, in a similar way, the mean arguments of Saturn, Jupiter, and Mars.

Place the thread of the centre at the mean motus of the Sun, on the zodiac of the first limb, and place the mean motus of the planet, marked on the second limb, under the same thread. Subsequently, move the thread to the origin of the second limb, and wherever the thread meets the first limb, that will be the required mean argument of the planet.

USE 7

[III. 7]

To find the mean centres of Saturn, Jupiter, Mars, Venus, and Mercury, in a similar way.

On the distance of the planets from the auges of the deferents, this distance being known as the 'centre'

Place the aux of the planet, as marked on the first wheel, in the direction of the origin of the first limb. Now place the thread over the mean motus of the planet on the second limb, and wherever it meets the first limb will be the required centre of the planet.

USE 8

[III. 8]

To equate the centres and arguments of Saturn, Jupiter, Mars, Venus, and Mercury; and to find in like manner the equations of the centre.

On equating the centres of the planets

Place the thread of the centre over the [mean] centre on the zodiac of the eighth sphere of the first limb, and place the mean argument, which has supposedly been determined, [under the thread] on the zodiac of the second limb, which advances against the natural

[1] *locus equatus*. The generic name 'equatorium' derives from the main purpose of the instrument, that of determining the *loci equati* of the planets.

[2] See III. 5 comm. The same figure occurs in III. 9 below.

signorum ab eodem filo. Deinde transfer filum ad centrum planete in circulo orbis sui, et statim cadet filum super centrum equatum in primo limbo, et super argumentum equatum in secundo, contra signa; et arcus interceptus in primo limbo, inter filum et locum priorem, puta finem centri medii, est equacio centri equata.

UTILITAS 9ª

De equacionibus argumentorum

Equacionem argumenti ad omnem longitudinem examinatam extrahere.

Quod si placuerit, posito[a] principio[b] limbi secundi contra principium limbi primi, transfer filum centri super centrum planete in primo limbo, et transfer almuri quod est in filo ad contactum ecentrici eiusdem planete. Deinde transfer filum centri super argumentum planete equatum, sicut exigit eius naturalis[c] successus in secundo limbo. Post hoc extende filum limbi super almuri, et quicquid fuerit inter filum limbi et caput limbi primi de gradibus altitudinis, ipsum est equacio argumenti precisa.

UTILITAS 10ª

De veris locis planetarum extrahendis

Verum locum Saturni, Iovis, Martis, Veneris, et Mercurii similiter invenire.

Quod si placuerit, pone filum centri super centrum medium planete in primo limbo. Deinde super[d] eodem filo pone medium motum planete in secundo limbo signatum et statim translato filo super centrum medium in circulo orbis planete, punctum illum secundi limbi qui sub filo ceciderit ad caput transferas primi limbi.

Deinde posito filo centri super equacionem examinatam in zodiaco octave spere, et ex ea parte qua illa equacio erat inventa per filum limbi, punctus ille secundi limbi qui sub filo centri ceciderit, ille

[a] H ponito [b] CAH principium [c] H natura vel [d] H sub

succession of signs from this same thread. Move the thread, then, to the centre of the planet on the circle of its sphere, and immediately the thread marks the equated centre on the first limb, and the equated argument on the second, against the signs. The arc along the first limb between the thread and its first position, taken to be that of the mean centre, is the equated equation of the centre.

USE 9

[III. 9]

To extract the equation of the argument at each longitude considered.

On the equations of the arguments

Having put the beginning of the second limb against the beginning of the first, move the thread of the centre over the centre of the planet on the first limb, and move the almuri to the point of intersection of the thread and the eccentric of the same planet.[1] Now transfer the thread of the centre to the equated argument of the planet, following the natural sequence [of signs] on the second limb. After this, draw the thread of the limb out over the almuri, and whatever lies between the thread of the limb and the origin of [the scale on] the first limb, in degrees of altitude, that same is exactly the equation of the argument.

USE 10

[III. 10]

To find the true place of Saturn, Jupiter, Mars, Venus, and Mercury, in a similar manner.

On extracting the true places of the planets

Place the thread of the centre over the mean centre of the planet on the first limb. Then set the mean motus of the planet marked on the second limb against the same thread, and at once, the thread having been moved to the mean centre on the circle of the sphere of the planet, move round to the origin of the first limb that point of the second limb which falls under the thread.

The thread of the centre having then been set on the zodiac of the eighth sphere at the equation found,[2] and [reckoning] up to the point where the equation was found (by means of the thread of the limb), it is true beyond question that the point of the second limb which

[1] The *almuri*, not mentioned at the beginning of Part III, is clearly distinguished from the thread of the centre, and yet it can hardly signify a radial metal pointer (as sometimes found on an astrolabe, and distinct from the alidade), which would have no function difficult of fulfilment by the thread itself. The word (from the Arabic *muri*—see the glossary) occasionally refers to a small pointer on the rete of an astrolabe. Meaning literally 'little thread', it must here refer to a marker or pointer on the thread—presumably a bead. Cf. the *margarita* or seed pearl of Apian's *Astronomicum Caesareum* and several earlier, but less well-known, instruments.

[2] *examinatum*. This must refer to the equation of the argument found in the last chapter.

procul dubio erit verus locus planete quesitus; cui si per 3[am] huius addideris motum octave spere, qui est circiter 10 gradus, habebitur verus locus ad nonam speram. Et hoc generale in omnibus.

UTILITAS 11[a]

De centro et argumento Lune equando

Centrum Lune et argumentum equatum ostendere.

Quod si placuerit, medium motum Solis in secundo limbo signatum pone directe sub capite primi limbi. Deinde transfer filum centri super medium motum Lune in secundo limbo signatum. Deinde filo non amoto, transferatur principium limbi secundi sub filo; et super quem numerum filum[a] cecidit in primo limbo super consimilem pone filum in secundo limbo; et tunc cadet filum super centrum Lune in primo limbo.

Deinde cum eodem centro non equato, move almuri ad ecentricum Lune, sicut dictum est conclusione nona huius; et statim pone argumentum Lune sub eodem filo, et tunc transferas filum centri super consimilem centrum in circulo orbis Lune primi limbi. Et locus secundi limbi qui sub filo ciderit erit argumentum equatum.

UTILITAS 12[a]

De vero loco Lune extrahendo

Verum locum Lune similiter invenire.

Quod si placuerit, pone medium motum eius, in secundo limbo signatum, in directo capitis primi limbi. Deinde almuri non variato, invenias equacionem argumenti eius examinatam per posicionem fili centri super argumentum eius equatum, et per extencionem fili limbi super almuri. Deinde considera qualis est numerus equacionis argumenti in circulo altitudinis, et super consimilem numerum in zodiaco orbis octave spere pone filum centri ex eadem parte qua cadit equacio, et punctus secundi limbi qui cadit sub filo erit verus locus Lune quesitus.

[a] C′ +prius

falls under the thread of the centre will be the true place of the planet, which we require. If, following Use 3, you add the motus of the eighth sphere, which is about 10°, you will have the true place at the ninth sphere.[1] And so for all [the planets named].

USE II

[III. 11]

To show the centre of the Moon and the equation of the argument.

On equating the centre and argument of the Moon

Set the mean motus of the Sun marked on the second limb directly under the origin of the first limb. Now move the thread of the centre over the mean motus of the Moon, marked on the second limb. Without the thread being moved, the beginning of the second limb is to be moved round until it is under the thread. Place the thread on the second limb over the same reading as is indicated by the thread on the first limb; and the thread will then fall on the [mean] centre of the Moon on the first limb.

With the same centre (not equated) move the almuri to the eccentric of the Moon, as explained in III. 9 above.[2] Immediately place the [mean] argument of the Moon under the same thread, and then move the thread of the centre to the same centre on the circle of the sphere of the Moon on the first limb. The point of the second limb falling under the thread will be the equated argument.

USE 12

[III. 12]

To find in a similar manner the true place of the Moon.

On working out the true place of the Moon

Set the mean motus [of the Moon], marked on the second limb, in the direction of the origin of the first limb. The almuri being unchanged,[3] you should find the equation of the [Moon's] argument, determined through the position of the thread of the centre over its equated argument, and by extending the thread of the limb over the almuri. Find the equation of the argument from the circle of altitude, and, setting this figure on the zodiac of the eighth sphere, draw the thread of the centre through it. The point of the second limb falling under the thread will be the required true place of the Moon.

[1] Cf. III. 5 above.
[2] This instruction is not required until III. 12 below.
[3] See Use 9.

UTILITAS 13[a]

De capite et cauda

Verum locum capitis geusahar reperire.

Quod si placuerit, per 4[am] huius, pone filum centri super medium motum capitis in secundo limbo in zodiaco, qui describitur secundum successionem signorum, et statim vide super quem numerum filum ceciderit in zodiaco descripto contra successum signorum; et iste gradus est verus locus capitis, et eius nadir est locus caude Draconis.

Per centra autem et argumenta scitur quantum distet planeta a longitudine longiori epicicli, puta per argumentum, et ab opposito augis per centrum.

UTILITAS 14[a]

De stacione et retrogradacione planetarum

Punctum vere stacionis prime et secunde, et tempus directionis et retrogradacionis ostendere.

Quod si placuerit, pone filum centri super centrum planete equatum in primo limbo, vel super centrum medium in orbe suo, et transfer almuri ad contactum ecentrici[a] planete. Deinde extende filum limbi super almuri, et move filum centri huc et illuc usque filum centri dividat filum limbi ad angulos rectos super almuri. Et si hoc fuerit versus 9 signa argumenti, est stacio prima, si versus 3 signa, erit stacio secunda precisissima.

Deinde circumvolve lento gressu filum centri circa limbum, secundum naturalem cursum planete in suo epiciclo. Et eius motus continue sequatur filum limbi semper extensum supra almuri. Et dum filum centri aliquid pertransit de limbo zodiaci, vide quantum pertransit filum limbi in circulo altitudinis; quia dum fuerit filum limbi insensibiliter motum,[b] dicetur stacionarius ad aspectum.

Si vero motus fili limbi fuerit sensibilis, et fuerit planeta inter secundam stacionem et primam, dicetur planeta directus.

[a] A centri [b] CAH motus

USE 13

[III. 13]

To find the true place of the head of geusahar.

On the head and tail [of the Dragon]

Using III. 4 above, place the thread of the centre on the mean motus of the head [of the Dragon] on the zodiac on the second limb, this being the zodiac described in the natural order of signs, and immediately take the reading indicated by the thread on the zodiac described against the succession of signs. That reading represents the true place of the head, and its nadir is the place of the tail, of the Dragon.

How distant a planet is from the furthest point of the epicycle is known from the centres and the arguments, for example by means of the argument, and [the distance] from the opposite aux [on the other side of] the centre.[1]

USE 14

[III. 14]

To show the point of the true station, first and second, and the time of direct and retrograde motion.

On the station and retrogradation of the planets

Set the thread of the centre over the equated centre of the planet, on the first limb, or over the mean centre on its circle,[2] and move the almuri to the point where it touches the planet's eccentric. Next, draw the thread of the limb over the almuri, and move the thread of the centre backwards and forwards until the thread of the centre crosses the thread of the limb at right angles at the almuri. If it should do so towards [3] signs of the argument, this is a first station, if towards [9] signs, it will be exactly a second station.[3]

Slowly now revolve the thread of the centre around the limb, in the direction of the natural course of the planet in its epicycle, and the thread of the limb is to follow its movements continuously, this thread being always drawn out over the almuri. And while the thread of the centre crosses the limb of the zodiac, observe how the thread of the limb intercepts the circle of altitude; for while the thread of the limb moves insensibly, [the planet] is said to be in[4] a stationary aspect.

If it should turn out that there is a perceptible motion of the thread

[1] This sentence seems unintelligible unless we assume that *centrum* at least is used in two different senses. The point being made is probably that the planets are by no means as simple as the Dragon. Simon Tunsted (version in MS. L) alters *opposito augis* to *auge deferentis*.

[2] That is, on one of the scales numbered here (10)–(15).

[3] The vagueness here is deliberate. See our commentary.

[4] Strictly speaking, 'near' should have been used, as is evident from what follows.

Si vero fuerit planeta inter stacionem primam et secundam, hoc est inferiori parte epicicli versus armillam, statim cum inceperit motus fili limbi esse velocior contra signa quam[a] motus fili centri cum signis dicetur planeta retrogradus. Quanto autem tempore erit planeta stacionarius, vel retrogradus, vel directus, faciliter patebit si ponatur filum limbi circulum[b] anni planete.

UTILITAS 15[a]

De mediis coniunctionibus et oppositionibus cognoscendis

Mediam coniunctionem et opposicionem Solis et Lune, necnon et quartas lunacionum, omni loco et hora ad omnem meridiem invenire.

Quod si placuerit, pone principium limbi secundi in directo radicis coniunctionis anni tui. Deinde transfer filum ad quam volueris coniunctionem vel opposicionem aut quartam tocius anni, et filum cadet in primo limbo super diem et horam coniunctionis vel cuiuslibet quarte ad quam operatus fueris, et similiter in zodiaco octave spere super illum locum in quo Luna fuerit ipsa hora in limbo secundo, si filo non moto transtuleris radicem Solis in limbo secundo ad principium limbi primi.

Et nota si posueris principium limbi secundi sub primo limbo in directo illius gradus contra quem stat finis lunacionis duodecime, tunc invenisti radicem lunacionum sequentis anni, nisi forte per embolismum impediatur, ut post dicetur.

Et hoc est generaliter verum in omnibus circulis mediorum motuum, et ideo, semel inventis radicibus planetarum, non oportet[c] amplius invenire novas radices, nisi oblivione et negligencia procurante.

UTILITAS 16[a]

De epactis et embolismis

Epactas cuiuslibet anni cum lunacione embolismali ostendere.

Quod si placuerit, dispone principium[d] secundi limbi pro radice lunacionum anni presentis. Deinde pone filum centri super luna-

[a] CAH quantum; C′ quam [b] C′H centrum [c] H oporteret
[d] CAH λ; L capud

of the limb, and the planet is between the second station and the first, then the planet is said to be direct.

If, in fact, the planet be between first and second station, that is, in the lower part of the epicycle towards the ring, as soon as the motion of the thread of the limb becomes faster against the signs than the motion of the thread of the centre with the signs, the planet is said to be retrograde. However, the length of time for which the planet is stationary, or retrograde, or direct, is made clear easily enough if the thread of the limb be placed over the circle of the year of the planet.

USE 15

[III. 15]

To find mean conjunction and opposition of the Sun and Moon, and also the Moon's quarters, at any place and hour, and at every meridian.

On discovering mean conjunctions and oppositions

Put the beginning of the second limb in the direction of the radix of conjunction for your year. Move the thread then to any conjunction, opposition, or quarter you please, of the whole [lunar] year,[1] and the thread will fall on the first limb over the day and the hour of the conjunction or quarter you wish to work with. Likewise, it will fall on the zodiac of the eighth sphere over the Moon's place at this same hour (on the second limb) if, the thread being kept fixed, you transfer the radix of the Sun on the second limb to the beginning of the first limb.

Note that if you place the beginning of the second limb below the first limb in the direction of that degree where the twelfth lunation ends, then you have found the radix of the lunations of the following year, unless by chance an embolism stands in the way [of the method],[2] as explained below.

And this result holds generally for all circles of mean motus, and thus once the radices of the planets are found, it is unnecessary to find new radices, except when forgetfulness and neglect make this necessary.

USE 16

[III. 16]

To show the epacts for any year, together with the embolismal lunation.

On epacts[3] and embolismal years

Place the beginning of the second limb in such a way as to show the radix of lunations of the present year. Then place the thread of

[1] Simon Tunsted (version of MS. L) replaces *tocius anni* with *coniunctionis*.

[2] *per embolismum impediatur*. For the significance of the remark and the meaning of 'embolism' see III. 16 below, and the commentary to it.

[3] *epacte* (nom. pl.)—intercalary days. From the Greek *ἐπακτή*, a verbal adjective from *ἐπάγειν*, 'to intercalate'. See III. 16 comm.

cionem duodecimam, et numerus dierum qui intercipitur inter filum et principium Ianuarii, si annus incipiat a Ianuario, erit numerus epactarum anni sequentis.

Si igitur per excrescenciam illorum dierum, contigerit lunacionem duodecimam distare a principio Ianuarii per 30 dies et amplius, ille annus erit embolismalis, hoc est continebit 13 lunaciones; quod cum contigerit non ponetur principium limbi secundi ubi[a] cadit finis lunacionis duodecime, sed ubi terminatur[b] lunacio terdecima. Verum[c] nota quod illius mensis dicitur esse lunacio in quo terminatur. Si igitur, in aliquo mense, contingat terminari duas lunaciones, scito quod annus iste est embolismalis, et posterior lunacio vocatur embolismus; ex quo patet quod post hoc capitulum faciliter sciri potest in quo mense cadet, primus, vel secundus, vel tercius, et sic de ceteris.

Invencio festorum mobilium et cetere proprietates kalendarii que oriuntur ex motibus Solis et Lune patent. Sed hec actenus.

UTILITAS 17ª

De vero loco Lune hora medie coniunctionis

Verum locum Lune ad horam[d] medie coniunctionis inquirere.

Quod si placuerit, pone medium motum Solis et Lune, in secundo limbo rote prime signatum, contra argumentum medium Lune, in zodiaco primi limbi signatum. Deinde transfer filum centri super consimilem numerum in circulo vero Lune, et ubi filum centri ceciderit in zodiaco limbi secundi, ipse est verus locus Lune. Differencia vero inter locum primum et secundum est longitudo inter Solem et Lunam, si operatus fueris[e] cum vero motu Solis.

UTILITAS 18ª

De tempore vere coniunctionis et opposicionis sciendo

Veram coniunctionem et opposicionem Solis et Lune ad minuta[f] horarum et dierum equatorum[g] cum suis noctibus invenire.

Quod si placuerit, primo invenias diem coniunctionis [vel][h] opposicionis, ad quem diem equabis motum Solis et Lune per premissam, et[i] non habuerint inter se distanciam. Prius invenisti et simul tempus medium et verum; que[j] si fuerit inter eos longitudo observa illam diligenter. Et accipe filum centri, et pone eum[k] super argumentum

[a] A nisi; C (?) [b] CA terminatus [c] H unde [d] H horas
[e] CA fuerit [f] CAH mora [g] CA equatis; H equatas
[h] C λ; AH et [i] CHL +si [j] C λ (erasure); L quod
[k] A λ; C′ ipsum

the centre over the twelfth lunation. The number of days between the thread and the first of January—if the year is to begin in January—will be the epact of the following year.

If, therefore, by the occurrence of those days, the twelfth lunation happens to be 30 days and more from the beginning of January, the year will be embolismal, that is, will contain [the beginnings or ends of] 13 lunations. When this happens, the beginning of the second limb is not to be put where the end of the twelfth lunation falls, but where the thirteenth lunation ends. Note that the month to which a lunation belongs is said to be that in which the lunation ends. If, therefore, in any month, two lunations should happen to end, you may know that the year is embolismal, and the last lunation is called embolismal. From this it is clear that, after this chapter, it may easily be decided in what month [a lunation] falls, whether the first, the second, the third, and so on.

The calculation of movable feasts, and other properties of the calendar arising from the motions of the Sun and Moon, are now evident. Let this suffice for the moment.

USE 17

[III. 17]

To inquire into the true place of the Moon at the hour of mean conjunction.

On the true place of the Moon at the hour of mean conjunction

Set the mean motus of the Sun and Moon,[1] marked on the second limb of the first wheel, directly against the mean argument of the Moon, which is marked on the zodiac of the first limb. Next, move the thread of the centre over the same figure on the true circle of the Moon, and where the thread of the centre meets the zodiac of the second limb, this will be the true place of the Moon. The difference, indeed, between the first position and the second will be the longitude between the Sun and Moon, if you work with the true motus of the Sun.

USE 18

[III. 18]

To find the true conjunction and opposition of the Sun and Moon in terms of minutes of equinoctial hours and days.

On obtaining a knowledge of the true time of conjunction and opposition

First find the day of the conjunction or opposition, and equate the motus of the Sun and Moon for this day, by means of the [explanation given] above, and there will be no distance between them. You found previously, at one and the same time, the mean and true [positions]: if there is any longitude between them, observe it carefully. Take the thread of the centre and put it over the equated

[1] Which are of course equal, by the definition of 'mean conjunction'.

Lune equatum; et transfer almuri[a] ad ecentricum motus horarum et Lune, qui dicitur *equator velocitatis Lune*. Deinde pone filum centri super longitudinem observatam prius in circulo illo qui dividitur per undecim, et statim cadet filum super horam et momentum coniunctionis certissime ubi situatur almuri, et in zodiaco limbi secundi super longitudinem, cum sua duodecima,[b] quodlibet signum computaveris pro uno gradu et quemlibet[c] gradum signi[d] pro duobus minutis.

Deinde equato Sole ad 9am speram secundum quod precessit, pone gradum Solis verum in directo diei et hore coniunctionis invente, vel opposicionis, in directo primi limbi et statim super consimilem locum graduum et signorum in circulo Iomyn, id est equacionis dierum, et statim dies et hora super quem cadit filum in limbo primo, ipsum est tempus equatum in diebus cum noctibus suis.

Et si placuerit seorsum scire quantum est tempus equacionis dierum, pone filum super consimilem numerum differencie primi loci fili ad secundum, super medietatem circuli altitudinis, ubi gradus resolvuntur in horas, et habebis propositum.

UTILITAS 19a

De velocitate Lune in una hora in omni loco epicicli

Motum Lune equatum in quotlibet horis in omni eius velocitate depromere.

Quod si placuerit, pone filum centri super argumentum Lune equatum, et transfer almuri ad ecentricum horarum. Deinde pone almuri super finem prime hore, et statim cadet filum in limbo secundo super motum Lune equatum in una hora, si gradum unum computaveris pro 2 minutis, et 15 gradus pro 30 minutis, et sic deinceps.

Item si posueris almuri super secundam vel terciam horam completam in eodem limbo, habebis motum Lune in tot horis, ita tamen quod unum signum computaveris pro uno gradu, et unum gradum pro 2 minutis, ut predictum est. Et in circulo diviso in partes undecim habebis differenciam motus Lune ad motum Solis, qui a quibusdam auctoribus *superfluum*[e] *Lune motus* vocatur.

[a] CAHL filum; C′ almuri [b] C′ +si [c] CAH quodlibet [d] A signum
[e] H super filum

argument of the Moon, and move the almuri to the eccentric of the motus of hours and of the Moon, which is called the *equator of the velocity of the Moon*. Then take the thread of the centre and place it over the longitude [difference] previously observed, using the circle divided into 11 [parts],[1] and immediately the thread will indicate the hour and moment[2] of conjunction precisely where the almuri is situated; and it will indicate, on the zodiac of the second limb, the longitude[3] ([on the scale] with an additional twelfth part, and counting one sign as 1°, and 1° as 2′).

Next, having equated the Sun at the ninth sphere as explained earlier, set the true [longitude] of the Sun [on the zodiac of the first wheel] in the direction of the day and the hour of conjunction or opposition already found, and [now set on the circle of the solar year] on the first limb. The equated time in days and nights is known at once, being where the thread falls on the first limb opposite the same [longitude] as before on the circle of iomyn.

If you wish to know separately the time [equivalent] of the equation of days, set the thread over the one half of the circle of altitude[4] at the figure expressing the difference between the first and second positions of the thread. There the degrees are converted into hours, and you have what was proposed.

USE 19

[III. 19]

To determine the equated movement of the Moon in any hour, for any velocity it may have.

On the hourly[5] velocity of the Moon, for any position on the epicycle

Set the thread of the centre at the equated argument of the Moon, and move the almuri to the eccentric of hours. Then set the almuri to the end of the first hour, and immediately the thread will indicate the equated motus of the Moon in one hour, at the second limb, each degree being counted as 2′, 15° as 30′, and so on.

Next, if you set the almuri to the end of the second or third complete hour, on the same limb, you will have the motus of the Moon in as many hours, so long as you count one sign as one degree, and one degree as 2′, as explained before. On the circle divided into eleven parts[6] you will have the difference between the motus of the Moon and that of the Sun, which by some authors is known as the *excess motus of the Moon*.

[1] Scale (26). [2] The interval, that is, between it and mean syzygy.

[3] The difference in longitude, that is, between mean and true syzygy.

[4] That is, over the 'scale of hours' (scale (8)), rather than the 'scale of altitude' (scale (67)), which together with it makes up a full circle.

[5] *in una hora*. This loose way of speaking, as though a velocity were a distance, was of course common before the seventeenth century. [6] Scale (26).

Patet igitur quod post hoc capitulum solum inveniri poterunt vere coniunctiones et opposiciones; et tempus Solis aut Lune perveniendi ad quodlibet punctum in orbe signorum signatum applica ut placet.

UTILITAS 20ª

De possibilitate eclipsis Solis et Lune inquirenda

Omnem coniunctionem et opposicionem eclipsabilem reperire.

Quod si placuerit, post opposiciones eclipsabiles signa verum locum capitis et caude in principio primi limbi, in principio opposicionis prime anni. Deinde pone principium limbi secundi directe contra gradum zodiaci limbi primi in quo fuerit prima opposicio eiusdem anni; quodsi iste locus non distaverit a capite vel a cauda per 12 gra., ipsa opposicio est eclipsabilis; si plus distet ne curaveris, sed transfer filum centri ad proximam opposicionem; et signa locum in zodiaco limbi primi et vide super quot gradus ceciderit filum centri in circulo capitis, et tot gradus computa (per 3ᵃᵐ huius) contra signa a loco capitis iam signato. Et si iste locus ut prius distaverit a filo citra 12 gra., erit eclipsis Lune; sin autem [non erit]. Sic procedas per singulas opposiciones anni.

Idem eciam facies in loco coniunctionis, si quesieris eclipsim solarem,[a] et si distaverint[b] dua loca predicta, scilicet capitis vel caude a loco coniunctionis signato, circa 6 sig., possibilis est eclipsis. Racio huius patet per 10ᵃᵐ communem proposicionem prime huius.

UTILITAS 21ª

De latitudine Lune

Latitudinem Lune per notam eius distanciam a capite vel a cauda cognoscere.

Quod si placuerit, pone filum centri super locum Lune in circulo geusahar, et cadet filum super gradus et minuta latitudinis in circulo latitudinis Lune certe.

[a] CA solaris [b] A distaverit

After this chapter alone it is evident that true conjunctions and oppositions may be found; and you are to add[1] the time at which the Sun or Moon arrives at any point marked on the circle of signs.

USE 20

[III. 20]

To find every conjunction and opposition at which an eclipse may take place.

On inquiring into the possibility of an eclipse of the Sun and Moon

For the oppositions at which an eclipse may happen, set the true place[2] of the head and tail [of the Dragon] at the beginning of the first limb, at [the time corresponding to] the beginning of the first opposition of the year. Then set the beginning of the second limb directly against the zodiac of the first limb at the first opposition for that year. If, however, the position be not more than 12° from the head or tail, an eclipse may take place at that opposition. If it be more [than 12°] distant, ignore it, and move the thread of the centre to the next opposition. Mark the place on the zodiac of the first limb, and take the reading indicated by the thread on the circle of the head [of the Dragon], reckoning the new position of the head against the signs. If, as before, this position is within 12°, there will be an eclipse of the Moon; but if to the contrary, [that is, if greater than 12°, there may be no eclipse].[3] Proceed in this way, taking the oppositions throughout the year one by one.

You will do the same with conjunctions, if you are seeking a solar eclipse. If the two aforesaid places (namely that of the head or tail [of the Dragon], and the marked position of the conjunction) are about 6 signs distant,[4] an eclipse is possible. The reason for this is evident from I. [19] of this work.[5]

USE 21

[III. 21]

To learn the latitude of the Moon from a knowledge of its distance from the head and tail [of the Dragon].

On the latitude of the Moon

Set the thread of the centre over the place of the Moon on the circle of geusahar,[6] and the thread will undoubtedly meet the circle of the latitude of the Moon at the degrees and minutes of [its] latitude.

[1] (?) *applica*. The last sentence is obscure. Perhaps there is an allusion to the need to add the time found in III. 18 to the time of a mean conjunction.

[2] As found, presumably, on scale (24).

[3] All MSS. lack this clause, which the sense requires. Cf. a similar construction in the middle of III. 24.

[4] See the commentary.

[5] The Latin text has a mistaken reference to I. 10. The present reference is not very apposite, but is the best that can be found.

[6] On the second face of the disc.

Et nota quod si motus geusahar fuerit a capite usque 6 sig., erit latitudo septentrionalis; si vero a cauda usque ad 12 sig., erit meridionalis.

UTILITAS 22ª

De diametro umbre cognoscenda

Medietatem diametri umbre Lune in omni eius elongacione ab auge epicicli cognoscere.

Quod si placuerit, revolve secundam faciem rote prime ita quod principium semicirculi eclipsis lunaris sit directe contra principium primi limbi, secundum successionem signorum. Deinde pone filum in zodiaco primi limbi, super gradus distancie Lune ab auge epicicli in eodem semicirculo primi limbi; et quicquid fuerit de filo inter duos exteriores ecentricos, ipsum erit semidiameter Lune. Residuum vero versus centrum ipsum erit semidiameter umbre Terre.

UTILITAS 23ª

De quantitate eclipsis Lune

Quantitatem eclipsis lunaris in motu et tempore, per quantitatem semidiametri umbre et Lune, et per notam latitudinem Lune demonstratorie ostendere.

Quod si placuerit, signa almuri in filo centri super locum quantitatis utriusque semidiametri, secundum premissam. Deinde transfer filum centri super tropicum eclipsis lunaris, ad minuta latitudinis Lune data. Deinde, filo non amoto, revolve rotam usquequo cadat filum limbi orthogonaliter super lineam minutorum casus et more contingendo almuri; et minuta linee sunt quantitas medietatis eclipsis necessario per demonstraciones Ptolomei, sicut patet conclusione ultima prime huius.

Quod si volueris scire tempus eisdem minutis correspondens, dispone almuri ad velocitatem motus Lune in una hora, sicut docetur 19ª huius. Deinde pone filum centri in circulo longitudinis, super tot minuta eclipsis accepta; et cadet almuri inter horas[a] super horas et partes horarum verissime.

[a] A +et

And note that if the [Moon's] geusahar-motus is between the head and 6^s, the latitude will be north; and if between the tail and 12^s, it will be south.

USE 22

[III. 22]

To discover the semidiameter of the shadow of the Moon for any value of its elongation from the aux of the epicycle.

On discovering the diameter of the shadow

Revolve the second face of the first wheel so that the beginning of the semicircle of the lunar eclipse[1] is directly against the beginning of the first limb,[2] [inscribed] according to the natural order of the signs. Then set the thread on the zodiac of the first limb, at the value of the angular distance of the Moon from the aux of the epicycle on the same semicircle of the first limb. Whatever length of thread lies between the two outer eccentrics,[3] this will be the semidiameter of the Moon. The remainder, indeed, towards the centre itself, will be the semidiameter of the Earth's shadow.

USE 23

[III. 23]

To demonstrate the quantity of a lunar eclipse, both in motion and time, by means of the size of the semidiameter of the shadow, and of the Moon.

On the quantity of an eclipse of the Moon

Set the almuri on the thread of the centre at the point corresponding to each of the two semidiameters, in accordance with the previous [chapter]. Then move the thread of the centre over the tropic of lunar eclipse, as far as the known latitude (in minutes) of the Moon. Keeping the thread in position,[4] turn the disc until the thread of the limb, whilst passing over the almuri, falls orthogonally over the scale of minutes of occurrence and duration. The minutes of the scale are necessarily the quantity of half-eclipse, from Ptolemy's proofs, as is clear in the last proposition of Part I of this work.[5]

If, however, you wish to know the time corresponding to these minutes [of arc], set the almuri at the velocity of the Moon (its motus in one hour), as explained in III. 19 above.[6] Then set the thread of the centre on the circle of longitude, over as many minutes [of arc] of eclipse as are [already] found. The almuri will indicate precisely the hours and parts of hours [required].

[1] The reference is to the entire half-disc, and not merely to one semicircular eccentric on it.

[2] That is to say, with the notation of II. 25–8 above, point *B* of the plate is moved to point *A* of the limb. (It was inscribed in a position 90° out of phase with this situation, and is illustrated accordingly in vol. iii.)

[3] Numbered here (48) and (49).

[4] With respect to the disc, not the limb.

[5] This is a mistake, if the text was written as presented here. The reference is I. 28, not I. 30.

[6] III. 18 would have been a better reference.

UTILITAS 24[a]

De minutis asus et mora eclipsis

Minuta casus et minuta more in eclipsi lunari seorsum cognoscere.

Quod si placuerit, pone filum centri cum suo almuri disposito ut pretactum est super latitudinem Lune in tropico eclipsis lunaris. Deinde revolve rotam usquequo filum limbi cadat orthogonaliter super lineam corde recte et verse contingendo almuri, quod est in filo centri. Et nota quod non plus[a] stabit orthogonaliter quam ipsum filum limbi steterit in linea cordarum super latitudinem Lune, si unum gradum eius computaveris pro 60 minutis, et terciam partem gradus pro 20 minutis. Deinde si filum limbi tetigerit intimum ecentricum erit mora; sin autem, non erit.

Quod si contingent, pone in filo limbi signum in loco contactus, et aliud signum in loco contactus cum almuri in filo centri. Deinde revolve rotam usque filum limbi directe iacuerit super lineam minutorum, et quicquid fuerit inter centrum rote et principium almuri erit minuta more veraciter, quod autem fuerit inter ambo signa erit minuta casus, modo procedendi in demonstracionibus Ptolomei et Albategni.

UTILITAS 25[a]

De digitis et parte Lune eclipsaris

Digitos diametri Lune eclipsatos una cum tota forma partis eclipsate et partis lucide de superficie eius, in omni momento duracionis eclipsis ostendere.

Quod si placuerit, formetur circulus in membrana secundum quantitatem diametri Lune cuius diametrum in 12 partes equales partire. Deinde fixa ipsa membrana in filo limbi, ita quod motu recto moveri possit super filum; et applicato almuri quod est in filo centri

‘C’ prius

USE 24

[III. 24]

To discover separately the minutes of [half the] partial and total phases of a lunar eclipse.

On the minutes of [half the] partial and total phases[1] of an eclipse

Set the thread of the centre, with its almuri set in the manner touched on earlier, to the latitude of the Moon, on the tropic of lunar eclipse. Next revolve the disc until the thread of the limb, while passing over the almuri, which is on the thread of the centre, falls orthogonally over the scale of corda recta and versa. And note that if the thread of the limb is orthogonal, it must be that it meets the scale of chords over the latitude of the Moon, each division of the scale being 60′, and a third part 20′, [which will be reckoned a minute division on the latitude scale]. If, then, the thread of the centre cross the inner eccentric,[2] there will be a period of totality, but if it does not do so, there will be none.

If, however, [the thread] meets [the eccentric], put a mark on the thread of the limb at the point of contact; and another mark where it touches the almuri, on the thread of the centre. Then revolve the disc until the thread of the limb lies directly over the scale of minutes, and whatever lies between the centre of the disc and the beginning (?) of the almuri will in fact be the number of minutes [of arc] of totality. What lies between the two marks will be the minutes [of arc] of the partial phase [of the eclipse], as in the proofs of Ptolemy and Albategni.

USE 25

[III. 25]

To show the eclipsed digits of the Moon's diameter, together with the shape of the eclipsed part and of the bright part of its surface, at any moment during the eclipse.

On the digits and the parts of the eclipsed Moon

A circular disc is to be made out of parchment, its diameter corresponding to the diameter of the Moon, and being divided into 12 equal parts. Fix, then, that same parchment on the thread of the limb, so that it may be moved on a straight line along the thread. The almuri (which is on the thread of the centre) having been set

[1] *casus* and *mora*. Both words, which generally mean nothing more than 'occurrence' and 'duration' respectively, amount to technical terms of medieval astronomy. *Casus* translates the Arabic *sukuṭ*, which in turn translates Ptolemy's ἔμπτωσις. *Mora* serves for the Arabic *mukth* and Ptolemy's μονή. Both terms, no doubt by virtue of their simple everyday meanings, are subject to some ambiguity. Thus referring to the figure of I. 22, Albategni uses the equivalent of *mora* to denote both $M_1 M_2$ (as here) and the difference between the shadow radius and UM_2.

[2] Numbered here (50).

ad medium circulum ecentricum illorum trium, scilicet ad equatorem umbre, circumvolvatur rota usquequo filum limbi cadat orthogonaliter super latitudinem Lune quam[a] habet in medio eclipsis, vel melius usquequo ipsum filum transeat per ipsam latitudinem in linea corde recte, et per latitudinem quam habebit in fine eclipsis super tropicum eclipsis lunaris. Et mota membrana circulari super filum limbi extensum, superextendatur filum centri super eius superficiem, ita tamen quod diameter membrane super respiciat centrum rote et almuri. Quod est in filo centri ostendet clarissime quot puncta diametri, et quanta pars superficiei, et secundum quam figuram fuerit eclipsata.

Et nota quod per istum modum harum trium operacionum componere poterimus tabulas eclipsis lunaris certissime, sive[b] compositas examinare.

UTILITAS 26ª

De corde recta et versa arcus dati

Cordam versam et rectam cuiuslibet arcus dati cognoscere.

Quod si placuerit, cordam rectam a radice fili limbi computa arcum duplicem arcui[c] tuo, et ad eius finem applica filum limbi, et pone ibi almuri in circulo medio communi circulo altitudinis et zodiaci limbi primi. Deinde remove filum super lineam cordarum, et filum cadet super cordam quesitum, si a radice fili limbi computaveris.

Si vero cordam versam[d] quesieris, computa ut prius arcum duplicem, incipiendo a capite limbi primi, et ad eius finem applicą filum limbi, et pone ibidem almuri. Deinde remove filum super lineam cordarum,[e] et cadet almuri super cordam versam si computaveris a capite limbi primi cordam illam. Hec[f] patent ex 9ª prime huius.

[a] CAHL quem [b] A λ; L sine [c] H arcus
[d] CA rectam; C′L versam [e] A cordam [f] A 3λ+ Hic patet pro

to the middle of the three eccentric circles,[1] namely the equator of the shadow, the [main] disc is to be turned until the thread of the limb falls orthogonally[2] over the latitude which the Moon has in mid-eclipse; or, better still, until the same thread indicates that same latitude on the scale of corda recta,[3] and indicates the latitude which [the Moon] will have at the end of the eclipse, where it crosses the tropic of lunar eclipse.[4] The parchment having been moved round with a circular motion over the extended thread of the limb, the thread of the centre is to be stretched out above its surface, so that the [graduated] diameter of the parchment is always directed towards the centre of the [main] disc and the almuri. That which is on the thread of the centre will show very clearly how many points[5] of the diameter and how much of the surface will have been eclipsed, and what will be the shape of the eclipsed Moon.

Note that by means of these three operations we may compile lunar tables with certainty, or test previous tables.

USE 26

[III. 26]

To discover the corda versa and corda recta of any given arc.

On the corda recta and versa of a given arc

For the corda recta,[6] reckon double your arc from the beginning of the thread of the limb, apply the thread to the end of this arc, and there set the almuri[7] on the circle common to the circle of altitude and the zodiac of the first limb. Then move the thread over the scale of chords, where [the almuri] will indicate the required chord, if you reckon from the beginning of the thread of the limb.

If you want the corda versa, count as before double the arc, beginning from the opposite end of the first limb, and apply the thread of the limb to the end of the arc, placing the almuri there. Next move the thread over the scale of chords, and the almuri will indicate the corda versa, if you count the chord from the opposite end of the first limb. These things are clear from I. 9 above.

[1] It is understood that this is to be done for a particular value of the argument.

[2] At right angles, that is, to the scale of corda recta.

[3] 'Taking one minute of latitude as one third part of a division on that scale' being understood.

[4] Eccentric semicircle (48).

[5] Digits.

[6] See our commentary for the idiosyncratic interpretation of this phrase, and of 'corda versa'.

[7] An almuri, that is, on the thread of the limb.

UTILITAS 27ª

De angulo qui est ex orizonte et ecliptica inquirendo

Cum notus fuerit gradus ascendens in hora eclipsis, et gradus nonagesimus ab ascendente, distanciam eiusdem gradus nonagesimi a puncto cenith per latitudinem Lune vel gradus medii celi in eadem hora cognoscere.

Quod si placuerit, per premissam, scias cordam rectam gradus Lune in altitudine vel altitudinis gradus medii celi, et in filo limbi pone ibidem almuri. Item scias cordam rectam distancie Lune ab ascendente, vel[a] gradus medii celi, et applica ipsam cordam ad arcum suum. Deinde a loco almuri egrediatur filum perpendiculariter, sicut dictum est conclusione penultima prime huius; et ubi filum absciderit lineam corde recte signa punctum, et iste est sinus altitudinis gradus nonagesimi ab ascendente, cuius arcum si abstraxeris de 90, residuum erit arcus quesitus, scilicet distancia gradus nonagesimi eiusdem a cenith capitis; et ipsa est quantitas anguli occidentalis proveniens ex concursu circuli altitudinis et ecliptice super punctum ascendentem.[b]

UTILITAS 28ª

De quantitate anguli secundi habenda

Quantitatem anguli occidentalis constituti super datum punctum ecliptice ex concursu circuli altitudinis cum ecliptica, post notam distanciam eiusdem puncti a cenith capitis, et[c] *per notam distanciam a cenith capitis*[d] *illius gradus qui ab ascendente distat per gradus 90 cognoscere.*

Quod si placuerit, pone almuri in filo limbi super cordam distancie gradus nonagesimi ab ascendente a cenith capitis. Deinde applica idem filum super arcum distancie puncti dati in ecliptica a cenith capitis. Deinde ab almuri ducatur linea perpendicularis ut prius, et ubi ceciderit filum super lineam cordarum nota locum, nam computando a radice fili ipsa erit corda recta arcus quesiti.[e]

[a] CA ut; C′ vel

[b] H +*Addicio marginis*: Sicut se habet sinus arcus . . . de 90 gra. remanet arcus altitudinis suprapropositi a cenith. (Cf. III. 27 seq. in the commentary below.)

[c] A 7λ

[d] H λ

[e] H +*Addicio quedam que fuit in margine*: Sicut se habet sinus arcus . . . ex concursu circuli altitudinis. (Cf. III. 28 seq. in the commentary below.)

USE 27

[III. 27]

Knowing the degree ascendent at the time of an eclipse, and the ninetieth degree from the ascendent, to discover the distance from the zenith of this same ninetieth degree by means of the altitude of the Moon, or of the degree of mid-heaven at the same hour.

On inquiring into the angle from the horizon and from the ecliptic

Find out the corda recta of the altitude of the Moon, from the previous chapter, or the corda recta of the altitude of mid-heaven,[1] and there set the almuri on the thread of the limb. Next, find the corda recta of the Moon's distance from the ascendent, or from mid-heaven, and apply that same chord to its arc. The thread[2] is then to be set at right angles through the almuri, as explained in the penultimate proposition of Part I. Mark the point where the thread intersects the scale of corda recta, and this is the sine of the altitude of the ninetieth degree from the ascendent. If the arc of the altitude be subtracted from 90, the remainder will be the arc required, namely the distance from the zenith of the nonagesimal degree.[3] And this is the angle to the west of [the circle of altitude], above the ascendent, between the circle of altitude and the ecliptic.[4]

USE 28

[III. 28]

To discover the size of the angle lying to the west [of the circle of altitude through the Moon] between the ecliptic and the circle of altitude, the zenith-distances of this point and of the point 90° from the ascendent being known.

On obtaining the size of the second angle

Set the almuri on the thread of the limb to the chord of the zenith-distance of the ninetieth degree from the ascendent. Then set the same thread to the zenith-distance of a given point of the ecliptic.[5] The [thread] is then to be drawn out at right-angles to the almuri, as before. Note where it meets the scale of chords, for, reckoning from the beginning of the thread, this will mark the corda recta of the required arc.

1 *medium celum*. This phrase denotes, as always here, the culmination of the ecliptic, that is where it crosses the meridian, although in a closely parallel passage in al-Khwārizmī (ed. Suter, p. 28) it refers to the highest point of the ecliptic, namely a point 90° from the ascendent. In the last sentence of this chapter such a point is referred to as *nonagesimus gradus*, a phrase used not uncommonly at the time, and still current in Kepler's day.

2 See our commentary.

3 *nonagesimus gradus*. To translate this as 'ninetieth degree' might lead to ambiguity. 'Mid-ecliptic' would suffice. See note 1.

4 The meaning of the Latin (especially *occidentalis*) can hardly be clear until after III. 28 has been read.

5 These two sentences are very loose and misleading. See our commentary. (Simon Tunsted here adds 'and there the almuri is placed'.)

UTILITAS 29ª

De diversitate aspectus in circulo altitudinis

Diversitatem aspectus in circulo altitudinis date ad tempus eclipsis lunaris cognoscere.

Quod si placuerit, applica filum centri ad lineam minutorum, et super punctum minuti 64 pone almuri.[a] Deinde revolve filum centri in una quarta circuli,[b] usque ad distanciam gradus Lune a cenith capitis. Deinde revolve rotam usque filum limbi, tangens almuri quod est in filo centri, cadet perpendiculariter super lineam minutorum casus; et minuta intercepta inter filum et centrum rote, ipsa minuta sunt diversitatis aspectus verissime.

Et nota quod hec operacio est inter cetera maxime precisa.[c]

UTILITAS 30ª

De diversitate aspectuum

Diversitatem aspectus in longitudine et latitudine apud omnem meridiem equaliter invenire.

Quod si placuerit, per antepremissam scias[d] quantitatem anguli occidentalis super punctum Lune[e] datum provenientem ex circulo altitudinis et ecliptica; et per premissam quantitatem diversitatis aspectus in[f] circulo altitudinis. Deinde pone almuri in filo centri super illam diversitatem aspectus in circulo altitudinis, et transfer filum centri in eadem quarta recedendo a linea minutorum casus secundum quantitatem arcus anguli dati.[g] Quo facto, non moto filo, revolve rotam usque filum limbi ceciderit super lineam minutorum orthogonaliter contingens almuri quod est in filo centri: et quot fuerint de minutis inter filum limbi et centrum rote, ipsum est diversitas aspectus in longitudine verissime.

Si vero computaveris arcum anguli predicti in quarta circuli a linea cordarum versus lineam minutorum, et per motum rote feceris filum limbi ab almuri predicto extendi orthogonaliter super lineam minutorum, quod fuerit inter ipsum filum et centrum rote erit diversitas aspectus in latitudine verissime.

Et nota quod per has quattuor operaciones expedite et precise facere poterimus tabulas diversitatis aspectus ad omnem regionem et ad omnem gradum zodiaci promptissime.

Causa operis patet in fine prime huius.

[a] H +vel melius super quantitatem 2 semidiametrorum inventam per 23
[b] H rote locate secundum cordas suas
[c] H +vel verius secundum *CB* (?) multiplicetur sinus distancie Lune a cenith capitis per cordam 64 min., et productum dividatur per 60, et exibit diversitas aspectus in *CAL* (?) concordis tabulis
[d] A λ   [e] AH linee   [f] A 3λ   [g] CA datus; C′ dati

USE 29

[III. 29]

To discover the parallax in the circle of altitude, at the time of a lunar eclipse, the altitude being given.

On parallax[1] in the circle of altitude

Apply the thread of the centre to the scale of minutes, and set the almuri to 64′. Then turn the thread of the centre into one quadrant, to the zenith-distance of the Moon. Now revolve the disc until the thread of the limb, which is touching the almuri which is on the thread of the centre, meets the scale of minutes of the partial phase at a right-angle. The number of minutes between the thread and the centre of the disc is undoubtedly the parallax.

Note that this operation is, like the others, extremely precise.

USE 30

[III. 30]

To find the parallax in longitude and latitude, applying equally to any meridian.

On parallax

Find out, using III. 28, the size of the western angle above the [ecliptic] place of the Moon, assumed given, between the ecliptic and the circle of altitude. From III. 29 find the parallax in altitude. Then set the almuri on the thread of the centre to the parallax in the circle of altitude, and move the thread of the centre into the same quarter, making the arc of the given angle with the scale of minutes of partial phase. Having done this, and not moving the thread, revolve the disc until the thread of the limb meets the scale of minutes orthogonally whilst passing through the almuri which is on the thread of the centre. The number of minutes between the thread of the limb and the centre of the disc is in fact the parallax in longitude.

If, however, you reckon the arc of the angle previously mentioned (within the quadrant) from the scale of chords towards the scale of minutes, and by moving the wheel you ensure that the thread of the limb is drawn out orthogonally over the aforesaid almuri and across the scale of minutes, then what is intercepted between the thread and the centre of the disc will most certainly be the parallax in latitude.

And note that by means of these four operations we may readily and accurately draw up tables of parallax for every region and for every degree of the zodiac extremely quickly.

The justification for this section is clear from the end of Part I.

[1] *diversitas aspectus.*

UTILITAS 31ª

De tempore apparentis coniunctionis cognoscendo

Tempus et horam visibilis coniunctionis eclipsabilis[a] *cognoscere.*

Quod si placuerit, adde vel subtrahe minuta diversitatis aspectus in longitudine a loco vere[b] coniunctionis. Adde[c] quidem si fuerit locus Lune inter ascendentem[d] et gradum nonagesimum, vel subtrahe[e] si fuerit ultra; et quod superest erit motus equatus cum quo operaberis sicut prius per operacionem 19 huius partis quarte, et habebis quesitum.

UTILITAS 32ª

De apparenti latitudine Lune cognoscenda

Apparentem latitudinem Lune in hora apparentis coniunctionis cognoscere.

Quod si placuerit, scias per 21am utilitatem precedentem veram latitudinem Lune ad idem tempus. Deinde si ipsa latitudo[f] meridionalis fuerit, adde eidem[g] diversitatem aspectus in latitudine, per 3am[h] huius. Si vero fuerit ipsa latitudo septentrionalis, subtrahe, et quod remanet erit apparens latitudo. Utrum autem latitudo[i] fuerit septentrionalis vel meridionalis cognoscere poterimus per eandem 21am huius partis.

UTILITAS 33ª

De quantitate eclipsis solaris cognoscenda

Quantitatem eclipsis solaris in motu et tempore pariter reperire.

Quod si placuerit, revolve rotam ad caput limbi primi, ita quod tropicus eclipsis solaris correspondeat medietati prime limbi secundum successionem signorum. Deinde filum centri pone super distanciam Lune ab auge; quo facto, applica almuri eius ad equatorem eclipsis solaris, et alterum almuri ad semicirculum orbis solaris. Deinde pone filum centri super latitudinem Lune visibilem in tropico eclipsis solaris et, filo non moto, revolve rotam usquequo filum limbi contingens ipsum almuri cadat perpendiculariter super lineam minutorum; et quod fuerit inter filum limbi et centrum rote sunt minuta eclipsis solaris certissime.

[a] CA eclipsis
[b] A vero
[c] H (and MSS. Joh. Gmunden) subtrahe; L minue
[d] CAL ascendens
[e] H (and MSS. Joh. Gmunden) adde
[f] CA longitudo; C′L latitudo
[g] CA eisdem; L sibi
[h] H 2am
[i] CAH longitudo; C′L latitudo

USE 31

[III. 31]

To discover the time and hour of a visible conjunction at which an eclipse is possible.

On finding out the time of appearance of a conjunction

Add the parallax in longitude to the true place of the conjunction, or subtract it from the same. Add,[1] if the place of the Moon be between the ascendent and the ninetieth degree; subtract, if it be beyond. The result will be the equated motus with which you will work, as you did before, in Use 19 of this fourth [*sic*] part; and you will arrive at the required result.

USE 32

[III. 32]

To discover the apparent latitude of the Moon at the time of visible conjunction.

On finding the apparent latitude of the Moon

Find out, by III. 21, the true latitude of the Moon at that same time. Then, if that latitude is south, add to it the parallax in latitude, with the help of III. 3. If, however, the latitude is north, subtract; and what results will be the apparent latitude. We may decide whether the latitude is north or south, moreover, by the same chapter III. 21.

USE 33

[III. 33]

To find the length of a solar eclipse, in motion as well as in time.

On discovering the length of a solar eclipse

Revolve the disc to the top of the first limb, so that the tropic of solar eclipse is brought into the first half of the first limb, following the natural order of the signs. Then set the thread of the centre over the distance of the Moon from the aux, and having done this, move its almuri to the equator of solar eclipse, and another almuri to the semicircle of the solar orbit. Now set the thread of the centre to the latitude of the visible Moon, on the tropic of solar eclipse, and, without moving the thread, revolve the disc until the thread of the limb, touching the [former] almuri, falls at a right-angle over the scale of minutes. What lies between the thread of the limb and the centre of the disc is in fact the number of minutes of solar eclipse.

[1] MS. L (Simon Tunsted's version) has 'Subtract'.

UTILITAS 34ª

De digitis et parte Solis eclipsata

Digitos diametri solaris una cum quantitate superficiei eiusdem eclipsata sensibiliter demonstrare.

Quod si placuerit, formetur membrana secundum quantitatem semidiametri contenti inter duo almuri in filo centri, hoc est secundum duplum eius quod cadit de filo centri inter equatorem eclipsis solaris et semicirculum corporis Solis. Et ipsa membrana fixa super filum limbi, revolvatur rota usquequo filum limbi cadat orthogonaliter super lineam corde recte, vel super[a] tropicum eclipsis solaris, ad minuta latitudinis visibilis Lune.

Et per motum ipsius membrane super ipsum semicirculum corporis solaris apparebit quantitas superficiei Solis, cuius semidiameter, si divisa fuerit in 6 partes equales, ostendet puncta eclipsanda. Aliter autem et melius apparebunt si utrique corpori Solis et Lune aptata fuerit membramina circularis secundum exigenciam semidiametri eiusdem, et membrana representans corpus Lune fixa fuerit in filo limbi, et altera membrana, cuius diameter dividitur in 12 partes equales, fixa fuerit super centrum rote. Et completum est.

Nota tamen quod[b] ex his que dicta sunt patet modus multiplicandi et dividendi, cum illa ars ad nichil deserviat astrologis nisi per tres sinus notos invenire quartum ignotum. Similiter extractio radicis, cum illa operacio solum adhuc inventa est ut, habitis duobus sinibus trianguli orthogonii, invenires[c] tercium sinum ignotum. Sed hec dicta sunt in fine primi huius expressius.

UTILITAS 35ª

De altitudine stelle super orizontem; et est hic subductio astrolabii

Elevacionem[d] stelle super orizontem accipere.

Quod si placuerit, teneas instrumentum contra stellam quousque radius lucis aut visualis transeat per foramina pinnularum *C* et *B*; et sit pinnula *C* propinquior corpori luminoso. Et tunc aspice super quem numerum graduum cadit filum limbi in circulo altitudinis, cuius computacio incipit a puncto *D* versus *A*, et iste est gradus altitudinis.

[a] H terminates Pt. III here. [b] A λ [c] CAH invenire [d] A Elevacio

USE 34

[III. 34]

To demonstrate the digits of the solar diameter together with the solar area visibly eclipsed.

On the digits and the eclipsed part of the Sun

A piece of parchment is to be cut equal to the semidiameter intercepted between two almuri on the thread of the centre, that is, equal to double that part of the thread of the centre which lies between the equator of solar eclipse and the semicircle of the body of the Sun.[1] Once that membrane is fixed on the thread of the limb, the disc is to be revolved until the thread of the limb falls orthogonally over the scale of corda recta, or over the tropic of solar eclipse, at the number of minutes of latitude of the visible Moon.

By the motion of this parchment over the semicircle of the solar body, the [eclipsed] area of the Sun will be apparent; and if its semidiameter be divided into six equal parts, it will show the number of digits eclipsed. This will be shown otherwise, however, and better, if a circular parchment be fitted [to represent] the body of the Sun, like that for the Moon, each in keeping with the appropriate semidiameter. The parchment representing the body of the Moon will be fixed on the thread of the limb, and the other parchment, whose diameter is divided into 12 equal parts, will be fixed over the centre of the [main] disc. [The arrangement] is then complete.

Note that from what has been said, a way of multiplying and dividing is evident, but this art serves astronomers[2] to no purpose, unless to find from three known sines a fourth, unknown, sine. Similarly with the extraction of a root; this operation was found merely in order that, having two sines of a right-angled triangle, one could find a third, unknown, sine. But these things were spoken of more expressly at the end of Part I of this work.

USE 35

[III. 35]

To obtain the elevation of a star above the horizon.

On the altitude of a star above the horizon; and here the astrolabe is introduced

Hold the instrument towards the star until a ray of light or a visual ray passes through the sighting-holes in the pinnules *C* and *B*; and let *C* be the pinnule nearer the luminous body. Note then the number of degrees indicated by the thread of the limb on the circle of altitude, counting from *D* towards *A*; and this is the altitude [of the star].

[1] Between, that is to say, the first and second almuri as they are positioned in the last chapter.

[2] *astrologi*. The modern distinction between astrologer and astronomer was not yet clearly established, and if we are to make the distinction, 'astronomers' is clearly the right word.

Nota sane, ex conclusione octava[a] primi, quod omnis divisio qui est unus gradus in limbo orbis octave[b] respectu filia centri, illa est dimidius gradus respectu fili limbi.

UTILITAS 36ª

De horis diei cognoscendis

Horas diei per notam Solis altitudinem super orizontem cognoscere.

Quod si placuerit, pone filum centri super gradum Solis in circulo directarum ascensionum, et transfer unum eius almuri ad contactum ecliptice. Iterum pone filum in circulo directo super nadir Solis et move aliud almuri ad contactum ecliptice. Quo facto, transfer almuri Solis super almucantarath altitudinis iam accepte, et vide super quem gradum filum limbi ceciderit. Et statim super eius oppositum move filum, et almuri nadir Solis cadet super horam diei quam queris. Cetera autem, ut arcus diei, altitudo meridionalis, azimuth, per se patent. Non potest autem almuri debite disponi[c] super eclipticam nisi quando capita Arietis sunt simul.

Item si placuerit hec eadem speculari in climate facto ad instar saphee, per solam altitudinem Solis acceptam, sine motu tanquam in quodam speculo omnia predicta relucent. Operacio autem que sit per sapheam habet suum tractatum ad singula capitula motus diurni.

UTILITAS 37ª

De horis noctis cognoscendis

Horas noctis per altitudinem stelle fixe cognoscere.

Quod si placuerit, pone filum centri super gradum Solis, cum revolveris duo capita Arietis limbi primi et secundi simul deinde, applicato almuri ad stellam, pone idem almuri super altitudinem

[a] A g; C (in later hand) +nona [b] A sui [c] CA disposicioni

Note well, from I. 8 above, that each degree-division on the limb of the eighth sphere with respect to the thread of the centre is half a degree with respect to the thread of the limb.

USE 36

[III. 36]

To discover the hours of the day, given the altitude of the Sun above the horizon.

On finding the hours of the day

Set the thread of the centre on the circle of right ascensions over the degree in which the Sun is found, and transfer one of its almuri to the point at which it meets the ecliptic. Again set the thread on the direct circle, over the nadir of the Sun, and move the other almuri to the point at which [the thread] meets the ecliptic. Having done this, move the almuri of the Sun[1] over the almucantar of the newly discovered altitude, and notice the reading indicated by the thread on the limb.[2] Immediately move the thread to the opposite point [of the limb], and the almuri of the nadir of the Sun will indicate the hour of the day which you are seeking. As for the rest, however, such as the arc of the day, the meridian altitude, [and] the azimuth, they are self-evident. But the almuri cannot be properly arranged over the ecliptic unless the heads of Aries are together.[3]

If it please you to observe these things on a climate made in the likeness of a saphea, given only the altitude of the Sun, all the aforesaid things appear as in a mirror, without moving anything.[4] The method of working with a saphea, however, has a treatise to itself, with individual chapters devoted to the diurnal motion.

USE 37

[III. 37]

To discover the hours of the night by means of the altitude of a fixed star.

On discovering the hours of the night

Set the thread of the centre over the position of the Sun, when you have brought the two heads of Aries on the first and second limb round together and when then, having applied the almuri to the star,

[1] As opposed to that of the nadir of the Sun.

[2] Simon Tunsted (MS. L) changes the text for the better here, substituting *limbi cecidit filum* for *filum limbi ceciderit*; but *in limbo* would have been still clearer.

[3] This means that the origin of the zodiac of the first limb (54) (at *D*, according to II. 22) is to be aligned with the head of Aries on the ecliptic circle of the plate.

[4] Note the same phrase in connection with the saphea in III. 40. The idea is perhaps that we need only look at the saphea to discover at once what we are seeking, without the use of a rete. This would be something of an exaggeration. It is just conceivable that some reference is being made here to the unusual way of calibrating the other astrolabe plate, which is a mirror image of the common plate. See II. 22–3.

stelle acceptam. Deinde, non amoto filo e loco, revolve clima quousque duo capita directe contra se fuerint, et statim almuri Solis cadet super horam noctis. Ex hoc patet qualiter noctis crepusculum, et aurora, et cetere proprietates noctis, accipi poterunt per almuri stelle; et almuri gradus Solis applica sicut placet.

UTILITAS 38[a]

De gradu ascendente habendo

Gradum ascendentem omni hora diei et noctis cognoscere.

Quod si placuerit, de die dispone almuri pro gradu Solis. Deinde pone filum in limbo exteriori super gradum Solis in circulo ascensionum climatis, et revolve clima quousque almuri Solis contingat orizontem orientalem. Deinde transfer filum super altitudinem acceptam, et ubi cecidit filum super gradus ascensionum eius climatis in primo limbo, ille procul dubio est gradus ascendens. Idem et equefaciliter cognoscere poteris per horam cognitam.

Si autem volueris hoc de nocte, dispone almuri pro nadir Solis et aliud pro stella. Deinde per proximam precedentem cognosce altitudinem nadir Solis, quod semper est in opposito Solis. Deinde pone filum super nadir gradum inter ascensiones huius climatis, et revolve clima quousque nadir contingerit orizontem. Postea transfer filum ad altitudinem nadir notam, et similiter[a] cadet hoc filum in primo limbo super gradum ascendentem.

UTILITAS 39[a]

De modo equandi duodecim domos

Principia duodecim domorum cognoscere.

Quod si placuerit, pone unum almuri super gradum nadir ascendentis in ecliptica et aliud super gradum ascendentem. Deinde pone

[a] A +cum

[you have] set that same almuri over the altitude found for the star.[1] Without moving the thread from its position, turn the climate[2] until the two heads [of Aries] are directly opposite one another, and the almuri of the Sun will immediately indicate the hour of the night. From this it is clear how dusk, and dawn, and other properties of the night, may be found by the almuri of a star: you may apply the almuri indicating the Sun's place as you please.

USE 38

[III. 38]

To discover the degree ascendent for any hour of the day and night.

On determining the degree ascendent

By day, set the almuri to the Sun's position. Place the thread on the outer limb over the position of the Sun on the scale of ascensions [appropriate to] the climate,[3] and revolve the climate until the almuri of the Sun meets the eastern[4] horizon. Then transfer the thread to the measured altitude, and where the thread falls over the degrees of [oblique] ascension for the climate on the first limb, that, without doubt, is the degree ascendent. You can discover the same thing equally easily from a knowledge of the hour.

If, however, you want to know the same thing by night, set one almuri towards the nadir of the Sun and the other towards [the position of] a star. Then, by III. 36, find out the altitude of the nadir of the Sun, which is always opposite the Sun. Set the thread to the nadir's position on the ascension [scale] for this climate, and revolve the climate until the nadir meets the horizon. Later transfer the thread to the known altitude of the nadir, and it will fall, as before, over the degree ascendent on the first limb.

USE 39

[III. 39]

To discover the points at which the twelve houses begin.

On the method of equating the twelve houses

Place one almuri over the nadir of the ascendent[5] on the ecliptic, and another over the degree ascendent. Then place the thread over

[1] That is, set it to the appropriate almucantar. There is a missing clause at this point, of the form: 'and have then revolved the plate, holding the thread in its new position, until the star again coincides with the position of the almuri.'

[2] *clima*—here meaning the astrolabe plate.

[3] That is, the scale of ascensions on the oblique circle (56) (see II. 18 above).

[4] This I believe to be a mistake for 'western', with serious consequences. But see the commentary.

[5] Simon Tunsted's version (MS. L) makes this the ascendent, not its nadir. See the three following footnotes, and III. 39 comm. and seq.

filum super ascendentem in circulo directo primi libri, et revolve clima usquequo orizon occidentalis[a] tangat almuri nadir ascendentis. Deinde move filum usque ad finem hore secunde, et ubi ceciderit filum in limbo primo inter gradus circuli directi, ipse gradus est inicium secunde domus.

Item move filum per alias duas horas inequales, usque almuri nadir tangat lineam quarte hore, et cadet filum super inicium tercie domus. Cum vero tetigerit lineam sexte hore, cadet filum super principium quarte domus. Ulterius, cum almuri in gradus ascendentis tetigerit inicium hore octave,[b] cadet filum super principium quinte domus; et sic deinceps, usque ad domum septimam. Et ultra earum[c] opposita sunt principia[d] domorum aliarum.

UTILITAS 40[a]

De alio modo cognoscendi ascensiones et reliquas domos

Alio modo contingit gradum ascendentem et domorum principia[e] invenire.

Quod si placuerit, et fuerit circulus ascensionum climatis descriptus in lamina climatis, tunc accipies ascendens hoc modo:

Pone caput Arietis quod signatur in climate directe contra caput Arietis limbi primi; et dispone almuri ad gradum Solis si sit de die, vel ad nadir Solis si sit de nocte et aliud almuri ad stellam, et predictum est. Deinde transfer filum ad gradum Solis inter ascensiones ipsius climatis, et filo immobiliter in ibi permanente revolve clima usquequo almuri gradus Solis tangat orizontem ex parte orientis. Post hec[f] pone almuri Solis super altitudinem acceptam. Deinde filo in ibi permanente, revolve caput Arietis climatis ad priorem disposicionem, hoc est contra caput Arietis primi limbi, et gradus ascensionum ville tue qui sub filo ceciderit ille est ascendens.

Eodem modo operaberis de nocte cum almuri Solis nadir.

[a] CAH occidentale [b] CA septime [c] CA eorum [d] A principalia
[e] A principalia [f] AL λ

the[1] ascendent on the direct circle [spoken of in] the first book, and revolve the climate until the western horizon meets the almuri of the nadir of the ascendent. Move the thread up to the end of the second hour, and where the thread falls on the first limb among the degrees of the direct circle, that place marks the beginning of the second house.

Next, move the thread through another two unequal hours, until the almuri of the nadir[2] touches the line corresponding to the fourth hour. The thread will then indicate the beginning of the third house. When the thread is at the sixth hour, it indicates the beginning of the fourth house. When, furthermore, the almuri of the degree[3] ascendent is at the beginning of the eighth hour, the thread will indicate the beginning of the fifth house. And so we continue, up to the seventh house. And beyond that, their opposites are the beginnings of the other houses.

USE 40

[III. 40]

It is possible to find in another way[4] the degree ascendent and the beginnings of the houses.

On another way of discovering ascensions, and the remaining houses

A circle of ascensions for the climate will have been described on the plate of the climate, and this being so, calculate the ascendent as follows:

Take the head of Aries marked on the climate and set it directly against the head of Aries on the first limb. Set the almuri to the degree of the Sun if the problem concerns daytime, or to the nadir of the Sun if it concerns night, [in which case] set the other almuri to a star, as explained before. Now transfer the thread of the position of the Sun on the [scale of] ascensions on the same climate. Keeping the thread fixed there, revolve the climate until the almuri of the degree of the Sun meet the eastern horizon. After this, set the Sun's almuri over the measured altitude. Keeping the thread there, revolve the head of Aries on the climate to [its] first position, that is, against the head of Aries on the first limb, whereupon the ascension ([on the scale] for your town) under the thread will be the ascendent.

You will work in the same way as regards the night, [but] with the two almuri of the Sun's nadir.

[1] MS. L reads differently here, and this might well be in part deliberate. (For *ascendentem . . . orizon* in the text read *gradum.*)

[2] MS. L would make this 'ascendent'.

[3] MS. L would have 'nadir of the degree' here.

[4] The title is misleading. The way is essentially similar to that already explained, but the instructions are repeated for towns where the scale (56) on the limb of the second face of the albion is inapplicable, i.e. where it applies to the wrong latitude.

Cetere autem domus inveniuntur omnino sicut in precedenti[a] capitulo dicitur; hec etiam omnia patent sicut in speculo in climate fundi facto[b] secundum operacionem saphee.

UTILITAS 41ª

De equacione planetarum per opus astrolabii

Vera loca omnium planetarum in longitudine et latitudine per notam eius altitudinem et azimuth invenire.

Quod si libuerit, de Sole per diem scias eius azimuth et altitudinem quam precise poteris per rectangulum vel per aliud instrumentum; et postea move filum almuri super consimilem[c] altitudinem et azimuth, et statim revolve filum almuri contra gradus et signa puta in circulo directo usquequo tangat eclipticam in illa medietate zodiaci inter duo solsticia, que congruit medietati[d] anni; et cum tetigerit eclipticam, ille gradus super quem ceciderit filum in circulo directo climatis vel limbi, ipse est gradus Solis.

Eodem modo de ceteris planetis per noctem, scias earum altitudines et azimuth et pone unum almuri super consimilem altitudinem et azimuth. Postea applicato altero almuri ad stellam notam positam contra certum gradum limbi primi, transferatur filum super altitudinem stelle acceptam; et considera quot gradus filum peragatur a situ illo in quo fuerat quando ponebatur super stellam. Deinde posito almuri altitudinis et azimuth in loco debito, retorqueatur filum contra signa circuli directi usque filum consimilem arcum transierit de gradibus limbi. Et necessario cadet almuri in zodiaco super gradum in longitudine et super gradum in latitudine, meridionali vel septentrionali, et ille est verus locus quem habet ille planeta pro hora altitudinis iam accepte.

UTILITAS 42ª

De horis diei et noctis cognoscendis per planetas

Horas diei et noctis per quemlibet[e] planetam super orizontem apparentem ostendere.

Quod si placuerit, scias verum locum planete in longitudine, secundum eius equacionem ad nonam speram, et scias eius latitudinem. Contra caput limbi primi, pone almuri unum super gradum Solis, si sit de die, vel super nadir eius, si sit de nocte, et aliud almuri

[a] C presidente; A precedente [b] CA facte [c] CA consimili
[d] A medietate [e] CAH quamlibet

The other houses are found exactly as explained in the previous chapter. On a climate[1] made in the manner of a saphea, furthermore, they all appear as in a mirror.[2]

USE 41

[III. 41]

To find the true places of all the planets, in longitude and latitude, from a knowledge of their altitude and azimuth.

On the equation of the planets, using an astrolabe

Find out the Sun's altitude and azimuth for the day, as accurately as you can, by means of a rectangulus[3] or other instrument. Subsequently move the thread with the almuri over this altitude and azimuth, and at once move the thread of the almuri round to the supposed position in the direct circle so that it meets the ecliptic on that half of the zodiac between the two solstices which corresponds to the half-year. And when [the almuri] meets the ecliptic, then the angular position of the Sun will be that angle indicated by the thread on the direct circle[4] of the climate or of the limb.

In the same way, for the other planets by night, find out their altitudes and azimuths and adjust one almuri to the altitude and azimuth [of one of them]. After this, the other almuri having been applied to a known star placed at a definite position on the first limb, the thread is to be transferred to the altitude of the observed star. See by how many degrees the thread has been moved since it was placed over the star. The almuri of altitude and azimuth having been then set in its proper place, the thread is again revolved against the signs of the direct circle until the thread has moved through a similar arc in terms of degrees of the limb. The almuri will necessarily indicate on the zodiac the longitude and latitude, north or south, and this will be the true place of the planet at the time of the newly observed altitude.

USE 42

[III. 42]

To show the hours of day and night by means of any planet visible above the horizon.

On discovering the hours of day and night by means of the planets

Find the true place of the planet in longitude, in accordance with its equation at the ninth sphere, and find its latitude. Against the top of the first limb set one almuri to the Sun's position, if it be by day, or to its nadir, if by night; and set the other almuri to the latitude

[1] The text adds *fundi*, perhaps alluding to the bottom of the inside of the mater.

[2] Cf. the note to III. 36 above.

[3] The *rectangulus* being Richard's own instrument, of course. See Section VI below.

[4] See III. 41 comm.

super gradum planete in longitudine et latitudine. Deinde cum almuri planete operare per omnia sicut cum stella fixa.

Cetera autem omnia que in astrolabio numerantur patent hic ad oculum. Ista eciam capitula de veris locis planetarum et horis perfecta facilitate patent in climate facto ad modum saphee.

Multa eciam alia per hec inquirenti et scienti patebunt; sed ignoscat Christus Iesus curiositati presumpte quantum in his declinavi a studio pietatis.

Explicit pars tercia tractatus Albionis

IV

Incipit pars quarta Albionis, adiecta pro errore vitando in composicione.

Albion, pro magna sui parte, non superficiem spere sed circulorum in spera descriptorum superficiem representat, super quam linee transeuntes, nunc a centro ecentricorum, nunc a centro orbis signorum tanquam radii visuales ab oculo, ipsos circulos debita proporcione conscindunt. Propter quod opus est ipsam superficiem vehementissime fore planam, alioquin de facili continget errare; in cuius remedium compendiosum erat quasdam tabulas calculare, quibus opus divisionis circulorum examinari posset ad unguem vel, si placuerit, eisdem tabulis ipsos circulos arte que sequitur dividere sine pluri.

Capitulum primum quarte partis

Hec prima tabula, que intitulatur *Tabula centri Saturni*, facta est hac arte:

Primo a sinistris descendendo, ponitur linea numerorum graduum 30 signa; et superius, ex transverso in singulis areis, ponuntur numeri singulorum signorum 12 (ut in prima area 0, in secunda 1, in tercia 2, et sic deinceps usque ad undecimam signam). Et in primo loco prime aree, contra unum gradum numeri, ponitur illud quod superest cum extracta fuerit equacio centri Saturni unius gradus de uno gradu, hoc est de 60 minutis. In secundo loco ponitur illud quod remanet cum extracta fuerit eius equacio centri duorum graduum de duobus gradibus; et sic deinceps usque ad complecionem 6 signorum.

and longitude of the planet. Proceed now with the planet's almuri exactly as for a fixed star.

All other things which are enumerated in [connection with] the astrolabe are here plain to see. Those chapters, furthermore, on the true places of the planets and the hours, may be put into effect with perfect ease, using a climate made after the manner of the saphea.

From all this, much else will emerge by way of inquiry and learning. But Christ Jesus forgive the presumptuous curiosity, inasmuch as we have turned aside to these things from the study of piety.

Here ends the third part of the treatise on the albion

IV

Here begins the fourth part of the albion, added in order to avoid error in construction.

ALBION, for the most part, represents not the surface of a sphere but a surface containing the circles which are described on a sphere. Lines on this surface cut the circles themselves in proper proportion, these lines crossing, now from the centre of the eccentrics, now from the centre of the circle of signs like visual rays from the eye. The surface itself must therefore be absolutely flat, or otherwise it will be easy to make a mistake. As a remedy for this it was advantageous to calculate certain tables, with which the work of dividing the circle could be verified to a nicety; or, if preferred, the circles could actually themselves be divided using these tables, by the following method, without more ado.

CHAPTER 1 of this fourth part

[IV. 1]

This first table, entitled *Table of the centre of Saturn*, was made in the following way.

First, coming down on the left is placed a column[1] of numbers of degrees, thirty to a sign. Above, at the head of separate [pairs of] columns,[2] the numbers of the 12 signs are placed (0 in the first space, 1 in the second, 2 in the third, and so on, up to eleven). In the first column and first row, against 1°, is placed what remains when the equation of the centre of Saturn [for a mean centre] of 1° is subtracted from 1° (60′). In the second row is placed what remains when the equation of its centre for 2° is subtracted from 2°. And so on, up to the completion of 6 signs.

[1] *linea.* [2] *ex transverso in singulis areis.*

Deinceps contra primum gradum septimi signi ponitur illud quod resultat cum addiderimus equacionem centri unius gradus ad 7 sig.[a] et 1 gra.; et sic ulterius usque ad complecionem 12 signorum.

Cum ista tabula examinari poterit divisio[b] circuli orbis Saturni; quia si divisiones concordaverint cum tabula, operatum est bene; sin autem, iuxta tabulam corrigatur; vel a capite dividi posset circulus iste opere expedito per dictam tabulam isto modo:

Ligetur[c] regula[d] in centro *G*, et extendatur ad zodiacum limbi primi super talem numerum graduum et minutorum qualis ponitur in prima area contra 5 gradus vel 6 de gradibus in linea numeri, et secundum illam quantitatem transeat una divisio per marginem et spacium. Iterum ponatur regula in zodiaco super talem numerum qualis reperitur in eadem area contra 10 gradus de gradibus qui sunt in linea numeri, et fiat altera divisio transiens per marginem et per spacium; et sic deinceps usquequo complete divisus fuerit circulus. Tandem dividere poteris marginem consimili modo ad singulos gradus per tabulam, vel in 6 aut 5 partes, cum circino. Et istud opus est expeditum, carens errore sensibili, licet non fuerit ita artificiale sicut primum opus[e] dividendi per circulos ecentricos, quem ad modum docetur parte secunda capitulo 3°.

Et hec est forma tabule continens tantummodo centrum equatum Saturni. Nota quod divisio omnium circulorum per tabulas est uniformis, et cum per tabulam unam circulus divisus fuerit absque errore in solis gradibus, ipse ostendet eandem proporcionem in omnibus fractionibus quam in integris representat tabula.

Tabula veri motus orbis Saturni.

[Vol. iii, p. 76]

CAPITULUM 2^m quarte partis

De tabula orbis Iovis.

Hec secunda tabula que intitulatur *Tabula centri Iovis* omnis calculata est et composita ex addicione et subtractione equacionis centri de gradibus equalibus, sicut precedens tabula centri Saturni; et eodem modo per eam examinari vel dividi potest circulus orbis Iovis, sicut predictum est de divisione circuli orbis Saturni. Et hec est forma tabule:

[a] CA gra.; C'L sig. [b] H omnino [c] H Signetur [d] A regulam [e] A operis

Now against the first degree of the seventh sign is placed what results when we add the equation of the centre for 1°, to 7^s and 1°. And so on, up to the completion of 12 signs.

The division of the circle of the sphere of Saturn may be verified with this table. For, if the divisions agree with the table, they have been well made; and if to the contrary, then [the divisions are] to be corrected from the table. Otherwise the circle may be freely divided from the beginning with the table in question in this way:

A rule is to be fastened at centre *G*, and extended to the zodiac o the first limb over the number of degrees and minutes to be found in the row [of the tables] opposite 5° or 6°. Make a division there through the margin and [adjoining] space. The rule is next to be set to the zodiac at the reading opposite 10°[1] in the table, and let another division be made crossing the margin and space; and so continue, until the circle is completely divided. Finally, you may divide the margin in like manner into single degrees (whether in groups of 6 or 5) by means of the table, with dividers.

This task is readily done, being free from sensible error, although not so accurate as the first method of dividing [the scales], namely by means of eccentric circles, as explained in II. 3.

Below is the form of the table containing only the equated centre of Saturn. Note that the division of all the circles by means of the tables is uniform, and that when a circle has been divided without error by means of one table, in degrees alone, it will show all fractions by due interpolation.[2]

IV. 1

Table of the motus of the circle of Saturn.

[See p. 76]

IV. 2

On the table of the circle of Jupiter.

This second table, which is entitled *Table of the centre of Jupiter*, is entirely calculated and constructed by combining (by addition or subtraction) the equation of the centre with an appropriate number of degrees, just as in the preceding table of the centre of Saturn. The circle of the sphere of Jupiter can be verified or divided in the same way by means of this table, as was explained in connection with the division of the circle of the sphere of Saturn. And this is the form of the table:

1 The author should have added 'or 12°'.

2 *ipse ostendet eandem proporcionem in omnibus fractionibus quam in integris representat tabula.*

Tabula veri notus orbis Iovis.

[Vol. iii, p. 77]

CAPITULUM 3^m

De tabula qua dividi potest orbis Martis.

Hec tabula, que intitulatur *Tabula Martis*, omnino calculata est et composita addicione et subtractione equacionis centri de gradibus equalibus, sicut precedens tabula Iovis; et eodem modo per eam examinari vel dividi potest circulus orbis Martis, sicut predictum est de tabula centri Iovis. Et hec est forma tabule:

Tabula veri motus orbis Martis.

[Vol. iii, p. 78]

CAPITULUM 4^m huius quarte partis

De tabula orbis Solis et Veneris.

Hec quarta tabula que intitulatur *Tabula centri Solis et Veneris* omnino calculata est et composita ex addicione et subtractione equacionis centri de gradibus equalibus, sicut precedens tabula Martis, et eodem modo per eam examinari vel dividi potest circulus orbis Solis et Veneris, sicut predictum est de tabula centri Martis. Et hec est forma tabule:

Tabula veri motus orbis Solis et Veneris.

[Vol. iii, p. 79]

CAPITULUM 5^m huius quarte partis

De tabula orbis Mercurii.

Hec quinta tabula que intitulatur *Tabula centri Mercurii* omnino calculata est et composita ex addicione et subtractione equacionis centri de gradibus equalibus, sicut precedens tabula centri Solis et Veneris; et eodem modo per eam examinari vel dividi potest circulus orbis Mercurii sicut predictum est de tabula centri Solis et Veneris. Et hec est forma tabule:

Tabula veri motus orbis Mercurii.

[Vol. iii, p. 80]

Table of the true motus of the sphere of Jupiter.

[See p. 77]

IV . 3

On the table by which the sphere of Mars may be divided.

[Explanation virtually as that of IV. 2, q.v.]

Table of the true motus of the sphere of Mars.

[See p. 78]

IV . 4

On the table of the sphere of the Sun and Venus.

[The explanation is omitted, being virtually identical to those above.]

Table of the true motus of the Sun and Venus.

[See p. 79]

IV . 5

On the table of the sphere of Mercury.

[Explanation omitted as before.]

Table of the true motus of the sphere of Mercury.

[See p. 80]

CAPITULUM 6^m huius quarte partis

De tabula qua examinatur divisio orbis Lune.

Hec tabula, que intitulatur *Tabula veri motus orbis Lune*, omnino calculata est et composita ex addicione et subtractione equacionis centri de gradibus equalibus, sicut docetur primo capitulo huius quarte partis. Et hec est forma tabule. Potest eciam vocari *Tabula centri equati Lune*.

Tabula veri motus orbis Lune.

[Vol. iii, p. 81]

CAPITULUM 7^m

De tabula veri motus Lune.

Hec tabula, que intitulatur *Tabula veri motus Lune* sive *equacionis argumenti*[a] calculata est et composita omnino sicut tabula veri motus orbis Solaris, per addicionem et subtractionem equacionis argumenti in longitudine longiori de gradibus equalibus; et per eam examinari poterit seu dividi ultimus et intimus circulus limbi primi, qui vocatur *circulus verus Lune*. Et hec est forma tabule:

Tabula veri motus Lune et equacionis argumenti pro hora coniunctionis.

[Vol. iii, pp. 82–3]

CAPITULUM 8^m

De tabula equacionis iomyn, id est diei naturalis.

Hec octava tabula, que intitulatur *Tabula iomyn equati*, facta hoc modo: disposita tabula cum 12 areis, et cum una linea numerorum pro gradibus descendendo a sinistris, et alia superius transversalis pro numero signorum 12, sicut in precedentibus. In primo loco prime aree ponitur id quod componitur ex equacione dierum addita ad unum gradum. In secundo loco, scilicet contra 2 gradus, ponitur id quod componitur ex equacione dierum correspondente duobus gradibus addita ad 2 gradus; et sic deinceps usque ad complecionem 12 signorum, ita quod in principio linee numeri contra[b] o erit 2 gra. et 10 min., et contra primum gradum Arietis 3 gra. 15 min.

Debet autem circulus iste per hanc tabulam dividi omnino sicut ceteri circuli, et erit inceptio huius circuli in directo 2[c] gra. et 10 min.

[a] CA argumentum   [b] H +cifram   [c] H 1

IV. 6

On the table by which the division of the circle of the Moon is examined.

[Explanation omitted. But notice the ending: '. . . it may also be called *Table of the equated centre of the Moon.*']

Table of the true motus of the sphere of the Moon.

[See p. 81]

IV. 7

On the table of the true [motus of the] Moon.

This table, which is entitled *Table of the true motus of the Moon*, or *Table of the equation of the argument*, is entirely calculated and drawn up as was the table of true motus of the solar sphere, by means of combining (by addition and subtraction) the equation of the argument at the greater distance,[1] with an appropriate number of degrees; and by its means the last and innermost circle of the first limb, which is called the *true circle of the Moon*, may be verified or divided. And this is the form of the table:

Table of the true motus of the Moon and of the equation [of the argument] for the hour of conjunction.

[See pp. 82–3]

IV. 8

On the table of the equation of iomyn, that is, of the natural day.

This eighth table which is entitled *Table of iomyn equated*, is made in the following way. A table is set out with 12 columns,[2] and with one line of number for degrees, down the left-hand side, and also another line across the top for the numbers of the twelve signs, as before. In the first row and column is placed what results when the equation of days is added to 1°. In the second row, that is, against 2°, is placed the sum of 2° and the equation of days corresponding to 2°; and so on, until the 12 signs are completed, so that at the beginning of the line of number, against 0, will be 2° 10′, and against the first degree of Aries 3° 15′.

The circle should be divided by means of this table just as are the other circles. The circle will begin in the direction of 2° 10′ Aries,

[1] *in longitudine longiori*. This refers to the only position significant at the time of conjunction.

[2] *area*.

Arietis, quia talis est equacio dierum correspondens fini Piscium, quod est vere principium Arietis.

Tabula iomyn sive circuli equantis motum diurnum.

[Vol. iii, p. 84]

CAPITULUM 9^m

De tabula latitudinis Lune ab ecliptica in septentrionem et meridiem.

Hec tabula, que intitulatur *Tabula latitudinis Lune*, communiter reperitur inter tabulas equacionum Lune, et est eius composicio per sinus, sicut componitur tabula declinacionis Solis; et per eam dividi debet semicirculus latitudinis Lune, sicut docetur capitulo 27° secunde partis huius. Et hec est forma tabule:

[Vol. iii, p. 85]

CAPITULUM 10^m

De tabula longitudinis cum sua parte duodecima, et de tabula coniunctionum duodecim.

Due breves tabule sequentes, quarum prima intitulatur *Tabula longitudinis cum sua duodecima* et secunda intitulatur *Tabula coniunctionum*, non sunt multum cum necessitate quia divisiones suorum circulorum sunt facilis operis; utraque tamen illarum componitur uno modo, scilicet per addicionem illius numeri qui ponitur contra unitatem duodecies replicati.

[Vol. iii, p. 86]

CAPITULUM 11^m

De tabula motus Lune in una hora ad augem, et ad longitudinem mediam, et ad eius oppositum.

Hec tabula, que intitulatur *Tabula motus Lune in una hora ad augem, et ad longitudinem mediam, et ad oppositum augis*, reperitur per[a] omnes tabulas Toletanas, sed iuxta calculacionem Albategni corrigitur; et per eam dividetur vel saltem examinari poterit divisio horarum in circulo coniunctionis, sicut docetur capitulo 17° secunde partis huius. Et hec est forma tabule:

[Vol. iii, p. 86]

[a] CH in

since such is the equation of days corresponding to the end of Pisces, which is the beginning of Aries.

Table of iomyn or of the equant of the daily motion.
[See p. 84]

IV. 9

On the table of the ecliptic latitude (north and south) of the Moon.

This table, which is entitled *Table of the latitude of the Moon*, is commonly found amongst tables of equations of the Moon, and it is drawn up by means of sines, just as is the table of declination of the Sun. By its means, the semicircle of the latitude of the Moon may be divided, as is explained in II. 27. And this is the form of the table:

[See p. 85]

IV. 10

On the table of longitude with its twelfth part, and on the table of twelve conjunctions.

The two short tables which follow, the first of which is entitled *Table of longitude, with its twelfth [part]*, the second *Table of conjunctions*, are not urgently needed, since the divisions on the corresponding circles are easily constructed.[1] Each is made, however, in the same way, namely by the addition, repeated twelve times, of the number in the first row.

[See p. 86]

IV. 11

On the table of the motion of the Moon in one hour, at the aux, at the mean distance, and at the opposite of aux.[2]

This table, entitled [as in this heading], is found in all Toledan tables, but is corrected according to a calculation by Albategni. By its means, the division of the hours on the circle of conjunction may be effected, or at least verified, as explained in II. 17. And this is the form of the table:

[See p. 86]

[1] The divisions are all equal.
[2] All positions being understood in relation to the epicycle at syzygy.

CAPITULUM 12m

[*De tabula stellarum fixarum.*]

Hec tabula, que intitulatur *Tabula stellarum fixarum*, verificata est pro modernis seculis ab Anno Christi 1327 per addicionem 17 graduum super eorum loca reperta per consideracionem Ptolomei in Alexandria, Anno Christi 138, sicut patet in 7° libro et in 8° *Almagesti*.

Et continet prima tabula gradum verum stelle, cum latitudine ab ecliptica. Secunda vero tabula continet gradum cum quo stella pervenit ad medium celi, et eius declinacionem a circulo equinoxiali. Et hec est forma tabularum:

[Vol. iii, p. 87]

CAPITULUM 13m huius quarte partis

De tabula qua examinari potest circulus orbis Mercurii.

Hec tabula, que intitulatur *Tabula medii motus Mercurii*, tabulata[a] est et composita modo communi tabularum expansarum ad singulos dies anni solaris pro medio motu planete diebus singulis extrahendo; et debet per eam dividi circulus involutus sicut docetur capitulo 19° et capitulo 20° secunde partis huius. Et hec est forma tabule (iste motus Mercurii est eius motus circa eius epiciclum, et dicitur motus argumenti):

[Vol. iii, pp. 88–9]

CAPITULUM 14m

De tabula medii motus Lune.

Hec tabula, que intitulatur *Medii tabula motus Lune*, calculata est et composita modo communi calculandi tabulas mediorum motuum planetarum ad singulos dies anni solaris; et per eam dividi debet circulus involutus sicut docetur capitulo 20° secunde partis huius. Et hec est forma tabule:

[Vol. iii, pp. 90–1]

CAPITULUM 15m huius quarte partis

De tabula orbis argumenti Lune.

Hec tabula, que intitulatur *Tabula medii motus argumenti Lune*, calculata est et composita communi modo tabularum expansarum ad dies singulas anni Solis[b] pro medio motu planete in suo epiciclo, et dicitur motus argumenti; et debet per eam dividi circulus involutus

[a] H calculata [b] H solaris

IV. 12

[*On the table of fixed stars.*]

This table, entitled *Table of fixed stars*, was verified[1] for modern times from the year 1327, by the addition of 17° to the star-positions determined by Ptolemy, in Alexandria in A.D. 138, which are to be found in *Almagest* VII and VIII.[2]

The first table contains the true degree of the star, with ecliptic latitude. The second has the mediation[3] and declination from the equinoctial. And this is the form of the tables:

[See p. 87]

IV. 13

On the table by which the circle of the sphere of Mercury may be checked.

This table, entitled *Table of the mean motus of Mercury*, was tabulated and drawn up in the manner usual with tables worked out for intervals of days (of the solar year), in order to extract the mean motus of the planet at daily intervals. By means of it, the spiral may be divided as explained in II. 19 and II. 20. And this is the form of the table. (This motus of Mercury is that about its epicycle, and is called the 'motus of the argument'.)

[See pp. 88–9]

IV. 14

On the table of mean motus of the Moon.

This table, entitled *Table of mean motus of the Moon*, was calculated and drawn up in the way usual with tables of mean motus of the planets, at intervals of a day of the solar year, and by means of it the spiral may be divided, as explained in II. 20. And this is the form of the table:

[See pp. 90–1]

IV. 15

On the table for the circle of the argument of the Moon.

This table, entitled *Table of mean motus of the argument of the Moon*, was calculated and draw up in the usual way of tables for intervals of single days of a solar year for the mean motus of a planet in its epicycle, known as the 'motus of the argument'. By means of it, the

[1] This does not mean that it was verified by observation.

[2] For the northern and southern hemispheres respectively.

[3] Literally 'the degree with which the star comes to the middle of the sky'. See Appendix 21.

sicut docetur capitulo 21° partis secunde huius. Et hec est forma tabule:

[Vol. iii, pp. 92–3]

CAPITULUM 16m

De tabula circuli directi.

Hec tabula ascensionum in circulo directo composita est per cordas, sicut docent canones; et debet per eam zodiacus dividi in secunda facie secunde rote, sicut docetur capitulo 22° secunde partis huius, de opere astrolabii. Et incipit hec tabula a principio solsticii hyemalis. Et hec est forma tabule.

[Vol. iii, pp. 94–5]

Item de capitulo 16° aliter tabula secunda est.

Hec tabula ascensionum signorum in circulo directo omnino non differt a priori nisi quod hec tabula incipit a puncto equinoxii vernalis. Non est igitur hec tabula multum necessaria nisi propter promptitudinem operandi. Et hec est forma tabule.

CAPITULUM [17m] ultimum huius quarte partis

Hec tabula, que intitulatur *Tabula ascensionum in circulo obliquo*, ubi videlicet est elevacio poli 51 gra. et 50 min. cuiusmodi est latitudo[a] civitatis Oxonie, calculata est et composita sicut docent canones in secundo libro *Almagesti*; et debet per eam dividi[b] circulus in limbo primo secunde faciei instrumenti, sicut docetur capitulo 18° secunde partis huius. Et hec est forma tabule:

[Vol. iii, pp. 96–7]

Explicit Albion

[a] H altitudo [b] H +secundus

spiral may be divided as explained in II. 21. And this is the form of the table:

[See pp. 92–3]

IV. 16

On the table of the direct circle.

This table of ascensions on the direct circle was drawn up by means of chords, as the canons explain. By means of it the zodiac on the second face of the second disc may be divided, as explained in II. 22, on the working of the astrolabe. The table begins from the winter solstice. And this is the form of the table:

[See pp. 94–5]

Further to 16. A second table.

This table of ascensions of signs on the direct circle differs nowhere from the previous one, except that this table begins from the vernal equinox. The table is therefore not much needed, unless it be to hasten the working. And this is the form of the table. [The table is omitted here.]

[IV. 17]

[*On the table of the oblique circle.*]

This table, entitled *Table of ascensions in the oblique circle*, namely that for an elevation of the pole of 51° 50′, the latitude of Oxford, was calculated and drawn up as explained in *Almagest* II [. 7]. By means of it the circle on the first limb of the second face of the instrument may be divided, as explained in II. 18. And this is the form of the table:

[See pp. 96–7]

Here ends the treatise on the albion

V

TRACTATUS RECTANGULI

AND

FOUR SHORT TRACTS CONCERNING CELESTIAL COORDINATES

SIGLA

A Oxford, Bodleian Library, MS. Ashmole 1796.
C Oxford, Corpus Christi College, MS. (D) 144.
D Oxford, Bodleian Library, MS. Digby 168.
F Cambridge University Library, MS. 1017 (Ee. 3.61).
H London, British Museum, MS. Harley 80.
L Oxford, Bodleian Library, MS. Laud Misc. 657.

For full discussion of the manuscripts, Introduction to the text, and Commentary, see vol. ii, pp. 287–308. The diagrams are printed in vol. iii, pp. 53–8. See also Plate XVI in vol. iii.

RECTANGULUS[a]

I

PROLOGUS[b]

In artem componendi instrumentum[c] rectanguli.

RECTANGULUM in remedium tediosi et difficilis operis armillarum eodem tempore quo composuimus[d] albion, hoc est sub annis Cristi 1326, concepimus ad rectificandum cursus et loca stellarum errancium[e] et fixarum, nodos eciam orbis noni[f] existentes in ecliptica apud equatorem et tropicos, una cum ceteris omnibus que investigari poterunt per armillas, astrolabium, aut turketum.[g] Cum racio fundamentalis in hoc solo theoreumate consistit Euclidis:[h]

> Inter parallelas omnium equidistancium linearum angulos relativos et magnitudines necesse est esse equales.

CAPITULUM PRIMUM

De[i] arte componendi tibiam primam.

Ad componendum igitur rectangulum, separemus laminam unam eris[j] quadrilateram, orthogonam sed oblongam, longitudinis quatuor digitorum, latitudinis autem[k] duorum, spissitudinis fere[l] semis, que sit *ABCD*; divisaque[m] linea[n] *AB* in 4 partes equales; sit una eius quarta *AE*; medietas quoque linee *AC* sit *AF*; a[o] quibus terminis *F* et *E* ducantur perpendiculares seu equidistantes super opposita latera que sint[p] *FG* et *EH*, intersecantes se super punctum unum.[q] Item[r] divisa linea *AE* in duo equalia ad punctum *K*, producatur equidistans a puncto *K* ipsi linee *AC* super lineam *CD* que sit *K*, acceptaque[s] linea *OK*, secundum quantitatem linee *KA*, et posito centro in puncto *O*, fiat circulus *OK*. Totumque superfluum quadrilateri *KF* preter semicirculum abradamus. Similiter et totum quadrilaterum *EG* abscindatur; factoque foramine triangulari, et decenter orbiculato cum suis marginibus in quadrilatero *FH*,[t] cuius basis erit in linea equidistanti[u] linee *CH*, et cuspis versus

[a] D RECTANGULUS DOMINI RICARDI DE WALLYNGFORD, ABBATIS SANCTI ALBANI (There is no early title on A, C, or L.)
[b] DL 6λ [c] F 2λ+RECTANGULUM [d] F +instrumentum quod nominatur
[e] HL erraticarum [f] S novi [g] AC torketum [h] H +Explicit prologus
[i] F 5λ+DE COMPOSICIONE TIBIARUM [j] F +aut alterius materie [k] F λ [l] L λ
[m] F 2λ+divisisque lineis [n] A λ [o] F 3λ+de quibus [p] AC sit
[q] CL +*I* (This is an interpolation in C.) [r] DL Iterum [s] L accipiatur
[t] H *FG* [u] ACF equidistante

THE RECTANGULUS

PART I

Prologue to the construction of the rectangulus.

WE designed the rectangulus to obviate the tedious and difficult work of making an armillary sphere, and we did so at the same time as we designed the albion—that is, in 1326. We conceived it as a means of determining the paths and places of the planets and fixed stars, and also the nodes of the ninth sphere in the ecliptic—at the equator and tropics—together with all the other problems which may be solved by the armillary, the astrolabe, or the turketum. The fundamental principle underlying the design is this one theorem of Euclid:[1]

Between parallels the relative angles as well as the lengths of equidistant lines are necessarily equal.

I. 1

How to make the first limb.[2]

To make the rectangulus we cut out a plate[3] of brass, rectangular but elongated, four digits[4] long, two wide, and about a half thick. This is *ABCD* (fig. I. 1). *AB* is to be divided into four equal parts, *AE* being one of the four parts. *AF* is a half of *AC*, and perpendiculars *FG* and *EH* are dropped from *F* and *E* on the opposite sides, intersecting in the point *I*. Next, divide the line *AE* into two equal parts at the point *K*,[5] draw a line *KL* through *K* parallel to *AC*, and with centre *O* on *KL*, at a distance from *K* equal to *AK*, construct a circle *OK*. We file away all the superfluous part *KF* outside the semicircle, and similarly the whole of the quadrilateral *EG* is to be cut away. A triangular aperture is made, and [the object] decently rounded off. The lower edge of the aperture in the quadrilateral *FH* having been made parallel to *CH*, and the aperture having its cusp at *M*, towards the point where *FI* and *KL* meet, and the

[1] The theorem is a conflation of Euclid I. 29 and I. 34.

[2] *tibia.*

[3] *lamina.* 'Block' might be thought a more appropriate word for a piece of the size given.

[4] A digit is a traditional measure equal to the thickness of a finger, and can be taken as about 0·8″, or 2 cm. I have somewhere seen a seventeenth-century work where it was taken to be three-quarters of an inch exactly.

[5] See our commentary below.

lineam *FI* in linea *KL* apud *M*, depositoque superfluo apud angulum *F*, talis apparebit[a] figura. Et tales tres laminas preparemus quarum[b] prima sit robustior, secunda parum subtilior et tercia mediocris, quas tres tibias nominamus.

CAPITULUM SECUNDUM

In[c] *artem componendi coxas.*

Capiamus iterum laminam alteram prioris spissitudinis sed quadratam, cuius latera sint equalia linee[d] *FC*, et sit[e] quadratum istud *FCBD*. Divisaque linea *FC* per equalia ad punctum *G*, producatur equidistans linee *FB*, que sit *GH*, in cuius medio *O*[f] posito centro fiat circulus equalis priori *OK*; ductaque contingente linea et equidistante linee *BD* super *CD*, que sit *LI*, totum superfluum quadranguli *GI*[g] abradamus preter semicirculum; et apparebit talis[h] figura, quam coxam vocamus.

Si igitur de circulo tibie abradamus medietatem spissitudinis eius in parte aversa[i] ab aspectu, et de circulo coxe abstulerimus medietatem spissitudinis eius in parte opposita aspectui, et applicemus coxam tibie in concavitatibus circulorum, et transierit unus paxillus cum equo per centrum utriusque circuli sicut in opere astrolabii satis docetur, apparebit tibia cum sua coxa, quasi pars una, excepto quod coxa moveri possit super circulum tibie motu elevacionis et depressionis tantundem. Possumus autem opere liniari decenter coxam cum suis lateribus perlimare, et compositorum talis apparebit figura. Deinde consimilem coxam aptabimus secunde tibie proporcionatam sibi in subtilitate si fuerit tibia secunda minor aut subtilior prima.

Tercie vero tibie non coxam aptabimus sed hallidadam cum pinnulis perforatis, coniungemusque hallidadam ipsi tibie cum paxillo et equo fortiter, ita quod motus eius elevacionis et depressionis sit difficilis sicut in astrolabio, sed multo difficilior et strictior sit motus duarum coxarum super suas tibias, ita[j] quod nullo levi pondere coxa super tibiam deprimatur. Eritque firmacio[k] hallidade ad terciam tibiam seu supremam cum paxillo et equo, sicut dictum est de tibiis atque coxis, et[l] firmacio equorum[m] ex parte aversa ab aspectu nostro[n] erit,[o] et erit hallidade cum antedescripta[p] tibia tercia talis figura.[q] Nec erit aliqua concavitas in circulo huius tercie tibie sicut in ceteris. De longitudine autem hallidade suo loco dicemus.

[a] L λ [b] H et [c] HL 2λ+DE ARTE; F 4λ+DE COMPOSICIONE COXARUM
[d] DL λ [e] C +qua [f] H λ [g] A *G* [h] L 2λ [i] L adversa
[j] ACDF 2λ+itaque [k] L formacio [l] F +erit [m] D λ [n] ACH 2λ
[o] F λ [p] ACFH λ [q] L +hic subscripta

superfluous metal in the neighbourhood of *F* having also been removed, this will be its appearance (fig. I. 2). We are to prepare three such pieces, of which the first is the most robust, the second is only a trifle more slender, and the third moderately so.[1] These three are called *limbs*.[2]

I. 2

How to make the joints.[3]

We are now to take another plate, of the thickness of the first, but square, whose sides are equal to the line *FC*. Let the square be *FCBD* (fig. I. 2a). The line *FC* is to be divided equally in the point *G*, and *GH* is produced parallel to *FB*. With centre *O*, the mid-point of *GH*, a circle is to be drawn equal to the former circle *OK*. A line *LI* being drawn on *CD*, parallel to *BD* and touching [the circle], all the superfluous part *GI* of the quadrangle is filed away, up to the semicircle. In shape it will have the appearance of the figure (fig. I. 2b), and we call it a *joint*.

If now from the circle on the limb we file away half the thickness (the half hidden from view), and remove the complementary half of the thickness of the circle of the joint, and if we apply the joint to the limb at the concavities of the circles, and push through the centre of each circle a peg with a horse, as is familiar from the astrolabe, then the limb with its joint will appear as a single object, apart from the fact that the hinge may be equally raised and lowered. We may, by adding a little ornamentation,[4] suitably file the joint and its sides so that the form of the result is as shown in the figure (fig. I. 2c). We then fit a similar joint to the second limb, of an appropriate thickness if the second limb was thinner or smaller than the first.

To the third limb we fit not a joint, but an alidade with perforated pinnules; and we fix the alidade firmly to this limb with a peg and horse, so that it offers resistance to being raised and lowered, just as in the case of an astrolabe. The two joints on the limbs are to be even firmer and more difficult to move, so that a small weight on a limb will not cause it to fall. The fixing[5] of the alidade to the third or uppermost support is with a peg and horse, just as was explained for limbs and joints, and the horse will be secured at a part hidden from view, while the shape of the alidade with the third limb will be as in the figure (fig. I. 2d). There will be no concavity in the circle of this third limb, as in the others. We shall have something to say of the length of the alidade, however, in the appropriate place.

[1] This refers to the possibility of reducing the half-digit dimension, as is evident from I. 2 below.

[2] *tibia.*

[3] *coxa* literally means 'hip', and remembering tha\ the hinged rod is here called the 'limb' (*tibia*), the analogy should be obvious.

[4] *opere liniari.*

[5] *firmacio.*

CAPITULUM 3m

In[a] *artem componendi columpnam seu*[b] *basim instrumenti.*[c]

Preparabimus[d] post hec columpnam unam[e] decentem ligneam vel eream, prout placet, ad supportandum perpendiculariter[f] instrumentum super planiciem orizontis, cum fuerit applicatum ad opus; que sit columpna *QP*, habens basim bone fixionis apud *P*. Et[g] in epistilo apud *Q* superemineat usus[h] paxillus perforatus[i] cum equo perpendiculariter, et in directum medietatis columpne que est *QP*. Deinde[j] fiat, in medio basis foraminis triangularis prime tibie, foramen aptum pro paxillo firmato in supremitate columpne, ita quod cum prima tibia[k] fuerit applicata ad columpnam et firmata cum equo possit tibia super columpnam moveri motu molaris, hoc est ab oriente per[l] meridiem in occidentem; et in applicacione prime tibie cum columpna talis apparebit figura. Consimiliter[m] fiet[n] paxillus perpetue firmitatis super caput tam prime coxe quam secunde, et foramina illis[o] apta in secunda tibia et tercia, sicut de prima tibia est dictum;[p] quc omnia cum fuerint invicem applicata, scilicet columpna, prima tibia, prima coxa, secunda tibia, secunda coxa, tercia tibia cum hallidada; et producta fuerit linea *PQ* usque in summitatem tercie tibie ipsa,[q] transibit per centra omnium paxillorum; et in parte opposita nostro aspectui tanquam una superficies apparebit, ut[r] presens[s] ostendit figura.[t]

CAPITULUM 4m

In[u] *composicione regularum ex ere*[v] *tibiis applicandarum; et de longitudine hallidade.*

Et cum preparate fuerint ipse[w] tres tibie, concavabimus superfluum[x] singularum, quod est apud *DI*, ita quod singulis tibiis applicari possint regule eree decenter planate, quante voluerimus longitudinis, latitudinis vero *HI*, eritque firmacio earum[y] fortis et ex parte aspectui nostro adversa,[z] ita quod in parte opposita nostro aspectui ipsius regule[aa] et tibie appareat superficies quasi una cum fuerint conclavate; oportetque hallidadam tante esse longitudinis quod perpendiculum ab eius fine descendens tangat fines omnium regularum. Deinde faciemus alias tres regulas ereas similes prioribus et equales in latitudine et spissitudine, sed habebunt longitudinem equalem totis tibiis cum suis regulis conclavatis, ita quod cum mote fuerint[bb] motu molaris super paxillum sicut et tibie,

[a] F 7λ+DE COMPOSICIONE COLUMPNE SIVE BASIS; HL 2λ+DE ARTE [b] DL VEL
[c] H INSTRUMENTORUM [d] F Preparemus [e] H λ [f] H λ [g] H 5λ [h] H autem
[i] F λ [j] L 2λ+Demenseat [k] S tibie [l] H 2λ+in meridiem et
[m] F et similiter [n] D fiat [o] L illius [p] CD predictum
[q] F 3λ+pertransibit [r] F 4λ [s] L subsequens [t] D +instrumenti tocius
[u] HL DE; F 11λ+DECOMPOSICIONE REGULARUM ET ALLIDADE [v] L LATONE [w] D iste
[x] H λ [y] DL eorum [z] ACDH adverse; L aversa [aa] A regulo [bb] AC fuerit

I. 3

How to make the column, or base of the instrument.

After this we shall prepare a suitable column of wood or brass, whichever we please, to support the instrument at right angles to the plane of the horizon, when fitted to the device. This column is *QP*, having a base well fixed to it at *P*. Surmounting the head[1] [of the column] at *Q* and perpendicular [to the top of the column] there is a perforated peg with horse, the peg pointing along the axis of the column *QP* (fig. I. 3a). Then let there be a hole made in the middle of the base of the triangular aperture in the first limb, suitable for taking the retaining peg at the top of the column, such that when the first limb is fitted to the column and made secure with the horse it will be possible to move the limb on the column with a rotatory motion,[2] that is, from the east, through the meridian, to the west. The first limb when applied to the column will appear as in the figure. In like manner, just as the peg makes the first joint permanently firm on the head [of the column], so with the second, and so for the apertures on the second and third limbs, as explained for the first limb. All these items, when added in turn, will amount to: the column, first limb, first joint, second limb, second joint, and third limb with alidade. And *PQ* will have been produced as far as the top of the third limb, and will pass through the centres of all the pegs. To our sight it will have the appearance of a single surface, as the present figure shows.

I. 4

On making the rules of brass which are to be applied to the limbs, and on the length of the alidade.

When the three limbs are prepared, we shall hollow out the superfluous part of each, namely *DI*, such that brass rules, suitably plane, may be applied to the individual limbs. The rules may be as long as we like, but of the breadth of *HI*, and firmly fitted to the part [of the limb] facing us, so that when they are riveted together the ends of the rule and of the limb appear as one surface.[3] It is necessary that the alidade be of such a length that the plumb-line[4] falling from its end touches the ends of all the rules. We should then make three brass rules similar to the first, and equal in breadth and width, and in length equal to the whole of the limbs with riveted rules, so that when they are moved on the peg with a rotatory motion in the same way as the limbs, as

[1] *epistilum*, deriving from *epistylium*, in turn from Gr. ἐπιστύλιον. The common English word is 'architrave', but 'epistyle' is known.

[2] *motu molaris*, the movement of a mill, a milling motion. The phrase occurs frequently below.

[3] This appears simply to mean that they touch perfectly.

[4] *perpendiculum*. This is actually the original meaning of the word, which could also be translated 'plummet'—cf. I. 5 below.

ut est dictum, perpendiculum hallidade earum fines contingat quemadmodum tibiarum. Propter quod nota quod oportet paxillos, super quorum centra tibie revolvuntur, tante esse altitudinis quod possint recipere[a] istas regulas et tibias ipsas[b] simul circumvolventes super eas, sicut tibie ut est dictum. Et cum coniuncta fuerit prima regula et prima tibia paxillo columpne talem[c] representabit figuram, et consimilem facient regule cetere et tibie coxis[d] iuncte. Parum quidem differunt coxe ab epistilo[e] columpne nisi quod hee mobiles hoc immobile perseverat.

Hec de partibus rectanguli dicta sufficiant.

CAPITULUM 5^m

In[f] divisione et mensuracione omnium[g] regularum.

Primam regulam prime tibie conclavatam, quam eciam ceteris omnibus esse expedit laciorem, marginabimus tribus marginibus in[h] parte opposita nostro aspectui, quarum due margines erunt prope latera latitudinis, et tercia in medio latitudinis eius erit. ('Marginem' hic vocamus[i] strictam superficiem inter duas lineas rectas equidistantes sed prope iacentes contentam.) Duoque maiora spacia inter[j] tres margines has contenta dividemus utraque in duo equalia per lineas prioribus equidistantes. Exibuntque iste linee marginum a radice paxilli super quem tibia circumfertur. Habebimus itaque in superficie regule tres margines pro divisionibus, et quatuor alia spacia pro litteris inscribendis, quorum duo spacia collateralia medie[k] margini assignentur; cetera duo[l] spacia suis marginibus in lateribus regule servient, ut[m] patet in presenti[n] figura. Ambas igitur margines exteriores a radice paxilli usque ad finem longitudinis earundem dividemus in 60 partes equales, divisione fideli; et vocabuntur hee sectiones 'gradus' seu 'partes cordarum', quarum parcium quamlibet dividamus in 60 minuta, vel in quotlibet longitudo permittet. Spacia eciam superioris et inferioris marginis dividamus per quinos gradus et[o] quinos,[p] pro litteris numerorum.

Mediam autem marginem dividemus per tabulam corde recte et verse,[q] sicut circuli solent per tabulas dividi; et ambo spacia dividantur per quinos gradus et quinos. In superiori quidem spacio scribantur littere numerorum graduum et signorum, et in[r] inferiori spacio similiter sed econtra.[s] Post hec dividemus omnes ceteras[t] regulas per duas margines tantummodo laterales[u] in 60 partes equales ut prius, nec

[a] L retinere [b] H ipsarum
[c] F 4λ+representabunt talem ut supra in hoc solis figurata est; L λ
[d] L 2λ+coxe coniuncte [e] H epistilio [f] DFHL DE [g] F λ [h] D λ
[i] D +unam [j] A in [k] A medio [l] L tres [m] F 5λ [n] A 2λ+subsequenti
[o] A 2λ [p] In A and C the diagram is numbered by sixes; H has no diagrams.
[q] F converse [r] H λ [s] F econverso [t] H reliquas [u] CD Litterales

explained, the plumb-line from the alidade touches the ends of the rules fixed to the limbs. Note that for this reason it is necessary that the pegs about the centres of which the limbs are revolved be of such a height that they can hold the rules as well as the limbs, moving round together as explained previously of the limbs. When the first rule and the first limb are joined together on the peg on the column, it will appear as in the figure, and similarly for the remaining rules and limbs connected to the joints. The joints indeed are little different from the head of the column: it is simply that the latter is fixed whilst they move.

These remarks on the parts of the rectangulus should be sufficient.

I. 5

On measuring out and dividing all the rules.

The first rule, pinned[1] to the first limb (this rule being broader than all the others), we mark out on its visible part[2] with three margins. Two of these will be along the edges, and the third along the middle of the length. (By 'margin' we mean the narrow surface between two parallel straight lines, placed close together.) We divide each of the two larger spaces between these three margins into two equal parts, by equidistant lines. These margins measure from the root[3] of the peg on which the limb is carried round. We shall have, therefore, on the surface of the rule, three margins for the divisions, and four other broader spaces for inscribing letters, two of which spaces are assigned, side by side, to the middle margin. The other two spaces, with their margins, are along the edges of the rule, as shown in the present figure. We therefore faithfully divide both outer margins into sixty equal parts, from the root of the peg up to the ends of the margins. These sections will be called 'degrees', or 'parts of chords', and we divide them into 60 minutes, or into however many the length permits. The upper and lower space of the margin we shall divide into groups of 5 degrees, so that they can be numbered.[4]

The middle margin, however, we divide by means of a table of right and versed chords,[5] just as circles are divided by means of tables. Both spaces are divided into groups of 5 degrees. In the space above are written the numbers of degrees and signs, and below likewise, although in the opposite direction. After this, we shall divide all the other rules into 60 equal parts, by two margins just as wide as before, and these

[1] This can scarcely refer to the riveted rule for several reasons. It seems, in fact, that the latter rules are henceforth often identified with the limbs to which they are attached.

[2] This means the part exposed to view, but it is not clear under what circumstances.

[3] See the commentary, and the final remarks to I. 5.

[4] *pro litteris numerorum.*

[5] This phrase is not to be taken at its face value. See the commentary. Here and elsewhere we translate 'right and versed chords' rather than the technical phrase 'corda recta and versa' to indicate looseness of meaning in the original.

habebunt iste regule mediam marginem cum spaciis alternatis. Propter quod omnes cetere regule esse poterunt stricciores quam illa que iam est descripta. Hallidada vero, quia nullam divisionem recipiet, potest esse quantum libet gracilis atque stricta.

Postremo in fine cuiuslibet regule filum cum[a] perpendiculo appendatur. Et nota quod superior corda, que vocatur 'corda recta', incipiet a paxillo versus finem regule; et inferior corda, que dicitur 'corda versa', a fine regule versus paxillum scribetur. Similiter in superiori[b] spacio medie marginis scribentur gradus trium signorum a paxillo versus finem; in inferiori spacio fiat econtra.

Explicit ars[c] componendi rectangulum[d]

II

INCIPIT ARS OPERANDI CUM ILLO[e]

PREFACIUM

RECTANGULUS omnium circulorum magnorum in spera solos semidiametros continet; que eciam etsi ut in spera non ab eodem centro procedant nil refert, nec ex hoc potest instrumentum hoc in aliquo argui falsitatis. Instructis quidem in disciplinalibus mathematicis satis constat. Et est[f] omni theoreumate certior proposicio hec[g] subiecta:

> Si ab aliquo puncto alterius parallelarum super alteram perpendicularis ducatur equalis in longitudine alteri linee inter easdem parallelas ex obliquo contente, necesse est perpendicularem parallelas exire, partemque eius inter parallelas contentam[h] esse cordam rectam arcus cuiuslibet anguli circa terminos oblique linee; residuum vero perpendicularis[i] extra parallelas pretensum[j] cordam esse versam illius arcus quo deficit arcus anguli acuti circa terminos oblique a quarta circuli est necesse.

Exempli gracia: Sint due parallele *AB* et *CD*, inter quas obliqua contenta sit *EF*, perpendicularisque ducta a puncto unius parallelarum *G*[k] sit *GH*,[l] equalis linee oblique *EF*, punctusque ubi intersecat parallelam *CD* sit *I*. Dicimus ergo quod *GI* minor est linea *EF*; et quod linea *GI* est corda recta arcus anguli *BEF*,[m] et similiter arcus anguli *FEG*, et quod

[a] L λ [b] S inferiori [c] F modus [d] F +*et sequitur modus operandi cum rectangulo*
[e] AC 2λ+*illo prefacio*; L *rectangulo* [f] L +in [g] H hic [h] A contentum
[i] L perpendicularum [j] A tensum; D protensum [k] F λ [l] A *GF*
[m] ACFS *AEF*

rules will be without a central margin with alternate spaces. On this account, both of the other rules may be narrower than that already described. The alidade, to be sure, carrying no divisions, may be as narrow and slender as you please.

Finally, hang a thread with a plummet[1] from the end of each rule. Note that the upper chord, known as the 'right chord', will begin at the peg and [increase] up to the end of the rule. The lower chord, known as the 'versed chord', will be inscribed from the end of the rule up to the peg. Similarly in the lower space of the middle margin will be written the degrees of three signs,[2] from the peg to the end; and in the lower space in the opposite sense.

Here ends the art of making the rectangulus

Part II

HOW TO USE THE RECTANGULUS

PREFACE

THE rectangulus contains only semidiameters of all the great circles on the sphere. Although they do not all proceed from nor refer to the same centre, it does not follow that for this reason I have designed the instrument wrongly. Indeed, those instructed in the mathematical disciplines are in agreement with it. And the following statement is a more certain proposition than any theorem:

> If from a point on one of two parallels a line is dropped perpendicular to the other, and equal in length to that part of a line obliquely contained between the same parallels, it is necessary that the perpendicular should go beyond the parallels, and that the part of it between the parallels should be the right chord[3] of whatever angle is at the ends of the oblique line. The residue of the perpendicular extending[4] beyond the parallels is necessarily the versed chord of that arc by which the acute angle at the ends of the oblique line is less than a right angle.

Example: Let *AB* and *CD* be two parallels, between which is to lie the oblique line *EF*, with *GH* equal to the oblique line *EF*, and perpendicular to the parallels on one of which *G* lies. *GH* is to meet *CD* in the point *I*. We say that *GI* is less than the line *EF* and that the line *GI* is the right chord[5] of the arc of angle *BEF*,[6] and similarly of the arc of angle *FEG*,

[1] *perpendiculum*. [2] That is, ninety degrees. [3] See n. 5 below.

[4] *pretensum*—cf. classical *praetentum*.

[5] Taking *EF* as the radius, $GI = \text{Sin}\,\widehat{BEF}$. On this meaning of *corda recta* (and *corda versa* below) see the commentary to I. 5 and II. Preface.

[6] Not *AEF*, as in S. (The end of the sentence supports the reading adopted.)

linea *IH* est corda versa arcus anguli *FEK*,[a] quo angulo angulus *BEF*[b] minor est recto. Hec ostendere instructis in sciencia corde et arcus, quibus solis dignum est hoc instrumento in consideracionibus suis uti, supervacuum iudicamus.

CAPITULUM PRIMUM HUIUS PARTIS DE NOMINIBUS PARCIUM RECTANGULI

Omnes partes rectanguli certis vocabulis[c] *nuncupare.*[d]

Prima pars que facta est ad modum columpne *basis rectanguli* dici potest. *Paxillus basis* dicitur ille clavus supereminens in summitate basis, perforatus pro repagulo qui *equus* vocatur[e] ut in astrolabio penitus[f] continetur. *Prima regula* vocatur illa lamina gracilis et oblonga que circa paxillum basis molariter circumfertur. Prima tibia dicitur pars illa que habet regulam sibi confixam,[g] prime regule in longitudine coequalem, que similiter movetur super paxillum basis motu molari. *Prima coxa* vocatur pars illa conclavata in qua iunctura prime tibie sicut in cardine suo movetur super paxillum motu elevacionis et depressionis, cum equo firmata[h] ut in astrolabio in parte a nostro aspectu aversa. Habetque hec coxa paxillum in summitate eius cum equo sicut epistilus basis[i] habet. *Secunda regula* et *secunda tibia* moventur super primam coxam circa paxillum sicut dictum est de prima regula et prima tibia respectu paxilli basis. *Secunda coxa* per omnia similis est prime coxe, in summitate secunde tibie situata. *Tercia regula* et *tercia tibia* sunt que super secundam coxam[j] moventur motu molaris.[k] Nec habet tercia tibia supra summitatem eius coxam, ut cetere, sed[l] loco coxe habet hallidadam. *Hallidada* est virgula longa firmata in summitate tercie[m] tibie, mota[n] super paxillum motu elevacionis et depressionis ut per foramina in eius tabellis altitudo siderum discernatur. *Perpendiculum basis* vocatur filum quod in superficie columpne appenditur cum modico pondere, ad rectificandum situm tocius rectanguli. Cetera autem perpendicula atque corde a suis regulis nominantur. *Paxillus* vocatur clavus rotundus coniungens[o] duas particulas instrumenti[p] invicem, ut altera parte quiescente altera moveatur. *Equus* vocatur repagulum confirmans laminas intra paxillum.

[a] A *FEG* [b] ACFS *AEF* [c] F λ [d] F nuncupantur [e] H perforatur
[f] ACFL plenius; S λ [g] L fixam [h] CFH firmatum; DLS firmato [i] L λ
[j] F λ [k] L molari [l] F +in [m] H λ [n] L movetur [o] F contingens
[p] ACF in; A +ad

and that the line *IH* is the versed chord of the arc of the angle *FEK*, by which angle *BEF* is less than a right angle. We deem it unnecessary to demonstrate these things to those who understand the science of chord and arc, who are fit to the use of this instrument.

II. 1. ON THE NAMES OF THE PARTS OF THE RECTANGULUS

To ascribe definite names to all parts of the rectangulus.

The first part, which is made after the manner of a column, may be called the *base of the rectangulus*. The nail projecting from the top of the base is called the *base peg*, this nail being pierced for a bolt, which is called the *horse*, as found at the middle of an astrolabe. The rather long and slender plate which is carried round the base peg with a rotatory motion is called the *first rule*. That part which has a rule fixed to it, equal in length to the first rule, and which likewise moves around the base peg, is called the *first limb*. That hollowed-out part, at which the junction with the first limb is as a hinge having a movement in elevation, is called the *first joint*. This is made firm with a horse, as in an astrolabe, [but] hidden from view. This joint has at its top a peg with a horse, just as has the capital of the base. The *second rule* and the *second limb* move on the first joint around the peg, just as was explained of the first rule and the first limb with regard to the base peg. The *second joint* is similar in all respects to the first joint, and is situated at the top of the second limb. The *third rule* and the *third limb* are what move with a rotatory motion on the second joint. The third limb has no joint at the top, as with the rest, but has an alidade in place of a joint. The *alidade* is a long rod fixed to the top of the third limb, moved on the peg with a motion in elevation so that the altitude of stars may be discerned through sighting-holes[1] in the alidade's small tablets.[2] The thread which is hung with a moderately sized weight from the surface of the column to level the rectangulus is called the *plumb-line of the base*.[3] The other plumb-lines and cords are named after the rules to which they belong. The round nail joining two parts of the instrument, one moving and one stationary, is called a *peg*. The bolt in the peg securing the plates is called a *horse*.

[1] *foramina*.

[2] *in tabellis*. The word 'pinnule' (*pinnula*) was usual for the corresponding part of an astrolabe, and is used elsewhere in *Rectangulus* (I. 2, II. 7, 9).

[3] Mentioned here for the first time.

CAPITULUM 2ma[a] DE ARTE COGNOSCENDI CORDAM RECTAM ET VERSAM ARCUS DATI

Cordam rectam et versam cuiuslibet arcus infra quartam circuli per arcum datum et quemlibet arcum per cordam datam cognoscere.

Quod[b] si placuerit, appende perpendiculum secunde tibie super hastam prime tibie ad quemvis arcum datum, ita quod super arcum illum orthogonaliter cadat filum; et quod est de hasta tibie inter filum et paxillum est corda recta arcus computati a paxillo versus finem haste; quod autem extra filum[c] fuerit computando a fine haste ad filum est corda versa arcus illius qui est similiter extra filum. Econtra facias[d] si per cordam scire volueris arcum suum. Et nota quod tunc demum cadit[e] filum perpendiculariter super partes regule seu[f] haste quando ipsum cadit super equales partes cordarum in utraque margine protractarum; et hec est causa precipua quare in singulis regulis corde arcuum duplicantur.

Sinus autem rectus et versus arcus maioris quarta circuli de facili ex premissis habetur, nam omnis sinus rectus arcus minoris quarta circuli est sinus rectus arcus residui de semicirculo. Et omnis sinus versus arcus maioris quarta circuli infra semicirculum maior est semidiametro per quantitatem corde recte illius arcus quo ille arcus datus est maior quarta[g] circuli, sicut[h] alibi est ostensum.

CAPITULUM 3m: DE[i] ARTE INVENIENDI MERIDIEM VILLE DATE[j]

In omni orizonte veram meridiem invenire.

Quod[k] si placuerit, pone instrumentum super basim[l] suam in aliquo loco eminenti, quo Sol aut stella fixa toto[m] die aut nocte possit per hallidadam intueri; erigaturque semper basis perpendiculariter super superficiem[n] orizontis. Deinde prime regule cuspidem apta ad meridiem prope verum;[o] qua ibidem immobiliter perseverante, move primam tibiam, cum toto residuo instrumenti, in ortu vel prope ad Solis aspectum, et eleva vel deprime hallidadam quousque[p] radius solaris transierit per utrumque foramen, si de die fueris operatus; vel usquequo[q] stella oriens per ipsa foramina possit[r] videri, si de nocte fuerit opus tuum; quod cum apprehenderis, appende filum appensum in cuspide prime tibie[s] perpendiculariter super gradus prime regule, et scias cordam versam et

[a] F 9λ (In a sixteenth-century hand the following title is added:) DE CHORDIS RECTIS ET VERSIS COGNOSCENDIS [b] L 3λ [c] L finem
[d] L facies [e] F cadet [f] F sum (This is altered in the margin to *sive*.)
[g] H +illius [h] L ut [i] F 6λ+DE VERA MERIDIE COGNOSCENDA
[j] H +VEL CIVITATIS [k] FL 3λ [l] AC basem [m] FH tota
[n] FHLS superficie [o] H 3λ (A space is left, suggesting an illegible exemplar.)
[p] CH usquequo; FLS usque [q] L usque [r] F +de nocte [s] F +aspectum

II. 2. ON DETERMINING THE RIGHT AND VERSED CHORD OF A GIVEN ARC

To find the right and versed chord of any arc less than a quadrant, given the arc; and any arc, given the chord.

Let the plumb-line of the second limb be stretched over the shaft[1] of the first limb, so that the thread falls at right angles over the arc in question. The chord of the arc is what lies along the shaft of the limb between the thread and the peg, measuring from the peg towards the end of the shaft. The distance beyond the thread, reckoning from the end of the limb towards the thread, gives the versed chord of the same arc. Working in the other direction you can find an arc from its chord. Note that the thread falls perpendicularly over the rule or shaft when it falls on equal chord-divisions on both margins of the limbs; and this was the chief reason for duplicating the scales of chords.

The sinus rectus and versus of an arc greater than a quadrant may easily be found from what has been said, for the sinus rectus of an arc less than a quadrant is the sinus rectus of the supplementary arc, i.e. the arc making up a semicircle. The sinus versus of an arc greater than a quadrant but less than a semicircle exceeds the semidiameter by the sinus rectus of the arc by which the given arc exceeds the quadrant, as is shown elsewhere.

II. 3. ON THE METHOD OF FINDING THE MERIDIAN AT A GIVEN PLACE

To find the true meridian on any horizon.

Put the instrument on its base in a high place, from which the Sun or a fixed star may be observed with the alidade throughout the day or night. The base is always to be erected perpendicularly to the plane of the horizon. Next, direct the point of the first rule to the approximate position of the meridian. Keeping it there, move the first limb with the rest of the instrument (if you are working by day) either to the rising Sun or to a near-by position, and either raise or lower the alidade until the solar ray passes through each sighting-hole. If you are working by night, move the alidade to the point where it is possible to see a star through both sighting-holes. When you have found this, draw out at right angles to the graduations of the first rule the thread attached to the tip of the first limb.[2] You then know the versed chord. Note the

[1] *hasta*. This word is confusing, but it is certainly synonymous with *regula* (as appears later in the paragraph), and we can only suppose that 'shaft of the first limb' refers to the *loose* rule next to that limb.

[2] That is, to the end of the riveted rule on that limb. Note that *perpendiculariter* here most emphatically does not mean 'vertically', but has the modern geometer's meaning of 'perpendicularly'.

arcum eius nota. Item cum Sol vel stella tetenderit[a] ad occasum, prima[b] regula semper immota,[c] move primam tibiam cum toto residuo quod est super ipsum ad Solem vel[d] stellam ut[e] prius, hallidada[f] existente in eadem disposicione in qua prius erat, quando per eius foramina[g] transiit[h] radius visualis; et observa tempus et horam in qua demum[i] poteris intueri eandem stellam, vel Solem, per ipsa foramina tabellarum, et tunc cognosce ut prius, per filum appensum in cuspide prime tibie, cordam versam et arcum distancie[j] a regula prius dicta; que si fuerit equalis[k] corde prius accepte, scias ipsam regulam veraciter in meridie esse sitam. Quod si fuerint diverse, coniunge utrosque arcus et, prima tibia nusquam mota,[l] per accessum vel[m] recessum regule[n] a prima tibia usquequo filum prime tibie cadere possit perpendiculariter super cordam versam medietatis iam compositi[o] arcus[p] ex utrisque; quo facto, scias cuspidem regule prime veraciter ad meridiem quesitam extendi. Hoc genus[q] querendi lineam meridionalem per Solem optimum est quando Sol est prope tropicos.

CAPITULUM 4^{m}: DE[r] ALTITUDINE STELLE SUPER ORIZONTEM ET AZIMUTH

Azimuth et altitudinem stelle apparentis super orizontem apprehendere.

Quod[s] si placuerit, pone primam regulam directe ad meridiem iam inventam, deinde totum residuum instrumenti move motu prime tibie ad stellam; et tunc eleva vel deprime hallidadam usquequo[t] radius stelle transeat per utrumque foramen. Quo facto, extende filum prime tibie orthogonaliter super primam regulam, et quicquid fuerit de regula intra filum versus paxillum ipsum[u] est corda recta distancie stelle a puncto veri orientis,[v] si fuerit stella in illa plaga que est inter meridiem et orientem; et quod de regula fuerit extra filum est corda versa distancie stelle a vera meridie. Similiter intellige[w] de azimuth occidentali, si fuerit stella in tali plaga; et sic[x] deinceps.

Item filum hallidade ubicumque ceciderit perpendiculariter super primam tibiam, quod fuerit intra filum est corda recta distancie stelle a cenith; quod vero de regula fuerit extra filum est corda versa elevacionis stelle a circulo orizontis; sed habitis cordis, per secundam huius de facili patent arcus.

Et nota quod azimuth proprie est longitudo stelle a meridie; altitudo autem est proprie latitudo stelle ab orizonte.

[a] H tendit [b] F 4λ [c] L immobiliter [d] L +ad [e] F 2λ+prima
[f] L hallidade; D +semper [g] DL foramen [h] F transierat [i] C demo; H demio
[j] S distantem [k] F equat [l] D motum; L moveatur [m] F et
[n] ACFH λ [o] H positi [p] F +prime regule [q] L opus
[r] F 7λ+AZIMUTH ET ALTITUDO STELLARUM COGNOSCENDA [s] FL 3λ [t] L quousque
[u] F et primum [v] D orizontis [w] H intelligi debet [x] H 2λ+sic de similibus

corresponding arc. Next, when the Sun or star is descending, the first rule being kept fixed, move the first limb with all above it to the Sun or star, as before, the alidade staying in the position held formerly, that is to say, when the visual ray passed through the sighting-holes. Now observe the time and the hour at which you are eventually able to view the same star, or the Sun, through the same sighting-holes of the tablets, and then decide as before on the versed chord, and on the arc of the distance to the rule previously mentioned, by means of the thread attached to the tip of the first limb. If this is equal to the chord formerly taken, you will know that the rule is truly placed in the meridian. If they are different, add together[1] both arcs and, not moving the first limb in any way, move the rule towards or away from the first limb until it is possible to make the thread of the first limb fall perpendicularly over [the rule] and at half the sum of the two arcs [on the numbering nearer the scale of] versed chord. Once this is done, you will know that the tip of the first rule points in fact towards the required meridian. This method of finding the meridian line by the Sun is best carried out when the Sun is near the tropics.[2]

II. 4. ON THE ALTITUDE OF A STAR ABOVE THE HORIZON, AND ITS AZIMUTH

To obtain the altitude and azimuth of a star appearing above the horizon.

Place the first rule in the meridian as now found, and move the rest of the instrument—by moving the first limb—towards the star. Next raise or lower the alidade until the ray from the star passes through each sighting-hole. Having done this, draw the thread of the first limb orthogonally over the first rule, and whatever is the reading of the rule (from the thread towards the peg) is the right chord of the distance of the star from the point of true east, if the star be in the quarter between east and the meridian. The reading of the rule beyond the thread is the versed chord of the distance of the star from the true meridian. Similarly, work out the westerly azimuth, if the star be in this quarter; and so for the rest.

Next, that which is within the point where the thread of the alidade falls perpendicularly on the first limb is the right chord of the distance of the star from the zenith. What is outside the thread on the rule is the versed chord of the elevation of the star from the horizon circle. Once the chords, are found however, it is an easy matter to find the arcs, using II. 2 above.

Note that the azimuth, strictly speaking, is the longitude of the star from the meridian, whereas the altitude, properly speaking, is the latitude of the star from the horizon.

[1] *coniunge.* [2] See commentary.

CAPITULUM 5m: DE[a] LATITUDINE VILLE TUE

Latitudinem regionis per stellam non occidentem deprehendere.

Quod[b] si placuerit converte primam regulam et totum eciam instrumentum fideliter ad septentrionem. Deinde cum fuerit aliqua stella perpetue apparicionis in directo instrumenti, hoc est in linea septentrionali, apprehende cum hallidada eius elevacionem ab orizonte; et nota cordam eius rectam et versam secundum doctrinam premissam, et cum iterum pervenerit stella illa ad lineam septentrionalem, iterum nota eius altitudinis cordam versam aut rectam. Oportet namque bis ipsam stellam transire per septentrionem, semel in maxima eius elevacione et semel in maxima eius depressione. Deinde illarum duarum cordarum versarum arcus quere, et medietatem excessus maioris arcus super minorem arcui minori adde, et arcus coniunctus erit latitudo regionis quesita, et eius corda versa erit corda latitudinis. Est quidem latitudo regionis distancia cenith capitis a maxima elevacione[c] capitis Arietis, cui equalis est altitudo poli septentrionalis. Si autem operatus fueris per cordas rectas, ut predictum est de cordis versis, arcus coniunctus erit elevacio capitis Arietis, seu[d] distancia poli septentrionalis a cenith capitis ville tue.

CAPITULUM 6m: DE LONGITUDINE ET LATITUDINE STELLE RESPECTU[e] CIRCULI EQUINOXIALIS

Latitudinem stelle a circulo equinoxiali una cum longitudine eiusdem a puncto equinoxialis existentis in medio celi deprehendere.

Quod[f] si volueris, converte primam regulam ad meridiem tibi notam. Deinde eleva sub eadem linea meridionali secundam tibiam, per motum prime coxe, usquequo[g] filum secunde tibie cadat perpendiculariter super primam tibiam in gradum et minutum corde verse altitudinis capitis Arietis, per premissam.[h] Quo facto, stabilietur in illo situ sub meridie secunda[i] tibia fixa semper. Deinde move terciam tibiam ad aspectum stelle cum hallidada, et eleva vel deprime hallidadam usquequo deprehenderis radium huius stelle. Quod cum perfeceris extende filum tercie tibie super terciam regulam existentem adhuc sub linea meridionali et, ut supra, quod fuerit de regula extra filum[j] ipsum est corda versa longitudinis stelle a puncto equinoxiali existenti in meridie. Quod autem de regula fuerit intra filum[k] ipsum est corda recta residui de quarta circuli existentis inter ortum et meridiem, vel inter occasum et meridiem.[l]

[a] F 4λ+DE LATITUDINE REGIONIS COGNOSCENDA [b] FL 3λ [c] A λ
[d] L 2λ+super distanciam [e] F 3λ+COGNOSCENDA [f] FL 3λ [g] H usque
[h] S premissa [i] AC 2λ+secundam tibiam [j] H 2λ+et primum
[k] H 2λ+et primum [l] F +si fuerint in illa plaga

II. 5. ON THE LATITUDE OF YOUR TOWN

To discover the latitude of a district by means of a star which is not setting.

Turn the first rule, as also the complete instrument, faithfully towards the north. When a star appears in line with the instrument, that is to say, in the northerly line, take its altitude above the horizon by means of the alidade. Note its right and versed chord, in accordance with the earlier explanation, and when once more that star returns to the northerly line, again note the right or versed chord of its altitude. For it is necessary that the star pass north[1] twice, once at its greatest elevation and once at its lowest point. Find then the arcs of two versed chords, and add to the lesser arc half the excess of the greater over the lesser arc. The result will be the desired latitude of the district, and its versed chord will be the chord of the latitude. The latitude of the district is indeed the distance of the zenith from the highest point reached by the first point of Aries, which is equal to the altitude of the north pole. If, however, you work with right chords, instead of versed chords, as explained, the arc obtained in the addition will be the elevation of the first point of Aries, or the distance of the north pole from the zenith of your town.

II. 6. ON THE LATITUDE AND LONGITUDE[2] OF A STAR WITH RESPECT TO THE EQUINOCTIAL

To discover the latitude of a star from the equinoctial circle, together with the angular distance[3] *of the same from the point of the equinoctial in the middle of the sky.*

Turn the first rule to the meridian which you have previously noted. Next raise the second limb under the meridian-line, by moving the first joint, until the thread of the second limb falls perpendicularly on the first limb at the degree and minute of the versed chord of the altitude of the first point of Aries (by the previous chapter). Once this is done, the second limb will be fixed steady in that position under the meridian. Next move the third limb, with alidade, to the direction of the star, and raise or lower the alidade until you find that star's ray. When you have done this, draw the thread of the third limb over the third rule, which is for this purpose under the meridian-line. As above, what is outside this thread on the rule is the versed chord of the longitude of the star from that point of the equinoctial which is on the meridian. On the other hand, what is inside the position of the same thread on the rule is the right chord of the complement of the angle between the meridian and the point of either rising or setting.

[1] That is, the meridian and its continuation to the northern point of the horizon.

[2] See the following note.

[3] *longitudo*. The whole expression is not equivalent to that subsequently abbreviated to 'mediation', for whose meaning see Appendix 21.

Item vide super quot partes et minuta signata in tercia tibia cadit filum allidade; et quod fuerit extra filum est corda versa latitudinis stelle a circulo equinoxiali,[a] bona[b] fide. Et quod de tibia est intra filum, ipsum est distancia stelle a polo mundi. Et nota quod, si cuspis hallidade deprimatur sub tercia regula, erit latitudo meridionalis. Si vero elevetur supra, erit latitudo stelle[c] septentrionalis.

CAPITULUM 7^m: DE[d] DECLINACIONE STELLE ET LATITUDINE AB ECLIPTICA

Declinacionem[e] gradus ecliptice existentis in medio celi a circulo equinoxiali per Solis aspectum deprehendere.

Quod[f] si placuerit, dispone instrumentum sub linea meridiana, sic quod elevetur secunda tibia secundum quod exigit altitudo capitis Arietis in meridie per antepremissam, scilicet quintam, huius. Deinde converte terciam tibiam cum allidada ad Solem, et tunc eleva terciam tibiam vel deprime usquequo[g] radius solaris transeat[h] per foramina pinnularum; sitque semper[i] allidada equidistans tercie tibie. Quo facto, absque motu elevacionis vel[j] depressionis tercie tibie, move ipsam terciam tibiam sub linea meridiana et extende filum tercie tibie super secundam tibiam perpendiculariter; et quod fuerit de secunda tibia extra filum ipsum est corda versa declinationis puncti ecliptice existentis in meridie a circulo equinoxiali. Vide ergo cui gradui[k] ecliptice debetur talis declinacio, et ipse est gradus medii celi vere.

CAPITULUM 8^m: DE VERO GRADU SOLIS[l] PER EIUS ASPECTUM HABENDO

Verum[m] locum Solis per solum eius aspectum apprehendere.[n]

Quod[o] si volueris,[p] per premissam invenies gradum[q] medii[r] celi, et terciam regulam dispones[s] sub meridie secundum quantitatem declinacionis eiusdem gradus ab equatore diei. Quo facto, ipsa tercia regula et toto quod sub ea continetur de[t] instrumento immobiliter existente in suo situ secundum premissarum[u] doctrinam, move terciam tibiam ad aspectum Solis. Quod si bene feceris, statim sine aliqua elevacione vel depressione hallidade a tercia tibia radius ipse solaris transibit per utrumque foramen. Tunc extende filum[v] tercie tibie super regulam terciam

[a] C equinoxii [b] AF 2λ [c] A λ
[d] F 7λ+DECLINACIONEM CUIUSLIBET GRADUS ZODIACI COGNOSCERE [e] L 14λ [f] FL 3λ
[g] A usque [h] C transiat [i] F λ [j] A et [k] D gradu
[l] F 5λ+COGNOSCENDO [m] L 8λ; S Certum [n] H *deprehendere* [o] L 3λ
[p] D placuerit [q] A gradus [r] D 2λ+me^r [s] DL disponas [t] LS in
[u] C premissorum [v] D +ipsius

Next see how many parts and minutes the thread of the alidade falls, along the third limb. What is outside the thread is the versed chord of the declination[1] of the star from the equinoctial circle, and of this you may be confident. What lies within the thread along the limb is the distance of the star from the pole of the world.[2] And note that if the point of the alidade be below the third rule, the latitude will be southerly, and if it be raised above [the rule], the latitude of the star will be northerly.[3]

II. 7. ON THE DECLINATION OF A STAR, AND LATITUDE FROM THE ECLIPTIC[4]

To discover the declination from the equinoctial circle of the degree of the ecliptic in the middle of the sky by means of an observation of the Sun.

Place the instrument under the meridian-line, the second limb being raised as it was before to give the altitude of the first point of Aries on the meridian, using [the information found] two chapters ago, that is, II. 5. Next, turn the third limb with alidade towards the Sun, and raise or lower the limb until the solar ray passes through the holes of the pinnules. The alidade is always to be parallel to the third limb. Having done this, by moving the third limb in elevation or depression, move that third limb under the meridian-line, and draw the thread from it over the second limb, and at right angles to this limb. That part of the second limb outside the thread will be the versed chord of the declination (from the equinoctial) of the point of the ecliptic which lies on the meridian. See, therefore, what degree of the ecliptic has this declination,[5] and this is indeed the degree of mid heaven.

II. 8. ON ARRIVING AT THE TRUE PLACE OF THE SUN FROM ITS ASPECT

To find the actual place of the Sun merely from its direction.

Find the degree of mid heaven from the previous chapter, and place the third rule under the meridian at an angle equal to its declination from the celestial equator. Having done this, move the third limb[6] to the direction of the Sun, keeping the third rule and all parts of the instrument below it fixed in the position decided by the previous instructions. If you do this properly, the same solar ray will at once pass through each sighting-hole without further raising or lowering the alidade.[7] Now draw out the thread of the third limb over the third rule,

[1] *latitudo.* [2] That is, the north polar distance. [3] See our commentary to I. 4.

[4] This title is quite erroneous. What follows it is correct.

[5] With the help of tables? The rectangulus could of course have been used.

[6] It must be understood to take with it the (parallel) alidade.

[7] The Sun must be on the ecliptic. Since the third limb passes through the meridian point of the ecliptic, and is constrained to move round a great circle of the sphere, that circle must be the ecliptic.

orthogonaliter, et quod fuerit extra filum de tercia regula erit corda versa arcus elongacionis veri loci vel gradus[a] Solis a gradu et puncto medii celi de ecliptica; et ipse punctus iam est notus. Erit ergo notus verus locus Solis per predictam huius.

CAPITULUM 9^m: DE[b] VERO LOCO LUNE COGNOSCENDO[c] IN LONGITUDINE ET LATITUDINE

Verum locum Lune una cum latitudine eius ab ecliptica cognoscere.

Quod[d] si placuerit, in hora qua simul videri possunt Sol et Luna super orizontem, secundum premissam et antepremissam dispone terciam regulam sub meridie secundum declinationem gradus medii celi in hora consideracionis tue, et scias memoriter tunc verum locum Solis. Deinde move[e] terciam tibiam ad aspectum Lune una cum hallidada equidistante; quod[f] si eius radii transeant per foramina pinnularum[g] scias quod Luna nullam habet latitudinem ab ecliptica; sin autem, eleva vel deprime hallidadam usquequo eius radius transeat per utrumque foramen. Quo facto, si extenderis filum hallidade orthogonaliter super terciam tibiam, quod de tibia fuerit extra filum erit corda versa latitudinis Lune ab ecliptica, et si hallidada fuerit supra terciam tibiam erit latitudo septentrionalis. Si vero cuspis eius sit sub tercia tibia, latitudo erit[h] meridionalis.

Item[i] extende filum tercie tibie super terciam regulam orthogonaliter, et quod fuerit de tercia regula extra filum est corda versa arcus distancie veri loci Lune a gradu medii celi; et ipse est notus et per consequens notus est verus locus Lune.

Et sic deprehendes vera loca omnium planetarum.

CAPITULUM 10^m: DE[j] VERIS LOCIS STELLARUM FIXARUM ET PLANETARUM IN LONGITUDINE ET LATITUDINE

Distanciam cuiuslibet stelle fixe,[k] *tam in longitudine quem in latitudine, a quolibet puncto in ecliptica designato, una*[l] *cum latitudine et gradu cum quo pervenerit ad meridiem demonstrare.*

Quod[m] si volueris,[n] de longitudine et latitudine eius[o] respectu ecliptice omnino, fac sicut dictum est de Luna et ceteris planetis erraticis. De gradu vero medii celi et de latitudine ab equinoxiali sic facies:[p]

Disposita secunda tibia sub meridie secundum altitudinem capitis

[a] A gradu [b] F 9λ+DE VERO GRADU LUNE COGNOSCENDO [c] CFHS COGNOSCENDE
[d] FL 3λ [e] L amove [f] A quod scilicet; F 2λ+ita videlicet quod
[g] F +ipsius allidade. Quod si sic moverit [h] C λ [i] A Tantum
[j] F 11λ+DE VERO LOCO STELLE FIXE COGNOSCENDO [k] L fixatam [l] CFHS uno
[m] L 3λ [n] F placuerit [o] H λ [p] F facias

and at right angles to it. What lies outside the thread along the third rule will be the versed chord of the arc of the elongation of the true place, that is the degree of the Sun from mid heaven. This same point is already known. The true place of the Sun will therefore be known by the last [chapter] of this [work].

II. 9. ON DETERMINING THE TRUE PLACE OF THE MOON IN LONGITUDE AND LATITUDE

To find out the true place of the Moon together with its latitude from the ecliptic.

At an hour when the Sun and Moon may be seen above the horizon at the same time, in accordance with the last two chapters place the third rule under the meridian in keeping with the declination of mid heaven at the hour of your inquiry, and commit to memory the true place of the Sun. Then move the third limb together with the parallel alidade to the direction of the Moon. If its rays should pass through the sighting-holes in the pinnules, you will know that the Moon has no latitude from the ecliptic; if the contrary, however, raise or lower the alidade until its ray passes through each sighting-hole. Once this is done, if you draw out the thread of the alidade orthogonally over the third limb, what lies along the limb beyond the thread will be the versed chord of the latitude of the Moon from the ecliptic. If the alidade be above the third limb, the latitude is northerly. If, on the other hand, its point be below the third limb, the latitude is southerly.

Next, draw out the thread of the third limb orthogonally over the third rule, and what lies along the rule beyond the thread is the versed chord of the arc of the distance of the true place of the Moon from mid heaven. This is known, and so consequently is the true place of the Moon.

In this way you discover the true places of all the planets.

II. 10. ON THE TRUE PLACES OF THE FIXED STARS AND PLANETS IN LONGITUDE AND LATITUDE

To show the distance of any fixed star whatever, both in longitude and latitude, from any designated point on the ecliptic, together with the latitude and degree with which it comes to the meridian.

For its latitude and longitude, both with respect to the ecliptic, do as you were told for the Moon and other wandering planets. For the degree of mid heaven and latitude from the equinoctial, do as follows:

Having positioned the second limb under the meridian according to

Arietis, sicut dictum est capitulo 6 supra, converte terciam tibiam cum hallidada ad Lunam, ita quod radius Lune per ipsa foramina transeat;[a] et tunc inter hallidadam et terciam tibiam sit distancia quam exigit latitudo Lune, si Luna latitudinem habeat. Deinde terciam regulam, equidistantem tercie tibie, move sub meridiem.[b] Cum igitur pervenerit[c] stella[d] ad meridiem, per verum locum Lune, deprehendes semper gradum medii celi et per consequens gradum cum quo stella pervenerit ad meridiem.

Omnia hec aliter cognosci poterunt sicut docet ars armillarum et turketi; cetera eciam[e] omnia que per astrolabium sciri possunt. Verum quia propter hec que[f] explicita sunt factum erat instrumentum rectanguli et inventum, quia eciam per ea que iam dicta sunt sufficienter docili discipulo patet via ad omnia[g] ea que per cetera instrumenta sciri possunt, ponimus istic[h] finem.[i]

[III]

CANON[j] TABULE CORDE VERSE

SEQUITUR[k] tabula pro rectangulo, brevis et utilis et labore non modico calculata per 11^{am} et 7^{am} conclusionem corde et arcus, mutata cum conclusione 7^{a} et 11^{a} ad talem litteram.

11^{a} sic: *Unius gradus et singulorum graduum corde verse arcum ostendere.*

7^{a} sic: *Diversarum cordarum arcubus cognitis, unam cordam omnibus communiter subtensam ostendere.*

Et istorum principia erant mihi difficilia[l] valde primo. Propositum[m] vero primo invenisse arcus ad singula minuta corde verse; et quia vidi quod labor foret mihi infinitus, nec multum utilis ad opus rectanguli, sed quod sufficeret tabula extensa ad 60 gradus, opus hoc brevius consumam; cuius operacio sic patet, et sufficientia[n] pulchra ⟨non⟩ minus.

Si arcum alicuius corde verse habere volueris, intra[o] cum gradibus corde verse, et in eius directo invenies arcum suum. Si vero in corda versa minuta[p] habueris, intra cum minutis in eandem tabulam et in eius directo invenies arcum suum, ita tamen quod gradus arcus[q] sint minuta, et minuta sint secunda, et sic deinceps. Tercio intra cum secundis corde verse,[r] et quod in eius directo inveneris sunt secunda, tercia, quarta, quinta, sexta, arcus[s] correspondentis eidem.[t]

[a] A transierit [b] CD meridie [c] C pervenit [d] AC stelle [e] C λ
[f] H +iam [g] H λ [h] F +hic Deo gracias; L +hic
[i] A +si Deus voluerit. Amen; CH Explicit ars operandi cum rectangulo anno Cristi 1326 feliciter; D +Explicit rectangulus Abbatis de Seynt Albon
[j] DL 4λ [k] DL Hec est [l] F difficillima [m] DF 2λ+Proposui autem
[n] L sufficit et erit [o] D +in hanc tabulam [p] DL fracciones [q] D λ
[r] D λ [s] L λ [t] L +arcui

the altitude of the first point in Aries, as is explained in II. 6, move the third limb with the alidade round to the Moon, so that the Moon's ray passes through the same sighting-holes. Then, if the Moon has any latitude, between the alidade and the third limb will be the latitude of the Moon. Now move the third rule, parallel to the third limb, into the meridian. When the star arrives at the meridian, you will always know —through the true place of the Moon—the degree of mid heaven, and consequently the degree with which the star arrives at the meridian.[1]

Everything here may be known otherwise, as the art of the armillary sphere and turketum teaches. Much else is also to be learned from the astrolabe. But we here draw to a close, since the rectangulus was invented and made for the purposes explained, and since, furthermore, enough has been said to permit a teachable pupil to discover for himself whatever may be known from the other instruments.

[III]

Canon to the table of versed chords.

A TABLE follows for use with the rectangulus. It is both short and useful, and was calculated with no little trouble by means of the *conclusiones* 11 and 7 of the [treatise on] chord and arc.[2] The order of these *conclusiones* is changed, thus:

11: *To show the arc of one degree, and of all integral degrees, of the versed chord.*

7: *The arcs of different chords being known, to show one chord subtended by all these arcs together.*

The underlying principles were at first very difficult for me. I had indeed proposed at first to work out the arcs for each minute of the versed chord; but because I saw that my labour would be endless, and not of much use with the rectangulus, and yet that a table with sixty entries would suffice, I brought this work to a close. The method is thus plain, as is its conciseness.

If you want the arc of any versed chord, enter [the table] with the number of degrees of the versed chord, and on the line opposite this number you will find its arc. If, however, you have minutes in the versed chord, enter with the minutes in the same table, and on the same line you will find the arc, but the degrees are to be treated as minutes, the minutes as seconds, and so forth. Thirdly, enter with the seconds of the versed chord, and what you find on the line opposite are seconds, thirds . . . sixths of the corresponding arc.

[1] That is, the mediation of the star.

[2] That the *Quadripartitum* is intended is certain. See our commentary.

Iste modus satis est ydoneus in communibus operacionibus, verumptamen quia, secundum quod probat Ptholomeus primo *Almagesti*, nulla[a] est proporcio corde ad cordam qualis est arcus ad arcum omnino, quod tamen supponit ista operacio, cuius pretextu errare contingeret aliquando in minutis et secundis. Ideo quando vis precise[b] operari, si habueris cum gradibus corde verse minuta, quotque[c] fuerint minuta tociens accipe illud quod est in directo graduum[d] cum quo intrasti, quod intitulatur 'pars sexagesima differentie', et similiter secundo, quot habueris secunda tociens accipe illud ita tamen quod illud quod intitulatur 'minuta' sint secunda, et sic deinceps. Et non errabis licet procedas[e] ad decimam. Ad capiendam vero partem sexagesimam differencie tociens quot habes minuta, reduc illam partem sexagesimam ad ultimum genus fractionis, et illud multiplica per minuta corde tue; et quod resultat erit illud quod debes addere arcui tuo. Et eodem modo si habueris secunda in corda,[f] multiplica eandem partem sexagesimam per secunda tua; quo facto, reducantur[g] ad priora genera fractionum si excedant 60. Et sic resultabit[h] quod erit addendum arcui tuo primo.

Si vero queras cordam versam per arcum tibi notum, intra cum arcu tuo, et si non inveneris precise intra[i] cum minuta propiori, et accipe differenciam illius ad arcum tuum cum quo intrasti;[j] et illam differenciam multiplica per 60 et multiplicatum divide per differenciam inter duos introitus. Et quod exit addendum est corde in lineam numeri in directo primi introitus. Et observa hic omnem regulam que in divisione et multiplicacione fraccionum per fracciones alibi observatur.[k] Si inveneris precise arcum tuum in tabula, tunc non oportet bis intrare.

Explicit[l] composicio rectanguli et operacio cum eodem 1326

Table of variant readings for the table of corda versa (page 432)

Amended			Reading in		Reading proposed
Row	Column (arc)	Column (diff.)	MS. F	MS. D	
9		q	25	35	35
18		t	31	21	21
23	m		25	25	35
26		q	40	40	38
38		q	16	16	13
39	q		52	52	51
39		q	22	22	21
47		s	0	0	3
48		q	25	25	45
52	s		15	17	17
53		q	38	38	28
54		m	29	29	58
56	q		0	1	1

[a] D (?); L non [b] D pace [c] DL quot [d] DL gradus [e] DL +usque
[f] L +tua [g] ACH reductantur [h] DL +illud
[i] DL 3λ; D +cum minori; L +intra cum minori [j] DL intras
[k] ACDFHL observantur [l] D 8λ; L 8λ+*Explicit operacio cum rectangulo*

This method is quite sufficient for ordinary purposes, yet since—as Ptolemy proved in the first book of the *Almagest*—ratio of chord to chord is not exactly the same as that of arc to arc, as was assumed in the above procedure, therefore there might happen to be some error in the minutes and seconds. On this account, when you wish to work accurately, if you have minutes as well as degrees in a versed chord, multiply the number of minutes by what you find on the line opposite the number of degrees you are working with, under the heading 'sixtieth part of the difference'. Secondly, multiply likewise by the number of seconds, but what are headed 'minutes' will now be seconds, and so on. You will not be in error, even if you go as far as [sexagesimal] tenths. To arrive at the product of the sixtieth part of the difference by the number of minutes, reduce that sixtieth part to the last [significant] place of sexagesimals,[1] and multiply that by the number of minutes in your chord. What results will be what must be added to your arc. In the same way, if you have seconds in your chord, multiply the sixtieth part by the number of seconds, and when this is done, carry one place forward anything in excess of sixty. In this way there results what must be added to your first arc.

If, moreover, you are looking for the versed chord corresponding to a known arc, enter [the table] with your arc, and if you fail to find it exactly, enter with the number of minutes nearest to it, and take the difference between that and your own arc. Multiply the difference by 60, and divide the product by the difference between the two entries [between which your arc lies]. What results is to be added to the chord in the line of number opposite your first entry. And here observe all the rules which are observed elsewhere in the division and multiplication of fractions by fractions. If you find your arc exactly in the table, then it is unnecessary to enter the table twice.

Here ends the treatise on the construction of the rectangulus and its use 1326

[1] Lit. 'fractions'.

[*Tabula corde verse*]

[sinus vers.]	arcus circuli magni g m s t q	differencia g m s t q	[sinus vers.]	arcus circuli magni g m s t q	differencia g m s t q
		10 19 9 42 22			1 5 47 34 41
1	10 19 9 42 22		31	61 5 47 34 41	
		4 54 14 50 22			1 5 1 42 28
2	15 13 24 32 44		32	62 10 49 17 9	
		2 58 15 47 43			1 4 34 12 5
3	18 11 40 20 27		33	63 15 23 29 14	
		2 50 40 29 5			1 4 1 46 34
4	21 2 20 49 32		34	64 19 25 15 48	
		2 31 1 32 28			1 3 6 50 23
5	23 33 22 22 0		35	65 22 32 6 11	
		2 17 8 17 25			1 4 0 1 37
6	25 50 30 39 25		36	66 26 32 7 48	
		2 5 20 1 58			1 1 2 0 29
7	27 55 50 41 23		37	67 27 34 8 17	
		1 59 26 23 40			1 1 48 53 13
8	29 55 17 5 3		38	68 29 23 1 30	
		1 47 0 15 35			1 1 22 47 21
9	31 42 17 20 38		39	69 30 45 48 41	
		1 54 34 43 30			0 54 57 27 0
10	33 36 52 4 8		40	70 25 43 15 51	
		1 35 51 52 5			1 7 35 30 25
11	35 12 43 56 13		41	71 33 18 46 16	
		1 39 0 43 38			0 59 5 29 34
12	36 51 44 39 51		42	72 32 24 15 50	
		1 34 13 47 51			1 0 2 5 9
13	38 25 58 27 42		43	73 32 26 20 59	
		1 30 51 48 11			0 59 22 21 3
14	39 56 50 15 53		44	74 31 48 42 2	
		1 29 19 33 32			0 59 33 40 19
15	41 26 9 49 25		45	75 31 22 22 21	
		1 23 19 44 4			0 59 46 19 47
16	42 49 29 33 29		46	76 31 8 42 8	
		1 15 57 51 51			0 58 3 18 39
17	44 5 27 24 20		47	77 29 12 0 47	
		1 27 43 21 8			0 58 59 49 45
18	45 33 10 45 28		48	78 28 11 50 32	
		1 20 31 56 59			0 57 57 48 44
19	46 53 42 42 27		49	79 26 9 39 16	
		1 17 38 50 0			0 41 43 27 32
20	48 11 21 32 27		50	80 7 53 6 48	
		1 16 8 38 11			1 14 43 54 49
21	49 27 30 10 38		51	81 22 37 1 37	
		1 14 42 9 6			0 56 40 56 28
22	50 42 12 19 44		52	82 19 17 58 5	
		1 13 24 6 30			0 57 45 6 28
23	51 55 36 26 14		53	83 17 3 4 33	
		1 12 10 8 28			0 59 3 7 42
24	53 7 46 34 42		54	84 16 6 12 15	
		1 11 6 3 31			0 57 4 38 49
25	54 18 52 38 13		55	85 13 10 51 4	
		1 10 1 56 38			0 57 27 45 57
26	55 28 54 34 51		56	86 10 38 37 1	
		1 8 55 26 1			0 57 23 42 2
27	56 37 50 0 52		57	87 8 2 29 3	
		1 8 18 46 57			0 57 20 22 18
28	57 46 8 47 49		58	88 5 22 51 21	
		1 7 32 43 14			0 56 26 19 1
29	58 53 41 31 3		59	89 1 49 10 22	
		1 6 18 28 57			0 58 10 49 38
30	60 0 0 0 0		60	90 0 0 0 0	

FOUR SHORT TRACTS CONCERNING CELESTIAL COORDINATES

From MS. Ashmole 1796

This manuscript is of importance primarily because it contains the writings to which we have given the title *Horologium*. The volume belonged to the sub-sacristan of the abbey of St. Albans, and it will be suggested in connection with those writings, on the design of an astronomical clock, that they were copied from Richard of Wallingford's unedited drafts. The following short pieces could equally well be by him, and might have been copied under the same circumstances. The evidence will be considered when we give a synopsis of the contents at the end of the Commentary on *Rectangulus*.

TRACT I (unfinished), fol. 39^r, lines 4–16

AD sciendum latitudinem stelle ab equinoxiali per eius latitudinem ab ecliptica, id est gradum cum quo celum mediat per verum eius gradum, aut est stella in eadem parte zodiaci cum circulo equinoxiali vel in diversa parte, si in eadem parte tunc fiat hec demonstracio sic:

Quarta spere in qua est stella, triangulus orthogonus equaliter *ABC*, et *A* caput Arietis vel Libre, et *C* polus zodiaci, et *D* polus mundi. Nam si *A* sit caput Arietis vel finis Virginis, tunc est *D* polus meridionalis. Et si *A* fuerit caput Libre vel finis Piscis, tunc est *D* polus septentrionalis. Et sit *T* locus stelle, et *E* verus eius gradus, et *TE* latitudo ab ecliptica. Per lineam *TE* mihi notam volo scire lineam *TF*, que est declinacio ab equinoxiali. Et per lineam *AE* volo scire arcum *ACG* cum quo gradu stelle pervenit ad meridiem.

Age ergo: *TFH* triangulus et *HIC* habent angulos duos oppositos preter rectos *TFI*. Proporcio lateris *HE* ad latus *EI* includens triangulum est sicut proporcio *HC* ad *TF* includens alterum angulum correlacium . . .

TRACT II follows *Rectangulus* immediately at fols. 165^v–166^v.

CUM HABUERIS tabulam fixarum factarum ad longitudinem et latitudinem ab ecliptica, et volueris per ipsam tabulam invenire longitudinem et latitudinem ipsarum stellarum ab equinoxiali, quia per ipsum gradum cum quo mediat celum et per eius latitudinem equinoxiali facilius erunt stelle situari in astrolabio et in quolibet alio instrumento, sic debes operari:

Primo accipe unam laminam plumbeam vel ligneam, bene latam et valde planam, et fac in illa lamina unum circulum, non nimis magnum sed mediocrem, super centrum *O*. Deinde divide ipsum circulum in quattuor partes equales per duas lineas orthogonaliter intersecantes se, scilicet *AB* et *CD*. Deinde divide ipsum circulum in 360 gradus equales, et postea incipe computare a *C* versus *B* 30 gradus, et hoc vocetur signum Arietis; et scribe sub illis 30 gradibus subtiliter '*Aries*'; et sic de omnibus signis—subsequentibus 30 gradibus pone '*Taurus*', et sic de aliis. Item infra circulum semper contra quinque gradus '5', et contra decem gradus scribe '10', et sic de aliis usque in finem signi.

Deinde protrahe lineam a *B* in continuum et directum ab *A* versus *Q* quantum volueris, et similiter a *B* versus *R*. Deinde divide *OR*[1] lineam intra et extra circulum in quot partes volueris, ponendo unum finem regule super *C* et alium super unum vel duos gradus, computando a *D* versus *B*; et sic potes dividere lineam *QR* in quot sectiones[2] volueris, sicut dividitur linea altitudinis in novo quadrante—non tamen oportet dividere eam ultra 90 partes. Similiter potes dividere lineam *OQ* in quot partes volueris, ponendo finem regule super *C* ut prius, et alium finem super quot gradus volueris, computando a *D* versus *A*.

Hoc facto, computa ab *O* versus *A* 24 gradus fere, id est 23 gradus et modicum plus quam dimidium gradus, id est 33 minuta; et ibi pone *H*, quod vocabitur polus declinacionis. Similiter ab *A* versus *Q* computa alios 23 gradus et 33 minuta, et ibi ponatur *F*. Dico ergo quod inter *H* et *F* sunt 90 gradus, quos tu debes subtiliter scribere sub linea *FH*, scilicet contra 5 sectiones pone '5', vel qualecunque volueris scribere. Item a puncto *B* versus *O* computabis alios 23 gradus et 33 minuta, et ibi pone *G*. Dico ergo quod inter *G* et *H*, quod est polus declinacionis, sunt 90 gradus precise. Deinde pone pedem circini in linea *BR* et alium pedem in *H*, et non cessa aptare circinum usquequo posses describere circulum transeuntem per *H* et per *C* [et] *D*. Centrum igitur illius circuli sit *L*. Dico ergo quod *L* equaliter distat ab *H* et *C* et *D*. Deinde super centrum *L* protrahatur linea dividens lineam *BR* ex utraque parte perpendiculariter, et sit hec linea *KM*. Hec est descripcio figure.

Si ergo volueris gradum cum quo stella celum mediat, per verum gradum eius, et similiter latitudinem eius ab equinoxiali, per gradum verum stelle, sic operaberis: vide in tabula tua latitudinem stelle ab ecliptica; et si stella fuerit septentrionalis ab ecliptica computa ab *F* versus *H* tot gradus eciam minuta si volueris; et similiter a *B* si volueris versus *H* computa tot gradus, et pone ibi punctum. Deinde totam illam lineam que est inter illa duo puncta divide in duas partes equales, et in puncto medio pone pedem circini aptati ad illa dua puncta, et contra illum gradum in quo stella est fac parvam et subtilem revolucionem

[1] MS.: *QR* [2] Over *partes*, which is struck through.

cum alio pede circini; nec oportet te facere totam circulacionem sed parvam partem eius versus illam partem in qua stelle verus gradus est; et vocetur illa circulacio parva *T*.

Deinde vide in tabula stellarum fixarum verum gradum ipsius stelle. Deinde pone pedem circini in linea *KM* et apta alium pedem ad verum gradum ipsius stelle in circulo tuo diviso per 360, usquequo posset alius pes transire per verum gradum stelle in circulo predicto et per polum[1] declinacionis *H*. Quo facto dico quod punctus in quo stat pes ipsius circini in linea *KM* equaliter distat ab *H* et vero gradu stelle et a nadir ipsius gradus. Deinde pede circini immobiliter stante in illo puncto in linea *KM*, vide ubi alius pes circini tangit parvum arcum quem fecisti, scilicet *T*, et illum punctum nota bene. His peractis pone pedem circini in *O*, quod est centrum circuli divisi per 360, et alium pedem pone in illo puncto quem bene notasti; et revolve pedem circini usque in lineam *OB* et vide quot gradus et minuta sunt inter circulum divisum et illum contactum. Et certe per tot gradus declinat illa stella ab equinoxiali. Habita igitur declinacione illius stelle ab equinoxiali, scribe talem numerum directe contra stellam illam, si vis facere tabulam novam ad longitudinem et latitudinem ab equinoxiali.

Deinde, si vis scire gradum cum quo stella mediat celum, protrahe unam lineam rectam a centro *O* per medium illius eiusdem puncti quem bene notasti, et sic in continuum usque in circumferenciam circuli divisi; et super quemcunque gradum illa linea[2] ceciderit ille est gradus cum quo stella mediat celum. Scribe igitur longitudinem istam et[3] id est cum quo mediat celum in tabula tua, directe contra stellam illam, et sic facies de omnibus stellis quotquot volueris habere verificatas certissime; et tunc potes faciliter ponere eas in astralabio vel quadrante, quia iste modus ponendi stellas, cum fuerint verificate, est facilior quam aliquis alius modus.[4]

* * *

Si autem fuerit meridionalis ab ecliptica tunc debes computare tantum a *G* versus *R* et ponere ibi punctum; similiter a puncto *F* versus *Q*, et ponere ibi punctum. Tunc in medio duorum punctorum pone pedem circini et revolve subtilem circulum per illa duo puncta, et sit ille circulus ut prius *T*. Ut prius, pone pedem circini in *KM* et revolve alium pedem a vero gradu stelle usque *H*, scilicet fac talem arcum qualis est *P*. Et ubicumque ille arcus abscindit alium, scilicet *T*, ibi nota locum stelle. Et tunc statim extrahere poteris, ut prius dictum est, gradum cum quo mediat celum et eius latitudinem ab equinoxiali, videlicet omnino

[1] MS.: *punctum* [2] MS.: *stella* [3] Superfluous *et* in MS.

[4] The star Alaioc is now taken as an example, and the entire method is rehearsed once more. We omit the example (for which see the discussion in vol. ii), and give the closing paragraphs next following.

ut prius. Cum igitur sic verificaveris omnes stellas quas habere volueris in astralabio, et scripseris eas in una tabula intitulata sic: *Tabula longitudinis et latitudinis ab equinoxiali circulo* tunc faciliter et certe ponere potes eas in instrumento tuo. Et iste est modus certissimus verificandi stellas secundum astronomos subtiles et sapientes. *Explicit.*

TRACT III, fols. 180rv (one diagram on fol. 179v belongs to this)

COGNITO VERO gradu stelle in ecliptica et latitudine eius ab ecliptica, docet Tholomeus in 8° libro *Almagesti*, cap. 5°, cognoscere gradum cum quo stella mediat celum et latitudinem eius ab equinoxiali[1] per unam demonstracionem que presupponit noticiam demonstracionis de cathis quam ponit in 12° cap., primo libro *Almagesti*. Ponam igitur in primo capitulum[2] Tholomei de verbo ad verbum, sed ubi inveneris aliquam litteram subtractam scias illam esse glosam linearum eiusdem textus. Deinde ad melius cognoscendum istam demonstracionem ponam *tractatum corde verse et recte* simul cum *tractatu de catis*, sine quibus non potest ista demonstracio componi ullo modo; et hoc totum supra huius conclusionem (?), quia utilis et mirabilis.

Describam itaque primum mediacionem celi earum, orbem descriptum super polos duorum orbium, orbis equacionis diei et orbis signorum, super quem circulum sint *ABGD*; et medietatem orbis equacionis diei, id est equinoxialis, super quem sint *ACG*, et sit polus eius *Z*, scilicet polus mundi; et medietatem orbis signorum super quem sint *BCD*, et polus sit *H*, scilicet polus zodiaci. Et ponam arcum orbis magni descripti super polos orbis signorum super quem sit *HT* et *KL*. Et imaginabor aliquam stellam fixam de quaquam (?) super notum punctum *T* quoniam loca earum, scilicet stellarum, non sunt nota nisi secundum consideraciones magnas, secundum quod descripsimus de eis apud similitudinem horum orbium descriptorum. Et describam arcum orbis magni descripti super polos equacionis diei, scilicet super polos mundi [et] super punctum *T*, qui est centrum stelle, super quem sunt *ZTMN*. Manifestum est igitur quod stella que est super punctum *T* mediat celum cum duobus punctis orbis equacionis diei et orbis signorum que sunt prius signata super punctum *M* et punctum *N*.

Et ostendam tibi istud, scilicet quis punctus est *M* in orbe signorum et quis punctus est *N*[3] in circulo equinoxiali. Et hec est una conclusio quam hic intendimus demonstrare. Et similiter ostendam arcum *TN* qui scilicet arcus ducetur (?) in declinacionem puncti *T* ab equinoxiali circulo, et hec est secunda conclusio quam hic intendit per id quod tibi narrabo. Et quia per illud quod declaravimus in principio libri *Almagesti*,

[1] MS.: *ecliptica* [2] MS.: capitulo [3] MS.: *M*

scilicet capitulo 12° primi libri de duobus orbibus magnis inter duos arcus *HA* et *AN* duorum magnorum orbium, protrahuntur duo arcus duorum magnorum orbium, scilicet arcus *HL* et arcus *ZN*, erit igitur proporcio corde dupli arcus *HA* ad cordam dupli arcus *AZ* aggregata ex duobus proportionibus, scilicet ex proporcione [corde] dupli arcus *HL* ad cordam dupli arcus *LT*, et ex proporcione corde dupli arcus *TN* ad cordam dupli arcus *NZ*. Unusquisque vero arcuum *AZ*, *ZN*, *KH*, *HB* est quarta circuli, ut supponitur. Et scimus ex[1] note superius, scilicet intra[2] stellarum, quod arcus *KT*, qui est latitudo stelle ab ecliptica, et arcus *KC*,[3] qui est eius longitudo, sunt noti.

Et scimus per id quod iam declaratum est de declinacione orbis signorum ab equinoxiali circulo arcum *ZH* et arcum *LK*. Ad habendum noticiam *KL*, scias ex prima cata consequencius, ex proporcione *PB* ad *AB*, ⟨que⟩ sit ex proporcione *PK* ad *KL* et ex proporcione *CL* ad *CA*.[4] Set *PB* est notum, quia est quarta circuli notum; et *PK* est notum, quia est quarta circuli; igitur notum erit sextum, notum quod est *KL*, igitur totum *TL* est notum, cum sit notum *TK*. Set nota quod arcus *KL* et arcus *LA* scientur per ascensiones *K* in circulo directo. Et ex eo declarabitur quod ex arcubus quesitis erunt arcus *HA* et arcus *AZ* et arcus *HL* et arcus *LT* et arcus *ZN* noti. Et propter hoc sciemus arcum *TN* residuum, qui scilicet est declinacio stelle ab equinoxiali circulo.

Et eciam quia proporcio corde dupli arcus *ZH* ad cordam dupli [arcus] *HA* agregetur ex duabus proporcionibus, ex proporcione scilicet corde dupli arcus *ZT* ad cordam dupli arcus *TN*, et ex proporcione corde dupli arcus *NL* ad cordam dupli arcus *LA*, erunt ex arcubus quesitis, cum quibus est sciencia. Propter illud quidem quod diximus hic arcus *ZH* et arcus *HA* et eciam arcus *ZT* et arcus *TN* noti; et propter illud quidem quod elevatur simul de orbe equacionis diei et de orbe signorum in spera recta ex arcu *KB* sciemus arcum *LA* et arcum *LN* residuum; et propter illud, sciemus ex toto arcum *NA* circuli equatoris et hoc *MB* orbis signorum.

Et nota quod statim in isto eodem capitulo demonstrat arcum stelle super orizontem et cum quibus gradibus oritur in quolibet orizonte de quo nichil ad presens.

Et nota quod hic vocat arcum duplum de cuius corda loquitur tantum arcum circuli maioris dupplicatum, ut si arcus *NA* sit 30 graduum erit eius duplum 60 graduum. Et si eiusdem dupli corda tota sit 150 minuta, erit eius medietas corda recta 30 gradus circuli. Hec satis manifeste patent primo libro *Almagesti*, capitulo 3°, vel ubi satis plane loquitur ostendo declinacionem zodiaci ab equinoxiali circulo.

[1] Lacuna? [2] Again a lacuna? [3] MS.: *KB*
[4] MS.: *CK* ad *CB*. See the commentary in vol. ii.

TRACT IV f. 189[rv]

SI HABUERIS declinacionem alicuius stelle a circulo equinoxiali, faciliter et certe investigare poteris latitudinem eiusdem stelle a via Solis, que ecliptica dicitur, per demonstracionem Geber in fine libri sui.

Item si certitudinaliter sciveris cum quo gradu zodiaci pervenerit stella ad medium celi civitate cuius meridies est tibi nota, certitudinaliter demonstrare poteris quis est gradus longitudinis eius vere in zodiaco ab Arietis capite. Est autem differencia inter gradum longitudinis stelle et gradum medii celi ista, quia linea transiens a polo mundi per centrum stelle necessario transibit per gradum medii celi eiusdem. Item[1] linea transiens a polo zodiaci per centrum stelle necessario transibit per gradum latitudinis eius in zodiaco qui est verus gradus eius. Est autem demonstracio convertendi longitudinem stelle in gradum medii et econtra et declinacionem in latitudinem eius.

Si neutra habeatur, primo per armillas inquiratur longitudo et latitudo stelle, sicut ars armillarum, qua habita faciliter reperies declinacionem et gradum medii celi sicut dicitur postea. Si tamen armillas non habueris, sed[2] solum instrumentum cognoscendi altitudinem graduum, tunc primo necesse est tibi scire gradum cum quo stella pervenit ad meridiem tuam, quod ita commode scies.

Primo oportet observare quoadusque stella illa pervenerit precise ad meridianam lineam, et statim in eodem momento investiga[3] per aliam stellam notam quis gradus zodiaci est in illo momento in medio celi, sicut docent canones vel quocumque alio vis ingenio. His habitis, vide eciam que fuerit altitudo illius stelle per quam operaris, dummodo fuerat in meridie. Deinde vide que est altitudo[4] capitis Arietis in illo loco, similiter que est altitudo illius gradus que simul est in medio celi cum stella ipsa, quod scies per declinacionem illius gradus ab equinoxiali circulo, sicut docent canones. Deinde operaberis altitudinem gradus ad altitudinem stelle, quod si stella fuerit altior gradu in medio celi et gradus fuerit septentrionalis, adde differenciam altitudinis super declinacionem gradus, et habebis stelle declinacionem ab equinoxiali, et cetera—satis docent canones, ideo supersedeo.

Habito itaque gradu medii celi et eius declinacione ab equinoxiali, si habebis alia duo, proponatur primo arcus declinacionis gradus medii celi, similiter et arcus distancie stelle ab equinoxiali, quibus ad invicem comparatis, dum pariter ad eundem polum declinaverit, minor de maiore dematur, et residuum erit eorum differencia. Hanc differenciam de 90 deme, et accipe eius cordam rectam et sit A; secundo accipe cordam residui declinacionis stelle ab equinoxiali de 90, et sit H; tercio deme totam declinacionem, que est 24 gradus, de 90, et accipe eius

[1] MS. adds *in* [2] MS.: *nisi* [3] MS.: *investigia* [4] *maxima* is understood.

cordam, et sit *C*; similiter quarto, cordam residui tocius declinacionis de 90 et sit corda *D*. His quattuor per ordinem dispositis, consequenter cordam tocius declinacionis, que est *C*, de 300 deme et sit residuum *E*, que dicatur *corda longior*. Deinde arcum declinacionis gradus medii celi de 90 deme, et illius residui quere similiter cordam, quam de 300 deme, et residuum vocetur corda sexta *F*.

His autem sex diligenter servatis, multiplica cordam tocius declinacionis *C* in directo tocius diametri, scilicet in 150 producta; que summa dividatur per cordam residui declinacionis gradus medii celi dempte de 90, et prodit corda declinacionis recte; cuius de 90 arcu dempto, et productum residui corda recte declinacionis vocabitur. Tunc ipsam recte declinacionis cordam multiplica per cordam distancie recte, et productum divide per sinus (?) distancie recte unum collectum per cordam longiorem. Et quod colligitur in cordam tocius declinacionis ductum dividetur per cordam sinus (?) declinacionis recte, quod igitur collectum fuerit ducetur in cordam longitudinis gradus medii celi stelle a capite Cancri aut Capricorni, cui propior fuerit aut aperte, ante aut retro, per ortum recti circuli[1] . . . per ortum circuli directi repertum.[2]

[1] I omit eight lines of text, the copy being too corrupt to be worth reproducing.

[2] Incomplete, as a result of missing folios.

VI

TRACTATUS HOROLOGII ASTRONOMICI

SIGLA

A Oxford, Bodleian Library, MS. Ashmole 1796.

B Brussels, Bibliothèque royale, MS. 2967.

C Cambridge, Gonville and Caius College, MS. 230/116.

For full discussion of the manuscripts, Introduction to the text, and Synopsis and Commentary, see vol. ii, pp. 309–70. The diagrams are printed in vol. iii, pp. 61–74. See also Plates XVII–XXII in vol. iii.

Since, to all intents and purposes, MS. A alone contains all of the following treatise, the spelling adopted here follows that of MS. A closely. In listing variants in Part I, especially in the difficult sections, I have erred on the side of triviality. The words of MS. A are not heavily contracted, but there are times when the copyist clearly cannot easily make out what he is copying, and on f. 175^{v}, for instance, words have been omitted, perhaps for this reason. This encourages the belief that the copy was made from informal drafts.

[TRACTATUS HOROLOGII ASTRONOMICI]

[PARS PRIMA]

CONCLUSIONES DIVIDENDI ROTAS PRO HOROLOGIO ASTRONOMICO PRO MOTIBUS PLANETARUM[a]

PRIMA CONCLUSIO

[1]

HABITA rota certi numeri dencium, unam[b] aliam rotam invenire que[c] habeat quotlibet dentes equales dentibus rote date.

Age:[d] Divide diametrum rote date in tot partes equales quot sunt dentes[e] equales in circumferencia, de quibus partibus diametri accipe quotquot partes[f] volueris; et super illis partibus[g] fac circulum cuius diameter[h] continet[i] omnes partes acceptas, et tunc circumferencia eius habebit tot partes equales alterius circumferencie partibus quot partes equales habet una diameter de partibus diametri alterius. Quia que est proporcio circuli ad circulum, illa[j] est proporcio diametri ad diametrum. Et[k] facere poteris protrahendo diametrum quantumcumque volueris longe, et accipiendo quotlibet partes equales.[l]

CONCLUSIO SECUNDA[m]

[2]

Habita rota que circumvolvitur perfecte in[n] certo tempore, unam[o] aliam rotam invenire que circumvolvetur[p] tractu[q] eiusdem in quotlibet partibus racionabilibus eiusdem temporis.

Verbi gracia: Data rota que circumvolvitur semel in die, et hoc precise, propositum est invenire aliam rotam que circumvolvitur[r] in 20 horis, hoc est in 20 vicesimis quarcis[s] partibus diei, vel in 20 tricesimis, vel in 100 undecimis, et sic de ceteris.

[a] Title in B and C only, the latter beginning '*Incipiunt canones*' [b] B λ [c] A qui
[d] B Sic age [e] B partes [f] B λ [g] B istas partes [h] A diametri
[i] B contineat [j] B eadem [k] C +sic [l] BC +placuerit
[m] B Alia conclusio [n] A λ [o] B λ [p] B circumvolvitur
[q] A tracte [r] A λ [s] B writes this and ensuing fractions in a modern style ($\frac{20}{24}$, etc.).

[A TREATISE ON THE ASTRONOMICAL CLOCK]

[PART ONE]

HERE BEGIN THE PROPOSITIONS DEALING WITH THE DIVISION OF THE WHEELS NEEDED FOR AN ASTRONOMICAL CLOCK SHOWING THE MOVEMENTS OF THE PLANETS

PROPOSITION 1

GIVEN a wheel of a definite number of equal teeth, to find another wheel having any desired number of teeth, [these being] equal in size to the teeth of the given wheel.

Method: Divide the diameter of the given wheel into as many equal parts as there are equal teeth on the circumference, and from these parts of the diameter take as many parts as you wish. Construct a circle with diameter equal to the number of parts chosen, whereupon its circumference will have as many parts equal to the parts of the other circumference as the one diameter has parts equal to those of the other diameter. This is the case, since the ratio of circle to circle is the ratio of diameter to diameter. And you may proceed in a similar way by producing the diameter as far as you like, and taking as many equal parts as you please.

PROPOSITION 2

Given a wheel which is turned completely in a definite period of time, to find another wheel which is turned by the movement of the same in any rational fraction of that time.

Example: Given a wheel which is turned once daily, and that precisely, it is proposed to find another wheel which is turned in 20 hours, that is to say in 20/24 of a day (or it might be in 20/30, or 100/11 of a day, and so on).

Age igitur: Divide rotam date revolucionis secundum quemcumque[a] numerum parcium equalium temporis mensurantis[b] ipsam revolucionem. Deinde rotam que trahi debet divide in quotlibet partes volueris equales dentibus rote prioris, secundum precedentem conclusionem; et applicetur rota rote, et habes[c] propositum. Quia cum rota trahens dividat tempus[d] sue revolucionis in partes equales numero dencium suorum, quot dentes | rota habuerit[e] tracta in tot partibus temporis eiusdem revolucionis complebit cursum proprie revolucionis; et hoc est propositum.

CONCLUSIO TERCIA[f]

[3]

Habita rota certe revolucionis, aliam rotam quotlibet[g] dencium cum illo[h] motu continuo trahere, ita quod in tota sua revolucione tantum unum dentem trahat de limbo rote tracte.

Age:[i] circa axem rote certe revolucionis fiat unus dens testudinatus, et applicetur illa axis ad rotam que trahi debet; et axis illa in tota[j] revolucione[k] rote sue continue trahet alteram rotam, et tum[l] tantum unum[m] dentem;[n] et hoc est propositum.[o]

Memorandum quod si fiat unus circulus 508 dencium qui semel in anno integro solari cum horis 6 complet cursum suum, et trahat alteram rotam 38 dencium, ille in anno 13 vicibus circumducetur et insuper quatuor signa 12 gradus 40 minuta 41 secunda in anno solari, et ille est medius motus Lune in anno solari secundum Alfunsum,[p] quia proporcio est probata.

Item memorandum quod si unus circulus motus lunaris in anno dividatur in 269 dentes, et moveat epiciculum[q] de 251 dentibus, tunc epiciculus movebitur in una lunacione integra semel et 25 gradus et

[a] C quemlibet; B quem [b] A mensuratis [c] BC habebis [d] B rotam
[e] C habuit [f] B Alia conclusio [g] B 2λ+quolibet dentes [h] B isto
[i] B Sic age [j] B 3λ+ista in illa [k] C +sua [l] A cum [m] C +eius
[n] B +trahet
[o] B does not follow our text beyond this point. C omits the remainder of the *conclusio*. (Omitting also the headings to 4 and 5, C at first appears to include them in 3.)
[p] A Afunsum [q] A epiculum

Method: Divide the wheel of the given period of revolution into any number of equal parts of the time measuring that revolution.[1] Then divide the wheel which is to be driven into as many parts as you wish,[2] equal in size to the teeth of the first wheel, this being done with the help of the previous proposition. And when the one wheel is applied to the other, you will have what was proposed; for, since the driving wheel divides the time of its revolution into parts equal to the number of its teeth, however many teeth the driven wheel has, in just so many parts of the aforesaid time of revolution it will complete the course of its own revolution. And this is what was proposed.

PROPOSITION 3

Given a wheel of a definite period of revolution, to drive continuously, by its movement, another wheel of any number of teeth, so that in its complete revolution it may drive only one tooth on the edge of the driven wheel.

Method: Make a snail tooth[3] about the axle of the wheel having a definite period of revolution, and let that axle be fixed to the wheel which is to be driven. The axle, during the complete revolution of its wheel, will drive the other wheel continuously, and then only one tooth. And this is what was proposed.[4]

It is to be borne in mind that if a wheel[5] be made with 508 teeth, this wheel completing its course once in a complete solar year including the six hours,[6] and if it drive another wheel of 38 teeth, then in one solar year the latter will be led round 13 times and a further $4^s\ 12^\circ\ 40'\ 41''$ (i.e. the mean motus of the Moon in a solar year, according to Alfonso), for this ratio has been worked out.[7]

Next, it is to be remembered that if a wheel of the annual lunar movement be divided into 269 teeth, and if it move an epicycle of 251 teeth, then in a complete lunation the epicycle will be moved through one

[1] This should not be 'any number', but rather 'a number equal to the denominator of the rational fraction mentioned in the heading'.

[2] (See the last note.) For 'any number' read 'a number equal to the numerator of the rational fraction'. This was probably thought to be obvious.

[3] That a *worm* is intended here emerges from later chapters. Cf. the 'snail' of later horology. See also the glossary.

[4] The foregoing proposition appears to have no relevance to the following worked examples, which seem rather to illustrate the first two propositions, if any so far.

[5] *Circulus* rather than *rota*. The apparently random choice of word is not noted subsequently.

[6] That is to say, a year of $365\frac{1}{4}$ days.

[7] It would be surprising if a simple ratio like 508/38 gave exactly the required figure. In fact it gives $13^{\text{rev}}\ 4^s\ 12^\circ\ 37'\ 54''$—a fairly good approximation.

49 minuta, qui est argumentum[a] in una lunacione integra; vel in 14 lunacionibus 1 gradus et 26 minuta, ut patet per canones.

Et memorandum quia primi numeri motus lunaris sunt magni valde, ideo si subduplantur idem proveniet, et commodius quam prius; scilicet si maior circulus solaris[b] habeat 254 dentes et circulus Lune habeat 19 dentes, tunc porcio 4 signorum 12 graduum at 49 minutorum et 49 secundorum habebit de illis 19 dentibus 7 dentes; et est quantitas dentis 18 gradus 57 minuta, vel in minutis 1137, igitur in fine certi[c] anni cum habebit 2 dentes, erit in circulo 37 gradus 54 minuta.

CONCLUSIO QUARTA

[4]

Si circulus certe revolucionis in partibus racionabilibus alicuius certi temporis, in tota sua revolucione, alterius circuli tantum moveat unum dentem, in tot revolucionibus integris quot habet dentes, tot de circulo tracto dentes movebit quota fuerit denominacio parcium racionalium eiusdem.

Verbi gracia: Circulus motus in 25 quartis decimis unius diei movet in sua revolucione unum dentem alterius circuli, ergo in 25 diebus movebit 14 dentes circuli tracti per eum; quia que est proporcio 25 ad 14 illa est proporcio 25 quattuor decimarum | ad integrum in conversa proporcione.

Similiter si in tribus quintis unius diei moveat unum dentem, in tribus diebus movebit 5 dentes, et sic de singulis, quia ter tria et bis tria idem sunt quod ter 5. Si ergo[d] in tribus quintis movet unum dentem, in ter tribus movebit tres dentes,[e] et in bis tribus duos; igitur in ter tribus et bis tribus movebit 5 dentes,[f] satis patet.

CONCLUSIO QUINTA

[5]

Habita rota certe revolucionis ut puta rota que in quotlibet partibus diei racionabilibus circuit precise zodiacum per numerum dencium

[a] A argumenti [b] A solari [c] A certum [d] C +semper [e] C λ [f] C λ

revolution and 25° 49′.[1] This is the argument in a whole lunation. In other words, in 14 lunations the epicycle will be moved through 1° 26′,[2] as is clear from the canons.[3]

It must be remembered that the figures given earlier for the lunar movement are very large, and that if they are halved, the same result is obtained, and more conveniently than before. If the larger (solar) wheel, that is to say, is to have 254 teeth, and the Moon's wheel 19, then a section of 4^{s} 12° 49′ 49″ will amount to 7 teeth out of the 19.[4] The size of a tooth is 18° 57′, or in minutes 1137, and therefore when, at the end of a year, two teeth remain (?), this amounts to 37° 54′ on the wheel.[5]

PROPOSITION 4

If a wheel, which revolves by a definite amount in rational parts of any fixed time, should move just one tooth of another wheel in a complete revolution, then (in as many complete revolutions as it has teeth)[6] it will move just as many teeth of the driven wheel as the denominator of the aforesaid ratio.

Example: A wheel, moved in 25 fourteenths of a day, moves in its revolution one tooth of another wheel. In 25 days, therefore, it will move 14 teeth of the wheel driven by it. The reason is that the ratio 25 to 14 is the same as the ratio of 25 fourteenths to unity, in converse proportion.

Likewise, if it move one tooth in three-fifths of a day, in three days it will move five teeth—and so for each case; for thrice three and twice three are the same as thrice five. If, therefore, in three-fifths of a day it moves one tooth, in thrice three days it will move three teeth, and in twice three two; and thus, clearly enough, in thrice three and twice three days it will move five teeth.

PROPOSITION 5

Given a wheel with a definite period of revolution, for example a wheel which moves around the zodiac exactly in any rational number of parts of a day, by virtue of the number of its teeth; to find another

[1] The ratio 269/251 gives a figure remarkably close to 1^{rev} 25° 49′, being in excess of this by only 1/251 minutes of arc.

[2] In excess, that is to say, of the integral number of revolutions (15).

[3] The intention was presumably to say 'as is clearly demanded by the canons'.

[4] He seems to be saying that if you count off seven teeth of the smaller wheel, you will have a good approximation to the vital 4^{s} 12° 40′ 41″. (This interpretation requires his 4^{s} 12° 49′ 49″ to be exactly 7/19 revolutions. This is only approximately true. More precisely, the answer given is 4^{s} 12° 37′ 54″.) The number is the number of teeth through which the smaller wheel moves in a year.

[5] I do not understand the relation of the year to two teeth. As already explained, seven is the number of teeth through which the smaller wheel moves in a year, over and above thirteen complete revolutions.

[6] As the section in parentheses stands, this does not make sense. What it should read, presumably, is: *in as many units of time as the numerator of the rational fraction.*

eius; aliam rotam invenire qui tractus ab illo[a] semel in anno circuit[b] zodiacum, qui est motus orbis Solis; vel qui semel in mense, qui est motus orbis Lune; vel semel in anno seu revolucione cuiuslibet[c] alterius planete, vel Capitis Draconis, et cuiuslibet orbis qui certo tempore[d] peragratur.[e]

Si vero per datam[f] rotam talis proporcionis in numero dencium non possit rota altera inveniri, eius[g] impossibilitatem[h] prompte poteris[i] deprehendere, et similiter econverso.

Age ergo: Accipe tempus revolucionis rote et vide quantum planeta movetur de zodiaco in illo tempore; et cum numero illius motus in gradibus et minutis[j] intrabis *Tabulam Proporcionis Zodiaci* divisi per quemcumque numerum infra 400; et cum ibidem[k] illum numerum inveneris in eius directo, invenies in linea numerum dencium quem oportet rotam tractam habere, qui debet trahi circa zodiacum in tempore revolucionis planete.

Quod si illum numerum in *Tabula*[l] *Proporcionis* invenire non poteris, scias quod non est possibile aliquam[m] rotam fieri dencium infra 400, que[n] tempore revolucionis planete per rotam datam trahi possit. Qualiter autem scietur quantum[o] planeta movetur[p] in toto tempore revolucionis date et divise per[q] partes eiusdem temporis racionales dicam tibi:

Accipe numerum denominacionis parcium racionalium per se, et numerum dencium per se, si fuerint infra integrum denominacionis, vel numerum qui[r] superabundat ab integro denominacionis per se, et cum illis duobus numeris intra[s] *Tabulam Proporcionis* distinctam per titulos. Videlicet, cum numero denominacionis intra tabulam consimilis tituli, et cum numero dencium si sit infra integrum, vel cum superhabundanti,

[a] C 4λ+que tracta ab illa [b] C circueat [c] C 7λ (haplog.) [d] C +zodiacum [e] C peragrat [f] C totam [g] C eiusdem [h] A in possibilitatem [i] A λ [j] A 4λ+et gradus et minuti [k] C λ (leaving a space) [l] C Tabulam [m] C aliam [n] AC qui [o] AC quod [p] C movet [q] A λ [r] C λ [s] C +in

wheel which, being driven by it, moves around the zodiac once in a year, which is the motus of the Sun's sphere, or once in a month, which is the motus of the Moon's sphere, or once in a year, or in the period of revolution of any other planet whatsoever, or of Caput Draconis, and of any sphere which is traversed in a definite time.

(If, in practice, it is not possible to find another wheel, by virtue of the given wheel's being of an unsuitable number of teeth, you will soon be able to detect this impossibility; and so conversely.)

Method: Take the time of revolution of the wheel, and see how much the planet moves through the zodiac in that time. Enter,[1] then, the *Table of Proportion of the Zodiac*[2] with the value of that movement, in degrees and minutes, it being possible to regard the zodiac as divided into any number of parts up to 400; and when you have found that number in its appropriate line, you will find on the line [of number] the number of teeth which the driven wheel (which is to be driven around the zodiac in the time of revolution of the planet) ought to have.[3]

If, however, in the *Table of Proportion*, you are unable to find that number expressing the planet's movement, you will know that it is not possible for another wheel to be made with fewer than 400 teeth and such that it may be driven by the given wheel in the time of revolution of the planet. On the other hand, just as it will be known how much the planet is moved in the entire period of a given revolution of the driving wheel, [that is], of the revolution divided into rational parts of that same time, I will instruct you as follows:

Take the denominator by itself, and the actual number of teeth,[4] if that number happens to be less than the denominator. (Otherwise take the number by which it exceeds the denominator.) With these two numbers, enter the *Table of Proportion*[5] in accordance with the headings. That is to say, with the number of the denominator, enter the table at the appropriate heading, and enter with the number of teeth in the line of number (if it be less than the denominator—otherwise enter with the

[1] I have kept, throughout, the archaic usage 'To enter a table with such and such a figure . . .'. The expression is still often used.

[2] The *Table of Proportion* to which he refers here (f. 181^v to f. 182^v and f. 84^r) has, as full title, *Table of Proportion, dividing a circle having 360 equal parts into any other number of parts, up to 400, for the wheels of a clock and the wheels of other instruments which are to be divided artificially.* It is simply a table giving the size of each tooth (expressed in degrees, minutes, seconds, thirds, and fourths) of a wheel divided into any number of teeth between 400 and 1. The line of number is the column (on the left of the tables) giving the number of teeth. The methods to be used in drawing up the table are given in Part 2, chapter 7, below.

[3] It is obviously being assumed that the planet's wheel is being driven by a single snail tooth on the axle of the first wheel.

[4] See the remark ending the synopsis of Prop. 5.

[5] He is now referring to a different *Table of Proportion*, which is to be found between f. 84^r and f. 88^r. The full title of this is *Table to discover what fractions (expressed in minutes, seconds, and thirds of a unit of 60 minutes) the numbers in the line of number are of the numbers in the headings.* In fact it allows us to express m/n as a sexagesimal fraction, where m/n is in its lowest terms, and $1 \leqslant m \leqslant 59$, and $1 < n \leqslant 59$. Actually the table given is not quite complete: it would have been unduly long had it been so.

intra in linea numeri; et in concursu invenies motum planete respondentem tempori revolucionis date.

Verbi gracia: Tempus revolucionis date est 20 tricesime sexte; igitur cum 20 | intrabis tabulam intitulatam titulo 36, et in concursu invenies id quod queris. Similiter si tempus revolucionis sit 30 vicesime quinte unius diei, cum 25 vicesime quinte sint integrum[a] racionalis denominacionis et superhabundant 5, cum illis 5 intrabis *Tabulam Proporcionis* sub titulo 25; et[b] concursu habebis quesitum.

Et nota bene[c] hanc conclusionem quia valet ad multa.

CONCLUSIO SEXTA[d]
(conversa precedentis)
[6]

Data rota quotlibet dencium, si ipsa possit trahi circa zodiacum in tempore revolucionis cuiuslibet[e] planete, per aliquam[f] rotam habentem aliquot dentes parcium racionalium revolucionis[g] diurne, et quot dencium erit rota eadem investigare.

Age ergo: Cum numero dencium rote date intra *Tabulam Proporcionis* incipientem a 400, et in eius directo accipe motum inventum in gradibus et minutis et cetera. Deinde cum illo numero graduum et minutorum intra tabulam medii motus[h] planete ad quem vis operari, quotiens oportet donec totum numerum inveneris signando dies et minuta diei secunda et tercia donec plene sciveris precisum tempus in quo planeta tot gradus et minuta et secunda et tercia peragratur de zodiaco per suum medium motum. Quo habito cum[i] minutis et secundis et terciis diei et ceteris[j] fraxionibus dumtaxat, intra *Tabulam Proporcionis* distinctam per titulos; et ubicumque inveneris illa minuta, secunda, tercia, cum ceteris fraxionibus, in tot dentes dividi debet rota trahens quot sunt

[a] C +sint integrum [b] C +in [c] C notanter
[d] A and C once again agree in the numbering of *conclusiones* [e] C cuius vis
[f] C aliam [g] C remancionis [h] A λ [i] C +gradibus [j] A λ

excess). At the point of meeting you will find the movement of the planet corresponding to the time of revolution given.[1]

Example: The period of the given revolution is 20 thirty-sixths of a day; thus with 20 enter the table headed 36, and at the point of meeting you will find what you are looking for.[2] Likewise, if the period of revolution be 30 twenty-fifths of a day, since 25 twenty-fifths are the integral part of the fraction, leaving behind 5 in the denominator, with that 5 enter the *Table of Proportion* under the heading 25; and at the point of meeting you will have what you were seeking.

Note this proposition carefully, for it is often useful.

PROPOSITION 6
(the converse of the last)

Given a wheel of any number of teeth, to investigate whether it be possible to drive that same wheel around the zodiac in the period of revolution of any one of the planets, by means of some other wheel having a number of teeth and having as period rational parts of the diurnal revolution;[3] and to find how many teeth that will have.

Method: Enter the *Table of Proportion*[4] with the number of teeth of the given wheel, beginning at 400, and from the appropriate line take the motus found, in degrees, minutes, and so on. Then, with that number of degrees and minutes, enter the table of the mean motus of the planet you are concerned with. Do this as often as is necessary, until you have found the entire number, by noting down the days, *minuta, secunda* and *tercia diei*[5] to the point where you know the precise time in which the planet traverses so many degrees, minutes, seconds, and thirds of the zodiac in its mean motion. When you have this, with the *minuta, secunda*, and *tercia diei* and the other fractions known exactly, enter the *Table of Proportion*,[6] under the several headings;[7] and when you have found those *minuta, secunda*, and *tercia* with the other fractions, the driving wheel ought to be divided into as many teeth as there are in the line of number,

1 I am unable to understand his meaning here, and the following examples only make the confusion worse. See commentary.

2 It is a curious fact that the author has chosen examples which will not be found in the *Table of Proportion*, since 20/36 and 5/25 are not in their lowest terms.

3 I can only make sense of this proposition if this is read 'having as its period an integral number of days'. See the comment at the end of the synopsis of Prop. 6. Notice that it is the *driven* (i.e. planet's) wheel of which the number of teeth is supposed known.

4 That is, the first of the two.

5 That is, sixtieths of a day, and so on. Retaining the Latin forms will help to prevent ambiguity later.

6 That is, the second *Table of Proportion*.

7 This free translation still fails to convey what I think was originally intended, namely 'search under the various headings, for the time just found, on each occasion looking only opposite the entry in the line of number corresponding to the whole number of days in which the driving wheel revolved once'.

in linea numeri sub illo titulo, vel in tot[a] quot resultant ex titulo et linea numeri si cum minutis, secundis, et terciis diei habueris unum diem integrum, vel ex titulo[b] bis aggregato si habueris duos dies.

Et si dictus numerus minutis et ceteris diei non poterit inveniri in *Tabula Proporcionis*, scias talem rotam tot dencium non posse moveri per aliquam rotam certe revolucionis parcium racionalium diei in integra revolucione illius planete. Et nota quod[c] ea que dico de rota habente certas partes revolucionis diei, idem debet intellegi de rota habenti certas partes racionales revolucionis cuiuslibet alterius temporis, visa igitur impossibilitate[d] illius,[e] accipiatur rota divisa in dentes alterius numeri, et operare sicut prius donec inveneris propositum destinatum.[f]

CONCLUSIO SEPTIMA

[7]

Habitis integris revolucionibus quibus[g] planeta certam partem zodiaci peragrat,[h] vel totum zodiacum et aliquot eius partes, numerum parcium racionabilium invenire in quot | dentes debet circulus dividi ad hoc quod deferat planetam in circulo signorum secundum eius medium motum, per alterum circulum divisum secundum numerum temporum integrorum.

Verbi gracia: Luna in 26 diebus pertransit secundum[i] eius medium motum [. . .],[j] et volo scire quot dentes habebit circulum ipsum deferens ad hoc quod moveat Lunam super zodiacum secundum eius medium motum continue.

Age ergo: Accipio residuum zodiaci non pertransitum quod est [. . .],[k] et per tabulas quero[l] in quanto tempore ipsum residuum peragratur, quod est [. . .].[m] Cum[n] ergo tempore intro in *Tabulam Proporcionis* distinctam

[a] C +dentes [b] C tali [c] A que [d] A in possibilitate [e] C +numeri
[f] C +et cetera [g] A quilibet [h] AC peragratur [i] C per
[j] A a space of approximately 17 letters length
[k] A a space of approximately 19 letters length [l] A quere
[m] A a space of approximately 28 letters length [n] C +illo

under that heading[1]—or, alternatively, as many as are together in that heading as the line of number, if you have a whole day in addition to the *minuta*, *secunda*, and *tercia*, and as many as are in the line of number together with twice as many as are in the heading, if you have two days.[2]

And if the said number of *minuta* and other parts of a day cannot be found in the *Table of Proportion*, you will know that a wheel with so many teeth cannot be driven by any wheel (assuming this to have a definite period of revolution of rational parts of a day)[3] in [the same period as that of] a complete revolution of the planet. (And note that what I am saying about a wheel with a period of revolution equal to definite parts of a day must be understood to apply equally to a wheel having a period of revolution of a definite number of rational parts of any other time whatsoever.)[4] Therefore, when the impossibility of working with this number is realized, a wheel divided into a different number of teeth may be taken, whereupon proceed as before, until you find what you are looking for.

PROPOSITION 7

Given that a planet traverses a definite part of the zodiac in an integral [number of days] or even [that it covers] the whole of the zodiac and some additional parts, to find the number of rational[5] parts in which a wheel should be divided up into teeth, so that it may carry the planet over the circle of signs in accordance with its mean motus, doing so by means of another wheel divided according to times.[6]

Example: In 26 days the Moon travels, in accordance with its mean motus, through [. . .].[7] I wish to know how many teeth to give to a wheel such that it will carry the Moon continuously over the zodiac, according to its mean motus.

Method: I take that part of the zodiac which is not traversed, that is to say [. . .], and I inquire of the tables how long it takes for this remainder to be traversed. This turns out to be [. . .]. Under the several headings I then enter the *Table of Proportions* with the time. I make a note of what

[1] I can only suppose that this is mistaken, and that it should read something like 'divided into as many teeth as are contained in the heading under which the time is eventually located'.

[2] I have restored the text as I think it was originally intended, but the method of proceeding where the length of time is greater than a day is quite erroneous.

[3] See the first footnote to this proposition.

[4] This is true in the abstract, but his method relies on the availability of tables giving the mean motus of the planet. These values were quoted in days and parts of a day, and thus the extension of the method to other time units (and hence to rational parts of a day, in the fullest sense) is a piece of wishful thinking.

[5] The word 'rational' could be omitted without loss of sense, or better still, altered to 'integral'. Or does *racionabilis* mean 'regular'?

[6] Unless this obscure phrase is explained satisfactorily, the problem itself cannot be understood. Is it simply a mistake for 'divided into an integral number of teeth'? In our commentary we suggest that it means 'moving an integral number of teeth in a day'.

[7] Numerical data are missing from this and the next three textual gaps.

per titulos, et signo titulum et lineam numeri, per quem titulum multiplico numerum 26 dierum integrorum, cui numero si addidero lineam numeri, ille erit numerus omnium dencium circuli[a] qui movetur[b] per rotam habentem dentes secundum numerum tituli.

Secundus casus: Item Luna in 30 diebus peragrat[c] totum zodiacum et ultra tot gradus [. . .],[d] cuius circuli porcionis quero tempus motus, cum quo ut prius ingredior *Tabulam Proporcionis* distinctam per titulos; et noto titulum in linea numeri, et cum titulo multiplico 30 de quo numero aufero lineam numeri et titulum quociens in tempore residui fuerit tempus integrum. Et quod remanet erit numerus dencium secure.[e]

CONCLUSIO OCTAVA

[8]

Habito certo tempore integro[f] quo planeta aliquam partem[g] zodiaci peragrat,[h] numerum certorum[i] dencium invenire quos[j] circulus ipsum deferens habebit ostendere.

Age ergo: Cum numero quem elegere[k] volueris, intra[l] *Tabulam Proporcionis* incipientis[m] a 400, et[n] in directo illius numeri electi accipe gradus, minuta, secunda, tercia, quarta, que tociens accipe dummodo[o] adequetur motui certe revolucionis date. Quo facto, nota quociens ipsum numerum[p] multiplicasti, quo in tempore dato tot dentes movebit de numero electo, quociens[q] et numerus contra illum acceptus multiplicatur.

Quod si numerus ille per suam multiplicacionem nunquam precise reddere voluerit partem zodiaci datam, scias quod numerus electus est inutilis. Elige ergo alterum donec invenies[r] numerum qui aliquociens acceptus reddit partem zodiaci datam precise. Et nota[s] quod si ille numerus propter eius grossiciem non vult reddere partem zodiaci datam | potes multiplicare loco suo quamlibet partem eius aliquotam[t] donec inveneris propositum.

Et tunc secundum quartam conclusionem,[u] sicut in numero integrorum accipit tot dentes, sic in partibus integri eiusdem denominacionis movebit unum dentem.[v]

[a] C circulo [b] C movebitur [c] C peragratur
[d] A space of approximately 19 letters length [e] C +Nota bene [f] A in tegno
[g] A partitem [h] AC peragratur [i] AC certarum [j] A quod [k] C eligere
[l] C +in [m] C incipiens [n] A +que [o] C donec [p] A +quod
[q] C λ [r] C inveneris [s] A λ [t] C aliquociens [u] A conclusioni
[v] C +Nota bene (Although there is no obvious evidence of a lacuna in A, C follows with a ninth *conclusio* beginning 'Habito certo numero'. Such a passage is to be found in A at f. 113^v, preceded by a long section which is nowhere in C. The *Ars componendi tabulam diversitatis aspectus in quovis climate* at f. 112^r must be irrelevant, but the passage beginning 'Memorandum quod si titulus et numerus dencium' looks like the second part of a *conclusio*. The new passage does not echo the contents of propos. 8, and is numbered separately. I take the following section to begin at 'Si unus circulus', f. 112^v.)

is in the line of number and the heading, and multiply the latter by 26, the number of whole days. If now I add to this number what I obtain from the line of number, I shall get the total number of teeth of a wheel which is moved by a wheel having as many teeth as are in the heading.[1] *Second example*: In 30 days the Moon traverses the entire zodiac and [. . .] degrees more. I now inquire as to the time which the Moon takes to move over this last part of a circle. With this, as before, I enter the *Table of Proportion*, following the headings; and I make a note of the heading and what is in the line of number. I multiply the heading by 30, and from the product I subtract what was in the line of number (and the heading as often as the whole time was contained in the time of the residue).[2] And what remains can safely be taken as the number of teeth.

PROPOSITION 8

Given a definite interval of time in which a planet traverses some part of the zodiac, to find the fixed number of teeth which the wheel carrying the planet will be required to have.

Method: Starting at 400, enter the first *Table of Proportion* with the number you choose, and from the line of the chosen number take the number of degrees, minutes, seconds, thirds, and fourths; and take them as many times over as is necessary to make up the movement of the given definite [time of] revolution. When this is done, note how often you have multiplied that number, how many of the teeth, whose number you hit upon, the driving wheel will move in the given time, and how often the number opposite that number of teeth is multiplied.[3]

But if that number, in its being multiplied, is never willing to make up exactly the given part of the zodiac, you will realize that the number chosen is of no use. In this case select another number, until you find a number which, [divided into 360° and] taken an integral number of times over, reproduces exactly the given part of the zodiac. And notice that if the number, on account of its size, is unwilling to produce the given part of the zodiac, you can multiply in its place any aliquot part of it a number of times, until you find what you are looking for.[4]

Lastly, by Proposition 4, just as in a whole number of days the device moves a definite number of teeth, so it will move one tooth in a rational fraction of that time.[5]

[1] Perhaps rather 'moving as many teeth daily as are in the heading'. MS. C has 'having such a number of teeth'. The extreme vagueness of the proposition is again evident here. See the commentary for a possible reconstruction of the writer's meaning.

[2] The section within brackets appears to be superfluous. See our commentary.

[3] The last part of the sentence seems merely to repeat the first half.

[4] This takes the reading of MS. A, and assumes that the author took *aliquotus* as a declinable adjective. The alternative reading is 'you can multiply in its place any part of it a number of times, until . . .'. Both readings make sense.

[5] This is a free translation of a very obscure sentence. Not only that, but the text so restored leaves much to the imagination. See the commentary.

CONCLUSIO NONA[a]

[9]

Memorandum quod si titulus et numerus dencium possint dividi per aliquid quod est commune[b] ad 60, tunc quantumcumque utrumque[c] minimus dividatur numerus dencium. Nux iuxta centrum manebit immutata,[d] et erunt illi dentes pars[e] diei eiusdem denominacionis, cuius est numerus quo dividitur titulus et summa dencium in commune.[f]

Probacio: Supponamus quod nux prope centrum habeat 28 dentes, vel dies, et titulus [35],[g] et summa dencium 725. Diviso titulo per [5],[h] est quociens 7. Item, diviso 725 per 5, est quociens 145. Dico ergo quod si nux tituli habeat 7 dentes, et summa circuli sit 145, quod numerus dencium nucis in centro erit 28 quinte unius diei, quia illud est pars | quinta 28 dierum, cuius est probacio: dividatur 28 per 5 et est quociens 5 et remanet 3, quod multiplica per 60 [et] sunt 180; quo iterum diviso per 5, est quociens 36, que sunt 3 quinte unius diei. Sed si quolibet 5 dierum accipiamus 5 quintas, fiunt 25 quinte, quibus additis tribus quintis diei, contentis in 36 minuta, fiunt 28 quinte. Patet ergo quod que est proporcio 28 dierum ad 725 dentes, illa est proporcio 28 quintarum unius diei ad 145 dentes, que similiter est quinta pars de 725. Et hoc est propositum.

[CONCLUSIO DECIMA]

[10]

Si unus circulus in certis eius revolucionibus alterius circuli certam porcionem circumducere debeat, utriusque circuli numerum dencium invenire.

Quod si volueris, accipe tempus unius revolucionis circuli pertrahentis, et vide quanta circumducti;[i] et per illam porcionem circuli tracti divide totum circulum 360 graduum, et numerum quociens nota. Deinde, quod remanet per 60 multiplica, et quod surgit per[j] habitam circuli porcionem divide, et numerum quociens serva. Et si quid ultra remanserit post divisionem,[k] per 60 multiplica, et collectum[l] divide sicut prius; et sic deinceps. Deinde vide quota porcio sit secundum quociens, et

[a] The section is in A only. [b] A cōe [c] A +ad [d] A in mutatus
[e] The passage could have been changed to 'et movebuntur illi dentes in numero parcium diei . . . cuius . . .'. [f] A cōe [g] A 45 [h] A 51 [i] A circumduci
[j] A proprius [k] A +ut [l] A collectui

PROPOSITION 9

It is to be borne in mind that if the heading[1] and the number of teeth may be divided by some number which is a factor of 60,[2] then the number of teeth of [driving and driven wheels is to be alike] divided by that factor. The nut[3] at the centre will remain unchanged, [but] the teeth of the [driving and driven wheels] will have as denominator [the factor aforementioned].[4]

Proof: Let us suppose that the nut nearest the centre has 28 teeth, or days,[5] that the heading is 35, and the total of teeth in the driving wheel 725. The heading being divided by 5, the quotient is 7. Next, 725 being divided by 5, the quotient is 145. I maintain, therefore, that if the nut [contains as factor(?)] the 7 teeth of the heading,[6] and the total in the circle be 145, then the number of teeth of the nut in the centre will be 28 fifths of a day, since that is a fifth part of 28 days. A check on this result is: 28 is divided by 5, the quotient being 5 and the remainder 3. Multiply this by 60, giving 180.[7] This being next divided by 5, the quotient is 36 *minuta diei*, which are three-fifths of a day. But if each of 5 days is divided up into five fifths, 25 fifths result, and adding these to three-fifths of a day, equivalent to 36 *minuta diei*, gives 28 fifths. It is clear, therefore, that whatever is the proportion of 28 days to 725 teeth, that is the proportion of 28 fifths of a day to 145 teeth, which likewise is one-fifth part of 725. And that is what was proposed.

[PROPOSITION 10]

In a certain number of revolutions, a wheel is to drive another wheel through a definite part of a revolution.[8] To find the number of teeth of each wheel.

Take the time of a single revolution of the driving wheel, and see by how much the driven wheel is to turn in that time.[9] Divide the answer into the complete circle of 360 degrees and note down the quotient. Next, multiply what remains by 60, and divide the answer by the given portion of the circle, keeping the quotient.[10] If there is any remainder after the division, multiply by 60, and divide the product just as before: and continue in this way. Next, see how big the second quotient is, and

[1] *titulus*. This must mean the heading of a table used in an earlier part of the section with which this was originally written.

[2] A purely conjectural interpretation of 'may be divided by some number *quod est commune* (?) *ad 60*'.

[3] See glossary. The use of the word to denote a small gear is still not entirely unknown in horological circles.

[4] This translation is purely conjectural.

[5] This apparently means that it rotates in twenty-eight days, at the rate of one tooth per day.

[6] Again conjectural. But 'if the nut of the heading has 7 teeth' would be unintelligible.

[7] This is to convert from days to *minuta diei*.

[8] *porcio*.

[9] From the tables of planetary motion—by implication.

[10] He has omitted to point out that the angle is to be converted to seconds before dividing.

tercium quociens de 60, et in tot partes quot[a] de circulo secundo remanserint, divide; et ille erit numerus dencium eiusdem porcionis. Porcio vero prima tot dentes continebit equales prioribus quota fuerit denominacio illius porcionis de 60, et si per numerum illorum dencium multiplicaveris primum quociens, et addideris illi numero numerum dencium prime porcionis invente, totum collectum dicet tibi quot dentes tractus circulus continebit.

Verbi gracia: Volo scire quot dentes circulus Solis habere debeat, ad hoc quod semel in anno circueat[b] circulum signorum, qui semel in die circumvolvitur circa polum mundi. Ad quod sciendum, intrabo in tabulas cum certo numero dierum, videlicet cum tribus, vel 6, vel quoquam[c] alio numero motus diurni, et extraho medium motum Solis in directo tot dierum—qui, exempli gracia, contra tres dies sunt 2 gradus 57 minuta et 24 secunda, cum paucis aliis fraccionibus, de quibus ad presens non curo; quod, reductum ad secunda, erit 10,644, qui sit *C*; per quod divido totum circulum 360 graduum. Similiter reductum ad secunda erit 1,296,000. Et exibit unus numerus quociens 121, qui sit *A*; et remanet porcio circuli 8,076,[d] qui sit *B*. Volo ergo scire quot dentes[e] porcio *B* habebit, equales dentibus prioris, quem[f] pertransit Sol in 3 diebus, scilicet *C*; ad quod sciendum, multiplica porcionem *B* in 60, et quod surgit divido sicut prius per porcionem *C*. Et exibit numerus quociens 45, et remanent tercia[g] 5,580, de quibus ad presens non curo. Dico ergo quod est porcio 60 ad 45 illa est proporcio porcionis *C* ad porcionem *B*.[h] | Quia[i] 60 continet 4 tercias de 45, ergo proporcio *C* ad *B* est sicut proporcio 4 ad 3, et, per consequens, si porcio *C* dividatur in 4 dentes equales, porcio *B* potest dividi in 3 dentes aliis coequales.[j] Sed totus circulus propositus continet 121 porciones *A*, inter se equales ipsi *C*; ergo si *A* multiplicetur per 4, numerus surgens erit numerus dencium omnium illarum porcionum equalium, scilicet 484, quibus si addantur 3 dentes porcionis *B*, erit numerus dencium tocius circuli motus Solis 487. Sed proprie[k] probata porcio *C* fuit motus Solis 3 dierum, ergo circulus motus diurni, qui est circulus circuli signorum, trahit in tribus diebus, de dentibus circuli Solis, 4 dentes; ergo in 363 diebus trahit 484 dentes circuli orbis Solis. Item, sicut in tribus diebus trahit 4 dentes, sic in uno die et dimidio trahit duos dentes. Et, per consequens, in 365 diebus et 6 horis trahet omnes 487 dentes; ergo in uno anno solari circulus motus diurni precise circumducit circulum motus Solis divisum in 487 dentes equales. Et hoc est propositum.

[a] A quod [b] A circuiat [c] A quovis [d] See the note to the translation
[e] A denteus [f] *angulum* is presumably understood [g] A tercie
[h] Letter rubbed [i] A scilicet [j] A quoequales [k] A proprius

the third quotient, and divide [the wheel] into as many parts as were left over from the second circle. This will be the number of teeth of the same portion. But the first portion will contain as many teeth equal to the first as is the denominator of that portion [turned into sixtieths], and if you multiply by the number of those teeth the first quotient, and add to that number the number of teeth of the first portion found, the whole result will tell you how many teeth the driven wheel contains.

Example: I want to know how many teeth the wheel of the Sun should have, in order that it go round the circle of signs once in a year, this circle revolving round the pole of the world just once in a day. To learn this, enter the tables with the number of days which the driving wheel is known to take for one revolution, namely with 3, or 6, or any other number for the daily motus, and extract the mean motus of the Sun on the line opposite so many days. For instance, opposite 3 days[1] is to be found 2° 57′ 24″, with other small fractions, which do not concern me at present. Reducing this to seconds gives 10,644, which is to be called *C*. By this I divide the whole circle of 360°. When likewise reduced to seconds, this is 1,296,000. The quotient, which is to be called *A*, is 121, and the remaining portion of the circle, which is to be called *B*, is 8,076. I want, then, to know how many teeth the portion *B* will have equal to the teeth of the wheel as a whole, which the Sun traverses in 3 days, and this number is to be known as *C*. To work this out, multiply the portion *B* by 60, and, just as before, I divide what results by the portion *C*. The quotient is 45, and 5,580 thirds remain, which do not concern me for the moment. The ratio of 60 to 45 is the same as the ratio of *C* to *B*. Since 60 contains 4 thirds of 45, the ratio of *C* to *B* is the same as the ratio of 4 to 3. As a result, if the portion *C* be divided into 4 equal teeth, the portion *B* is to be divided into 3 teeth equal to the others. But the entire wheel proposed contains 121 portions (*A*), equal amongst themselves to *C*. Thus if *A* be multiplied by 4, the result will be the number of teeth in all those equal portions, namely 484. If the three teeth of the portion *B* be added to them, the number of teeth of the entire wheel of the Sun's motus will be 487. But the correctly estimated portion *C* was the motus of the Sun for 3 days, and therefore the wheel of the diurnal motus, which is the wheel of the circle of signs, in 3 days drives 4 teeth of the Sun's wheel. In 363 days it therefore drives 484 teeth of the wheel of the orb of the Sun. Next, just as in 3 days it drives 4 teeth, in $1\frac{1}{2}$ days it drives 2 teeth. As a consequence, in 365 days and 6 hours it drives all 487 teeth; and therefore in one solar year thewheel with a diurnal movement drives round the wheel of the Sun's motus, divide into 487 teeth. And that is the proposition.

1 In the tables of mean motus for the Sun.

[CONCLUSIO UNDECIMA]

[11]

Habito certo numero electo, secundum quem volueris zodiacum dividere, quod in eius directo inveneris secundum quantitatem unius dentis exscribe,[a] cui adde partem eius sexagesimam continue tociens donec adequetur alicui numero medii motus in tabula medii motus planete. Quo facto, vide quociens ipsam partem sexagesimam addidisti, et certus est quod in tanto tempore quantum fuerit in linea numeri, medii motus circulus dentatus movebit unum dentem de numero electo. Et tot sexagesimam alterius dentis quociens partem sexagesimam addidisti [...][b] et per consequens in tanto tempore sexagies[c] accepto movebit 60[d] dentes et tot quociens pars sexagesima addebatur.

Si vero per talem addicionem partis sexagesime non vult numerus ille equari alicui numero medii motus, vide per quantum tempus deficit a concordia unius cum altera, et cum fraccionibus tocius temporis preter integra[e] intra in *Tabulam Proporcionis* distinctam per titulos. Et cum ibidem similiter tempus in fraccionibus inveneris, scias quod in tot integris et in tot partibus racionabilibus quot sunt in[f] linea numeri sub titulo movebit illum dentem cum suis partibus sexagesimis additis.

Et per consequens, secundum conclusionem quartam premissam, in tot diebus vel integris movebit tot dentes quot sunt sexagesima cum[g] partibus sexagesimis aliquociens additis;[h] ut si duodecies[i] additur movebit 60 et 12 dentes, scilicet 72.

Et nota hoc pro[j] opere subtilissimo.[k]

[a] A quam scribit [b] The sense suggests a substantial lacuna, in A and C.
[c] A form of *sexagesies*. [d] C 67 (?) [e] A 2λ [f] A λ [g] C in [h] A λ
[i] A duo dentes [j] A per [k] C +et cetera

[PROPOSITION II][1]

Having chosen some definite number according to which you wish to divide up the zodiac,[2] which number you will find on the line in the first *Table of Proportion* according to the size of a single tooth, there written down, add to that part of the zodiac a sixtieth part,[3] and do so repeatedly, until a figure is reached which equals some value of the mean motus in the table of mean motus for the planet in question. When this is done, see how many times you have added that sixtieth part, and you will know that the toothed wheel of the mean motus moves one tooth of the chosen number in as many units of time as were in the line of number. As many sixtieths of a tooth as the number of sixtieths added, [and one further tooth, together make up the figure obtained for the mean motus of the planet].[4] Consequently, in the time it takes the planet to move sixty times the mean motus arrived at above, the wheel will move sixty teeth, and as many more as the number of sixtieth parts added.

If, indeed, the number refuses to be equated with any value of the mean motus by the addition of sixtieth parts, see what the time difference is between the one and the other,[5] and with the excess time expressed as a fraction of the whole, enter the second *Table of Proportion*, under the heading.[6] And when you have expressed the excess time as a fraction, you may be sure that the wheel moves one tooth, and its added sixtieth parts, in as many whole days as the planet takes, together with the number of rational parts of the same quantity as are indicated by the line of number and heading conjointly.[7]

Consequently,[8] according to the fourth proposition above,[9] in as many days as the planet takes, or units [of time], the wheel will move as many teeth as there are sixtieths added. Thus when 12 sixtieths are added, it will move 60 and 12 teeth, that is to say, 72.

And note this result for the construction of the most subtle [sorts of] device.

[1] It appears that there might be some heading or preamble missing. The section is in MS. A only.

[2] The sense later requires that the wheel of mean motus be divided into *n* parts. Is *zodiacum* in error for *rotam*?

[3] Bear in mind that division by sixty is particularly easy in the sexagesimal system.

[4] This long restoration seems to be required by the trend of the argument.

[5] This could mean a number of things. Narrowing these down to two, I suggest that the sentence might also have read 'see how long it takes the planet (or wheel?) to traverse the amount by which the mean motus is exceeded'. Reasons for thinking both unsatisfactory, however, are given in the commentary below.

[6] That is to say, using the tables inversely, the problem being to determine an integer (which will occur in the line of number, when found) which divides the number in one of the headings, giving the fraction found.

[7] For comment on this passage see the synopsis and commentary in vol. ii. There might be a passage missing.

[8] Here we have a very elementary worked example, which is an illustration of the first, rather than the second, paragraph.

[9] I cannot see any reason for citing this proposition in a worked example.

[CONCLUSIO DUODECIMA]

[12]

Ad componendum verum motum Solis in rotis. Accipe motum Solis diurnum, qui est 59 minuta 8 secunda 19 tercia 37 quarta secundum Alfunsum, et divide eum secundum quemcumque numerum volueris, et quociens erit tota pars diei motus diurni quota est ille numerus divisor. Deinde cum illo numero intrabis in *Tabulam Proporcionis* vel cum duplo, vel triplo, quadruplo, quintuplo, et sic de ceteris, usque ad complecionem numeri divisoris. Et si in *Tabula Proporcionis* inveneris numerum cum quo intravisti,[a] scias quod numerus quem in linea numeri inveneris in eius directo in *Tabula Proporcionis* erit numerus dencium tocius circuli; et quantitas unius dentis erit numerus quociens qui in capite tabule ponitur.

Et si forte non possis invenire numerum cum quo velles intrare in *Tabula Proporcionis*, tunc cuilibet illorum numerorum[b] adde integrum motum Solis diei diurnum, et fac inde tabulam et cum toto collecto et singulis intra iterum *Tabulam Proporcionis*. Et si consimilem numerum inveneris, scias quod linea numeri in eius directo ostendet tibi numerum dencium rote tue.

Verbi gracia: Vellem scire quis numerus dencium divideret equaliter rotam[c] ad motum Solis diurnum. Ad quod faciendum divide motum Solis diurnum per 58, et exibit numerus quociens 0 gradus 1 minuta 1 secunda 10 tercia 41 quarta, de qua radice constituo tabulam expansam usque ad 58 expansas lineas numeri. Et cum singulis numeris in *Tabulam Proporcionis* intravi, et quia per tabule introitum non invenero aliquem numerum equalem alicui illorum in *Tabula Proporcionis*, expando[d] unam aliam tabulam 58 linearum addendo singulis numeris

[a] A intrasti [b] A numerum [c] A rotum [d] A expanda

[PROPOSITION 12]

To represent the true motus of the Sun by means of wheels:[1]

Take the Sun's diurnal motus, which according to Alfonso is 59′ 8″ 19‴ 37^{iv}, and divide it by any number you like.[2] The answer is that part of a circle which results when the circle is divided by a number we shall here call the 'divisor'.[3] Next, enter the first *Table of Proportion* with that answer, or double, or three, or four, or five times it, and so on, until you find a number opposite some divisor. If, in the *Table of Proportion*, you find the number with which you entered, you will know that the number in the line of number will be the number of teeth required for the whole circle,[4] and the size of a single tooth will be the number at the head of the table.[5]

If by chance you are not able to find in the *Table of Proportion* the number with which you wanted to enter, then to any of those numbers,[6] add the whole diurnal motus of the Sun. Next make a table, and with the individual multiples [of the original sexagesimal fraction], and with these figures [added on to the diurnal motus], enter the *Table of Proportion* once more. As before, if you find the number, you know that the line of number shows you the number of teeth of your wheel, on the same line.[7]

Example: If I wanted to know what number of teeth should divide a wheel equally to represent the diurnal motus of the Sun I should act as follows. Divide the diurnal motus of the Sun by 58, yielding the quotient 0° 1′ 1″ 10‴ 41^{iv}. With this radix I draw up an expanded table, with up to 58 'lines of number' set out. I entered the first *Table of Proportion* with the individual numbers, and because, in entering the table of submultiples which I compiled, I did not find any number equal to any in the first *Table of Proportion*, I drew up another table of 58 lines, the lines being lines of the table already spoken of, increased by the whole

[1] The substance of the following passage is repeated in Part II, section 3, which, however, contains much else besides. The two sections are not arranged as parallel texts, as elsewhere, since here the emphasis is on the theory underlying the author's practice, whereas later it is practice which is of paramount importance.

[2] This 'any number you like' is the number of teeth on the wheel which moves at a constant rate, viz. one revolution per day. If we were indeed to use any number, this would mean adding an extra gear to the day-axle of the main clock. This might be avoided by using an existing gear with the right velocity, but the following trial-and-error procedure would then be jeopardized. In fact 'any you like' might merely indicate that this is indeed a trial-and-error method.

[3] A free translation, which seems to give the sense correctly. The 'divisor' is the number of teeth on a wheel which moves with the Sun's mean motus.

[4] The 'whole circle' is that referred to in the last note.

[5] The author envisages writing down the original sexagesimal fraction of the daily motus, and successively adding it to itself, to double itself and so on, thus making a table. This section will be made clearer by referring to the actual tables of Part II, section 3.

[6] That is, the multiples of the smallest sexagesimal fraction.

[7] Without the following examples it would scarcely have been possible for a contemporary to have understood this section. The fact is that the line of number does not give the number we are really anxious to know.

supradictis integrum motum Solis diurnum. Deinde cum singulis eorum intravi *Tabulam Proporcionis* quousque invenero numerum equalem illi numero cum quo intravero, et directe contra 6 in linea numeri tabule expanse per 58 continentur 1 grad. 5 min. 15 sec. 23 ter. 40 quar. Et in *Tabula Proporcionis* contra lineam numeri 331 invenitur idem numerus. Scio ergo quod si rota divisa fuerit per 331 equaliter mensurabit motum Solis diurnum in perpetuum. Quia si dividas zodiacum per 331 idem numerus exibit super quam racionem calculatur | *Tabula Proporcionis*. Et consimiliter reperies in subduplis numeris, vel subtriplis, et sic de ceteris abbreviando dentes, vel in duplis, vel in quadruplis, et sic de aliis multiplicando dentes.

Verbi gracia: In effectum ponendo in opere facio unam rotam que complet motum suum precise in tempore diurno, et circa centrum eius figo unum circulum divisum in 29 dentes cui adiungo unum alium circulum mobilem habentem 29 dentes et tres, quod est subduplum ad 58 [et] 6. Quando igitur circulus motus diurni perfecerit revolucionem suam revolvet secum totum circulum 32 dencium mediante circulo fixo habente 29 dentes tantum exceptis tribus dentibus de illis 32, quos cum secum attraxerit tunc perficiet illum circulum 32 dencium in parte revolucionis sue nove. Et per consequens oportet circulum motus diurni complere revolucionem suam et tres dentes[a] nove revolucionis, antequam semel revolverit integrum habentem 32 dentes, qui mediante nuce sua in eius centro confixa tantum unum dentem trahet de circulo diviso in 331 dentes.

Eodem modo per omnia facies de motibus aliorum omnium planetarum.

[CONCLUSIO TERCIADECIMA]

[3]

Nota pro *Tabula Proporcionis* in simplici, si in aliquo tempore parciali motus rote principalis moveat unum dentem de rota planete, ad sciendum in quot temporibus integris moveat quot dentes tali utendum regula:

Quod si in 3 quartis unius diei moveat unum dentem, tunc in tribus diebus movebit 4 dentes. Similiter, si in 5 octavis unius diei moveat unum dentem, per 5 dies movebit 8 dentes; et sic de omnibus aliis numeris.[b]

[a] A +dentes   [b] A numerus

of the Sun's diurnal motus. I then entered the first *Table of Proportion* with the individual numbers of the newly formed table until finding one of them there,[1] and this was opposite 6 in the line of number of the original table of 58 lines, namely $1^\circ\ 5'\ 15''\ 23'''\ 40^{iv}$.[2] In the first *Table of Proportion*, opposite 331 in the line of number, the same number is found.[3] I know, therefore, that if the wheel is divided uniformly into 331 parts, it will measure the diurnal motus of the Sun perpetually.[4] This is the case because, if you divide the zodiac by 331, the *Table of Proportion* gives the answer calculated above. You will find an exactly similar number if you divide both sets of teeth by two, three, and so on, or if you double them, triple them, and so on.[5]

Example [continued]: Putting my calculations into effect, I make a wheel which moves round exactly once per day, and about its centre I make a circle of 29 teeth. To this I annex another wheel of 29 plus 3 teeth, that is to say, a half of 58 and 6. When, therefore, the circle with a diurnal movement performs its revolution, it moves by means of the attached circle of 29 teeth the whole circle of 32 teeth apart from 3 of these 32, which are moved during the next revolution. In consequence, the circle of diurnal movement must complete its revolution and three teeth of the new revolution, before revolving once the wheel of 32 teeth, this latter being in the middle of the nut fixed at its centre, and driving in each revolution one whole tooth of the circle divided into 331 teeth.[6]

You proceed in the same way for the motions of all the other planets.

[PROPOSITION 13]

Using the simple *Table of Proportion*, knowing that in some part of the time of movement of the main wheel it moves one tooth of a planet's wheel, in order to discover in what times it will move just how many teeth, such a rule is to be used:

If in three quarters of a day it moves 1 tooth, then in 3 days it will move 4 teeth. Likewise, if in five-eighths of a day it moves 1 tooth, in 5 days it will move 8 teeth; and so on for all other numbers.

[1] Literally: until finding a number equal to the number with which I enter.

[2] He is, in effect, saying that $1\frac{6}{58} \times 59'\ 8''\ 19'''\ 37^{iv}$ is equal to $360^\circ/331$. In fact, the first is equal to $1^\circ\ 5'\ 15''\ 23'''\ 43^{iv}$, and the second is $1^\circ\ 5'\ 15''\ 24'''\ 28^{iv}$—an extremely good fit. (Notice that the entry for $360^\circ/331$ in the *Tables of Proportion* is fully accurate.)

[3] Not strictly true. See the preceding footnote.

[4] The text probably does not differ much here from the original, but it certainly leaves much unspoken. What we want to know is *how to drive* this wheel, or how to use it to drive another, which in turn measures out the Sun's motion. In fact the example continues, although this point is only made clearer by amplification of the text, at the end of this section. For an explanation, see the commentary.

[5] He is simply making the point that it is gear *ratios* which matter, and not the actual numbers of teeth.

[6] In brief: the 29-tooth wheel revolves once per day, it is meshed with a 32-tooth wheel, on whose axle is a single tooth, which advances in each of its revolutions 1 tooth of the 331-tooth wheel, on yet a third axle. This last wheel represents the Sun's mean motus. See also Part II, section 4, for a slightly different arrangement.

[CONCLUSIO QUARTADECIMA]

[14]

Item pro *Tabula Proporcionis* in composicione[a] tali utendum[b] est regula: Si in uno die et 3 quartis moveat unum dentem, ergo in 4 diebus movebit 7 dentes, quia 4 et tres faciunt 7 similiter. Si in uno die et octavis unius diei moveat unum dentem, ergo in 8 diebus movebit 11 dentes, quia 3 et 8 faciunt 11; et [sic] de aliis.

Nota pro regula brevissima ad cognoscendum precisum tempus in quo aliquis planeta movetur unum dentem de dentibus equalibus in sua rota divisis tali utendum[c] est arte:

Primo scias numerum illorum dencium quotlibet fuerint infra 400, et cum illo numero intra in *Tabulam Prime Proporcionis*, et quod in eius directo inveneris de gradibus, minutis, secundis, terciis, et quarcis, accipe, et illa est | quantitas unius dentis. Deinde vide in quanto tempore ille planeta movetur tot gradibus, minutis, secundis, terciis, quarcis, et istud tempus serva in minutis diei, secundis, terciis, et quarcis diei, cum quo tempore intrabis *Tabulam Secunde*[d] *Proporcionis* (intra minutis, secundis, terciis, quarcis),[e] et in quacumque tabula inveneris illum numerum temporis, scias quod titulus eiusdem tabule superius et linea numeri a dextris dicet[f] tibi in quanto tempore parcium vulgarum planeta pertransit unum dentem, prout satis docetur in canone supradicto.[g]

[CONCLUSIO QUINTADECIMA]

[15]

Quomodo autem invenitur illud tempus quod iubet[ur] servare, videlicet in minutis diei, secundis, terciis, et quarcis,[h] scias hoc modo: Cum illo numero prius invento in *Tabula Prime Proporcionis*, contra numerum dencium equalium,[i] intrabis in tabulam medii motus planete, factam ad modum tabularum Alfunsi, et ubicumque inveneris illum numerum, in gradus, minuta, secunda, et cetera,[j] si precise inveniri poterit in eius directo, in linea numeri est tempus quesitum in minutis et secundis diei. Si vero precise illum numerum non inveneris, tunc accipe minorem propiorem, tamen quem extrahe de gradibus et minutis tuis in *Tabula Prime Proporcionis*, et nota in lineam numeri medii motus planete quid est, aut minuta[k] diei, vel secunda, secundum communem

[a] A composicionis [b] A utandum [c] A utandum [d] A Secundo de
[e] The bracketed phrase appears to be redundant [f] From *dicent*? [g] A supradictum
[h] Is there a substantial lacuna here? [i] A begins a new sentence here.
[j] A begins another sentence here. [k] A minuti

[PROPOSITION 14]

Next, for the compound *Table of Proportion*, a rule like this is to be followed:[1]

If one tooth moves in a day and three quarters, then in 4 days 7 teeth will move, since 4 and 3 make 7. If one tooth moves in a day and an eighth, in 8 days 11 teeth will move, since 3 and 8 make 11; and so in general.

For the shortest rule to learn the exact time it takes any of the planets to move one tooth of the equal teeth into which its wheel is divided, a method like this is to be used:

First, knowing how many teeth there are in the planet's wheel, this being fewer than 400, enter the first *Table of Proportion*[2] with that number. Whatever you find, on the line of the number, in degrees, minutes, seconds, thirds, and fourths, will be the size of a single tooth. Next find out in how much time that planet moves so many degrees, minutes, seconds, thirds, and fourths. With this time, in *minuta*, *secunda*, *tercia*, and *quarta diei*, enter the second *Table of Proportion*,[3] and in whichever table you find the time, you may know that the title above the table and the line of number on the right will give you the time (as a vulgar fraction) for the planet to make one tooth—in keeping with what is taught in the canon above.

[PROPOSITION 15[4]]

This is how you may go about finding the time in *minuta*, *secunda*, *tercia*, and *quarta diei* [which a planet takes to cover an angular distance of one tooth of a wheel divided uniformly into any given number of teeth]:

With the number already found in the first *Table of Proportion* opposite the number of equal teeth, enter the table of mean motus of the planet, this table being drawn up after the fashion of the Alfonsine tables. Wherever you find the number, in degrees, minutes, seconds, and so on, if you find it exactly on some line or other, then in the line of number of the planetary tables is the time sought in *minuta* and *secunda diei*. If you fail to find the number exactly, then take the next lesser number, this being taken out of your number of degrees, minutes, etc., in the first *Table of Proportion*; and note what there is in the line of number of the table of the mean motus of the planet, whether in *minuta* or *secunda diei*, in accordance with the common method of working with the Alfonsine

[1] *In composicione* (compound) is taken to be contrasted with *in simplici*. The latter indicates that simple fractions are being considered, the former that we are considering mixed, or compound fractions.

[2] Lit. *Table of First Proportion*. This is the first occasion on which the tables are distinguished, but hereafter the distinction is adhered to.

[3] Lit. *Table of Second Proportion*.

[4] Although made into a new section, what follows is merely an elaboration of the immediately preceding paragraph.

modum operandi per tabulas Alfunsi. Deinde quod remanserit post subtraccionem quere iterum in eiusdem tabulis vel minus eo quod subtrahes, ut prius de residuo; et non cessabis sic intrare cum residuo in tabulas medii motus planete donec consumpseris totum numerum quem primo invenisti in *Tabula Prime Proporcionis*. Et cum quolibet introitu notabis lineam numeri quid significat in minutis, secundis, terciis, et quarcis, diei, in quo planeta complet tantum motum quantum invenisti in *Tabula Prime Proporcionis* contra numerum dencium equalium rote diei; cum quo intrabis *Tabulam Secunde Proporcionis*, ut predictum est, et invenies id quod queris.

[CONCLUSIO SEXTADECIMA]

[16]

Ad inveniendum in quot dentes debet circulus dividi ad hoc quod in certa parte temporis [parcium vulgarum][a] certam eius partem vulgarem pertranseat, tali utendum arte:

Accipe partes vulgares temporis tui ut verbi gracia tres quintas unius diei, vel anni, vel cuiuslibet temporis quod per 60 potest dividi; et cum illo tempore intra | in *Tabulam Proporcionis* distinctam per titulos, sub denominacione temporis tui; et quod in directo linee numeri inveneris, de minutis, secundis, terciis, quarcis, extra scribe et vide quantum de zodiaco planeta movetur in tot fraccionibus temporis. Et illum motum bene nota, quem quere in eadem *Tabulam Proporcionis* distinctam per titulos, ita quod minuta *Tabule Proporcionis* predicte utaris pro gradibus, et secunda pro minutis, et sic de ceteris. Quem motum si sub aliquo titulo inveneris, scias quod unum signum[b] 60 graduum esset divisum per numerum tituli; tunc tempore dato moveret de illo signo tot dentes quota est linea numeri contra quem invenitur ille motus planete.

Si autem non poteris illum motum in tota *Tabula Proporcionis* invenire, adde sibi motum unius diei, vel anni, aut cuiuslibet alterius temporis cuius accepisti sexagesimam partem, et cum illo tempore aggregato intra

[a] See the translation. [b] A gradum

tables. Then look in the same tables for what was left after you had done the subtraction, or a part of this, as before; and continue to enter the table of mean motus of the planet with what was left over, until you have accounted for the entire number which you originally found in the first *Table of Proportion*. And whatever you began with, you must note the line of number which shows in how many *minuta*, *secunda*, *tercia*, and *quarta diei* the planet covers as great an arc[1] as you found in the first *Table of Proportion* against the number of equal teeth of the day-wheel.[2] With this time, enter the second *Table of Proportion*, as explained earlier, and you will find what you are looking for.[3]

[PROPOSITION 16]

This is the rule to be used to discover into how many teeth a wheel must be divided such that in a definite rational part of the time[4] the circle will turn through a definite rational part:

Take rational parts of your chosen time as, for example, three-fifths of a day, or of a year, or of any time which may be divided by 60; and with this time enter the second *Table of Proportion* in accordance with the headings, the headings corresponding to the denominator of your time fraction. Write down what you find on the line opposite the appropriate figure in the line of number, in minutes, seconds, thirds, and fourths, and see through how much of the zodiac the planet moves in this time. Make a careful note of the motus, and look it up in the *Table of Proportion*, following the headings, transposing it so that you treat the minutes of the aforesaid *Table of Proportion* as degrees, the seconds as minutes, and so on for the rest. If you find this motus under one of the headings, you may be sure that it is the answer to the division of one sign[5] of 60 degrees by the number in the heading; and then, in the given time, it should move, in that sign, as many teeth as the number in the line of number, opposite which was found the motus of the planet.

If, on the other hand, you are unable to find the motus in the entire *Table of Proportion*, add to it the motus for one day, or year, or any other time of which you know the sixtieth part, and with the aggregate time[6]

[1] Lit. 'completes as much motus'.

[2] For *rote diei*. Is the *diei* out of place here?

[3] This last sentence, which suggests that the author wants to express the time arrived at as a rational fraction of a day, appears to be totally disconnected from the rest of the section. It is, in fact, merely a reaffirmation of the last sentence of Proposition 14.

[4] The text adds 'of rational (vulgar) parts', but these words seem superfluous.

[5] Signs of 60 degrees were known as *signa physica*. as opposed to the *signa communia* of 30 degrees. (The emendation of the text at this point is confirmed by the next paragraph.)

[6] The sentence is put together very loosely. Thus 'with the aggregate time' (*cum illo tempore aggregato*) should rather read 'with the motus corresponding to the aggregate time'.

Tabulam Proporcionis sicut prius. Et si inveneris tantum motus contra aliquam lineam numeri, scias quod unum signum 60 graduum dividatur in tot dentes quot est titulus tabule sub quo invenisti motum tuum. Scias quod in tempore iam accepto ex aggregato movebit[a] tot dentes de illo signo quota est linea numeri contra quem invenitur ille motus. Et nota quod si cum motu tuo habueris signa, quociens habueris signa tociens aggregabis numerum tituli ad lineam numeri.

Si vero adhuc istum motum invenire non poteris in *Tabula Proporcionis*, adde sibi motum 2 dierum, vel tres, vel 4, vel quotlibet[b] volueris, donec habueris hanc proporcionem. Quod si volueris addere plures dies, elige tibi aliud tempus cum quo operare sicut prius.

Et nota quod si titulum sub quo invenisti medium motum tuum multiplicaris per 6, numerus surgens erit numerus dencium tocius circuli. Et unus circulus temporis iuxta axem habebit titulum temporis, et alius habebit lineam numeri si intrasti cum simplici, vel lineam numeri cum titulo tociens accepisti integrum tempus cum fraccionibus. Et nota quod fixus circulus semper habebit dentes secundum numerum tituli tantundem.

[PARS SECUNDA]

[II. 1]

Quantitas horologii pro sonitu unius cimbe. Dividatur rota pro ponderibus[c] dialis in 72 dentes[d] cuius centrum distabit a basi 13 dentes rote eiusdem,

[a] From *movebuntur*? [a] A quolibet
[c] A proponderis (See the note to the translation.)
[d] Not *dentibus*. Our authority for this expansion is from unabbreviated passages on f. 81^{v} and elsewhere.

enter the *Table of Proportion* as before. If you find that motus against some figure in the line of number, you may be sure that one sign of 60 degrees may be divided with as many teeth as are in the heading of the table under which you found your motus. You may be sure that in the time now taken in aggregate, as many teeth will move in that sign as the number in the line of number against which that motus was found. And note that if with your motus you have a number of signs, you must add to the line of number a corresponding multiple of the number in the heading.

If you are still unable to find the motus in the *Table of Proportion*, add to it the motus for two days, or three, or four, or any number you like, until you have the figure you are looking for. And when you wish to add more days, choose another time with which to operate, as before.

Lastly, notice that if you multiply by 6 the heading under which you found your mean motus, the number resulting will be the number of teeth of the whole circle.[1] And one time-wheel[2] next to the axis will have as many teeth as were in the heading[3] [where we were expressing the time as a sexagesimal], working with the simple [version of the *Table of Proportion*], and the other will have [as many as were indicated in] the line of number, together with the heading as many times over as you took the whole time with the fractions. And note that the enmeshed (?) wheel[4] will always have as many teeth as the number in the heading.[5]

[PART II]

[II. 1]

The quantities for a timepiece capable of ringing a bell.[6]

The wheel of the weight for the day movement[7] is divided into 72

[1] There being six signs of 60° (*signa physica*) in the circle. [2] *circulus temporis.*

[3] *habebit titulum temporis.* This last section is difficult to understand. It is discussed in the commentary.

[4] *fixus circulus*—presumably meaning the 'affixed' wheel.

[5] This sentence presumably repeats the beginning of the last.

[6] It appears to be taken for granted throughout that the reader knows the general plan of a striking clock. Note that *cimba* is not from the classical *cymba*, but from *cymbalum* meaning 'cymbal' or 'bell', as in Vitruvius (Gr. *κύμβαλον*). *MLWL2* has *cymba* meaning 'monastery bell or chime'. The relative dimensions of the bell recommended here are given at the end of the section. There is evidence in II. 3 that the words *clok* and *cimba* were interchangeable, countering arguments that the absolute dimensions of the bell were small (as the gongs or bells associated with early striking jacks). We deduce some probable dimensions for the St. Albans bell, at p. 272, vol. iii. In the 1360s Edward III's three royal clocks had three bells together weighing 13,228 lb., the bell of the Westminster clock alone accounting for 9,261 lb.

[7] *rota pro ponderis dialis*, the words as they appear in the manuscript, are difficult to translate. The English 'dial' is from the adj. formed from *dies* in the phrase *rota dialis*, or 'day wheel'. The Norwich rolls (see p. 316, vol. ii) show that *diale* was already a noun in horological parlance by 1324/5. We shall later see it used as an adjective of other nouns. *Ponderis* is gen. of *pondus* (weight). The *pro* is not superfluous, for a similar phrase (with the same spelling)

et corda sex dencium a media linea tocius operis, cuius axis in longitudinem infra tenonem continet cordam[a] 15 dencium. Finis axis rote eiusdem extra tenonem pro nuce dialis continet cordam 2 dencium 30 minutorum. Summa eiusdem nucis 4.[b] A tenone ex parte dialis ad medium rote ponderis continetur[c] corda 2 dencium. Longitudo calige dialis, cum capite, continet cordam 11 dencium. Latitudo laminis predicte calige continet cordam 10 dencium 40 minutorum.[d] Diameter rotarum strob continet cordam 21[e] dencium, cuius centrum distat a basi corda 13[f] dencium, et a media linea corda 6 dencium 25[g] minutorum. Longitudo pinnularum strob continet cordam[h] unius dentis 30 minutorum. Spacium inter rotas strob continet cordam 2[i] dencium 30 minutorum. A tenone ex parte dialis ad medium inter rotas strob continetur[j] corda 8 dencium. Summa dencium rotarum strob 30. Longitudo axis strob infra tenonum continet cordam 15 dencium. A

[a] A corda [b] A (margin) +vel 3 [c] A continet
[d] Both figures are over an erasure. [e] Over an erasure. See the translation.
[k] Possibly over an erasure. [g] The second digit is over an erasure. [h] A corda
[i] There is a rubbed word here, with a figure '2' replacing it in the margin.
[j] A continet

teeth. The centre of the wheel is separated from the base by a distance of 13 teeth of that same wheel, and is a chord of 6 teeth from the centre line of the entire device, the length of its arbor being a chord of 15 teeth beyond the tenon.[1] The end of the arbor of the same wheel, outside the tenon, for the nut with a daily motion, contains a chord of 2 teeth 30 minutes. The number of teeth on this wheel is 4.[2] From the tenon on the dial side[3] to the middle of the weight-wheel[4] is a chord of 2 teeth. The length of the barrel[5] which has the daily motion, together with its head, is a chord of 11 teeth. The breadth of the plate of the aforesaid barrel contains a chord of 10 teeth 40 minutes. The diameter of the strob wheels[6] contains a chord of 21[7] teeth. Its centre is a chord of 13 teeth from the base, and a chord of 6 teeth 25 minutes from the median line. The length of the strob pins[8] is a chord of 1 tooth 30 minutes. The space between the strob wheels contains a chord of 2[9] teeth 30 minutes. From the tenon on the dial side to the middle of the strob wheels is a chord of 8 teeth. The strob wheels have 30 teeth. The length of the strob arbor beyond the tenon contains a chord of 15 teeth.[10] From

occurs in connection with the striking movement. We therefore amend the following word to *ponderibus*.

[1] *infra tenonem. Tenon* is not in the usual dictionaries. *OED* supposes the English word 'tenon' to come from the French *tenir*, but our word has the appearance of an intermediary between the Latin *tenere* and the English word. At first sight the word of the text refers to that part of the axle (or 'arbor') which enters the bearing, but, considering the section as a whole, the tenons appear to be the pillars carrying the bearings. The name was taken, presumably, from the fact of their holding the bearings.

[2] Reading *Ssm*^{a} as *summa*. Note that a marginal gloss here suggests 3 as an alternative to 4.

[3] *ex parte dialis.*

[4] *rota ponderis.* Does this refer to the wheel of 72 teeth?

[5] *Caliga* must surely refer to the weight drum, which would slide over the axle, and which might then have a ratchet control for winding. See the remarks on the striking *caliga*, n. 11, p. 477 below.

[6] *rotarum strob.* The word *strob* is not in any dictionary I have seen. The best explanation I can offer was first suggested to me by Dr. Avraham Eran, of Jerusalem. He pointed out that according to the *glossarium* of L. Diefenbach, a *stroba* was a crossbow (*Armbrust*). The similarity between the crossbow and the verge of a clock, with its crosspiece for the weights, is striking. In the translation, 'strob' is treated as an English word. Apart from *stroba*, words with which it is cognate might include the following: Ger. *stob*, p.p. stieben, to start, or set suddenly in motion; O.E. *strop*, from Lat. *struppus*, strap, noose: Lat. *strobilus*, pine-cone, from Gr. στρόβιλος, in turn from a verb meaning 'to twist'. (The word 'pinion' comes, curiously enough, from a word for the same object.) The scribe is likely to have heard the word spoken, and is therefore unlikely to have miscopied from the obvious substitute 'strok' (M.E.). Our last possibility is that the word is a shortened form of 'stourbill', a stirred or disturbed rod. This word was used in the fifteenth century in connection with the Westminster clock, as Mr. Anthony Turner has pointed out to me. (See *OED* under 'Nut'.)

[7] The word is over an erasure.

[8] *Pinnulae* are the sighting vanes of an astrolabe or quadrant (class. Lat. for feathers, fins, or vanes). Our interpretation of the dimensions of the escapement, and of its general form, suggests that *pinnula* should be translated 'pin'. Certainly in II. 7 below, *pinnulum* (amended to *pinnulam*) is a word used of a cylindrical rod.

[9] Altered from a figure which was perhaps originally 3.

[10] Since 15 is the depth of the case, we are confirmed in the view that there is only a single pallet. Admittedly it is possible to bend a verge round the arbor of an escape wheel. Dondi drew it so, and a clock working in the Jagiellonian Museum at Cracow is built in this way.

tenone ex parte dialis ad medium nucis in axe strob continetur[a] corda 2 dencium. Diameter nucis predicte continet [corda] 2 dencium 40 minutorum. Spissitudo eiusdem nucis continet cordam unius dentis 20 minutorum. Summa dencium eiusdem nucis 8. In hasta, corda semicirculi strob ab angulo ad angulum continet cordam 3[b] dentis 20 minutorum. Semidiameter[c] semicirculi eiusdem continet cordam unius dentis 40 minutorum. Medietas haste strob pro ponderibus continet diametrum rote et cordam 12 dentis. Quadratura plumborum pro hasta strob continet cordam 6 dencium. Diameter rote dialis continet diametrum rote ponderis et cordam 7 dencium, cuius centrum distat a basi diametro[d] rote ponderis et corda 3 dencium 20 minutorum. Summa dencium eiusdem rote 120.

Diameter rote in centro dialis, pro mensuracione sonitus, continet cordam 13 dencium 30 minutorum. Summa dencium rote predicte 10. Diameter rotarum pro ponderibus[e] sonitus continet cordam 13 dencium 30 minutorum. Summa dencium rotarum predictarum 12. Longitudo dencium rotarum predictarum continet cordam unius dentis et 30 minutorum. Spacium inter rotas predictas continet cordam 2 dencium.[f] Corda semicirculi[g] in hasta malleoli ab angulo ad angulum continet cordam 3 dencium 20[h] minutorum. Semidiameter[i] semicirculi haste predicte continet cordam unius dentis 40 minutorum. A tenone ex parte dialis ad medium rotarum sonitus continet cordam 2 dencium. Longitudo axis rotarum sonitus continet cordam infra tenonem 15 dencium. | Longitudo tocius calige sonitus continet cordam 10 dencium. Latitudo lamine predicte continet cordam 11 dencium 30 minutorum.[j] Centrum rotarum ponderis sonitus distat a basi diametro rote et corda 3 dencium,

[a] A continet

[b] Cf. p. 474 n. *i*. A marginal '3', verified by the following sentence, replaces an erased numeral, which ultra-violet light shows to have originally been '3'!

[c] A semidiametri [d] A diametrum [e] A ponderis

[f] There is a '+' sign over the '2', and '30' (minutorum?) is added in the margin.

[g] A semidiametri

[h] A marginal '20' replaces an erased '30', visible with ultra-violet light.

[i] A semidiametri

[j] Both figures are over erasures.

the tenon on the side of the dial to the middle of the nut on the strob arbor is a chord of 2 teeth. The diameter of the aforesaid nut contains a chord of 2 teeth 40 minutes. The breadth of the same nut is a chord of 1 tooth 20 minutes. This nut has 8 teeth. On the verge, the chord of the strob semicircle from one angle to the other[1] is 3 teeth 20 minutes. The semidiameter of this semicircle is a chord of one tooth 40 minutes. The middle of the strob verge[2] for carrying weights contains one diameter of the wheel[3] and a chord of 12 teeth. The length of the rectangular lead [weights] for the strob shaft[4] contains a chord of 6 teeth. The diameter of the journal wheel contains the diameter of the weight-wheel[5] and a chord of 7 teeth.[6] Its centre is, from the base, one diameter of the weight-wheel and a chord of 3 teeth 20 minutes. This wheel has 120 teeth.

The diameter of the wheel in the centre of the dial [day wheel?], for measuring out the strike [of the bell], contains a chord of 13 teeth and 30 minutes. This wheel has 10 teeth. The diameter of the wheels for the weight serving the strike is 13 teeth 30 minutes. The aforesaid wheels have 12 teeth. The length of the pins[7] of the said wheels is 1 tooth 30 minutes. The space between the same wheels is a chord of 2 teeth (30 minutes).[8] The chord between the angles of the semicircle on the shaft carrying the hammer[9] is 3 teeth 20 minutes. The semidiameter of the semicircle on the aforesaid shaft is 1 tooth 40 minutes. From the tenon on the dial side to the middle of the striking-wheels[10] is a chord of 2 teeth. The length of the arbor of the striking wheels contains, beyond the tenon, a chord of 15 teeth. The length of the entire barrel for the strike[11] is 10 teeth. The breadth of the plate[12] mentioned previously is 11 teeth 30 minutes. The distance of the centre of the weight-wheels for the strike[13] from the base is one diameter of the [standard] wheel and

1 *ab angulo ad angulum*. This is a crucial phrase in interpreting the form of the escapement. See our commentary.

2 'Middle' was probably intended to emphasize that the *quadratura plumborum* (see below) was being ignored.

3 This presumably refers to the wheel of 72 teeth, in terms of which all our measurements are made. Its diameter is approximately 23 teeth, or, more exactly, $22^{t}\ 55'$.

4 *Quadratura plumborum* might just possibly refer to a square or rectangle carrying the weights. With a beam of traditional shape (¬_⌐), the inner rectangle might have been called a *quadratura plumborum*, and the dimension 6, which is rather small, might then apply to it.

5 *rota ponderis*. The diameter in question is that referred to in the last note but one.

6 Is this a mistake for 4? See our notes to the diagram on p. 329, vol. ii.

7 *dentes*.

8 Added in the margin.

9 *hasta malleoli. Malleolus* generally signifies a small hammer, by contrast with *malleus*.

10 *rote sonitus*. This refers to the two wheels (without teeth) carrying the pins which move the hammer.

11 *caliga sonitus. Caliga*, a soldier's boot (class.), might possibly by the Middle Ages have come to be equated with a corresponding piece of leg armour. The word could certainly apply to an ordinary stocking (*MLWL2*). This all suggests a loose wooden barrel, by an obvious transference of meaning, and it is hard to see why such should have been used if it was not meant to allow for a ratchet ('spring click') for winding.

12 This must refer to the disc on the end of the barrel which prevents the rope from coming off.

13 *centrum rotarum ponderis sonitus*.

et a media linea corda 10 dencium 35 minutorum. Latitudo cimbe continet cordam 14 dencium, profunditas eius continet cordam 12 dencium. Altitudo tocius operis, in postis a basi, continet diametrum rote et cordam 24 dencium et 40 minutorum. Latitudo tocius operis, infra tenonem, continet diametrum rote et cordam[a] 13 dencium.[b] Spissitudo tocius operis infra tenonem continet cordam 15 dencium. Latitudo omnium laminarum de quibus fit domus continet cordam 3 dencium [·].[c] Post hec applicetur una lamina pro facie instrumenti, videlicet pro horis naturalibus cognoscendis, que habeat in longitudine[d] a basi diametrum rote bis et cordam unius dentis 30 minutorum,[e] et in latitudine diametrum et cordam 19[f] dencium 30 minutorum.[g] Et cetera de diali, et aliis sicut placet.

[II. 2]]

Pro motu Solis medio, hoc est numerus dencium omnium rotarum, videlicet, strob 30, in cuius axe fiat rota 8 dencium; cui aptetur rota ponderis 72 dencium; in cuius axe fiat nux 4 dencium; cui aptetur rota dialis 120; cui conclavetur rota pro medio motu Solis 115 dencium; cui correspondet 8, super hasta fixa in altera fine rote 24 dencium; cui correspondet testudinatus[h] 1 dentis; in cuius axe fiat rota 18 dencium, cui correspondebit rota 17 dencium, et cetera.

[a] A cod'
[b] The sentence is struck out by a later reader. See the note to the translation.
[c] The number of minutes is erased, but is shown by ultra-violet light to have been 30 originally.
[d] A longitudo
[e] A later hand has struck through 'cordam . . . minutorum'.
[f] Over an erasure.
[g] A later hand has struck through '30 minutorum'.
[h] A testud' (Elsewhere the MS. has 'testitunatus'.)

a chord of 3 teeth, and from the median line it is a chord of 10 teeth 35 minutes. The breadth of the bell is 14 teeth, and its depth 12 teeth.[1] The height of the entire device, from the base to the posts,[2] is one diameter of the wheel and a chord of 24 teeth 40 minutes. (The breadth of the whole thing contains one wheel diameter and a chord of 13 teeth.)[3] The depth of the whole device beyond the tenon is 15 teeth. The breadth of all the planks[4] which go to make up the clock case is 3 teeth [minutes erased]. After all this,[5] a plate is fitted as the face of the instrument, namely so that the natural hours may be known,[6] and the length of this from the base should be 2 wheel diameters and a chord of one tooth 30 minutes.[7] Its breadth should be one diameter 19 teeth and 30 minutes. As for the dial and the other parts, you may decide for yourself.

[II. 2]

For the mean motus of the Sun, the numbers of teeth required for each wheel are as follows:

The strob wheel is to be 30. On the arbor of this there is to be a wheel of 8; with which the weight-wheel of 72 teeth is to engage, on whose arbor there is to be a nut of 4 teeth; to which is to be fixed the day wheel of 120; to which a wheel of 115 teeth, for the mean motus of the Sun, is to be connected. To be meshed with this is a wheel of 8, on a shaft at the other end from a wheel of 24 teeth. This engages with a snail-wheel of 1 tooth, on the arbor of which there must be a wheel of 18 teeth, which will mesh with a wheel of 17 teeth, and so on.

[1] This tells us a great deal about the word *cimba*. Both from its association with the word for the cymbal, a relatively flat object, and judging from the very flat bells illustrated in early clocks, one might imagine the *cymba* to have been highly flattened in shape. The dimensions quoted show that this was not so. Besides *cymba*, there were at least a dozen medieval Latin words for bells of different sorts. Cf. p. 473 n. 6, above, and note the word *clok* in II. 3 below.

[2] *postis* (abl. plural). The classical Latin ablative from *postis* (post, door-post) would have been *postibus*, but there seems to have been a medieval preference for *posta* or *postella* as nom. sing.

[3] As explained in the note to the text, this sentence has been struck out by a later reader; and rightly so. Taken at its face value, it contradicts the last sentence of the section. Lest anyone should be tempted to suppose that the *depth* of the clock is being referred to, and that a depth of thirty-six teeth will allow the striking movement to be placed in a more conventional position (i.e. end-on with the going train), we point to the next sentence, and also to the fact that several other measurements are incompatible with the idea.

[4] *lamina*. This is a conjectured translation; but the clock-house is not likely to have been made of costly metal plates.

[5] That is, presumably, after assembling the case.

[6] *hore naturales* (unequal hours).

[7] '30 minutes' erased.

[II. 3]

Pro sonitu[a] *unius clok,* hoc erit numerus dencium omnium rotarum, videlicet: magna rota pro pondere, 100 dencium; sibi correspondet rota 20 dencium, cui, super axem rote predicte, conclaventur due rote, quarum[b] numerus dencium [sit] 10; quibus aptetur una hasta, cum malleolo, in qua eciam figatur unus semicirculus, qui correspondet duabus rotis conclavatis ad mensuracionem sonitus. Deinde applicetur una cimba meliori modo quo placet. Deinde aptetur[c] unum doleum magnum[d] rote, ita ut tociens vertatur doleum quociens vertitur magna rota, quod eciam habebit in circumferencia 25 divisiones, et in longitudinem, videlicet infra caput,[e] 24 divisiones, in quibus, secundum computacionem congruam, inponantur[f] 24 dentes; cui applicetur aliud doleum, minus doleo predicto quantitate[g] unius divisionis in circumferencia, et infra caput[h] divisionum [tot] quot[i] et aliud continet, in quibus divisionibus secundum computacionem congruam inponantur 24 dentes, in cuius capite, versus diale, fiant 24 foramina, quibus

[a] A sonitus
[b] A quorum
[c] A apptetur (Reading this as *applicetur* would not materially change the sense.)
[d] See the note to the translation.
[e] A capit' (This is less likely to be an abbreviation for *capita* than a mistake for *caput*; but see below, where *infra capitis* occurs.)
[f] A imponatur
[g] A quantitatus
[h] A capitis
[i] There is probably a lacuna here.

[II. 3]

In order to make a clock[1] *strike*, these are the numbers of teeth required for all the wheels:

The large wheel for the weight is to have 100 teeth; a wheel of 20 teeth meshes with it, and to this wheel,[2] which is above the arbor of the first wheel, two wheels are to be riveted having 10 teeth. With these wheels a shaft engages, together with a hammer. A semicircle is also to be fixed on the shaft, engaging with the two riveted wheels, and serving to mete out the strike.[3] A bell is next added, in any suitable way.[4] After this, a large barrel[5] is to be fixed to the wheel, in such a way that the barrel turns just as often as does the large wheel. The barrel should have 25 divisions[6] on the circumference, and be 24 divisions in length, namely below the head. On these divisions, according to a suitable calculation, 24 teeth[7] are placed. Linked with [the barrel] is another barrel, smaller than the first in circumference by an amount of one division, but [in length] below the head having as many divisions as there are on the first barrel, [that is to say], the divisions on which, according to a suitable calculation, the 24 teeth are placed.[8] At the head of the barrel against

[1] *clok. MLWL2* gives *cloca* (bell) as occurring *c.* 550, *c.* 1000, 1274, 1610. There are several cognate words cited before the time of our MS. It is generally held—and no doubt correctly—that the English word 'clock' derives from a word for 'bell'. There is no good reason for assuming that an essential intermediary between the Latin and English was a Dutch word (as the *OED* suggests). The reason for this belief is bound up with an erroneous view of the origin of the clock in England. See p. 370, vol. ii. Note the Middle English *clogan*, and the Irish *clog* (as in the phrase *clog an eadhachta Phatraic*, the bell of St. Patrick's will, said by some to be still preserved in Belfast). When Alfred the Great translated Bede, he rendered *campana* as *cluggan*.

[2] This passage is crucial to the understanding of the text as a whole. See the commentary in vol. ii.

[3] *qui correspondet duabus rotis conclavatis ad mensuracionem sonitus*. There seems to be no possibility that this refers to an arrangement for altering the 'measure of the duration' of the strike.

[4] *meliori modo quo placet.*

[5] *Doleum* is not a word of classical Latin. *MLWL2* gives its meaning as 'cask', 'tun' (13th and 15th cent.), and 'prison' (14th cent.). Cf. also classical *dolium*, a large jar or vase, and, according to *MLWL2*, occurring *c.* 1282 with the meaning 'wooden casing round millstone'. By reason of the later mention of its use to cut off the strike, a *doleum* is here taken to approximate in function to our so-called 'locking-plate'. The commonest type of early locking-plate (or count-wheel) comprises a flat ring, toothed on the inside and slotted on the outside, and held by four projections at the end of a cross. The chiming-barrel of a clock which plays tunes on a ring of several (often eight) bells, after the manner of the more familiar musical box, is obviously closer in appearance, if not in function, to one of the *dolea* (i.e. that with twenty-four suitably disposed *dentes*, or what I take to be pegs) of the text.

[6] We are not told what sort of divisions, obviously because this is, up to a point, immaterial, and no doubt on the original clock the divisions were not the standard tooth in size.

[7] The word is *dentes*, but 'pegs' would probably have been a better translation.

[8] An alternative interpretation would be that teeth were also placed on the twenty-four divisions of the second barrel.

aptetur unum springum pro clausura sonitus, et cetera. Et hec sequens est forma doleorum predictorum.

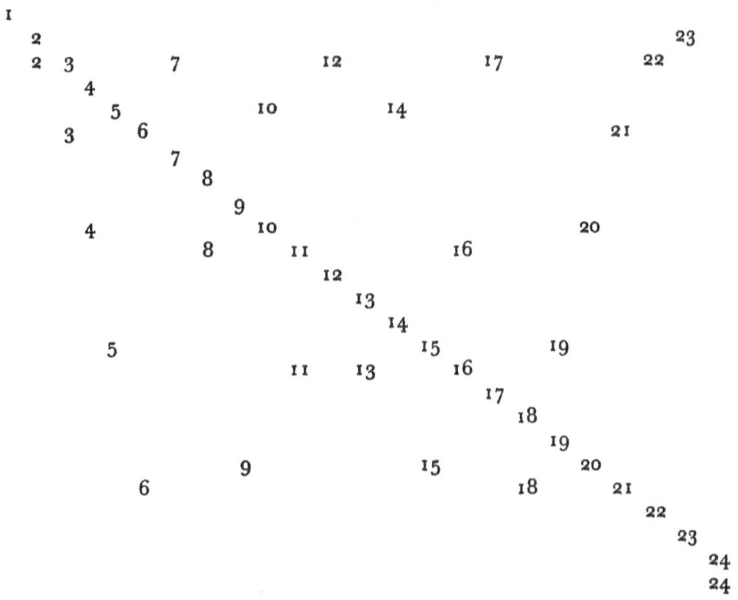

[II. 4. 1]

Pro composicione motus Solis in rotis.

Inicium operis est ut operarius habeat tabulam proporcionis pro omnibus gradibus zodiaci, et est composicio illius tabule [facilis?], cuius composicionis titulus erit *Tabula dividens circulum habentem 360 partes equales in quotlibet alias partes equales infra 400 partes*;[a] et incipit linea numeri istius tabule a 400 partibus usque ad unitatem. Primo pro composicione in linea numeri a 400 usque ad 360 partes: quod zodiacus non potest dividi per 400, quia maior est zodiaco, ideo multiplica zodiacum per 60, tunc erunt minuta; postea divide illa minuta per 400, et exibit primus numerus quociens in minutis 54, et nichil remanet. Iterum multiplica zodiacum per 60, et divide per secundum numerum in linea numeri, videlicet per 399, et exibit numerus quociens in minutis 54, et numerum remanentem iterum multiplica per 60, et iterum divide per lineam numeri, videlicet per 399; et exibit in secundis 8. Iterum numerum remanentem multiplica per 60, et divide per lineam numeri ut prius, et exibit 7 in terciis. Iterum numerum remanentem multiplica

[a] This is precisely the first part of the rubric to the tables on ff. 181^{v}–182^{v}, 84^{r}. The continuation is there *pro rotis horologii et aliorum instrumentorum artificialiter dividendis*. (The MS. has *dividendum*.)

the dial[1] 24 slots[2] are made. With them[3] a spring[4] engages to shut off the strike,[5] etc. And the following is the form of the aforementioned barrels.[6]

[II. 4. 1][7]

To represent the motion of the Sun on a wheel.

The first thing the maker[8] should have is a table of proportion for all degrees of the zodiac. The composition of this table is [simple?], its title being *Table dividing a circle, having 360 equal parts, into any other number of equal parts up to* 400.[9] The line of number of this table goes from 400 to unity. First for the composition of the table between 400 and 360 in the line of number: since the zodiac cannot be divided by 400, this being greater than the number of degrees in the zodiac, therefore multiply the zodiac by 60, turning it into minutes. Subsequently, divide the number of minutes by 400. The answer is in the minutes column, namely 54, and there is no remainder. Next, multiply the zodiac by 60, and divide by the second number in the line of number, namely by 399, giving 54 in the minutes column. Multiply the remainder by 60, divide by the line of number, namely by 399, giving 8 for the seconds column. Multiply the remainder by 60, divide by the line of number, and 7 goes in the thirds. Again multiply the remainder by 60, divide by the line of number, and 13 goes in the fourths. Discard whatever remains without any qualms, for it is not necessary to calculate beyond fourths. And so the process goes on for all entries in the line of number as far as 360, always dividing the zodiac in minutes by the parts in the line

1 The spelling of the phrase *versus diale* has not been changed, since *MLWL* gives *diale* as the nominative form, and since this would also have been the accusative form, assuming a neuter noun.

2 *foramen*.

3 I take it that this does not refer to the slots, but to the twenty-four teeth mentioned earlier in the Latin sentence. See the commentary.

4 *unum springum*. No dictionary I have seen records this Latin form, which clearly comes from the Old English *springan*, through the Middle English noun *spring* or *springe*, denoting a bent or sprung twig or sapling, especially as in a snare for game. Many early locking-plates are, of course, held by a spring-operated catch, which tends to confirm the idea that the function of the *dolea* is to control the strike in some such manner as does a locking-plate.

5 *pro clausura sonitus*. *Clausura* in classical Latin meant, e.g., 'lock', 'bar', 'bolt'. *MLWL* gives a meaning of 'fence' or 'fencing'.

6 Is this a mistake for *doleum*, in the singular? Only one *doleum* has the special form. MS. A has no diagram which would have served to illustrate the form of the *doleum*, but collating the MS. shows that the table of f. 181r follows here. See the commentary and notes on the collation.

7 Reasons for placing this here are given in the commentary.

8 *MLWL* cites *operarius* (with genitive), a worker of something (6th cent.). It also points out that *operaria* meant 'forge' (*c.* 1315).

9 It is strange that he should now introduce the table as though this were the first time the reader had encountered it. It is scarcely possible that the sequence of what we have called parts I and II was originally inverted. It is not unlikely, however, that Part II was *written* before part I, and it is also conceivable that we are dealing with what were intended to be two different treatises.

per 60 et divide per lineam numeri ut prius, et exibit 13 in quarcis. Et quicquid remanet abice, et non cura, quia ultra quartas non oportet calculare. Et sic fiat processus per omnes partes in linea numeri usque 360, dividendo semper zodiacum in minutis per partes linee numeri, in minutis, secundis, terciis, quarcis, quia in gradibus nichil resultat.

Sed cum perveneris ad 360 partes in linea numeri divide per zodiacum; et exibit unus gradus, nichilum in minutis, secundis, terciis, quarcis. Postea, per numerum sequentem in linea, videlicet 359, divide zodiacum; et exibit in numero quociens in gradus 1; et numerum rema[nen]tem multiplica per 60, et iterum divide per lineam numeri, et exibit nichilum in minutis. Iterum multiplica numerum remanentem per 60 [et divide per lineam numeri], et exibit in secundis 10. Et sic procede in terciis et in quarcis. Et sic fiat tota *Tabula Proporcionis*, dividendo zodiacum per omnes partes linee numeri a 400 usque ad unitatem.

[II. 4. 2]

Et si aliquod instrumentum pro vero motu Solis componere volueris, fiat operacio rotarum per divisionem inferius declaratam, | videlicet, quod si fiat mocio per instrumentum manuale sive per horologium, idem perveniet. Pro instrumento autem moto per horologium, preparentur[a] omnia que pertinent ad motum dialem, preter diale. Et loco dialis[b] fiat opus ad modum talis figure precedentis.

[II. 4. 31]

A est circulus maximus divisus in 360 partes equales; *B* circulus ecentricus divisus in 331 dentes;[c] *C* nux correspondens circulo ecentrico, divisa in 8 partes; et in alia extremitate lancee nucis est unus circulus divisus in 16 partes, et ille idem circulus correspondet nuci transversali divise in 2 partes, ad modum instrumenti torcularis. Et super axem eiusdem nucis figitur circulus *F*, divisus in 32 partes, cui correspondet circulus *E* fixus, divisus in 29 partes equales. Et figitur circulus *E* super

[a] A preparantur

[b] The three contractions of the word are alike (*dial'*). For the case endings of the word see note 3 to the translation opposite.

[c] The word *partes* was first written, and then erased.

of number, [dividing it] into minutes, seconds, thirds, and fourths; for the number of degrees in the quotient is nil.

But when you come to 360 parts in the line of number, divide by the zodiac.[1] The answer is one degree, and no seconds, thirds, or fourths. Afterwards divide the zodiac by the number coming next on the line, namely 359. The number of degrees in the quotient is 1. Multiply the remainder by 60, and again divide by the line of number, giving no minutes as a result. Again multiply the remainder by 60 [and divide by the line of number], resulting in 10 seconds. Carry on in the same way for the thirds and fourths. And in this way the entire *Table of Proportion* is drawn up, dividing the zodiac by each element of the line of number between 400 and 1.

[II. 4. 2]

If you want to make an instrument for the true motion of the Sun, make a device incorporating wheels, divided after the manner set out below; that is to say, if the motion be effected by a manual instrument or by a timepiece, the result is the same.[2] For an instrument driven by a timepiece, however, one must prepare everything which relates to the daily movement, quite apart from the dial itself. At the dial, there is to be a device such as is illustrated in the preceding figure.[3]

[II. 4. 31]

A is the largest circle, divided into 360 equal parts; *B* is the eccentric wheel divided into 331 teeth; *C* is the nut meshing with the eccentric wheel, being divided into 8 parts. At the other end of the shaft[4] on which the nut is situated is a wheel divided into 16 parts, which meshes with the transverse nut divided into 2 parts, after the fashion of the screw[5] of a wine-press. On the arbor of this same nut, a wheel *F* of 32 parts is to be fixed, with which a wheel *E* engages, this being divided into 29 equal parts. The wheel *E* is fixed on a beam *G*. *H* is a plate, on which

[1] i.e. 'divide the zodiac by the number of degrees in the zodiac'.

[2] This is the first (and last) occasion on which the author hints at the existence of manual instruments. It would have been arguable, had he not mentioned them, that he arrived at the idea of the astronomical clock independently of a knowledge of earlier geared equatoria.

[3] There is an empty space left for a figure on the previous page, that is, f. 168^r. In fact, however, f. 167^v is occupied by a figure which is only a trifle too large to fit this space. (Most of f. 167^v is blank.) This suggests that there is no missing figure, and that MS. A is at this juncture bound correctly.

[4] *lancea* (class. Lat., hence our 'lance'). Notice that the word used earlier in the text (as it is here arranged) was *hasta*. The change is no doubt an indication that the technical vocabulary of horological Latin was not sufficiently standardized. The word *lancea* occurs again.

[5] *instrumentum*. 'Screw' is a very free translation, but the author is obviously referring to this essential part of the wine-press, rather than the wine-press as a whole. This gear would now of course be called a worm-wheel. The word is not a synonym of *scru*, which is used elsewhere apparently with the meaning 'hole' or 'bearing'. See the note to II. 7. 2, below.

trabem *G*. *H* est lamina super quam dividitur circulus signorum, et circulus anni. *I* corpus Solis. *K* due parve rote super quas movetur axis tocius operis. *L* caliga super quam corpus Solis. *M* axis tocius operis.

[II. 4. 32]

. . . Summa tocius 331; cui rote[a], videlicet 331, aptetur finis unius haste 8 dencium. In altero[b] fine, versus centrum, fiat alia rota 16 dencium cui correspondebit una nux 2 dencium; super cuius axem figatur una rota 32[c] dencium, cui correspondebit una rota fixa circa centrum tocius operis, super unam trabem ferream | immobiliter fixam, 29 dencium. Et nota quod hasta predicta et rota cum nuce debent figi super unam laminam conclavatam calige[d] moventis corporis[e] Solis in fine interiori versus rotam obliquam, que quidem caliga se extendet a rota predicta usque ad laminam divisam; et ibi figatur corpus Solis et cetera.

[II. 4. 41]

Post divisionem circulorum per litteras, fiat divisio illorum per tabulas illis constitutas. Et primo fiat divisio laminis *H*. Dividatur extimus locus laminis in limbo in 360 partes equales, et erunt illi[f] gradus gradus equinoxiales. Deinde fiat spacium aliquantulum latum pro nominibus mensium et pro numero. Deinde aliud parvum spacium pro diebus anni. Deinde aliud spacium pro nominibus signorum, et pro numero in eodem spacio, et aliud spacium pro gradibus signorum. Deinde per tabulas ascensionis in circulo directo, et per gradus equinoctialis in extimo loco laminis, dividatur circulus signorum; et per eundem circulum signorum sic divisum, et per almanak de novo factum, fiat divisio circuli dierum anni. Et erit ille circulus, si placet, inter gradus equinoctiales et circulum signorum. Deinde, infra istos duos circulos, fiat circulus Capricorni; postea circulus equinoctialis; deinde circulus Cancri;

[a] A +uno [b] A rota
[c] A 31 (See the translation for an explanation.) [d] A caligis
[e] A corpus (The amendment makes a radical change in interpretation, but is certainly required.)
[f] A ille

is marked a circle of signs and a year circle.[1] *I* is the model of the Sun. *K*[*K*] are two small bearings[2] in which the arbor of the entire device[3] moves. *L* is the tube[4] on which is fixed the model of the Sun.[5] *M* is the axis of the whole thing.

[II. 4. 32]

. . . The sum of all this is 331. With this wheel, that is to say, the wheel of 331, the end of a shaft of 8 teeth is meshed. At the other end, and towards the centre, there is to be another wheel of 16 teeth, meshing with a nut of 2 teeth, on whose axis is fixed a wheel of 32 teeth.[6] With this there meshes a wheel of 29 teeth fixed at the centre of the entire device, and on an immovable iron beam. And notice that the aforesaid shaft,[7] and the wheel with a nut, must be fixed over a plate which is riveted to the moving tube of the model[8] of the Sun, [the attachment being] at the inner end, towards the oval wheel,[9] the tube extending from that wheel as far as the divided plate. And there, among other things, the model of the Sun is to be fixed.

[II. 4. 41]

After the division of the circles by letters,[10] let them be divided by means of the tables drawn up for them. First let the plate *H* be divided. The outermost edge of the plate, on the limb, is to be divided into 360 equal parts—equinoctial degrees. Next there is to be a narrow[11] space for the names of the months and for the number,[12] and then another small space for the days of the year.[13] After this, another space for the names of the signs, and the number can go in the same space; and then another space for the degrees of the signs.[14] By means of the tables of ascension on the direct circle, and the equinoctial degrees on the outermost part of the plate, the circle of signs is divided; and by means of that same circle of signs, thus divided, and by means of an almanac which is up to date,[15] the circle of days of the year should be divided. And if it please you, that circle will lie between the equinoctial degrees and the circle of signs. Below those two circles the circle of Capricorn is to be drawn; after that, the circle of the equinox; next, the circle of

[1] *circulus anni* (i.e. a calendar ring).
[2] *due parve rote.*
[3] That is, the main arbor.
[4] *caliga.*
[5] At the end of the pointer? Cf. the end of II. 4. 32, and the commentary.
[6] The figure 31 in the text is corrected to 32 in order to give the correct (annual) rotation of the wheel of 331 teeth. II. 4. 31, above, supports this amendment.
[7] *laminam conclavatam.* (Presumably not an error for *conclamatam* or *concavatam.*) *Conclavare* meant 'to nail together'.
[8] *corpus.*
[9] *rota obliqua.*
[10] i.e. after giving each a letter.
[11] *aliquantulum latus*—'a little broad'.
[12] Perhaps originally meant to be in the plural, *pro numeris*, 'for the numbers of the days of the month'.
[13] i.e. for the day divisions.
[14] i.e. the degree marks.
[15] *de novo factum.* Cf. the parallel text, where the almanac of Profatius is mentioned.

postea ecliptica. Et fiat regula que habebit cursum suum super exteriorem partem axis tocius operis; et extendet extremitas una regule ad corpus Solis, et ibi attachietur, ad demonstrandum diem, et gradum Solis, et locum Solis in ecliptica.

[II. 4. 42]

Divisio laminis exterioris.

In facie operis dividatur exterius circulus laminis in 360 partes equales, | et erunt gradus equinoxiales, vel gradus equales. Deinde, per *tabulam ascencionis signorum in circulo directo*, dividatur circulus signorum, videlicet zodiacus. Deinde, per eundem circulum signorum, fiat divisio dierum per almanac Profacii,[a] sive de nova calculacione; et scribantur nomina signorum.[b] Postea protrahatur circulus Capricorni,[c] et ceteri tres circuli, et fiat postea inscripcio ecliptice ita quod caput Capricorni in ecliptica[d] sit contra caput Capricorni in circulo signorum, et caput Cancri, et Arietis, et Libre, contra[e] capita sua | in circulo signorum.

[II. 4. 43]

In facie instrumenti componitur unum almicantaraz perforatum, et subtus almicantaraz una regula demonstrans gradum Solis et eius nadir: et figitur almicantaraz super pinnulam[f] exeuntem a centro operis inferiori, et nullo modo movetur. Postea componitur sub regula una lamina fixa super caligam correspondentem rote diurne, per quam quidem rotam movetur totum opus. Et protrahitur super laminam subtus regulam, primo, zodiacus divisus per gradus ascencionis in circulo directo, et stelle ibi situantur prout componuntur in rethe astralabii:[g] et in limbo illius laminis, extra almicantaraz fixum, protrahuntur duo circuli coniuncti cum spaciis pro numeris et tam pro nominibus mensium quam pro nominibus signorum. Et dividitur intimus circulus

[a] A Prefacii
[b] A sanctorum
[c] A Capricornii (This spelling is adhered to throughout.)
[d] A ecliptice
[e] A qua
[f] A pinnulum
[g] The spelling is characteristic of the MS.

Cancer; and after that, the ecliptic.[1] There is to be a pointer[2] carried on the outer part of the main axis of the entire device; and one end of the pointer extends to the model of the Sun, and there it[3] is to be attached,[4] to show the day, the Sun's degree, and the place of the Sun in the ecliptic.

[II. 4. 42]

The division of the outer part of the plate.

On the face of the device, the outer circle of the plate is to be divided into 360 equal parts, these being equinoctial, or equal, degrees. Next, with the help of the *Table of ascensions of the signs on the direct circle*,[5] the circle of signs, namely the zodiac, is to be divided. Next, on that same circle of signs, let the day divisions be made, using the almanac of Profatius or making a fresh calculation; and the names of the signs are to be written in. After that, the circle of Capricorn must be drawn,[6] together with the other three circles;[7] and afterwards, let the inscription of the circle of the ecliptic be such that the first point of Capricorn on the ecliptic lies opposite the first point of Capricorn on the circle of signs, and let the first point of Cancer, of Aries, and of Libra be against their first points in the circle of signs.

[II. 4. 43]

A pierced almucantar-plate[8] is put on the face of the instrument, and, underneath it, a pointer showing the degree of the Sun and its nadir. The almucantar-plate is fixed by a rod[9] going from the lower centre of the face of the instrument. It does not move. Under the pointer is placed a plate, fixed on the tube on the journal wheel, by which wheel, indeed, the entire device is moved. Drawn on the plate under the pointer is, first, the zodiac, divided by degrees of right ascension, and then the stars are put on it, just as they are placed on the rete of an astrolabe. On the limb of this plate, outside the set of fixed almucantars, two circles are drawn, joined together with spaces for numbers, and both for the names of the months and the names of the signs. The inner of these two circles

[1] These are the circles normally drawn on the rete of an astrolabe, before parts of them are later cut away.

[2] *regula.*

[3] Presumably the Sun's model.

[4] *attachietur.*

[5] Cf. *Albion* IV. 16, and the commentary on that chapter and table.

[6] Capricorn is the sign which the Sun enters at the winter solstice.

[7] Namely, the circles of Aries, Cancer, and Libra. By 'circle of Aries' he means, presumably, the quadrant of the ecliptic circle covering the season of spring; and similarly for the other three 'circles'.

[8] *unum almicantaraz.* (Not illustrated.) This is an unusual arrangement, as explained in our commentary.

[9] *pinnula.* (Altered in keeping with the earlier, and more common, usage.) The translation 'rod' is confirmed by the figure on f. 176^r. This has significance for the understanding of the escapement mechanism, in connection with which the word *pinnula* is used. See p. 475 n. 8.

illorum duorum circulorum per *tabulam ascencionis signorum in circulo directo*, et per gradus equales prout dividitur zodiacus in rethe astralabii. | Et protrahuntur 12 linee a centro poli mundi, per capita 12 signorum, usque ad extimum circulum, qui dicitur circulus 12 mensium: et dividitur extimus circulus illorum duorum circulorum in omnibus diebus anni, non tamen per gradus equales, sed per gradus zodiaci et per almanak que dicitur *quarta tabula Solis*. Et in extimo loco scribitur numerus dierum cum nominibus mensium; et protrahuntur in facie illius laminis infra zodiacum, et extra stellas[a] de prima magnitudine, prout situantur in rethe astralabii.

[II. 4. 5]

In eadem facie componitur secunda lamina lacior prima fixa super axem predictum, videlicet prime laminis; et describitur per circulos sicut est[b] limbum prime laminis, videlicet per duos circulos coniunctos cum spaciis pro numeris, et pro nominibus signorum. Et dividuntur illi duo circuli in secundo limbo, extimus circulus per gradus equales, et intimus circulus per gradus zodiaci, divisus per gradus equales et per tabulam ascencionis. Et inter istos duos limbos sic divisos componitur una lamina in qua situatur Caput Draconis et Cauda; et correspondet Caput Draconis gradibus zodiaci in extimo limbo, et movetur illa lamina Capitis Draconis inter duas primas laminas fixas super caligam super quam figitur circulus diurnus . . .

[II. 4. 6]

Deinde fiat divisio rote oblique in fundo instrumenti, videlicet rota 331, tali modo: sit *A* in figura presenti rota diurna, super quam infigitur rota obliqua, et primo accipias 2 gradus Tauri incipiendo ab Ariete de gradibus equinoxialis et a duobus[c] gradibus Tauri protrahes lineam ad centrum. Deinde accipe 9 grad. et 50 min. de illis 60 in diametro, et iuxta illos gradus fiat punctus in linea protracta a duobus[d] gradibus Tauri, qui sit *F*. Postea protrahatur linea ab 28 grad. Geminorum ad centrum, et de diametro diviso in 60 ut prius 17 grad. 20 min., et fiat punctus in dicta linea, qui sit *G*. Deinde in Leone 27 grad., et protrahatur ad centrum, et 9 grad. de diametro, et fiat punctus in linea qui sit *H*. Deinde in Sagittario 28 grad. 30 min., et protrahatur ad centrum, et de diametro accipe 14 grad., et fiat punctus et sit *I*. Deinde

[a] A stelle [b] A et [c] A duo [d] A duo

is divided by means of the *Table of ascensions of the signs on the direct circle*, and by equal degrees, just as the zodiac on the rete of an astrolabe is divided. Twelve lines are drawn from the centre of the pole of the world, through the first points[1] of the twelve signs, as far as the outer circle, known as the 'circle of the twelve months'. The outermost of these two circles is divided into the days of the year, not by equal degrees, however, but by the help of the zodiacal degrees together with [that table of] an almanac which is known as the *fourth table of the Sun*.[2] In the outermost place are written the numbers of the days[3] with the names of the months; and they are drawn on the face of this plate below the zodiac, and outside the stars of first magnitude, just as on the rete of an astrolabe.

[II. 4. 5]

On the same face a second plate is put, wider than the first, and fixed on the aforementioned axis, namely that of the first plate. It is marked out with circles just as the limb of the first plate, namely with two circles joined with spaces for numbers and for the names of the signs. These two circles on the second limb are divided, the outermost circle into equal degrees, the innermost by degrees of the zodiac, the latter being divided with the help of the equal degrees and a table of ascensions. Between the two limbs so divided, a plate is put bearing the head and tail of the Dragon. The head of the Dragon reaches as far as the degrees of the zodiac[4] on the outermost limb,[5] and the plate carrying it moves between the first two plates which are fixed on the tube on which the journal wheel is also fixed.

[II. 4. 6]

Next to be divided is the oval wheel[6] in the depths[7] of the instrument, that is, the wheel of 331 teeth. This is how to proceed: let *A* in the present figure[8] be the journal wheel, on which the oval wheel is fixed. Join to the centre the point which is two equinoctial degrees (measured away from Aries) from the first point of Taurus. Next, counting the diameter as 60°, take a point *F* at a distance of 9° 50′ in the direction of the second degree of Taurus. Take now a point *G*, measuring 17° 20′ from the centre,[9] in the direction of the 28th degree of Gemini. Join 27° Leo to the centre, and take a point *H* at a distance of 9° from the centre. Join 28° 30′ Sagittarius to the centre, and let *I* be a point at 14°

[1] *per capita.*

[2] An almanac gives the Sun's longitude for the days of the years in the leap-year cycle, thus requiring four parallel tables. The cycle was sometimes called *ciclus compoti.*

[3] i.e. the dates.

[4] *correspondet . . . gradibus zodiaci.*

[5] i.e. on the lower of the two plates.

[6] *rota obliqua.* This is also called the 'eccentric wheel' (*circulus ecentricus*). A table for dividing the wheel is included at II. 4. 8, below.

[7] *in fundo.*

[8] f. 82r.

[9] Lit. 'as before'.

separetur spacium in diametro ex parte Cancri, 50 grad. de diametro, et fiat punctus et sit *K*. Item separetur alius punctus ex parte Capricorni, 46 grad. 50 min., et sit ille punctus *L*.

Deinde ponatur pes circini in puncto *I*, et extendatur usque *K*, et fiat porcio circuli usque ad circulum *A*. Ponatur idem circinus in puncto *G*, et extendatur usque *L*, et porcio circuli ex parte Capricorni. Deinde restringatur circinus, et ponatur in puncto *F*, et coniungantur [porciones] circuli *K* [et] *L* ex parte Piscium. Ille idem circinus ponatur in puncto *H*, et coniungantur porciones circuli *K* et *L* ex parte Libre. Postea duplentur porciones propter spissitudinem.

Postea fiat divisio isto modo: aptetur regula[a] protendens a centro instrumenti, ita quod acuitas regule iaceat super dictum circulum obliquum. Et alia regula in facie instrumenti conclavata motui Solis; et ponatur illa regula in facie instrumenti super principium Arietis, vel super inicium diei mensis sibi correspondentis in tabula pro eadem rota calculata. Et primo videas quod caput Arietis superius et inferius sint in locis suis, videlicet quod una regula iaceat super unum caput et alia super aliud.

Deinde prospice tabulam quot dies et quot minuta diei correspondent 6 dentibus, et fiat ibi nota; deinde quot respondent 12 dentes, et fiat nota. Et erunt in tota divisione 55 spacia et 1 dens post; quodlibet[b] spacium debet dividi per 6; et erunt in toto 330 dentes cum uno dente . . .

[II. 4. 7]

Sequitur composicio tabularum motus Solis in die naturali correspondentis tabule proporcionis per quas quidem invenitur proporcio dencium rotarum *F* et *E*; et est composicio talis:

[a] A r^{a} (The contraction is used throughout, as is r^{u} for *respectu*.)
[b] A quotlibet

from the centre. Next, a division is to be made in the neighbourhood of Cancer. Let *K* be a point at 50° of the diameter measuring from the centre towards the first point of Cancer. Another point, *L*, is to be marked at 46° 50′ measuring towards Capricorn from the centre.

The foot of a pair of dividers is now to be placed at the point *I*, and these are to be extended as far as *K*. An arc of a circle is then drawn going as far as the circle *A*. The same dividers are now to be used from centre *G*, with radius *GL*, to draw the Capricorn part of the circle.[1] Next the dividers must be closed up somewhat, and from centre *F* the parts of the circles *K* and *L* are to be joined in the region of Pisces. The same dividers, [that is, with the same radius], are to be placed with centre *H*, and the parts of the circles *K* and *L* in the region of Libra are to be joined up.[2] Next, all arcs are to be duplicated, to form [a margin of] sufficient breadth.

The whole thing is to be arranged as follows. A pointer must be fitted at the centre of the instrument,[3] such that the pointed end of this lies on the oblique circle referred to above. Another pointer on the face of the instrument is connected to the motion of the Sun; and this pointer is placed on the face of the instrument, and on the first point of Aries,[4] or on the beginning of the day of the month corresponding to it in the table calculated for that wheel.[5] First you should see that the upper and lower initial points of Aries are in their places, namely that one pointer lies on one, and the other pointer on the other.[6]

Next look up the table to see how many days and how many *minuta diei* correspond to 6 teeth,[7] and note down the answer; then how many correspond to 12 teeth, and note this down. There are, in the entire division of 331 teeth into groups of 6, 55 spaces, with 1 tooth left over; that is to say, altogether there are 330 teeth plus 1 tooth . . .

[II. 4. 7]

Here follows[8] the method of drawing up the tables of the motus of the Sun in a natural day, based on[9] the [Sun's] table of proportion,[10] by which a ratio between the numbers of teeth in two wheels (*E* and *F*) is found:

1 Roughly speaking, this extends from Libra to Pisces, through *L*.

2 The result of this construction is a pseudo-ellipse. Cf. the construction advocated in *Albion* II. 6.

3 Lit. 'stretching forth from the centre of . . .'.

4 *principium Arietis*, not *caput Arietis*, as before.

5 That is, the pointer should point to the day when the Sun first enters Aries.

6 In short, we have two pointers, one traversing the calendar scale, the other the zodiac scale.

7 It is never made clear why we should do this.

8 Beginning here, the text repeats the substance, but not the words, of Part I, prop. 12.

9 *correspondens* (*tabule*).

10 We shall qualify this table as the Sun's, to distinguish it from the first and second *Tables of Proportion* used so extensively elsewhere.

Accipe motum Solis in die naturali, quondam est 59 min. 8 sec. 19 ter. 37 quar., et divide per quemcumque numerum volueris infra 60 vel extra. Et sit gracia exempli numerus dividens 58, [et sit] numerus dividendus motus Solis in die naturali. Et limetur una tabula ab unitate usque ad 58, videlicet in gradibus, minutis, secundis, terciis, quartis, quia ulterius non respondebit *Tabule Proporcionis*: et scribatur | 58 pro titulo tabule, [et] in linea numeri ab unitate usque ad 58. Et in capite tabule, supra unitatem in linea numeri, scribantur in suis locis gradus, minuta, secunda, tercia, quarta. Deinde per 58, qui est titulus tabule, dividantur 59 min. de motu Solis in die naturali, et numerus quociens erit 1 et numerus remanens erit 1. Illa unitas in numero quociens scribatur in tabula tua in linea minutorum, edirecto unitatem in linea numeri. Deinde numerum remanentem, videlicet 1, multiplica per 60; et post multiplicationem adde 8 in secundis de motu Solis in die naturali; et totum collectum divide per titulum tabule, videlicet per 58, et erit numerus quociens unitas. Scribatur illa unitas in linea secundarum, et numerum remanentem iterum multiplica per 60; cui adde 19 in terciis de motu Solis in die naturali; et totum collectum divide per 58, sicut prius, et erit ille numerus quociens 10 in linea terciorum. Et numerum remanentem iterum multiplica per 60, et addantur 37 quarta[a] de motu Solis, et dividatur collectum per 58; eritque numerus quociens 41, et scribatur 41 in linea quartarum. De numero remanente non est curandum si fuerit infra 30; et quia nichil resultat in gradibus scribatur 0.

Deinde ulterius procedendo in composicione dupla 41 in quarcis et scribe numerum provenientem in sequenti spacio videlicet 22. Deinde adde 41 ad 22, et numerum provenientem scribe in sequenti spacio, videlicet 3. Ad illum numerum 3 adde 41, et provenient 44; et sic procede usque ad finam tabule, semper addendo primum numerum in quarcis omnibus aliis. Et sic procede in terciis, secundis, minutis, et gradibus. Sed videas bene quam resultat 60 in quarcis tunc adde unitatem terciis, et quam resultat 60 in terciis tunc adde in secunda, et sic in minutis.

[a] A quarte

Take the motus of the Sun in a natural day, which motus was at one time $59' \, 8'' \, 19''' \, 37^{iv}$, and divide it by any number you please, whether this be less than or greater than 60. Take this example: the divisor shall be 58, and the number to be divided shall be the motus of the Sun in a natural day. A table is to be bordered[1] from unity up to 58, namely in degrees, minutes, seconds, thirds, and fourths—and no further, since anything more would fail to correspond with the Sun's *Table of Proportion*; and '58' is written up as the heading of the table, the numbers 1 to 58 being written in the line of number.[2] At the top of the table, against unity in the line of number, are to be written in the appropriate places the numbers of degrees, minutes, seconds, thirds, and fourths. To find them, the 59 minutes of the Sun's motus in a natural day are divided by 58, which is the heading of the table. The quotient is 1, with 1 the remainder. That unit in the quotient is written in your table in the line of minutes, opposite the figure 1 in the line of number. Subsequently, multiply the remainder, namely 1, by 60; and after doing so, add the 8 seconds of the motus of the Sun in a natural day. Divide the total by the heading of the table, namely 58. The quotient will be unity. That unit is to be written in the line of seconds, the remainder being multiplied by 60 and added to 19, the number of thirds in the Sun's daily motus. Dividing by 58, as before, gives the quotient 10, which is put in the line of thirds. Again multiply the remainder by 60, the 37 fourths of the Solar motus being added, and the total being divided by 58. The quotient will be 41,[3] and 41 is accordingly written in the line of fourths. As for the remainder, it is not necessary to bother about it if it comes to less than 30.[4] And because there were no degrees in the answer, we there write 0.

After all this, it is necessary to proceed with the table, first by doubling the 41 in the column of fourths. Write the resulting number, namely 22, in the next space. Add 41 to 22, and write the result, namely 3, in the next space. Add 3 to 41, giving 44; and carry on in this way up to the end of the table, adding the first number in the fourths to every other.[5] And do the same for the thirds, seconds, minutes, and degrees. But remember that for every 60 which results in the fourths, you must add 1 to the thirds, and for every 60 in the thirds, you must add 1 to the seconds, and similarly for the minutes.

[1] *limetur*. This use seems to be unknown elsewhere; but *limes* can denote a border or boundary. There is, alternatively, a possible connection with *limbus*, or even with the English verb 'limn', to illuminate (of a manuscript).

[2] See pp. 498–9, and also pp. 524–6 for a summary of the tables generally.

[3] It is in fact 40, with 57 remainder. See the next footnote.

[4] Note the precaution. Taking 41 instead of 40 above, we notice that the writer follows the unspoken half of his rule, viz. that above 30 we are to go to the next higher figure.

[5] He does not appear to be unduly worried by the accumulation of fractions below the fourths. In the chosen example, by the time 58/58 is reached by his method, the accumulated error is only about one-fourth. In some cases, however, the error could be relatively serious —if ignoring fourths can ever be taken as serious for 14th-century astronomy.

Et si vis probare tabulam tuam, contra 5 in linea numeri, vel contra 10, vel contra 20, vel contra quemcumque numerum volueris, multiplica omnes numeros quociens in suis locis scriptis contra unitatem in capite tabule, per 5, vel per 10, vel per 20, vel per quemcumque numerum linee numeri volueris; et numerum provenientem in gradibus, minutis, secundis, terciis, quarcis, scribe contra illum numerum linee numeri sic multiplicatum. Et sic procede quousque perveneris ad complecionem tabule tue per 58, et erit in fine tabule contra 58 precise motus Solis in die naturali, videlicet gradibus, minutis, secundis, terciis, et quarcis, si bene feceris.

Si autem ista tabula de motu Solis simplici non sufficiat, nec potest inveniri aliquis numerus in ista tabule correspondens *Tabule Proporcionis*, fiat una tabula alia eiusdem quantitatis in gradibus, minutis, secundis, terciis, et quarcis, et super eandem lineam numeri; et erit illa eadem tabula prius facta, cum addicione integri motus Solis in die naturali. Et fiat tali modo:

Accipe in tabula tua de motu simplici omnes numeros quociens, in gradibus, minutis, terciis, quartis, contra 1 in linea | numeri, quibus adde integrum motum Solis in die naturali, videlicet gradus gradibus, minuta minutis, et sic de aliis. Et omnes illos numeros provenientes scribe in secunda tabula contra unitatem in linea numeri in suis locis, videlicet per gradus, minuta, secunda, tercia, et quarta. Deinde fiat composicio ulterius. Accipe numerum in quarcis in tabula simplici, et adde numerum in quarcis in tabula composita;[a] et numerum provenientem scribe in tercio[b] spacio sequenti. Iterum accipe numerum in quarcis in tabula simplici, et adde secundo numero in quarcis in tabula composita; et numerum provenientem scribe in tercio spacio. Et sic per totam composicionem tabule composite in quarcis. Et sic procede in terciis, secundis, minutis, et gradibus, et semper quando ex addicione alicuius numeri de tabula simplici ad aliquem numerum in tabula composita resultat. Plus 60 si fuerit in quarcis, adde unitatem terciis, et sic in secundis, minutis, et gradibus.

Et est probacio composicionis istius tabule talis: Adde motum Solis integrum cuicumque numero volueris in tabula simplici, et numerus proveniens erit numerus tabule composite contra illum eundem numerum in linea numeri.

Et sequitur composicio inicialis tabule simplicis et composite Solis. Et est generalis modus faciendi tabulas correspondentes *Tabule Proporcionis*, et eciam pro mocione aliorum planetarum, prout patet in rubrica, *Ad componendum verum motum Solis in rotis*, et [in] capitulo, videlicet, de divisione duarum parvarum rotarum cum nuce.

[a] A compositi (The manuscript is consistent in using this word to qualify abl. and gen. forms of '*tabula*', and the use is not noted again below.)

[b] Crossed through in the MS.

If you wish to verify your table, opposite 5 in the line of number, or 10, or 20, or any other number you like, multiply the quotient written at the top of the table, by 5, or by 10, or by 20, or by any number you please; and write the result, in degrees, minutes, seconds, thirds, and fourths, against the number in the line of number which was used in the multiplication.[1] Carry on as far as 58, when, at the end of the table, you should get exactly the Sun's daily motus, in degrees, minutes, seconds, thirds, and fourths, if you have done it properly.

If, on the other hand, this table of the simple motus of the Sun should not suffice, and it proves to be impossible to find any number in it corresponding to the Sun's *Table of Proportion*, let another table be made, of the same quantity, in degrees, minutes, seconds, thirds, and fourths, and on the same line of number; and it will be the same table as before, apart from the addition of the whole of the Sun's diurnal motus. This second table is to be made as follows:

Take the quotient in your table, in degrees, minutes, seconds, thirds, and fourths, which comes opposite 1 in the line of number, and add to it the whole solar motus for one natural day; that is to say, degrees to degrees, minutes to minutes, and so for the rest. Write all the resulting numbers in the second table in the proper places against 1 in the line of number, namely in degrees, minutes, seconds, thirds, and fourths. Now for the next step in drawing up the table, take the number of fourths in the simple table and add the number of fourths in the compound table. Write the result in a third space, following the other two. Next take the number of fourths in the simple table and add it to the second number of fourths in the compound table. Write the number which results in the third space. And so for the completion of the entire compound table, so far as the fourths are concerned. Carry on in the same way for the thirds, seconds, minutes, and degrees, invariably adding the number in the simple table to the number in the compound table. If you get more than 60 in the fourths, add a unit to the thirds, and similarly for the seconds, minutes, and degrees.

This is how to verify the table: add the entire daily motus of the Sun to any number you choose from the simple table, and the number resulting will be the number in the compound table against that same number in the line of number.

What follows is the beginning of the first table, simple and compound, of the Sun. This is indeed the general method of making tables based on the table of proportion, not only for the Sun but also for the motion of the other planets, as is clear from the rubric, *To represent the true motus of the Sun by means of wheels*, and also from the chapter on the division of two small wheels and a nut.

[1] It goes without saying that there should be agreement with the number written there.

Tabula simplex vel composita

	per 58					idem cum integro				
	gr.	m.	s.	t.	q.	gr.	m.	s.	t.	q.
1	0	1	1	10	41	1	0	9	30	18
2	0	2	2	21	22	1	1	10	40	59
3	0	3	3	32	3	1	2	11	51	40
4	0	4	4	42	44	1	3	13	2	21
5	0	5	5	53	25	1	4	14	13	2
6	0	6	7	4	6	1	5	15	23	43

[II. 4. 8]

Post composicionem tabularum correspondentium *Tabule Proporcionis*, restat componere unam tabulam per quam dividatur circulus ecentricus, sive circulus 331 dencium correspondens nuci. Et est composicio talis:

Limetur una tabula que continet 331 spacia, et erit ista tabula pro Sole tantum. Et in prima linea scribantur 331 tantum ordinate et successive per suos limites; et vocentur dentes circuli ecentrici. In secunda linea dies, et tercia linea partes diei. Et est composicio talis: Scribe numerum dencium prout hic incipiuntur, ab unitate usque ad complecionem 331. Deinde in secundo limite unitatem, pro uno die, in tercio loco 6, pro partibus dierum. Deinde multiplica numerum dierum in primo limite per quemcumque numerum dencium volueris, et edirecto illius numeri dencis scribe numerum resultantem. Postea per illum eundem numerum dencium multiplica partes diei in primo limite, | et divide per 58. Et numerum quociens adde numero dierum: et sic fiat de omnibus numeris usque ad complecionem tabule.

Tabula ad dividendum circulum ecentricum in 331 dencium

dentes	dies	partes diei	gr.	m.	s.	t.	q.
1	1	6	1	5	15	23	40
2	2	12	2	10	30	47	20
3	3	18	3	15	46	11	0
4	4	24	4	21	1	34	40
5	5	30	5	26	16	58	20
6	6	36	6	31	32	22	0
7	7	42	7	36	47	45	40
ultimus numerus istius tabule scribatur hic inferius							
331	365	14	359	59	55	33	40

Line of number	Simple table 58					Compound table with whole [motus added]				
	°	′	″	‴	iv	°	′	″	‴	iv
1	0	1	1	10	41	1	0	9	30	18
2	0	2	2	21	22	1	1	10	40	59
3	0	3	3	32	3	1	2	11	51	40
4	0	4	4	42	44	1	3	13	2	21
5	0	5	5	53	25	1	4	14	13	2
6	0	6	7	4	6	1	5	15	23	43

[II. 4. 8]

After drawing up the tables based on the Sun's table of proportion, it remains to compose a table enabling us to divide the eccentric [i.e. oval] wheel, that is, the wheel of 331 teeth, meshing with the nut. This is how we go about it:

A table is drawn up with 331 spaces, this table being for use with the Sun, and nothing else. In the first column the numbers from 1 to 331 are written in sequence,[1] in the appropriate spaces in the table;[2] and these are called 'teeth of the eccentric wheel'. In the second column are days, and in the third are parts of a day. And this is how the table is filled in: Write out the number of teeth, as we have started them here, from 1 to 331. Next, write 1 in the second column for 1 day, and in the third column write 6, for the parts of a day. Multiply the number of days on the first row by any number of teeth you like, and on the line opposite this number of teeth write the number which results. Subsequently, multiply the number of parts of a day on the first row of the table by that same number of teeth, and then divide by 58. Add the quotient to the number of days; and in this way we go on for all numbers, up to the end of the table.

Table to divide the eccentric wheel into 331 teeth[3]

Teeth	Days	Parts[4] of a day	°	′	″	‴	iv
1	1	6	1	5	15	23	40[5]
2	2	12	2	10	30	47	20
3	3	18	3	15	46	11	0
4	4	24	4	21	1	34	40
5	5	30	5	26	16	58	20
6	6	36	6	31	32	22	0
7	7	42	7	36	47	45	40
the last number of this table is written here below:							
331	365	14	359	59	55	33	40

[1] *in prima linea scribantur 331 tantum ordinate et successive.*

[2] *per suos limites.*

[3] The full table is given elsewhere in the manuscript See pp. 524–6 below.

[4] Notice that these parts are 58th parts of a day.

[5] Notice that the figure 43 (not 40) was arrived at above. (This is not a copyist's error.) It is different again from the figure in the fourths column (28) opposite 331 in the (first) *Table of Proportion*, which in turn differs from that quoted in I. 9 in the thirds column (24).

Verbi gracia: Accipio ultimum numerum tabule dencium, qui est 331, et multiplico per unitatem in primo limite dierum; et resultant[a] numerus 331. Deinde, per illum eundem numerum dencium multiplico 6 in primo limite, que sunt partes diei, et numerum resultantem divido per 58. Et numerus quociens inde proveniens addetur diebus, et scribetur in limite dierum; et erit numerus 365, autem remanens est 14, et debet scribi in limite parcium dierum. Et sic fiat composicio omnium talium tabularum, videlicet, multiplicando dies et partes diei per numerum dencium edirecto, et dividendo per titulum tabule simplicis et composite.

Et si volueris componere tabulam correspondentem dentibus, diebus, et partibus diei, in gradibus, minutis, secundis, terciis, et quarcis, accipe numerum inventum in tabula simplici vel composita per tabulam proporcionis; et scribe in superiori parte tabule per suos limites. Et illum eundem numerum multiplica per quemcumque numerum dencium volueris, et numeros provenientes in suis locis divide per 60, videlicet, reducendo quarta[b] in tercia, tercia in secunda, secunda in minuta, minuta in gradus. Et numeri provenientes scribendi sunt contra numerum dencium per quem primo multiplicati; et sic fiat de omnibus numeris dencium usque ad totam complecionem tabule.

Si autem leviori modo volueris componere, fiat composicio per modum tabule simplicis et composite; primo, videlicet, in quarcis, postea in terciis, secundis, minutis, et gradibus; et idem proveniet.

[II. 5. 11]

Postea figitur alia rota magna, super quam figitur corpus Solis; et dividitur ille circulus deferens corpus Solis in 160; inter quas rotas fixas componitur una rota magna mobilis, et defert corpus Lune. Et est latitudo illius rote per spacium unius pollicis in limbo; et in interiori parte deferentis Lune figitur una rota continens 118 dentes. Deinde procedit una hasta fixa super deferentem Solis cum duabus rotis; et aptatur extima rota haste, videlicet 8 dencium, circulo[c] deferentis Lune, [diviso] in 118 dentes; et alia rota, in alia extremitate haste iuxta caligam, videlicet 8 dencium, aptatur circulo testitudinato 2 dencium.[d] Et est ille circulus testitudinatus, prospecta facie instrumenti, ex parte dextra haste predicte. Figitur autem super axem circuli testitudinati una rota versus[e] partem inferiorem instrumenti 12 dencium, et aptatur illa

[a] This use of the plural was commonplace.
[b] A quartas (The same ending is used for the four words denoting sexagesimal parts.)
[c] A circuli
[d] A unius dentis (See the note to the translation.)
[e] Written above *super*, which is deleted.

Example: I take the last number under the heading 'Teeth' in the table, namely 331, and multiply by unity, which appears at the top of the column of days. The result is 331. Next, by that same number of teeth, I multiply the 6 at the top of the table, giving the parts of a day; and I divide the result by 58.[1] The quotient is added to the days, and is written at the end of the days column—in fact it will be the number 365, although there will be a remainder of 14, which must be written in the 'parts of a day' column. In this way, all such tables are made, namely by multiplying the days and parts of a day by the number of teeth on the line, and by dividing by the heading of the simple and compound table.

If you want to compile a table corresponding to teeth, days, and parts of a day, in degrees, minutes, seconds, thirds, and fourths, take the number found in the simple or compound table, with the help of the Sun's table of proportion. Write it in the upper part of the table, in the appropriate spaces.[2] Multiply that same number by any number of teeth you please, and divide the result by 60, that is to say, by reducing fourths to thirds, thirds to seconds, seconds to minutes, and minutes to degrees. The resulting numbers are to be written against the number of teeth first used in the multiplication. And we go on in this way until the table is finished.

On the other hand, if you want to compile the table more easily, go about it in the way used for the simple and compound table—first, that is, for the fourths, then for the thirds, the seconds, the minutes and the degrees—and the same sort of table results.[3]

[II. 5. 11]

Next, another large wheel is fixed, to which the model of the Sun is attached; and the wheel carrying the model of the Sun is divided into 160. Between the wheels thus far put into position, a large movable wheel is placed, carrying the model of the Moon. The breadth of that wheel at its rim is an inch. Within the wheel carrying the Moon a wheel of 118 teeth is fixed. From there proceeds a shaft, fixed on the wheel carrying the Sun, and bearing two wheels. The outermost wheel of the shaft, namely that of 8 teeth, engages with the wheel of 118 teeth carrying the Moon. The other wheel at the other extremity of the shaft, next to the tube, which wheel is also of 8 teeth, engages with a snail wheel of 2 teeth.[4] The snail-wheel, when you are facing the instrument,[5] lies on the right-hand side of the aforesaid shaft. However, on the arbor of the snail-gear a wheel of 12 teeth is fixed (towards the lower

[1] See p. 499 n. 4.

[2] *per suos limites.*

[3] *et idem proveniet.*

[4] We here follow II. 1 and the figure of f. 176r. The original reading here ('one tooth') would give a completely erroneous motion.

[5] *prospecta facie instrumenti.*

rota ad unam rotam 24 dencium fixam super rotam 44 minutorum de conclusione:[a] et movetur illa rota 44 minutorum super crucem fixam in medio instrumenti, et continet 198 dentes, et illo eodem circulo 198 dencium aptatur unus[b] axis fixus super unam extremitatem | crucis, 6 dencium. Et super illam axem, [cum nuce] 6 dencium, figitur una rota 59 dencium, cui aptetur una hasta habens in summitate unam rotam testitudinatam 2 dencium, et concordat cum rota 59 dencium. In alia extremitate haste est unus circulus 12 dencium. Facie autem instrumenti prospecta,[c] ponitur hasta cum duabus rotis in sinistra parte operis: et aptatur inferior pars haste ad unam rotam testitudinatam 12 dencium fixam super extremitatem calige deferentis corpus Solis . . .

[II. 5. 12]

Memorandum quod in isto instrumento Sol movetur quolibet die 24 horas, et in toto anno nichil de excessu pro anno bisextili: et in 4 annis deficit unus dies bisextilis.

Et movetur Luna per caligam Solis, ita quod in qualibet lunacione movet Sol precise Lunam, per rotas sibi aptatas, 29 dies 12 horas, videlicet per circulum divisum in 118 dentes. Et quolibet die capit 4 dentes per circulacionem Solis a puncto in predictum.

Quantitas unius dentis de circulo 44 minutorum est 7 min. 20 sec. 30 ter. unius hore. Et movet Sol in 4 diebus et 22 horis 7 minutis 20 secundis 30 terciis unum dentem de illo circulo magno 44 minutorum, quia hec est sexta pars unius lunacionis [cum 44 minutis 3 secundis 3 terciis].[d] Quia ille magnus circulus 44 minutorum movetur per nucem 6 dencium sibi coaptatam; quare nux in 29 diebus 12 horis precise movetur a puncto in punctum.

[a] The contraction could equally well be expanded *composicione*.
[b] A circulus (The word is pointed below, as though to erase.)
[c] A perspecta
[d] A in 44 minutis . . . (See the translation.)

part of the instrument), and this wheel engages with a wheel of 24 teeth, fixed on the wheel of the surplus 44 minutes.[1] The wheel of 44 minutes is pivoted in a cross-piece fixed in the middle of the instrument, and it has 198 teeth. The wheel of 198 teeth engages with a wheel of 6 teeth fixed on an arbor at one extremity of the cross. And on this arbor, with the wheel of 6 teeth, is fixed a wheel of 59 teeth which is to engage with a shaft having at its end a snail-wheel of 2 teeth, meshing with the wheel of 59 teeth. On the other end of the shaft is a wheel of 12 teeth. Looking at the face of the instrument, the shaft with the two wheels is placed on the left-hand side of the device. The lower part of the shaft engages with a snail-wheel of 12 teeth,[2] fixed on the end of the tube carrying the model of the Sun.

[II. 5. 12]

It is to be borne in mind that, in this instrument, the Sun moves twenty-four hours in any day, and in a whole year nothing to make up for the excess length of the bissextile year: and thus in four years it will be one bissextile day slow.[3]

The Moon is moved by the Sun's tube, so that in any lunation the Sun moves the Moon, by means of the wheels connecting with it (namely by means of the wheel divided into 118 teeth), in exactly 29 days 12 hours. And in any day, this advances the wheel by 4 teeth, the Sun's wheel moving round one complete revolution.[4]

The size[5] of a single tooth of the wheel of 44 minutes is $7^m\ 20^s\ 30^t$.[6] In $4^d\ 22^h\ 7^m\ 20^s\ 30^t$, the Sun moves one tooth of this great wheel of 44 minutes, since this is one-sixth part of a lunation,[7] not forgetting the $44^m\ 3^s\ 3^t$.[8] Since the great wheel of 44 minutes is moved by the nut of 6 teeth meshed with it, the nut, in exactly $29^d\ 12^h$, is moved round one complete revolution.

1 This peculiar name for the wheel of 198 (or 196) teeth is explained at p. 507 n. 1. The *de conclusione* of the text supports the interpretation there offered.

2 Evidence will shortly accumulate to show that this is a crown wheel with spiral teeth. The use of the adjective *testitudinatus* (or its orthographic equivalents!) is therefore not by itself sufficient to indicate the form of a wheel. One must also take into account whether the wheel is designated as *rota, circulus,* or *nux,* and even then there is always room for doubt. For the form of our new snailed crown-wheel see Plate XXI (from ff. 176^v–177^r of MS. A) and the lower figure on p. 72, vol. iii. It is later (II. 7. 2) called *quasi testitudinata.*

3 Notice that no special provision for adjusting the instrument is mentioned.

4 *a puncto in predictum,* lit. 'from the point to the aforesaid point'. On several occasions below the phrase *a puncto in punctum* is used with the same meaning.

5 Not the angular size, but rather the time to traverse.

6 The words *unius hore* which follow suggest that minutes, etc., are meant in the modern sense, and not what we are here calling *minuta diei* (etc.). The sense of this phrase, as of the whole paragraph, is very complicated however. For a discussion see our commentary. The paragraph provides a clue to the peculiar name 'great wheel of 44 minutes'. See p. 507 n. 1.

7 A better reason would have been 'since it was designed to do so'.

8 See text for the amendment made. This is the only way I can understand the original: we are to use the exact figure of $29^d\ 12^h\ 44^m\ 3^s\ 3^t$ rather than the rough figure of $29^d\ 12^h$.

[II. 5. 13]

Ad faciendum motum Lune medium.

Primo fiat composicio Solis secundum eius verum motum, prout satis evidenter probatur in suo loco; et super infimam extremitatem calige Solis fiat unus circulus 12 dencium, ad modum circuli testitunati. Deinde aptetur unus circulus super unam hastam 12 dencium, correspondens ad dictum circulum testitunatum 12 dencium. Deinde super aliam extremitatem haste fiat unus nux testitunatus 2 dencium; et aptetur illa nux 2 dencium ad unam | rotam divisam in 59 partes equales, et habeat ille circulus 59 dencium[a] axem[b] divisam in partes equales, ad quam quidem nux[c] 6 dencium aptetur [et?] una rota divisa in 196 partes equales. Et habebit illa rota 196 dencium unam caligam curtam super quam componatur una rota 24 dencium, cui aptetur alia rota super caligam Solis 12 dencium. Et super finem axis circuli 12 dencium fiat nux testitunatus 2 dencium, cui nuci aptetur alius circulus super hastam [8][d] dencium; et super aliam extremitatem haste fiat alter circulus 8 dencium; et fiat illa hasta super caligam Solis. Deinde aptetur illa ultima rota 8 dencium ad circulum divisum in [118],[e] super quem circulum erit corpus Lune.

Patet namque quod si Sol moveat quolibet die a puncto in punctum per motum sibi ordinatum, movebit precise 4 dentes de circulo Lune diviso in [118][f] dentes; et per consequens sequitur quod in 29 diebus 12 horis circuibit totum circulum Lune, quia tociens sunt dentes in circulo videlicet quater 29, et 2 dentes. Et quia una lunacio est 29 dies 12 hor. 44 min. 3 sec. 3 ter., de 44 min. 3 sec. 3 ter. mocionem explanabo.

Patet ex predictis quod, in qualibet lunacione, axis circuli divisi in 59 dentes movet de circulo 44 minutorum 6 dentes, [scilicet] de circulo

[a] A dentes
[b] In error? See n. 3 to the translation opposite.
[c] A nucem
[d] Required by the diagram of f. 176[r].
[e] After erasure, only the initial '1' remains. The diagram of f. 176[r] indicates the correct reading of '118'. For the probable original reading see the following footnote.
[f] A 160

[II. 5. 13]

To represent the mean motus of the Moon.

First let a device be made to represent the moving Sun, in accordance with his true motion, as is sufficiently clearly explained[1] in the relevant place.[2] At the far end of the Sun's tube, let there be a wheel of 12 teeth after the style of a snail-wheel. Fitted to a shaft there is then to be a wheel of 12 teeth, engaging with the snail-wheel of 12 teeth already mentioned. On the other extremity of the shaft, there is to be a single snail-nut of 2 teeth, which is to be meshed with a wheel divided into 59 equal parts. This wheel of 59 teeth should have an arbor[3] divided into equal parts[4] and fixed to the arbor there is to be a certain nut of 6 teeth meshing with a wheel divided into 196 equal parts. The wheel of 196 teeth will have a short tube, on which is placed a wheel of 24 teeth, and into this is meshed the other wheel of 12 teeth, on the Sun's tube. On the end of the arbor of the wheel of 12 teeth, let there be a snail-nut of 2 teeth,[5] with which is to be meshed another wheel, on a shaft of [8][6] teeth; and on the other end of the shaft there is to be yet another wheel of 8 teeth. This shaft is to be carried on the tube of the Sun.[7] Next, mesh the last wheel of 8 teeth with the wheel divided into 118, on which wheel will be the model of the Moon.[8]

For it is clear that if the Sun move round once, in any day you care to name, with the motion ordained for it, it will move exactly 4 teeth of the wheel divided into 118 teeth.[9] As a result, it follows that the entire wheel of the Moon will go round once, since there are 4 times 29, and 2 teeth, in the wheel.[10] And since one lunation is $29^d\ 12^h\ 44^m\ 3^s\ 3^t$, I will explain the motion during the $44^m\ 3^s\ 3^t$.

It is clear from what has already been said[11] that, in any lunation, the arbor of the wheel divided into 59 teeth moves 6 teeth of the wheel of

[1] *probatur.*

[2] *in suo loco.*

[3] *axis.* Is this a mistake for *circumferencia*? See the next footnote.

[4] Is he here hinting at a lunar scale, running from 1 to 29½ (the number of days in a lunation)? The wheel of 59 teeth does in fact revolve once in 29½ days exactly. The diagram at f. 176r gives no hint of such a scale.

[5] The diagram on f. 176r represents this clearly enough as a worm pinion (Plate XVII).

[6] Required by the diagram of f. 176r (Plate XVII).

[7] In other words, the angular velocity of the 12/2.8/8 . . . train is the (diurnal) velocity of the Sun, 360° in 24^h.

[8] *corpus Lune.*

[9] The manuscript has 160, but although there is a wheel of 160 teeth (see the beginning of II. 5. 3, and elsewhere), the author probably intended the figure 118, a wheel of which moves very nearly 4 teeth per day.

[10] This reading gives a wheel of 118 teeth, which is indeed the wheel of the Moon.

[11] Perhaps he is referring to the theory of Part I, and in particular to Proposition 10. It is certainly far from clear otherwise.

diviso in 196 dentes. Et est quantitas unius dentis 7 minuta hore 20 secunda 30 tercia, ex qua sequitur quod in 33 lunacionibus movebit 24 hore 13 min. 40 sec. 39 ter., quia in qualibet lunacione movebit 6 dentes; unde sequitur quod in 33 lunacionibus movebit totum circulum 196 dencium, videlicet 24 horas et, de nova revolucione, 2 dentes pro [13][a] minutis et ceteris; quia si multiplicentur 44 min. 3 sec. 3 ter. [per 196] usque ad tercias, et acciperes sextam partem, et illam sextam partem reduxeris ad minuta, que est quantitas dentis,[b] post multiplicacionem 196 per quantitatem unius dentis pervenient inde 24 hore 13 min. 39 sec., per addicionem 2 dencium de nova revolucione, et cetera.

[II. 5. 2]

Et aptetur inferior pars haste ad unam rotam testitudinatam 12 dencium fixam super extremitatem calige deferentis corpus Solis, sub qua quidem extremitate calige Solis figitur una hasta cum duabus rotis, qua inferior rota haste habet 15 dentes, et aptatur uni circulo testitudinato unius dentis fixo super crucem in medio instrumenti: et per illum motum retrahitur Sol quolibet die per unam dictam in suo cursu.

In superiori parte haste componitur una rota 9 dencium, et habet dentes bene longas, et aptatur illa superior extremitas haste uni circulo diviso in 219 dentes. Et est circulus quodammodo ecentricus in duabus partibus, et concentricus in aliis duabus partibus. Et figitur ille circulus 219 dencium super circulum diurnum tocius operis, et figitur ille circulus diurnus super interiorem extremitatem calige deferentis[c] duas primas laminas stellarum et graduum.

[a] A 3 [b] A +et [c] A deferencium

44 minutes,[1] that is to say, of the wheel divided into 196 teeth. The size of a single tooth is $7^{m}\ 20^{s}\ 30^{t}$,[2] from which it follows that in 33 lunations it will move $24^{h}\ 13^{m}\ 40^{s}\ 39^{t}$,[3] since in any lunation it will move 6 teeth. From this last fact it follows that in 33 lunations it will move the entire circle of 196 teeth, namely 24^{h}, and 2 teeth of the next revolution for the 13 minutes,[4] etc. This is so, since if $44^{m}\ 3^{s}\ 3^{t}$ be multiplied by 196 as far as the thirds, and you take one-sixth part, and you reduce that sixth part to minutes,[5] this being the size of a tooth,[6] after multiplying 196 by the size of a single tooth the result turns out to be $24^{h}\ 13^{m}\ 39^{s}$, through the addition of 2 teeth of the new revolution, and so on.[7]

[II. 5. 2]

The inner part of the shaft is fitted to a snail of 12 teeth at the end of the tube carrying the model of the Sun. Under this, the end of the tube for the model of the Sun, a shaft with two wheels is fixed. The inner[8] wheel of the shaft has 15 teeth, and engages with a snail-wheel of one tooth fixed on the cross in the middle of the instrument. By the movement [of this system of wheels] the Sun holds back in its course a certain amount.[9]

On the outer part of the shaft a wheel of 9 teeth is placed, and this has rather long teeth. This upper end of the shaft engages with a wheel divided into 219 teeth. The wheel is eccentric, in a sense, in two of its parts, and concentric in the other two parts.[10] This, the wheel of 219 teeth, is fixed on the journal wheel[11] of the main device,[12] and this journal wheel is fixed inside the tube carrying the first two plates, of stars and of degrees.

[1] What appears to be perfectly obvious to the author, but can hardly be so to the reader, is the relation between the wheel of 196 teeth, and 44 minutes. The period of rotation of the wheel being 963⅔ days provides no answer. In fact, the evidence of this and other sections suggests that the name arises from the fact that the wheel of 196 teeth is simply the first and largest in the train, the aim of which is to convert a periodicity of $29^{d}\ 12^{h}$ into the exact period of a lunation, namely $29^{d}\ 12^{h}\ 44^{m}\ 3^{s}\ 3^{t}$.

[2] See the preceding footnote. On his working, the figure should read $7^{m}\ 20^{s}\ 49^{t}$.

[3] On his working the figure should be $24^{h}\ 14^{m}\ 41^{s}\ 38^{t}$.

[4] The manuscript has 3 minutes, which is surely a mistake.

[5] Or 'continue the division as far as the minutes'.

[6] This makes no apparent sense. Perhaps there is a lacuna here.

[7] As always, the worked example is tantalizingly sketchy. The general plan seems to be that of comparing the theoretical requirements of the clock with the motions actually achieved. The author points out that in 33 lunations the wheel of 196 teeth revolves once, plus very nearly 2 teeth; and that this is more or less the motion theoretically required. He does not make a straight statement of the errors inherent in his design, as did Giovanni de' Dondi, for example. But then, Dondi's errors were greater!

[8] *inferior*. The wheels on the shaft are not permanently one lower than another, since the whole assembly moves round the wheel of 219 (corresponding to the wheel of 331 in the alternative scheme of II. 4. 3).

[9] *unam dictam*. Is there a word missing here?

[10] An earlier version of the wheel of 331. See the commentary.

[11] But note that the day in question is sidereal, and not solar.

[12] *tocius operis*.

[II. 5. 3]

Et nota quod in extimo loco limbi deferentis corporis Lune, ex parte dextra corporis Lune facie instrumenti circumspecta, componitur unus axis; et in fine remociori limbi[a] figitur super axem una rota 8 dencium; et aptatur circulo deferentis corporis Solis, diviso in 160 partes. Et in alia extremitate componitur una parva rota testitudinata 2 dencium,[b] et aptatur ad unam rotam 8 dencium fixa super alium parvum axem in alia extremitate secundi axis fixi. Super limbum deferentem Lune figitur una parva rota testitudinata 2 dencium,[b] et aptatur illa parva rota 2 dencium[b] ad unam rotam fixam super exteriorem extremitatem axis Lune. Et dividitur illa rota super axem Lune in 10 dentes equales. Deinde componantur omnia ista et coniungantur; et patebit, per mocionem circuli diurni, motus Solis, et Lune, et capitis Draconis, et stellarum fixarum in locis eorum.

[II. 5. 4]

Memorandum quod quia[c] Sol in 32 diebus pertranseat 29 dentes de circulo diviso in 331 dentes, tunc pertransit 29 triginta duas partes unius dentis in uno die. Sed 331ª pars circuli est 1 gradus 5 minuta 15 secunda 24 [tercia 28] quarta; hoc 29 vicibus est 31 grad. 32 min. 26 sec. 49 ter. 32 quar., cuius pars 32 est 2 min. 2 sec. 21 ter. 23 quar. 22 quin., quod multiplicatum per 29 facit 59 min. 8 sec. 20 ter. 17 quar. 38 quin., quod plus est motu Alfonsi in uno die. Ergo annus secundum hunc motum minor est quam annus solaris Alfonsi qui constat ex 365 diebus 34 min. 33 sec. 9 ter. 58 quar. Et per consequens ad servandum proporcionem oportet[d] similiter ponere motum Lune plus velocem quam ponit Alfonsi; sed tam Alfonsi quam secundum Ptolomeum. Motus elongacionis Solis a Luna ponitur in die secundum Alfonsum, 12 grad. 11 min. 26[e] sec. 41 ter. 38 quar.; secundum vero Ptolomeum 12 grad. 11 min. 26 sec. 41 ter. 30 quar. 18 quin. Si ergo predictus motus Solis in die addatur ad summam elongacionis in die in qua ambo conveniunt, erit quod surgit motus Lune proporcionatus predicto motui Solis, quod est 13 grad. 10 min. 35 sec. 1 ter. 37 quar. 56 quin.; et nota hic[f] bene hoc secundum elongacionem Ptolomei.[g] Secundum Alfonsum est 13 grad.

[a] A +et
[b] A unius dentis (See the commentary for a justification of the alteration.)
[c] A si [d] Reading obscure [e] A 46 [f] A hec [g] A percul

[II. 5. 3]

Notice that on the outermost edge of the limb carrying the model of the Moon, on the right-hand side of the Moon looking at the face of the instrument, an arbor is placed; and at the end away from the limb, fixed to this arbor, there is a wheel of 8 teeth. This engages with the wheel carrying the model of the Sun, divided into 160 parts.[1] At the other end of the arbor is placed a small snail-wheel of 2 teeth,[2] engaging with a wheel of 8 teeth, fixed on another short arbor, and on the other end of this second arbor. On the limb carrying the Moon is fixed a small snail-wheel of 2 teeth,[3] and this small wheel of 2 teeth engages with a wheel fixed on the outer end of the Moon's arbor. This wheel on the Moon's arbor is divided into 10 equal teeth. And then all this is to be put together and assembled; and, by the movement of the journal wheel, the movement of the Sun, the Moon, the head of the Dragon, and the fixed stars in their places, will all be apparent.

[II. 5. 4]

It is to be remembered that since in 32 days the Sun traverses 29 teeth of the wheel divided into 331 teeth, then it traverses $\frac{29}{32}$ of a tooth in a day.[4] But $\frac{1}{331}$ of a circle is $1^\circ\ 5'\ 15''\ 24'''\ 28^{iv}$;[5] 29 of these make $31^\circ\ 32'\ 26''\ 49'''\ 32^{iv}$,[6] and a 32nd part of a tooth is $2'\ 2''\ 21'''\ 23^{iv}\ 22^{v}$. Multiplying this by 29 gives $59'\ 8''\ 20'''\ 17^{iv}\ 38^{v}$, which is more than the motus in one day, according to Alfonso. Thus a year, corresponding to this motus, is less than an Alfonsine solar year of $365^{d}\ 34^{m}\ 33^{s}\ 9^{t}\ 58^{q}$. As a result, if we keep the right ratios, we are obliged to make the Moon move more rapidly than does Alfonso. This is what Alfonso did for Ptolemy. The motus of elongation of the Sun from the Moon according to Alfonso is, in a day, $12^\circ\ 11'\ 26''\ 41'''\ 38^{iv}$; while according to Ptolemy it is $12^\circ\ 11'\ 26''\ 41'''\ 30^{iv}\ 18^{v}$. Now if the aforesaid daily motus of the Sun be added to the total daily elongation for the day in which they come together, the result will be the motus of the Moon corresponding to the aforesaid motus of the Sun. This (i.e. the Moon's motus) is $13^\circ\ 10'\ 35''\ 1'''\ 37^{iv}\ 56^{v}$, using Ptolemy's figure for the elongation, and $13^\circ\ 10'\ 35''\ 1'''\ 55^{iv}\ 36^{v}$ using Alfonso's. According to this supposition,

1 That is to say, a wheel fixed to the *caliga Solis*.

2 See the commentary for the reason for altering the text.

3 See the commentary.

4 The truth of this is evident only to those with a knowledge of the train under discussion: $(1^{d})/29.32/2.16/8.331/(365\frac{7}{29}{}^{d})$. (See II. 3, commentary and synopsis.) The whole trend of these opening sentences suggests not an oval gear of the kind we have suggested elsewhere, but a gear whose teeth are moved uniformly in time. This implies a non-uniform scale of days.

5 The last figure is supplied on the strength of the statement immediately following. This is the only occasion on which this calculation is obviously done correctly (omission of the 28^{iv} is obviously a scribal error). For the incorrect versions, on which we may perhaps blame the author's carelessness in copying his own writings, see p. 499 n. 5.

6 This is a misleading statement, since it is not used in the calculation.

10 min. 35 sec. 1 ter. 55 quar. 36 quin. Nota bene. Scito ergo pro certo, secundum hanc supposicionem, quod quanto tempus motus Lune contra certos dentes fuerit minus quam tabula proporcionis in tempore, tanto verius infra [26][a] quarta zodiaci, cui correspondet in tempore pro uno die duo quarta[b] fere tanto melius quia 2 quartis diei respondent 26 quarta[b] zodiaci. Nota hoc infallibiliter.

[II. 6. 11]

Et dividitur illa lamina Draconis per dentes perforatas in 177 partes equales, quibus aptatur una parva rota 6 dencium, super finem cuiusdam haste. Et super aliam extremitatem illius haste figitur una rota 12 dencium, et aptatur illa rota 12 dencium ad unam rotam fixam super caligam Solis, in parte anteriori continentem 19 dentes. Postea sequitur caliga Solis, super quam in parte anteriori fixa est rota 19 dencium correspondens Capiti[c] Draconis.

[II. 6. 12]

Motus Capitis Draconis procedit a summitate calige Solis, videlicet de una rota divisa in 19 dentes, et movetur secundum mocionem Solis, hoc est semel in anno a puncto in punctum motu rapto,[d] cui coaptatur una rota 12 dencium, et movetur illa rota 12 dencium quolibet anno 19 dentes. Est autem alia rota 6 dencium, in alia extremitate haste, et movetur quolibet anno 9 dentes et dimidius; et hoc bene patet per computacionem mocionis. Et sunt in toto circulo Capitis Draconis 177 dentes, patet namque ex predictis quod quolibet anno movetur de illa rota 9 dencium et dimidii. Et est quantitas unius dentis de zodiaco 2 grad. 2 min. 2 sec. 2 ter.; unde patet quod si multiplicetur motus parvi circuli, capientis 9 dentes et dimidium in anno, per 19, exibunt de magno circulo 180 dentes[e] et dimidium; unde patet per unam calculacionem, quod in 19 annis revolvitur magna rota a puncto in punctum et, de revolucione nova, 3 dentes et dimidius. Tunc sequitur quod in 19 annis movetur per totum zodiacum cum 7 gradibus, et hec est precisa mocio Capitis in 19 annis preter 20 minuta; et de illis non est multum curandum. Attamen patet quod in 57 annis deficiet per unum gradum equalem, set leviter instrumentum[f] potest emendari.

[a] A 20 [b] A quarte [c] A capitis [d] A rapti [e] A dentis
[f] A intuenti

you may therefore be quite sure that the Moon will take less time to cover a certain number of teeth than the *Table of Proportion* would lead one to suppose. The maximum discrepancy[1] will be 26^{iv} of the zodiac, corresponding to almost 2 *quarta diei*, since 2 *quarta diei* correspond to 26^{iv} of the zodiac. Take careful note of this.

[II. 6. 11]

The Dragon plate is divided, by means of perforated teeth,[2] into 177 equal parts, engaging with which is a small wheel of 6 teeth, on the end of a shaft. And on its other extremity a wheel of 12 teeth is fixed to the shaft, and this wheel of 12 teeth meshes with a wheel fixed to the Sun's tube, containing 19 teeth on its foremost side. After this, the tube for the Sun follows, at the front part of which tube is the wheel of 19 teeth, relating to the head of the Dragon.

[II. 6. 12]

The movement of the head of the Dragon is taken off the top of the Sun's tube, namely from a wheel divided into 19 teeth. It moves in accordance with the Sun's motion, that is, once in a year, but with a reduced velocity.[3] Meshing with the wheel of 19 teeth is a wheel of 12 teeth, and this moves 19 teeth in a year. The wheel at the other end of the shaft, however, has 6 teeth, and this moves $9\frac{1}{2}$ teeth in any year; and this is perfectly evident in computing the motion. There are, all told, 177 teeth on the wheel of the head of the Dragon, and yet it is clear from what has been said before that, in any year, $9\frac{1}{2}$ teeth of this wheel move. One tooth, measured along the zodiac, is $2^\circ\ 2'\ 2''\ 2'''$; whence it is clear that, if the movement of the small wheel, which moves $9\frac{1}{2}$ teeth in a year, be multiplied by 19, $180\frac{1}{2}$ teeth of the large wheel will result; whence it is evident from calculation that in 19 years the large wheel will turn one complete revolution, and $3\frac{1}{2}$ teeth of the new revolution. It follows from this that in 19 years it moves through the whole zodiac and a further 7 degrees, and this is the exact motion of the head [of the Dragon] in 19 years less 20 minutes—and one need not worry about these. It is obvious that in 57 years it will be short by 1 equal degree,[4] but the instrument can easily be corrected.

[1] In one year (in the case of the Moon, of course).

[2] *per dentes perforatas*. Elsewhere in the text the verb *perforare* and its derivatives are used strictly to imply the existence of *holes*. This fact, together with the diagram of f. 176r, which represents the Dragon's head as projecting from the edge of the Dragon plate, suggests the use of an unusual form of perforated gear. This is drawn in vol. iii, p. 72, fig. II. 6.1. It has the advantage that it will not allow end-play in the engaging nut. The phrase comes again, in relation to the Dragon train on p. 517, but now in relation to another wheel. This may be a mistake, but if not, and if we accept the drawing on f. 176r in relation to *that* passage (but not this), a 'perforated wheel' is simply a crown-wheel.

[3] *motu rapto*, lit. 'with robbed motion'.

[4] An ordinary degree, as opposed to the 'degrees' on certain astrolabe scales.

[II. 6. 13]

Post motum Lune fiat motus Capitis hoc modo, videlicet, super caligam Solis figatur una rota 38 dencium cui respondebit [rota] unius haste 24 dencium. In altera fine haste[a] fiat rota 6 dencium cui respondebit una rota 177 dencium, in cuius centro figatur una caliga protendens se a rota 177 dencium usque ad laminam divisam; et in illo fine figatur Caput et Cauda Draconis. Et nota quod uterque finis haste Capitis debet figi super caligam diurnam.

[II. 6. 2]

Et quia in tabulis Tholetanis continetur quod cursus equalis Capitis Draconis in una hora est 8 secunda, et sunt hore in anno solari 8,765 hore 49 min. 15 sec. 59 ter. 12 quar.; tunc sequitur quod motus Capitis Draconis in anno, secundum quod movetur in una hora 8 sec., est 19 grad. 28 min. 46 sec. 34 ter. 8 quar. Et sunt hore in 19 annis 166,550 hor. 36 min. 3 sec. 44 ter. 48 quar., unde sequitur quod medius motus Capitis Draconis in 19 annis, secundum quod movetur una hora 8 sec., [est] | 370 grad. 6 min. 44 sec. 48 ter. 32 quar.

Est autem medius motus Draconis in anno, secundum Alfonsum, 19 grad. 20 min. 28 sec. 26 ter. 7 quar. 10 quin. Et per ipsum Alfonsum, in 19 annis, 367 grad. 29 min. 0 sec. 16 ter. 13 quar. Et si dividatur rota Capitis Draconis in 177 dentes, erit quantitas unius dentis 2 grad. 2 min. 2 sec. 2 ter. 2 quar. Quantitas unius anni de rota divisa in 177 dentes erit 19 grad. 19 min. 19 sec. 19 ter. 19 quar. Et si illa divisio moveatur secundum mocionem Solis, prout pretactum est in mocione Capitis Draconis, in instrumento erit quantitas 19 annorum 367 grad. 29 min. 0 sec. 16 ter. 13 quar. Et si fiat divisio rote Capitis Draconis in 176 dentes, erit quantitas unius dentis 2 grad. 2 min. 43 sec. 38 ter. 10 quar. Quantitas unius anni: 19 grad. 25 min. 54 sec. 32 ter. 35 quar. Et si illa eadem divisio, videlicet 176 dencium, fiat per motum Solis, erit quantitas 19 annorum 369 grad. 12 min. 16 sec. 19 ter. 5 quar.

Motus Capitis Draconis secundum Prefacium Iudeum, in anno, est 19 grad. 21 min. 21 sec. 32 ter. 7 quar. In 19 annis: 367 grad. 45 min. 49 sec. 10 ter. 13 quar. Rota 176 dencium excedit: excedit Prefacium in 19 annis, 1 grad. 26 min. 27 sec. 8 ter. 52 quar. Item rota 176 dencium est minus mocionis 8 secundorum in una hora; in 19 annis 0 grad. 54 min. 28 sec. 31 ter. 27 quar. Item rota 177 dencium est minus Prefacii 0 grad. 38 min. 43 sec. 3 ter. 6 quar. Illa eadem rota 177

[a] A hastis

[II. 6. 13][1]

After the motion of the Moon, let the motion of the head of the Dragon be represented in this way: On the Sun's tube[2] is to be fixed a wheel of 38 teeth, with which meshes a wheel of 24 teeth, fitted on a shaft. At the other end of the shaft there is to be a wheel of 6 teeth, with which a wheel of 177 teeth meshes. In the centre of this a tube must be fixed, reaching from the wheel of 177 teeth up to the divided plate.[3] The head and tail of the Dragon are to be fixed on the tube, [on the wheel of 177 teeth]. And note that both ends of the Dragon shaft[4] must be fixed on the journal tube.[5]

[II. 6. 2]

Since the Toledan tables have it that the uniform course of *Caput Draconis* in an hour is $8''$, and since in a solar year there are $8{,}765^h\ 49^m\ 15^s\ 59^t\ 12^q$, therefore it follows that the motus of *Caput Draconis* in a year, if it moves $8''$ in an hour, is $19^\circ\ 28'\ 46''\ 34'''\ 8^{iv}$. The number of hours in 19 years is $166{,}550^h\ 36^m\ 3^s\ 44^t\ 48^q$, whence it follows that the mean motus of the head of the Dragon in 19 years, assuming that it moves $8''$ in an hour, is $370^\circ\ 6'\ 44''\ 48'''\ 32^{iv}$.

The mean motus of the Dragon in one year, according to Alfonso, is $19^\circ\ 20'\ 28''\ 26'''\ 7^{iv}\ 10^{v}$. And by this same Alfonso, in 19 years it is $367^\circ\ 29'\ 0''\ 16'''\ 13^{iv}$. If the wheel of the head of the Dragon be divided into 177 teeth, the size of each tooth will be $2^\circ\ 2'\ 2''\ 2'''\ 2^{iv}$. The length of a single year on the wheel divided into 177 teeth will be $19^\circ\ 19'\ 19''\ 19'''\ 19^{iv}$. If this division[6] is to move in keeping with the motion of the Sun, as touched upon previously in connection with the motion of the head of the Dragon, on the instrument, 19 years should amount to $367^\circ\ 29'\ 0''\ 16'''\ 13^{iv}$. If the wheel of the head of the Dragon be divided into 176 teeth, the size of a tooth will be $2^\circ\ 2'\ 43''\ 38'''\ 10^{iv}$. The size of a year will be $19^\circ\ 25'\ 54''\ 32'''\ 35^{iv}$. And if this same division, namely into 176 teeth, be made in terms of the motus of the Sun,[7] 19 years will amount to $369^\circ\ 12'\ 16''\ 19'''\ 5^{iv}$.

The annual motus of the head of the Dragon, according to Profatius Judeus, is $19^\circ\ 21'\ 21''\ 32'''\ 7^{iv}$. In 19 years: $367^\circ\ 45'\ 49''\ 10'''\ 13^{iv}$. The wheel of 176 teeth moves more than this—in fact by an amount $1^\circ\ 26'\ 27''\ 8'''\ 52^{iv}$ in 19 years, assuming that Profatius is right. The wheel of 176 teeth moves rather less than $8''$ per hour—by an amount $0^\circ\ 54'\ 28''\ 31'''\ 27^{iv}$ in 19 years. The wheel of 177 teeth falls short of the figure quoted by Profatius by an amount $0^\circ\ 38'\ 43''\ 3'''\ 6^{iv}$. That same

[1] The ground covered here is repeated in II. 7. 5, but our opening sentence shows that this is not a misplaced fragment. (In the later chapters, the Moon is dealt with *after* the Dragon.)

[2] See the synopsis.

[3] *laminam divisam*—viz. the astrolabe plate which serves as the face of the clock.

[4] That is, the shaft carrying the wheels of 24 and 6 teeth. The arrangement is clear on the diagram of f. 176^r (Plate XVII).

[5] See the synopsis for the meaning.

[6] i.e. of a wheel divided in this way (see below).

[7] *per motum Solis.*

dencium est minus mocionis 8 secundorum in una hora; in 19 annis 2 grad. 39 min. 37 sec. 41 ter. 25 quar. Rota 176 dencium excedit Alfonsum in 19 annis, 1 grad. 43 min. 16 sec. 2 ter. 52 quar.

Et est circulus 176 dencium in medio inter Alfonsum et numerum mocionis 8 secundorum in una hora, set propius est numero mocionis 8 secundorum in una hora, per o grad. 48 min. 47 sec. 31 ter. 25 quar. Et ideo per illam rotam 176 dencium erit tota nostra operacio pro Capite Draconis.

[II. 7. 1]

Composicio superioris instrumenti.

Prima rota est rota diurna, et dividitur in 360 dentes, cui conclavatur una rota ecentrica, pro motu Solis, 331 dencium, cui applicatur una nux, ad modum clave, 8 dencium super finem unius haste. Et in altero fine figitur una nux 16 dencium, cui respondet una nux testitudinata 2 dencium. Iuxta centrum instrumenti, et in alio fine axis eiusdem nucis, figitur rota 32 dencium, cui respondet rota fixa circa centrum tocius operis, super trabem ferream transversantem totum instrumentum. Et figitur | illa trabes super tabulas horologii. Hec de motu Solis.

[II. 7. 2]

Post motum Solis sit motus Lune isto modo:

Super unam laminam longam ferream, que ducit hastam Solis, que eciam conclavatur calige Solis super interiorem finem, conclavatur una rota quasi testudinata. Et est rota bene lata ad latitudinem[a] fere

[a] A latitudo

wheel of 177 teeth falls short of 8″ per hour by an amount $2° 39' 37'' 41''' 25^{iv}$ in 19 years. The wheel of 176 teeth exceeds Alfonso's figure for 19 years by $1° 43' 16'' 2''' 52^{iv}$.

The motion of the wheel of 176 teeth lies between Alfonso's figure and the figure of 8″ per hour. It is more correct than the later figure by a margin of $0° 48' 47'' 34''' 25^{iv}$.[1] And with a discussion of this wheel of 176 teeth, we have explained all the results of our calculations, as far as they relate to the motion of the head of the Dragon.[2]

[II. 7. 1]

Assembling the above instrument.

The first wheel is the journal wheel, being divided into 360 teeth. This is riveted to[3] an eccentric wheel of 331 teeth, for the motus of the Sun. To this is applied a nut of 8 teeth, after the fashion of a club,[4] on the end of a shaft. At the other end is affixed a nut of 16 teeth, engaging a snail-nut of 2 teeth. In the centre of the instrument, and at the other end of the arbor for that nut, a wheel of 32 teeth is fixed, meshing with a wheel[5] fixed around the middle of the whole device, on an iron beam crossing the whole instrument. This beam is fixed on the frame[6] of the horologe. And so much for the motus of the Sun.

[II. 7. 2]

After the motus of the Sun, the motus of the Moon is to be represented, and in this way:

On a long strip of iron plate, which leads round the pointer [carrying] the Sun, and which also is riveted to the Sun's tube at the inner end,[7] is riveted a wheel which is rather like a snail-wheel.[8] The wheel is quite

[1] The meaning of this statement, confirmed by the figure quoted, is that, if T, A, W refer, respectively, to the motions of the head of the Dragon in 19 years according to Toledo, Alfonso, and the wheel of 176, then $(W-A)-(T-W) = 0° 48' 47'' 34''' 25^{iv}$.

[2] A totally different translation might have been given of this sentence: 'Thus by this wheel of 176 teeth shall we represent the motion of the head of the Dragon.' This translation, which gives *operacio* the sense of *opus*, is ruled out by the obvious superiority of the wheel of 177 teeth. (See synopsis.)

[3] *conclavatur.* This is our first (and last) encounter with the *rota diurna . . . 360 dencium.*

[4] *ad modum clave.* The last word is unlikely to be a mistake for *clavis* ('key'). In likening the nut, on its shaft, to a club or cudgel (classical meaning), he is in any case singling it out for comment. Is this meant to convey the idea that the nut was *tapered*? Tapered medieval maces—even with leaves as on a pinion—still survive. But this idea is not borne out by the diagrams (see commentary), and it seems more probable that the phrase is simply meant to convey the excessive *length* of the nut. Whether or not the nut was tapered makes no difference to the working out of the problem.

[5] Namely the wheel of 29 teeth.

[6] *super tabulas.* See p. 517 n. 1.

[7] *super interiorem finem.* (The plate seems to correspond with the row of dots near the base, on f. 176r (Plate XVII). The word translated 'pointer' is now *hasta*, not *regula*.

[8] *una rota quasi testitudinata.* Cf. the note to the translation of II. 5. 11.

duorum pedum. Et est numerus dencium eiusdem rote 12, et sit illa rota ad modum istius figure, cui respondet una parva [rota] 12 dencium, et figitur super hastam prope finem superiorem. Et prope finem alterum eiusdem haste figitur una nux testudinata 2 dencium, cui respondet una rota 59 dencium; et in alio fine axis super quam figitur illa, figitur alia rota parva 6 dencium (et nota quod hasta correspondens magne rote testudinate et rota 59 dencium figuntur super trabem transversalem ferream fixam super tabulas) cui correspondet rota 44 minutorum. Et est numerus eiusdem rote 194, super quam figitur una rota circa centrum 24 dencium.[a] Et ista rota non caliga nec figitur super caligam alterius, sed pendet supra caligam Solis; cui rote, videlicet 24 dencium, aptetur[b] una rota parva 12 dencium; et alio fine axis eiusdem[c] rote figitur nux testitudinata 2 dencium, cui respondet una rota 16 dencium fixa super finem inferiorem unius haste. Et in superiori fine eiusdem figitur rota 16 dencium, cui respondet rota deferens corpus Lune 255 dencium. Et figitur, portans duas rotas 16 dencium, cum scru, super unam laminam compositam super caligam Solis. Et rota 255 dencium pendet super caligam Solis sine aliqua caliga; sed habet in uno[d] diametro curtam laminam conclavatam,[e] in cuius fine prominent duo tenacula. Apposita ista rota, videlicet 255 dencium, ad predictas rotas prominet finis calige Solis, super quam figitur, cum tenaculis, circa centrum, una rota perforata 38 dencium; et in illa rota incipit motus Capitis.

[II. 7. 3]

Tunc additur unum firmaculum circa centrum, includens omnes caligas et rotas circa centrum fixas. Tunc rota perforata apponitur parva

[a] A dentes [b] A figitur [c] A eius [d] A unius [e] A conclavata

wide, being almost 2 feet at its widest. It has 12 teeth, and is made in the manner depicted in the accompanying figure, [f. 177^r (Plate XXI)]. It engages with a small wheel of 12 teeth, which is fixed on a shaft near its upper end. Near the other end of this same shaft is fixed a snail-nut of 2 teeth, engaging with a wheel of 59 teeth; and at the other end of the arbor on which the latter is fixed, another small wheel of 6 teeth is fixed. And notice that the shaft linking the large snail-wheel and the wheel of 59 teeth are fixed on a transverse iron beam, fixed on the tables.[1] With the wheel of 6 teeth meshes the wheel of 44 minutes. The number of teeth of this wheel is 194; and on it, and about its centre, is fixed a wheel of 24 teeth. This wheel is neither with a tube,[2] nor is it fixed on the tube of any [other wheel], but is suspended above the tube of the Sun. Engaging with this wheel, namely that of 24 teeth, is a small wheel of 12 teeth; and at the other end of the arbor of the wheel is fixed a snail-nut of 2 teeth, engaging with a wheel of 16 teeth, fixed on the lower end of a shaft. At the upper end of this shaft is fixed a wheel of 16 teeth, engaging with the wheel of 255 teeth carrying the model of the Moon. [This shaft], carrying the two wheels of 16 teeth, is fixed with a 'screw',[3] over the plate placed on the Sun's tube. The wheel of 255 teeth has no tube, and is placed on the tube of the Sun; but on one diameter it has a short strip riveted,[4] at the ends of which project two bolts.[5] The end of the Sun's tube (near the centre of which is fixed, with fasteners, a perforated wheel of 38 teeth)[6] protrudes, when the wheel of 255 teeth is fitted to the aforesaid wheels. On the wheel of 38 teeth begins the motus of the head [of the Dragon].

[II. 7. 3]

Next is added a support[7] at the centre, holding all those tubes and wheels which are fixed at the centre. The perforated wheel of 38 teeth

[1] Since the beam is depicted on f. 176^r, we can see that 'table' corresponds to the side-pieces of the frame—probably not themselves of iron, if they were in the shape of very large flat plates. But this is not at all obvious.

[2] *ista rota non caliga.* Might this once have been *ista rota non habet propriam caligam?*

[3] *cum scru.* The idea of fixing a moving shaft or arbor with a screw, in an unspecified way, is very puzzling until we look into older meanings. Diez (*Etymologisches Wörterbuch der romanischen Sprachen*, 4th ed., 1878) derives 'screw' from the Latin *scrobem*, acc. of *scrobs*, a ditch, trench, or hole. John Mincheu (*The Guide into the Tongues*, 2nd ed., 1627) derives the word from French *escroue*, which is in any case a reasonable intermediary, and he adds this very enlightening definition: 'a scrue, the hole or hollow thing wherein the vice of a presse, &c doth turn'. It seems highly probable that *scru* is to be translated 'bearing', for which, as it happens, we have no other word in the text.

[4] *conclavata.*

[5] *tenacula.* I used the word 'riveted' for *conclavata*, but 'rivets' might be equally appropriate here. See the diagram of f. 176^r (Plate XVII).

[6] See p. 511 n. 2. Previously it was the *last* wheel (of 177) in the train which was said to have perforated teeth. Now, and subsequently, it is the wheel of 38.

[7] *firmaculum.*

rota fixa super finem unius haste; et continet 24 dentes. Et in alio fine haste sit rota parva 6 dencium. Tunc apponitur una lamina firmans superiorem finem haste predicte; et alius finis haste figitur in caliga diurna, inter rotam perforatam et finem lamine continentis superiorem finem haste predicte. Et firmatur illa lamina cum tenaculis super caligam diurnam. Tunc apponitur una rota cum caliga, et firmatur cum tenaculis, sed pendet super caligam diurnam. Et est numerus dencium eiusdem rote 166, cui correspondet nux[a] 6 dencium fixa[b] super caligam diurnam. Tunc apponitur una lamina cum caliga, et figitur superior finis cum tenaculis lamine fixe super diametrum rote 255 dencium. |

[II. 7. 4]

Tunc apponitur una[c] longa lamina portans Solem circa diale, et in alio fine habet laminam ob umbratam Lunam in eclipsi; et circa cuius centrum figitur una rota 48 dencium, et ista, cum sibi appositis, vertit[d] Lunam in concavitate[e] quadam. Et ista lamina longa firmatur cum tenaculis ad laminam fixam super caligam Solis, que ducit hastam correspondentem rote 255 dencium.

His apponitur una lamina cum plurimis firmaculis, in cuius dorso componitur una rota 48 dencium; et correspondet rote fixe in longa lamina circa centrum instrumenti. In alia parte lamine, super finem axis rote 48 dencium, figitur una parva rota 18 dencium, cui correspondet alia parva [rota] 18 dencium. Et illa rota non figitur, sed habet in medio quadrature[f] foramen que recipit hastam in cuius fine superiori ponitur Luna.

[II. 7. 5]

Et ita fit ut Luna possit ascendere et descendere tempore quo obviat Capiti vel Caude[g] Draconis; et ideo adiungitur haste[h] portanti[i] Lunam quasi regula perforata 18 foraminibus, cui aptatur una quasi semirota 18 dencium in eius dorso, in cuius alia parte componuntur due lamine perforate utraque earum 5 foramina. Et lamina contingens omnia ista firmatur super caligam fixam super laminam firmatam lamine[j] fixe super diametrum rote 255 dencium.

[a] A nuci [b] A fixi [c] A +una [d] A vertunt [e] A concavitatem
[f] A quadrat' [g] A Cauda [h] A hasta [i] A portantem [j] A +lamine

is then engaged with a small wheel fixed on the end of a shaft; and this has 24 teeth. At the other end of the shaft there is to be a small wheel of 6 teeth. Then a plate is added, securing the upper end of the aforementioned shaft; and the other end of the shaft is fixed on the diurnal tube, between the perforated wheel and the edge of the plate containing the upper end of the aforesaid shaft. And this plate is made secure with bolts, and pivoted on the diurnal tube. A wheel with tube is then added, and fixed with rivets, but it is suspended over the diurnal tube. The number of teeth of this wheel is 166. It engages with the nut of 6 teeth, fixed on the diurnal tube. Now is added a plate with a tube, and the upper end is fixed, with fasteners, to the plate fixed on the diameter of the wheel of 255 teeth.

[II. 7. 4]

A long plate, [i.e. a strip of metal], is next added, carrying the Sun around the dial, and at the other end it has a plate representing the shaded part of the Moon during an eclipse.[1] At its centre[2] is fixed a wheel of 48 teeth; and that, together with its associated wheels, turns[3] the Moon in a concavity. The long plate is secured with fasteners to the plate fixed on the Sun's tube, which drives the shaft meshing with the wheel of 255 teeth.

To the wheel of 255 a plate is added, with many supports,[4] on the back of which is a wheel of 48 teeth. This engages with the wheel fixed on the long plate around the centre of the instrument. On another part of the plate, on the end of the axis of the wheel of 48 teeth, is fixed a small wheel of 18 teeth, with which meshes another small wheel of 18 teeth. This wheel is not fixed, but has a hole in the middle of its cross,[5] which takes the shaft at the upper end of which the Moon is placed.

[II. 7. 5]

So that the Moon may ascend and descend[6] in time with its encounters with the head or tail of the Dragon, the shaft carrying the Moon is connected to something like a perforated rule with 18 slots, engaging with something like a half-wheel of 18 teeth on its curved part. On another part of this, two perforated plates are put, each with 5 slots. And the plate touching all these is fixed on the tube which is fixed on the plate secured to the plate fixed on the diameter of the wheel of 255 teeth.

[1] This 'long plate' must refer to the curved strip which passes outside the model of the Moon to the right of the figure on f. 176r (Plate XVII). The crescent-shape might be a side-view of the eclipse plate, unless it is meant to represent the Sun itself, which might well be the case.

[2] This presumably means the centre of the instrument generally. See our commentary.

[3] *vertit* (or *vertunt*).

[4] *firmaculum*.

[5] I take this to mean that we are dealing with a four-spoked wheel.

[6] This is likely to mean no more than 'move one way and another'. It can only be for lunar eclipses since the Sun is in front of the Moon and the head of the Dragon.

Tunc apponitur una lamina deferens Caput et Caudam Draconis, et in circumferencia exteriori fit ad modum magne pellvis abscise[a] fundo. Et prope Caput in lamina transversali iuxta Caput figuntur due lamine. Utraque earum continet 5 dentes, et simili modo fit iuxta Caudam; que dum proveniunt ad duas laminas perforatas fixas super ligniculum, in cuius dorso est semirota, expellunt Lunam ultra Caput et Caudam. Et figitur ista lamina, cum Capite et Cauda, super unam caligam exeuntem a rota divisa in 166 dentes.

[II. 7. 6]

Tunc apponitur diale, in cuius circumferencia gradus et nomina signorum ecliptice. Et in alio circulo inscribuntur dies mensium et nomina earum, tunc ecliptica, tunc stelle quedam tam maioris quam minoris[b] magnitudinis. Et figitur diale super caligam diurnam exeuntem a rota diurna. Tunc apponitur almicantaraz, tunc unum suppositorium ad supportandum finem axis portantem totum instrumentum. Et completum est, si Deus voluerit. Amen.

[SUPPLEMENTUM]

Memorandum quod si circulus magnus in quo fixus est Sol habuerit 360 dentes, tunc nux per quam[c] movetur corpus Lune habebit 18 dentes; et erit idem quod modo que nunc habet 160 dentes | et nux eius 10 dentes. Item, si niger circulus in quo fixa est Luna habuerit 354 dentes, et nux movens eum habuerit 12 dentes, idem erit sicut modo; quia nux habet modo 8 dentes et niger circulus 118.

Item si nux illius parvi circuli, qui circulus habet 59 dentes, habuerit 11 dentes, tunc circulus 44 minutorum habebit 363 dentes, et idem erit sicut modo; quia nux habet nunc 6 dentes et magnus circulus 44 minutorum habet 198. Si circulus dierum tortuosus habuerit 487 dentes, et nux iuxta centrum eius habuerit 9 dentes, tunc nux iuxta limbum circuli dierum habebit 12 dentes; et tunc complebit cursum suum in 365 diebus et 6 horis precise.

[a] The MS. originally read *abscise*, but an 's' is erased, leaving *ab cise*.
[b] A minorius
[c] A quem

A plate is now added, carrying the head and tail of the Dragon, and its outer circumference is made in the shape of a large *pellvis*, cut off at the base.[1] And near the head [of the Dragon] on the transverse plate[2] two plates are fixed, each with 5 teeth; and similarly next to the tail. When they come into contact with the two perforated plates,[3] fixed on the little piece of wood[4] (the half-wheel being on the back of this), these move the Moon outwards when it passes beyond the head and tail. This plate is fixed, with the head and tail, *Caput* and *Cauda*, on a tube on which is pivoted the wheel divided into 166 teeth.

[II. 7. 6]

Next, a dial is added, on the circumference of which are degrees and names of the signs of the ecliptic. In another circle the days of the months and their names are inscribed, then the ecliptic, then some of the stars, of both greater and lesser magnitudes. The dial is fixed on the diurnal tube, stemming from the journal wheel. The almucantars are now added, and a support for the end of the arbor carrying the whole instrument. And the work is finished, if God be willing.

[SUPPLEMENTARY CHAPTER]

It is to be remembered that if the large wheel, on which the Sun is fixed,[5] has 360 teeth, then the nut by which the model of the Moon is moved will have 18 teeth. The same result will obtain if, as at the present time, [the Sun's wheel] has 160 teeth, and the nut 10 teeth.[6] Next, if the black wheel on which the Moon is fixed has 354 teeth, and the nut moving it 12 teeth, the same result is obtained as at present, when the nut has 8 teeth and the black wheel 118.

If the nut of this small circle (which has 59 teeth) is to have 11 teeth, then the wheel of 44 minutes will have 363 teeth, with the same result as at present; for the nut now has 6 teeth and the large wheel of 44 minutes 198. If the oval wheel[7] of days has 487 teeth, and the nut at its centre 9 teeth, then the nut near the limb of the wheel of days will have 12 teeth;[8] and it will then complete its course in 365 days and 6 hours precisely.

[1] Perhaps this means 'basin' (*pelvis*). Alternatively, this might mean that the dragon plate was shaped like a supposed dragon skin (*pellis*). Cf. Apian's drawings for the sort of thing he may have had in mind.

[2] The basin?

[3] The half-wheel and rack?

[4] *ligniculum*. Presumably a support for the half-wheel.

[5] See II. 5. 3, and the figure depicting the *rotation* of the Moon.

[6] It was this last pair of figures which we encountered in II. 5. 3. (This is dealing with only the first and last wheels of the train.) The statement is false. See our synopsis.

[7] *circulus tortuosus*. This is the third characterization of this oval wheel.

[8] The 9 and 12 should clearly be interchanged, but the text is not altered, for reasons given in the commentary.

Memorandum pro motu Capitis in novemdecim annis: si circulus fixus iuxta centrum Solis habuerit 19 dentes et nux mota super ipsum habuerit quot volueris dentes, ita quod alta nux movens Caput habuerit totidem dentes, tunc si circulus Capitis Draconis habuerit 361 dentes, tunc precise Capitis Draconis complebit limbum zodiaci, et similiter revolucionem propriam, in 19 annis; quod fieri non potest, quia in 19 annis pertransit ultra zodiacum 7 grad. 46 min. 45 sec., secundum tabulas Arsachelis. Si ergo dividamus zodiacum ita quod quilibet gradus dividatur in duo, erunt tot tricesima[a] [secunda] sive dimidii gradus 720. Similiter, si numerus 361, qui componitur ex 19 multiplicato[b] in se, multiplicetur per 2, hoc est duplicetur, erunt tot 722; a quo, si extraxeris quindecim partes, scilicet 7 integra et dimidium, remanebunt tot 707. Si ergo dederimus circulo Capitis 707 dentes, complebit in 19 annis omnes dentes suos et ultra 15 dentes. Sed quando usque complet de propriis dentibus 15 dentes, tunc complet de limbo zodiaci 15 dimidios graduum et 15 minuta qui sunt 7 grad. et 45 min., quod patet per multiplicacionem et divisionem; quia est proporcio 707 ad 15, illa est proporcio 720 ad 15 integra et 16 minuta illarum divisionum zodiaci et 18 secunda; quod insensibiliter differt a motu Capitis in 19 annis, quia 15 dimidii graduum faciunt 7 gradus integros et 30 minuta, quibus additis minutis et 18 secundis fiunt 7 grad. et 46 min. et 18 sec.

Hec est regula: Vide quantum circulus ductus superfluit in tempore motus circuli trahentis, in quantum superfluit in aliquot huius temporibus, ita quod illud superfluum sit pars aliquota primi superflui. Deinde ipsum ultimum superfluum multiplica per quotlibet partes, et nota quod tales partes habet primum superfluum. Et illum | numerum multiplica per numerum temporum revolucionis circuli trahentis; et de illo[c] collecto subtrahe partes secundi superflui, et residuum erit numerus circuli ducti.

Verbi gracia: Superfluum epiciculum Lune in una lunacione est 25 grad. 49 min., et superfluum in 14 lunacionibus est 1 grad. et 26 min., quod continetur[d] in primo superfluo 18 vicibus. Si ergo primum superfluum multiplicabis per 18 partes, et totum collectum per 14, erunt partes 252; de quo numero si subtraham secundum superfluum, quod est unitas, remanet 251, qui est numerus dencium epiciculi. Et numerus dencium circuli trahentis est 269, scilicet plus per 18. Et hec regula infallibilis.[e]

[a] A trisesime (The word *tot* is here used in this way for the first time. It is used twice in the next sentence in places where it indicates that the answer to a calculation follows. It might have been copied from *totum* or *tota*.)

[b] A multiplicatum

[c] Word illegible in MS.

[d] A continent

[e] A infallibilia A short tract of 17 lines follows, beginning 'Proprietates signorum'.

It is to be borne in mind that the head of the Dragon will move round in 19 years; if the wheel fixed at the centre of the tube for the Sun has 19 teeth, and if the net engaging it has any number of teeth whatsoever[1] and the upper nut moving the head of the Dragon has the same number of teeth, and if there are 361 teeth on the wheel of the head of the Dragon, it will complete exactly one circuit of the limb of the zodiac, and likewise its own revolution, in 19 years. But this cannot be, since in 19 years it traverses 7° 46′ 45″ beyond the complete zodiac, according to the tables of Arzachel. If, therefore, we divide the zodiac so that each degree is divided into two, the number of groups of thirty seconds or half degrees will be 720. Similarly, if the number 361, which is the square of 19,[2] be multiplied by 2, that is to say doubled, the answer will be 722. From this, if you extract 15 parts, that is to say 7½, 707 remain. If, therefore, we give 707 teeth to the circle of the head of the Dragon, in 19 years it will complete a circuit of those teeth, and a further 15 teeth besides. When it completes 15 teeth more, then it completes, from the limb of the zodiac, 15 half degrees and 15′, which is 7° 45′, as is clear from multiplication and division; for the ratio of 707 to 15 is the same as the ratio of 720 to 15^{p} 16′ 18″.[3] This differs insensibly from the motus of the head of the Dragon in 19 years, since 15 halves of a degree make 7° 30′, which, when the minutes and the 18″ are added, gives 7° 46′ 18″.[4]

Rule: See by how much the angle turned through by the driven epicyclic wheel exceeds one revolution in the time of a revolution of the driving wheel. Let the excess in any integral number of revolutions be expressed as an integral multiple of the original excess. Next, multiply this last excess by as many parts as were in the first. Multiply this, the number of parts, by the number of times the driving wheel revolves. Subtract from the answer the number of parts in the second excess, [i.e. subtract 1], and the remainder will be the number of teeth of the driven wheel.

Example: the epicyclic excess[5] of the Moon in a lunation is 25° 49′, and the excess in 14 lunations is 1° 26′, which is contained in the first excess 18 times.[6] If, therefore, you multiply the first excess by 18 parts, and the whole answer by 14, there will be 252 parts. If I subtract the second excess, which comprises one [of these parts], from this number, 251 remain. This is the number of teeth of the epicycle. The number of teeth of the driving circle is 269, that is to say, 18 more. And this is an infallible rule.

[1] *quot volueris dentes.*
[2] *qui componitur ex 19 multiplicatum in se.*
[3] The last figure should read 33″.
[4] Wrong, but see our synopsis.
[5] *Superfluum epiciculum.*
[6] *18 vicibus* (18 × 1° 26′ = 25° 48′).

APPENDIX: A SUMMARY OF THE TABLES

On f. 181^r of MS. A there comes an unusual table which is reproduced in Plate XVIII, and the significance of which (it concerns the method of arranging one of the *dolea* for the striking) is discussed in the commentary to II. 1–3.

Immediately after this there follows the preamble to the next table:

> Hec sequens tabula est tabula proporcionis, dividens circulum habentem 360 partes equales in quotlibet alias partes equales infra 400 partes, pro rotis horologii et aliorum instrumentorum artificialiter dividendum.
>
> (The following is a table of proportion, dividing a circle of 360 equal parts into any number of parts below 400, for the wheels of a clock, and of other instruments which are to be divided artificially.)

Rather more than ten sides of tables, with titles, follow (ff. 181^v–182^v, and 84^r–87^v). These are not worth reproducing, as being simply tables to obviate the long division required to decide the number of degrees, minutes, seconds, thirds, and fourths (i) in each tooth of a wheel with any number of teeth from 1 to 400 (the table ends 'factum est, Deo gracia', which perhaps tells us something of the medieval attitude to long division); (ii) in nearly all proper fractions of the form m/n degrees, where $1 \leqslant m \leqslant 59$ and $1 < n \leqslant 59$.

The heading of the first table repeats all but the first four words of the preamble. The title of the second table is:

> Tabula inveniendi proporcionem ad 60 minuta qualis est proporcio linee numeri ad titulos tabularum scriptarum.
>
> (Table to find the ratio, expressed in terms of a unit of 60 minutes, of the number in the left-hand column (line of number) to the number in the heading.)

The title is repeated as a running headline. On the eleventh side, after a short table, there comes a canon to further short tables, probably unconnected with our text. The canon begins: 'Pro latitudinibus trium superiorum intra tabulam bipartialem cum argumento equato et quod in eius . . .' This might be a fragment of a longer treatise on planetary latitudes.

From f. 88^v to 90^r there are conventional sexagesimal multiplication tables, accompanied by an explanation on f. 89^v. This fragment uses the word *geuzahar* for *Caput Draconis*, which the clock text never does, although the word is used in the treatise on the albion.

Elsewhere in MS. A (ff. 114ʳ–116ᵛ) there are further tables, the first covering three and a half sides, the rest being shorter. The columns are headed *Dentes*, *Dies*, *Partes*, *Gradus*, *Minuta*, in the first table, and *Dentes*, *Gradus*, *Minuta*, *Secunda*, *Tercia*, *Quarta* in the rest.

(i) This is the table of which a specimen section was given on f. 170ʳ (II. 4. 8), the line of number running from 1 to 331. This was perhaps mistakenly intended to assist in the division of the Sun's wheel (see commentary to II. 4. 31), although the tables say nothing of the fact.

(ii) Line of number: 1, 10, 20, 30,..., 200, 206 (teeth). Heading: *Pro Capite.*

(iii) Line of number: 1, 10, 20,..., 250, 255 (teeth). Heading: *Pro Luna.*

(iv) Line of number: 1, 10, 20,..., 190, 199. Heading: *Pro Saturno.*

(v) Line of number: 1, 10, 20,..., 220, 224. Heading: *Pro Iove.*

(vi) Line of number: 1, 10, 20,..., 230, 235. Heading: *Pro Marte.*

The last table is followed by this explanation:

Per istas tabulas debent dividi omnes rote pertinentes ad instrumentum de motibus omnium planetarum. Et sunt Lune dentes 225 [*sic*], Capitis 206, Saturni 199, Iovis 224, Martis 235, Solis 331. Nota quod motus accessionis et recessionis octave spere in Anno Domini 1250 fuit 9 graduum 4 minutorum 4 secundarum, et movetur per annum 31 secundarum, unde in quolibet anno sequenti tot secunde addende sunt octave spere.

By means of these tables it is possible to divide all the wheels relating to the instrument showing the movements of all the planets. The wheels are to have, for the Moon 225 teeth [for 255], for the head [of the Dragon] 206, for Saturn 199, for Jupiter 224, for Mars 235, for the Sun 331. Notice that the motus of accession and recession of the eighth sphere in A.D. 1250 was 9° 4′ 4″, and it moves 31″ per annum; wherefore in any subsequent year an appropriate number of seconds must be added to the [motus] of the eighth sphere.)

This note is of some intrinsic interest, although it has nothing to do with the clock as such. It might have come from Maudith's work (see Appendix 25). The present figures can be verified by computing the value of what is usually called the *equation* of the motus of the eighth sphere, for A.D. 1250. (March then began 35 days before the first day of A.H. 648.) The periodicity of trepidation is scarcely ever stated explicitly, and the figure obtainable from Thebit's writings varies according to the method used to extract it. To summarize briefly, Thebit's equation may be reasonably expressed

$$10^\circ\,45'.\sin(1^\circ\,34'\,02''+360^\circ.\,T/4181{\cdot}5),$$

where T is the epoch, expressed in Arabic years from the Hegira.

(Arzachel's expression, deriving from a modified theory, uses a periodicity of perhaps exactly 4,200 Arabic years.) With the parameters of the above formula (following Thebit, that is), the equation at 31 December 1250 is 9° 2′ 59″. This is reasonably close to the 9° 4′ 4″ quoted, but not calculated, it seems, by our author. The eighth sphere does not move at 31″ per annum. The so-called 'small circle of the eighth sphere', having a periodicity of about 4,000 years, moves considerably more than this. The statement can only refer to the rate of change of the equation discussed above. The rate of change of the equation per annum is not a constant quantity, but is given (with the above example) by

$$(10^\circ\,45'.2\pi/4181{\cdot}5).\cos(1^\circ\,34'\,02''+360^\circ.T/4181{\cdot}5).$$

(The writer should have known the first fact, but not, of course, the second.) At the end of the year 1250, the rate of change was 31″·385, and 31″ was presumably a figure tabulated by a Latin author.

The last set of tables, on ff. 185^r–186^v, expresses $12^s.p/q$ (answer in signs, degrees, and minutes), with p in the line of number, and q the heading (but not filled in). The range of the tables is:

$$p = 1, 2, 3, \ldots, q$$

and

$$q = 4, 6, 8, 12, 13, 15, 16, 17, 18, 115, 140, 78, 24 \text{ (in this order)}.$$

They are clearly for the purpose of marking the position of each tooth on a wheel of a specific number of teeth, using a graduated circular scale, and are probably compiled in connection with II. 4. They use the method recommended there, of dividing by q and filling in the table by successive addition. As a result of the use of this method, there are accumulated errors, but never of more than one minute.

VII

ECCLESIASTICAL AND SPIRITUAL WRITINGS

INTRODUCTION

A NUMBER of short pieces, all nominally composed by Richard of Wallingford during his tenure of the abbacy of St. Albans, are preserved in the *Gesta Abbatum* of MS. Cotton Claudius E. iv (ed. H. T. Riley, vol. ii, 1867). It is extremely unlikely that the abbot actually composed all these short pieces, and indeed he might have composed none of them. These writings are, in the order in which they appear in the *Gesta*: constitutions for the dependent cell at Redbourn; a formula for the admission of a monk or nun at St. Mary de Pré; two legal documents, both being plaints to the king, and written in French; a supplication to the Pope, written in 1332 and relating to the inquisition into Richard's incapacity; and finally, four prayers, three of them personal, and the fourth a general confession of sins. All these writings are printed below, in substantially the same form as they appear in Riley's edition of the *Gesta*, but with the orthography of the manuscript restored in preference to Riley's largely classical version. Translations into English are also given, even of the letter to the Pope, which is ostensibly from the abbot and convent, and which the author of the *Gesta* held to be no more than an outline of the contents of the actual supplication. In addition to all these writings, evidence is given for the quondam existence of others concerning monastic life, and a document relating to Saint Julian's Hospital is printed from another source.

The few documents printed here are included both with a view to completeness and to give some idea of a life from which mathematics and astronomy were remote. Once he had become abbot of St. Albans, the demands made on Richard of Wallingford by his Order were considerable, whether we consider the burden of church services, the maintenance of monastic discipline, the need to safeguard the financial interests of the abbey, especially by litigation, or Church government at the level of the General Chapter. The life of an abbot demanded a variety, not only of practical, but of intellectual, skills. The Benedictine Rule required that each monk give an appreciable part of his day to study, and life in a monastery of Black Monks resembled in a number of respects that in the universities, although by the fourteenth century the intellectual, and even the spiritual, leadership of the country had long since passed from the monasteries to the friars and secular masters at the University of Oxford. These men were in turn largely responsible for having attracted a proportion of the monastic population to the universities, thus underscoring the fact that the monasteries had lost their once prized self-sufficiency. On the other hand, the worst excesses of scholasticism were not likely to be found in the relatively small monastic communities, where there was no great need for unnecessary subtlety of the sort contrived only to shake off an opponent in disputation, and where the whole exercise must have seemed pointless in the light of the severely practical and ritualized life of the community. In a fourteenth-century abbey, the defence of privilege must have seemed far more important than the defence of a thesis. Richard of Wallingford nevertheless adapted himself successfully and easily, so far as we can tell, to the routines of St. Albans, and no doubt his astronomical clock was to some extent compensation for what he had put behind him. He was fortunate in that one of the marks of an English abbot—as opposed to his continental counterpart, whose power was increasingly held in check after reforms begun by Innocent III—was the autocratic manner in which he was able to control the revenues of the abbey. In the documents reproduced below, not only does the feudal character of the abbot emerge clearly, but also his autonomy, even in regard to statutes of the General Chapter. So much for his character *ex officio*.

As a more personal testimony we close with his four prayers.

I. *Tractatus de Statutis Generalium Capitulorum Ordinis et Legatorum Otonis et Ottoboni*

We are told by the author of the *Gesta* (ii. 196) that Richard compiled such a treatise, with the *incipit* 'Ascultate, obedientie filii', and that before making his visitation of the monastery and its dependent cells upon his confirmation as abbot he enacted that the statutes should be observed. (Note that the Rule of Saint Benedict begins similarly: 'Ausculta O fili . . .' Cf. Proverbs 4: 20.) The compilation is presumably to be identified with that which in the list of Richard's writings (ibid., p. 207) is called *Decretales et Constituciones Capitulorum Provincialium*. At *Gesta*, ii. 469–82, there is a series of thirty-five statutes alleged to be a conflation of Richard's statutes with those of Michael of Mentmore. In the first reference, Richard is said to have had about sixteen titles. Curiously enough, Abbot Thomas de la Mare later (1351) gave constitutions under nineteen headings (*Gesta*, ii. 418–20), which could explain the 'circiter sexdecim'.

Before the Fourth Lateran Council of 1215 there was no legislative authority to control the monastic body in England. The Lateran Council created a legislative body for the Black Monks, and gave to certain officials the duty of ensuring that its enactments were carried out. It was decreed that the superiors of houses of Black Monks in each province should meet in chapter every three years, to legislate for the reformation of their order (Knowles, i, ch. 2). From the Council of 1215 to the constitutions of Benedict XII (1336, the year of Richard's death) there were two separate chapters of English Black Monks, one for the Province of Canterbury and the other for the Province of York. After 1336 there was to be only a single provincial chapter. Supplementary decrees came from papal legates mentioned in the rubric above. From 1234 the legate

Cardinal Otho held a series of visitations, on one occasion (1238) presenting a set of statutes written by Pope Gregory IX. The other legate mentioned in the title, Ottoboni, was less active by far, made fewer personal visitations, and was generally prepared to leave the enforcement of his decrees to the bishops and provincial chapters. The character of the deliberations of the provincial chapters may be assessed from Pantin, i–iii. They were concerned to abolish private property, to ensure the moral safety of those leaving the walls of the monastery, especially on visits to houses in the neighbourhood and to the universities; and among their many other points of deliberation, reference was repeatedly made to the problem of diet. The autonomy of the English abbeys before Innocent III was not easily forgotten, and the edicts of the provincial chapters were a long series of compromises.

II. *Super Prologum Regule Sancti Benedicti*

This work is known to have existed from *Gesta*, ii. 207 (the short list of Richard's writings), and from a treatise on the Benedictine Rule to be found in British Museum, MS. Add. 6162 (*inc.*: 'In omnibus omnes . . .', ff. 40–2; see especially f. 42ᵛ, 'Sed Venerabilis pater abbas Ricardus Sancti Albani videtur nontire (*sic*) super Prologum Regule quod nulla sit differencia . . .'). For the contents of the latter manuscript see Pantin, ii, p. xviii. Richard's commentary on the Rule of Saint Benedict, or at least on its prologue, is paralleled by the commentaries of two other Englishmen, John of Beverley, a monk of Durham, and Uthred of Baldon. Uthred's treatises survive. See W. A. Pantin, 'Two treatises of Uthred of Baldon on the monastic life', in *Studies in Medieval History presented to F. M. Powicke* (Oxford, 1948), pp. 363–85, especially p. 367.

III. *Privilegia Monasterii Sancti Albani*

Known only from the list of writings, *Gesta*, ii. 207, this was apparently a short table of privileges ('tabulam breviter compilavit').

IV. *Constituciones pro fratribus apud Redburnam commorantibus*

The following text is taken from MS. Cotton Claud. E. iv, f. 204^{v} (previously numbered f. 197^{v}), for which see *Gesta*, ii. 202–5. A few lines of this are reproduced in Pantin, i. 213, since they contain a reference to the General Chapter which met at Abingdon in 1329.

Considerable attention had always been paid to mental relaxation within the monasteries, which was achieved at first by freedom for two or three days from all community acts other than the principal Mass

(Knowles, i. 284). Country walks were occasionally permitted *ad hominem*, but were in some houses by custom permitted to all as of right. At length the larger houses established manors as retreats, and these often grew to the proportions of small monasteries, with their own chapels and refectories, for example. Redbourn was a case in point. A cell of Saint Amphibalus was established there in 1178, as the result of the discovery of what were supposedly the remains of the Saint and his fellow martyrs. Richard's constitutions are, despite the antiquity of the establishment, the earliest known, and indeed they are the earliest to have survived for any such monastic retreat. After his death the place continued to thrive. His constitutions were consciously aimed at preventing too great a relaxation of discipline. In Thomas de la Mare's time, life there became easier. By then the priory had grown to the point where it was reasonable to say Mass on Sundays and feast days. Abbot Thomas himself preferred Redbourn to the other manors, staying there often, rebuilding a chapel there, and making a number of gifts to the place. A lengthy account in the *Gesta Abbatum* (ii. 397–401) of his relations with the priory leaves little doubt, however, that until he and Abbot Michael had made certain concessions of fuel and food, Redbourn was a somewhat spartan establishment. Needless to say, the quality of life there is not readily gauged from a set of rules: in due course, Abbot John Whetehamstede changed the regulations by what seems to us remarkably little. The convent opposed these changes, as an encroachment on its liberties, and the abbot gave in with good grace (*Registrum J. Whethamstede*, ed. H. T. Riley (Rolls Series), ii. 211–12).

From the text of the constitutions below, it will be seen that three monks of the cloister in turn were to spend a month at Redbourn. The convent in Richard's time at St. Albans probably numbered about a hundred, and if we are to assume that there was no overlapping of periods of retreat, this entails that each monk could hope for a vacation to Redbourn rather more often than once in three years.

The importance of the obedientiaries is in evidence in two places in the constitutions. By the fourteenth century, the distinction between the possessions of the members of the convent and the barony of the abbot was firmly drawn. The Rule of Saint Benedict of course forbade monks to regard themselves as in possession of property. The obedientiaries were the officials who, in the economically desirable annual audit, as in other matters of finance, relieved the conventual conscience.

No mention is made in the earliest monastic regulations of the deliberate relaxation of the regime of silence, prayer, work, and seclusion (Knowles, ii. 245). One of the most important opportunities for recreation in the life of a monk was provided by periodical blood-letting, and the statutes of the General Chapters, for example, are frequently concerned with *minutio*, with the consequent suspension of duties, with diet, and with conversation and rest. There are in Pantin, vol. i, for instance, five documents, apart from Richard of Wallingford's, containing legislation on these matters (documents of 1221, 1249, 1278, 1287, and 1293, for which see pp. 237, 36 sqq., 100, 258, and 262 respectively). In later documents of this sort, references to bleeding become fewer, whilst references to hunting become rather more numerous. The best-known allusion to the monk's weakness for the chase is in Chaucer's General Prologue to the Canterbury Tales, wherein the monk, who 'lovede venerie', and who disagreed strongly with the dictum that huntsmen are not holy men, kept fine horses and swift greyhounds to indulge in his favourite pastime of hunting the hare. The Ellesmere illustration shows him with horse and hounds. For earlier references to hunting, see Knowles, i. 108, and for a history of the wider question of the circumstances under which meat might be eaten see ibid., i. 17–21.

In the following version of the text, capitals are used liberally, as seems to be the custom with this sort of document.

De Constitucionibus Ricardi Abbatis pro fratribus apud Redburnam commorantibus

Inprimis, tres fratres claustrales qui inibi vice sua succedunt, per mensem integrum duntaxat morentur: qui, eundo et redeundo, nonnisi cum honesta vectura incedant, per Cantorem et Camerarium ab Obedienciariis in termino procuranda. Nec licebit fratri, durante termino more sue ad Monasterium, sine petita licencia et obtenta, nisi in comitiva sui Prioris, redire, et hoc cum honesta vectura: nisi super hoc secum, ex certa causa et necessaria, dispensetur. Quod similiter intelligi volumus de hiis fratribus qui illuc de claustro ad visitandum confratres de licencia certa procedunt.

On the Constitutions of Abbot Richard for the brethren staying at Redbourn

In the first place, three brethren of the cloister at a time, chosen in turn, shall stay there for at least one complete month. Both in going and in returning they shall travel in none but a decent conveyance,[1] to be obtained at each end [of the stay] from the Obedientiaries by the Cantor and Chamberlain. It is unlawful for a monk during his stay to return to the Monastery without permission duly sought and obtained, unless he be in the escort of the Prior, and then in a decent conveyance. This is the case unless he get dispensation for established and necessary reason. We wish it to be known that brethren leaving the cloister to visit their fellows likewise need permission.

[1] According to *MLWL*, *vectura* is ambiguous as between a mount and a wagon. Compare, therefore, Pantin, i. 240 (York, anno 1221): 'Item prelati et obedienciarii regulariter licenciatis visitare amicos suos inveniant omnia necessaria in sumptibus et equis, ita quod proficiscentes revertantur die statuto, nisi racionabilis causa subfuerit.'

Item, fratres ibidem commorantes ad Matutinas de nocte in ecclesiam, hora competenti, conveniant, pulsato classico, ut moris est. Et mane Horas Canonicas simul dicant, et Missam de die singuli audiant; nec cuiquam liceat prius interiorem portam egredi, quam divum[a] pensum fuerit, ut premittitur, persolutum. Nec liceat fratri sacerdoti per quatriduum a celebracione cessare, nisi de consilio Prioris. Quod si quis alicuius premissorum transgressor fuerit assuetus, admonitus prius, et iterum, si non emendaverit, terminum solacii sui perdat; nichilominus, pro suo excessu, in Capitulo Monasterii objurgandus.

Item, fratres post Horas et Missam, ad hauriendum aerem exeuntes, simul ordinate incedant ad loca vicina privata a concursu publico, consueta, tempore debito ad prandium redituri. Nec licebit alicui, infra spacium unius milliaris, neque ad vicinas domus, ad quemvis amicum, divertere, sine petita a Priore licencia, et obtenta, honesta causa primitus assignata; nec quovis extra septa,[b] ante vel post prandium, progredi, nisi de licencia et sciencia Prioris, ut antiquitus fieri consuevit; sub pena claustrum temere egredientibus infligenda.

Item, nulli fratrum ibidem in termino commoranti liceat, durante termino suo, ad alia loca ultra milliare a Prioratu distantia, sine

[a] Riley suggested *divinum*.

[b] MS. *cepta*

Next, the brethren shall assemble together at Matins in the church at night in the customary manner, summoned by the ringing of bells. And in the morning they shall say the Canonical Hours together, and they shall hear the Mass for that particular day. It is unlawful for any to go out of the inner gate before the divine office has been performed, neither may brethren who are priests cease from celebration for a period of four days, unless on the advice of the Prior. If any be an habitual transgressor in any of the aforesaid respects, he shall first be admonished and then, should he fail to mend his ways, he shall lose his time for recreation; notwithstanding this, he shall be rebuked for his excess in the Chapter of the Monastery.

Next, the brethren going out to take the air[1] after the Canonical Hours and the Mass will go together in an orderly manner to some customary near-by place away from the public throng, until the time when they are due to return for the main meal of the day. No monk may turn aside within the space of a mile, or visit near-by houses, to see any friend, without the permission of the Prior, sought for good reason and obtained beforehand. As is the ancient practice, no monk shall go beyond the boundaries, whether before or after the meal, unless with the knowledge and permission of the Prior. A punishment is to be inflicted on those so leaving the cloister.

Next, during the term of their stay, none of the brethren may travel more than a mile's distance beyond the

[1] Compare the Christ Church rule of 1298, which would have the monks walk in the fields *propter aerem graciorem* (Knowles, i, p. 284 n. 1).

competenti vectura procedere, aut extra alicubi pernoctare, sine licencia Abbatis, vel eius, cui, Abbate absente, competit licencias tales dare; nisi forte quis in comitiva Prioris, in negociis domus arduis devillantis, exiverit. Nec extra Cellam, quamvis rogati, comedant sine licencia in quovis loco ad quem de licencia Prioris divertunt, sub pena in Capitulo quinquagesimo primo Regule nostre expressa.

Item, nulli eorum liceat, sine licencia et voluntate Prioris, ante communem refectionem jantari, vel in tempore regularis jejunii, sine eius dispensatione, cenare. Super quo Prioris conscientiam oneramus, cum quibus et quando debeat et oporteat dispensare. Eleemosynaque reliquiarum, prout honestis[a] solet pauperibus erogari, fideliter ipsis pauperibus mendicantibus per Priorem et eius ministros distribuenda, cotidie colligatur. Propter quod volumus, et districte precipimus Sub-coquinario domus nostre, quod, omni necessitatis excusacione postposita, cotidie ipsis fratribus de coquina tam integre[b] serviatur, acsi in mensa Refectorii residerent; alioquin ipso die non extra Refectorium, nec ad descum, ipsis servientibus presideat, donec eis

[a] *honestius* Riley

[b] Marginal correction from the *ingre* of the text.

Priory without a suitable conveyance, nor may any pass the night away from the Priory without the Abbot's permission, or that of one to whom, in his absence, the Abbot has granted the right to to give such permission. The exception is when, perhaps, the monk leave in the company of the Prior in the course of enacting the difficult business of the house outside the town. Although they be invited to do so, the brethren shall not eat outside the Cell except at places which the Prior has given them permission to visit; and the relevant penalty is stated in chapter 51 of our Rule.[1]

Next, it is lawful for none, without the wish and licence of the Prior, to breakfast before the common repast, or to eat without the Prior's dispensation at the time of a rule of fasting. In this matter we leave it to the conscience of the Prior as to those who are to have dispensation, and when. The small alms, which are customarily given to the deserving poor, are to be collected daily, to be faithfully distributed to the poor who beg, by the Prior and his ministers. On this account we wish and most rigorously instruct the Sub-kitchener of our house that, putting aside all pleas of necessity, the brethren should have as much food daily supplied from the kitchen as if they were sitting at the table of the Refectory. And then, each day, and within the Refectory and not at a desk, let him preside over the

[1] The Rule is edited by Abbot Cuthbert Butler: *Regula Monasteriorum S. Benedicti* (Freiburg-im-Breisgau, 1912; 2nd ed. 1927, repr. 1936). This edition has many excellent critical notes, but often amends the original in small ways without recording the fact. An edition made from the so-called 'Authentic Text', with some revision of vulgarisms, and with the orthography of the current Roman missal and breviary, and accompanied by a fine English translation, is by Abbot Justin McCann: *The Rule of St. Benedict* (1952).

fuerit de porcione annone sue, et de arreragiis, plenarie satisfactum; quod similiter extendi volumus ad defectus communis servicii, si quos fratres in Infirmitorio patiantur.

Item, licet Obedienciarii, racione obediencie sue, possint ex permissione in negociis cure sue ad loca non multum distancia proficisci, licencia singulis vicibus non petita; secus autem, si ad loca procul distancia prodeant; tamen illicenciati in huiusmodi locis vicinis prandere aut pernoctare non possunt, nec debent, in quibus verisimile est quod commode possent prandio et Hore Completorii interesse; quanto magis neque Prioratum de Redburne possint sine licencia declinare, ut ibidem se reficiant, vel pernoctent. Quibus negociis officii sui euntibus non licet extra viam ad loca impertinencia proposito vie sue divertere. Loca autem vicina nostra interpretacione vocamus maneria de Kyngesbury et Childewik, Nova Terra et Apsa, Sopwelle et Grangia, personatus de Langlee, et Redburne, et similia.

Item, cum[a] ex Canone, gravi pena vallato, monachis interdicatur venacio, Priori et fratribus inhibemus ne canes venaticos habeant ibidem, aut teneant; neque venacioni intersint, aut a proposito, causa venandi, per campos aut sylvas discurrant, sub pena Canonis prenotata; nec etiam

[a] *eum* Riley

servers, until they have served the brethren their rightful portion of the produce and any arrears, to their satisfaction. We wish this rule to be extended likewise to supply any deficiencies in the common service, if there be brethren ill in the Infirmary.

Next, although by reason of their [vow of] obedience Obedientiaries may without permission travel to places not far distant to transact business in their charge, and without their having sought permission on each occasion, it is otherwise should they travel to distant places. Nor are they permitted to dine in near-by places or to stay away for the night, nor should they do so, for in these circumstances they should be in a position to return for dinner[1] and Compline. Even more important, they may not quit the Priory of Redbourn without permission to eat out or pass the night in the neighbouring places. It is not lawful for those going out of the Monastery to leave the road in order to visit places not on the route proposed for the transaction of their proper business. We take 'places in our vicinity' to mean the Manors of Kingsbury and Childwick, Newland, Apsa, Sopwell, and Grange, the parsonages of Langley and Redbourn, and similar places.

Next, from the Canon, and enforced by liability to a serious penalty, the brethren are forbidden to hunt, and we forbid either the Prior or the brethren to use or to own hunting-dogs. They shall not engage in the chase, or course through the fields and woods with the intention of taking game, under the aforementioned penalty of the Canon. And they shall

[1] On the distinction between *prandium* and *refectio* in the Rule of St. Benedict see McCann, ed. cit., p. 188.

ad quecunque loca, recreacionis gracia, procedentes inordinate incedant: nec hayas nec clausuras vicinorum diripiendo transiliant. Quod si quis suo inordinato incessu vicinorum aut extraneorum transeuntium offendat aspectum, ut scandalum suscitetur, a carnibus vel piscibus ipso die abstineat, prout dieta requirit. Adjicientes, quod nullus personas inhonestas, seu suspectas, ad convescendum, seu confabulandum, loco aut suspecto tempore introducat; aut extra ad tales accessum seu familiaritatem habeat cum eisdem, sub pena superius annotata.

Item, ordinavit Abbas memoratus quod cum per Capitulum Generale, ultimo apud Abyndonam celebratum, anno Christi MCCCXXVIII° [provisum est] quod per totam Quadragesimam nullus minucionem in termino acciperet, nisi propter actualem necessitatem duntaxat, ut in Constitucione plenius continetur; tum quia multum erat contra antiquam consuetudinem Monasterii Sancti Albani, ubi ab antiquo consuetum erat, quod minuti irent ante prandium extra portas, et toto [tempore][a] post prandium similiter, usque ad Completorium; ordinavit Abbas viam mediam, isto modo: Quod minuti in termino per totam Quadragesimam usque ad Primam Dominicam in Passione, die Dominico comederent in Refectorio; et post prandium irent in Infirmaria spaciatum; die Lune et die Martis comederent in oriolo; et ante pran-

not proceed to any places of recreation in a disorderly manner, nor shall they jump over their neighbours' hedges or fences, damaging them in the process. If, by virtue of unruly behaviour, any give rise to offence in the eyes of his neighbours or those living beyond the neighbourhood, so as to occasion a scandal, he shall abstain from flesh or fish, according to what is the food for the day. Let none introduce dishonourable or suspicious persons to eat or converse at a place or time which is suspect. None should approach such persons, or make himself familiar to them, under the aforementioned penalty.

Next, the Abbot draws attention to the last General Chapter, held at Abingdon in 1328,[1] in which it was ordained that for the whole of Lent there should be no blood-letting, except inasmuch as this is absolutely necessary, as is fully explained in the Constitution. Since, however, this ordinance is quite contrary to the ancient use of the Monastery of St. Albans, where from ancient times it was the custom of those whose blood had been let to go outside the gates before the main meal, and to go out likewise for the whole of the time available after the meal and before Compline, the Abbot decrees a compromise of the following sort: For the whole of Lent up to the first Sunday of the Passion, those monks who have been bled should go to the Refectory; and after the meal they should go to walk in the Infirmary; on Monday and Tuesday they should eat in the ante-chamber,[2] and they should not

[a] In excerpting this passage, Pantin (vol. i, p. 213) suggested *cito* to replace *toto*

[1] The General Chapter met in March 1328 (O.S.), or 1329 (N.S.).

[2] *MLWL* gives the meanings of '*oriolum*' as 'Gallery', 'upper room', or 'porch'. The meaning here is certain, from *Gesta*, i. 314: 'atrium nobilissimum in introitu, quod porticus vel oriolum appellatur'.

dium irent non extra portas, nisi usque ad Sanctum Germanum et gardinum: post prandium autem in gardinum duntaxat; nec luderent ad billos, nec aliquid aliud, nisi tantum in confabulacione sancta, et haurirent auram puriorem. Qui vero die Jovis et Sabbato ex cursu essent de minutis, in termino comedant in oriolo; et usque ad Sanctum Germanum irent ante prandium; et post prandium tantum in Infirmariam, de licencia presidentis. Diebus autem feria quarta et sexta feria, nullus iret in solacio, nec comederet extra Refectorium: nec aliquis gradus currat in tabula a Prima Dominica Septuagesime usque ad Pascha, neque a Dominica Prima Adventus usque ad Natale. Ista de Quadragesima sunt concessa, sub condicione quod nulla fiat insolencia, sub pena perdendi totum solacium antedictum.

go outside the gates before the meal, except as far as Saint Germain and the garden. After the meal, however, they should go only to the garden, and they should not play at bills[1] or any other game, nor should they engage in anything other than holy conversation and breathing pure air. Those being bled on Thursday and Saturday will eat in the antechamber, and they may go as far as Saint Germain before the meal, and only to the Infirmary afterwards with the permission of the monk in charge. On the fourth and sixth days of the week, however, none shall go for recreation, or eat outside the Refectory, nor shall any run a single pace on the *tabula*[2] from the first Sunday of Septuagesima until Easter, or from the first Sunday in Advent until Christmas. The concessions relating to Lent are on condition that nothing give rise to immoderate behaviour, under penalty of losing the whole of the recreation aforesaid.

V. *Forma recipiendi et profitendi fratres et sorores apud Pratum*

It appears that no certain rule was observed at Saint Mary de Pré, before the time of Richard of Wallingford, whose formula for admission is set forth in the following document (*Gesta*, ii. 214).

The priory at Saint Mary de Pré (or 'des Prez', or similar plural form meaning 'of the meadows') was founded by Abbot Warin as the consequence of a dream had by a man of Walden (*Gesta*, i. 199–200). The man's vision was of Saint Amphibalus, who supposedly revealed the whereabouts of relics of himself and his companions, and asked that a shrine be built there. Warin built a church, and houses for women lepers, who were to be veiled, and obey a rule. Matthew Paris states that in the thirteenth century the inmates were exceedingly poor (*Chronica Majora*, ed. H. R. Luard, vol. v, Rolls Series, 1882, p. 452). They were not numerous, being perhaps about ten monks, rather fewer nuns, and in Richard's time three or four sisters. This last group of inferiors was

[1] Possibly skittles, or even a primitive version of cricket.
[2] A running track? Cf. *tabula rotunda*, a tournament.

finally excluded by Abbot Thomas de la Mare, who also ruled that henceforth only literate nuns were to be admitted (*Gesta*, ii. 401–2). This perhaps had something to do with the fact that his sister Dionisia had previously become a nun at the priory (*Gesta*, ii. 373). The *Gesta Abbatum* does not lead one to suppose that the priory was still a refuge for lepers in the fourteenth century.

The procedure for the reception of brethren in the Rule of Saint Benedict was cap. LVIII. The new monk was 'not to be granted an easy admittance'; but after the initial hurdles were overcome, and he was seen to be in earnest, he was to make a written promise: 'Suscipiendus autem in oratorio coram omnibus promittat de stabilitate sua et conversatione morum suorum, et obedientiam . . .' (ed. McCann, p. 130). This is echoed by Richard's formula.

Recipiendus frater ad Prata debet coram Domino Abbate et Archidiacono Monasterii, et aliquo spectate fidei seculari, jurare fidelitatem et devocionem, atque gratitudinem Monasterio Sancti Albani; et quod nunquam erit in consilio, vel opem prestabit, clam vel palam, ut aliquis frater aut soror, sive monialis, recipiatur in loco illo, nisi per Abbatem, prout semper moris fuit. Quando vero professionis votum solemniter debet emittere, suam professionem propria manu scribet; cuius tenor talis esse debet:

> Ego frater N., sacerdos, promitto stabilitatem et conversionem morum meorum ad castitatem, et voluntariam paupertatem, et obedienciam, secundum Regulam Sancti Benedicti, coram Deo et Sanctis eius, in hac ecclesia, que constructa est in honorem Sancte Marie, in presencia Domini N. Abbatis, et in presencia Domini N. Archidiaconi, Monasterii Sancti Albani.

The brother to be received at St. Mary de Pré should appear before Dom. Abbot and the Archdeacon of the Monastery, and a secular of tried faith, and should swear fealty and devotion, and gratitude to the Monastery of St. Albans. No brother, sister, or nun, is ever to be received by any but the Abbot, which has always been the custom, and no persuasion or force shall be used, openly or secretly. When solemnly making public his vow, the monk should write his declaration in his own hand, and the tenor[1] of it should be as follows:

> I, brother N., priest, promise steadfastness and the conversion of my ways to chastity and freely chosen poverty, and obedience, in keeping with the Rule of Saint Benedict, in the presence of God and his Saints. This I do in the church which was built in honour of Saint Mary, and in the presence of Dom. N., Abbot, and in the presence of Dom. N., Archdeacon, of the Monastery of St. Albans.

[1] i.e., the literal content. The word is used in the modern legal sense.

VI. [*Charta confirmatoria de LX solidis Hospitalis Sancti Juliani*]

As part of his so-called 'Auctarium Additamentorum', included as an appendix in his monumental volume *Matthaei Paris Monachi Albanensis Angli, Historia Major* (2nd ed., 1684), William Wats printed a number of documents relating to the Hospital of Saint Julian. One of these is a letter from Richard of Wallingford, printed again here from the source manuscript Cotton Nero D. i, f. 193, compared with Wats's version, which is followed in all respects but that of orthography (ibid., pp. 1158–9).

Wats prints a series of similar documents, from which the early history of the Hospital may be pieced together. All are taken from the same Cottonian manuscript, which is written in a fourteenth-century hand. The Hospital was founded specifically for lepers ('pauperes Christi, videlicet Lazaros') by Geoffrey, Abbot of St. Albans (1119–46), on a piece of land called 'Kyngesho' near Watling Street. (This land, we are told, was held by a glazier (*vitriarius*) by the name of Alulfus, but without the permission of the convent.) For its upkeep the hospital was assigned a number of different tithes and other income. Henry II later confirmed the foundation charter, and granted a penny a day in perpetuity to the hospital. Richard's visitation of the dependencies of the monastery, shortly after his inauguration, resulted in his withdrawing the most significant of the hospital's tithes, namely tithe of rent of the vill of St. Albans, this tithe being worth 60 shillings. His letter shows that in 1329, having—as he says—examined the charters, he was prepared to concede the right to the relevant tithe.

It is clear from a number of references in the *Gesta Abbatum* that the lepers were discontented with their treatment at the hands of successive abbots, and from Abbot Michael's constitutions for the Hospital (anno 1344; *Gesta*, ii. 483–502, in which MSS Nero D. i and Claudius E. iv are collated) it is equally clear that, in the eyes of at least one abbot, discipline had reached a very low ebb. These constitutions nevertheless give a very good idea of life within a fourteenth-century colony of 'lepers'.

Omnibus ad quos presentes litere pervenerint, Ricardus miseracione divina Abbas Monasterii Sancti Albani, eternam in Domino salutem. Noveritis nos inspexisse quasdam chartas, fratribus infirmis hospitalis Sancti Juliani iuxta Sanctum Alba-

I, Richard, by the divine compassion Abbot of the Monastery of St. Albans, greet in the Lord all to whom this letter comes. You will be aware that we have examined certain charters, granted by our predecessors to the infirm brethren of the Hospital of Saint Julian near St. Albans,

num, eis a predecessoribus nostris confectas super diversis possessionibus eis ab eisdem predecessoribus concessas; inter quas possessiones reperimus dictis fratribus infirmis esse concessos sexaginta solidos, annuatim percipiendos de reditu ville Sancti Albani; prout in eisdem chartis plenius continetur. Quam quidem concessionem reditus antedicti, approbamus et confirmamus per presentes; eisque sex libras pro duobus annis preteritis, quibus dictus reditus fuerat eisdem fratribus subtractus, persolvimus in die confectionis presencium, nolentes quod per nos aut successores nostros, reditus memoratus fratribus antedictis de cetero subtrahatur. In cuius rei testimonium, sigillum nostrum presentibus est appensum. Datum apud Sanctum Albanum, die Sanctorum Septem Dormiencium, Anno Domini MCCCXXIX°.

and concerning sundry possessions granted to them by those same predecessors, among which possessions we discover the said infirm brethren to have been granted 60 shillings annually from the income of the vill of St. Albans, as is more fully explained in those charters; which grant of the aforesaid income we approve and confirm by these presents, and we pay to the brethren on the day of writing the six pounds withheld from them over the past two years, unwilling as we are that the income mentioned should be deducted either by us or our successors from the remainder of what is due to the aforementioned brethren. In testimony of this act, our seal is affixed to these presents. Given at St. Albans on the festival of the Seven Sleepers, [27 July], A.D. 1329.

VII. *Billa super extorcionibus per villanos Sancti Albani Monasterio ibidem illatis*

This plaint made against the townsmen by Richard of Wallingford is to be found in *Gesta*, ii. 227–8, and is printed here with Riley's translation from the French of the original, with small and insignificant changes. This and the following document are the only known examples of writings purporting to be by Richard of Wallingford which were written in French, and it is possible that they were drafted by one of his obedientiaries, or by his attorney John of Bledelawe (cf. *Gesta*, ii. 237–8).

An outline of the trial of strength between the townsmen of St. Albans and the abbots, lasting several years, and more or less decided by the firm actions of Richard of Wallingford, is given in our brief account of his life, vol. ii (pp. 9–10), the details having been drawn from *Gesta*, ii. 215–60. The year 1327 was one of widespread disturbance, consequent on the weak government of Edward II. There were risings in Bury St. Edmunds, London, Abingdon, and Oxford, for example, and it might well have been that Richard of Wallingford left Oxford for St. Albans in the mistaken belief that the monastery there would be a more peaceful place. At St. Albans, however, a commune had been set up which demanded that two burgesses represent it in parliament, and that it be given a number of other rights (of common, of multure, and so forth). The outcome of the following plaint represented the new abbot's first great victory: the justiciars made inquisition into the charge, found a number of the townsmen guilty, and committed many to prison.

A study of the disturbances throughout the country as a whole will be found in N. M. Trenholme's 'The risings in the English monastic towns in 1327', *American Historical Review*, vi (1901), 650–70.

Billa super extorcionibus per villanos Sancti Albani Monasterio ibidem illatis

Come apres le coronement nostre Seignur le Roi, que Dieu garde, fist crier sa pees par my toute son roialme, les gentz de la ville de Seint Alban sei entrelierent par serment, et a force et armes, countre la pees nostre Seignour le Roi, l'Abbeie de Seint Alban, et les moignes et les servauntz leinz demorantz, assegerent, et diverses assautz de nuytz et jours fairent; issint que moigne ne sergeaunt, ne nul autre del' Abbeie, pout issir ne entrer, ne viande a eux venir, par quaraunte jours continuauntz; et manacerent de mettre lour Abbeye en feu et en flamme; par quel ensegement, plusours de ceux furent en despeir de vie; par quele duresse, l'Abbe lour graunta divers choses prejudiciels a sa esglise, et encountre l'estat nostre Seignour le Roi. Et firent que les gentz l'Abbe n'entrerent la ville; c'est assavoir, Hughe de Ware, taunt q'il feseit fyn a eux par vint livres, dount il paia quaraunte solz; et puis firent crier sa pees par my la ville pur salvementz aler et venir: et Johan de Maideford fist sa fyn a eux en la [mesme] manere pur cent solz, dount il paya deux

Bill relating to extortions by the townsmen of Saint Albans on the Monastery there

Whereas after the coronation our Lord the King, whom God preserve, caused to be proclaimed his peace throughout all his realm, the people of the town of Saint Albans combined by oath, and by force of arms, against the peace of our Lord the King. They did besiege the Abbey of Saint Albans, and the monks and the servants therein dwelling, and did make divers assaults by night and day; so that neither monk nor servant, nor any other person of the Abbey, could go forth or enter, nor could meat to them come, for forty days continuous. And they menaced to set their Abbey on fire and in flames, through which besieging, many within were in despair of their lives; by reason of which duress, the Abbot did grant unto them divers things prejudicial to his church, and against the estate of our Lord the King. And they did cause that the people of the Abbot could not enter the town; that is to say, Hugh de Ware, until he had made fine to them for twenty pounds, of which he did pay forty shillings, upon which they did cause their peace to be proclaimed throughout the town for him safely to go and come; also, John de Maideford made his fine to them in the [same] manner for one hundred shillings, of which

marcs; et Johan de Mundene, baillif lo Roi, pristrent et enprisonerent, tant q'il lour eust done scyssaunte acres de sa terre, ou il eust este decolle. Et Johan Poleyn, Simond de Nortone, William de Okhurste, et William de Hoo, gentz l'Abbe, assegerent en la ville de Seint Alban, en la meson Adam le Hussher, a force et armes, tant q'ils feseient fyn a eux a lour volentez, ou ils eusent decollez; et puis feserent crier lour pees, acrochauntez a eux roial poer. Et autres des sergeantz l'Abbe, et de ses tenauntz, de jour en autre pristrent, encrochauntz a eux roial poer, tauntque gree fust fait a eux a lour volente. Et unqore, les dites gentz de la dite ville de Seynt Alban fount tieles duresses, manaces, bateries, ensegementz, de jour en autre, a les gentz le dite Abbe, et coillettes et taxacions, en meyntenaunce de lours mauveistez, issint que l'Abbe ne pust trover de lui servir. Dount l'Abbe prie la grace nostre Seignour le Roy d'une lettre de sa Targe a Sir Roger de Grey, et ses compaignons, denquere especialment de cestes choses, et autres tortz a l'esglise faitez, et sur ceo mettre remedie.

he paid two marks; and John de Mundene, the King's bailiff, they did take and imprison, until he had given them sixty acres of his land, or he would have been beheaded. Also, John Poleyn, Simon de Norton, William de Okhurste, and William de Hoo, people of the Abbot, they did besiege in the town of Saint Albans, in the house of Adam le Hussher, by force of arms, until they had made fine to them at their will, or they would have been beheaded; and then did cause their peace to be proclaimed, usurping to themselves royal power. And others of the servants of the Abbot, and of his tenants, from day to day they did take, usurping to themselves royal power, until it had been done to them at their will. And further, the said people of the said town of Saint Albans do inflict such duresses, menaces, batteries, besiegings, from day to day, upon the people of the said Abbot, and collections and taxations, in maintenance of their misdeeds, that the Abbot cannot find wherewith to serve him. Wherefore the Abbot prayeth the favour of our Lord the King of a letter under his Privy Seal to Sir Roger de Grey, and his fellows, to make inquisition especially as to these things, and other wrongs against the church committed, and to apply remedy thereto.

VIII. [*Billa Domino Regi de redditu viginti solidorum*]

A further plaint to the king from Richard of Wallingford, also written in French, is to be found in *Gesta Abbatum*, ii. 267–8. As with the foregoing document in French, Riley translated it, and his version is printed here.

The suit now concerns a mill at Little Langley, on which the Abbot demands payment of a rent of 20 shillings due from Isabella, Queen Dowager. For an endorsement on the petition, an injunction to the

Chancellor, the Bishop of Winchester, instructions to the justiciars, and the verdict of the inquisition (in the abbot's favour), see *Gesta*, ii. 269–78.

Billa, sive Supplicacio, Domino Regi . . .

A nostre Seignour le Roy, et a son Conseil, prie son Chapelein, l'Abbe de Seint Alban, que come il eit suy par peticions en divers Parlementz pur vintz soldre de rente du molyn du Petite Langele, que devynt en la manere nostre Seignour le Roy, come del purchace Alianore, jadys Reyne d'Engletere, ael nostre Seignour le Roy q'ore est; dont ses predecessours furent seisiz de an en an, de temps le Roy Henry le Secounde tantques a temps la dite Alianore; et des arrerages de quele rente le Roy, piere nostre Seignour le Roy q'ore est, fist gree tantqes al an de son regne tierz; de queu temps encea mesme la rente lui ad este detenue, si come piert par enquest sur ceo pris par petition de Parlement, et returne en Chauncellerie;—q'il pleise a nostre dit Seignour le Roy, et a son bon Conseil, ordeigner que le dit Abbe puisse aver bref au gardeyn du manoir de Petite Langlee, q'ore est en la mayn la Roigne Isabelle, q'il paie desore au dit Abbe la rente avantdite; et ensement, autre brief as Tresorer et Barouns del Escheker, que les arrerages de mesme la rente del tierz an avantdite soient au dit Abbe alowez al dit Escheker, en les dettes q'il doit a nostre Seignour le Roy illoeques.

Bill, or Supplication, to the Lord King . . .

Unto our Lord the King, and to his Council, prayeth his Chaplain, the Abbot of Saint Albans, that whereas he has sued by petitions in divers Parliaments for twenty shillings of rent from the mill of Little Langele, that came to be in the manor of our Lord the King, as of the purchase of Alianor, formerly Queen of England, grandmother of our Lord the King that now is; of which his predecessors were seised from year to year, from the time of King Henry the Second until the time of the said Alianor; and the arrears of which rent the King, father of our Lord the King that now is, duly discharged until the third year of his reign; from which time downward the same rent has been withheld from him, as appeareth by Inquisition thereon taken by petition of Parliament, and returned into Chancery; that it may please our said Lord the King, and his good Council, to order that the said Abbot may have a writ unto the keeper of the manor of Little Langlee, which is now in the hand of Queen Isabella, that he pay from henceforth unto the said Abbot the rent aforesaid; and also, another writ to the Treasurer and Barons of the Exchequer, that the arrears of the same rent from the third year aforesaid be unto the said Abbot allowed at the said Exchequer, in the debts which he oweth there unto our Lord the King.

IX. *Littera supplicatoria Domino Pape*

As explained elsewhere (vol. ii, p. 11), a Papal Bull of November 1332 was addressed to the Bishop of Lincoln, enjoining him to investigate

the extent of Richard's incapacity, consequent upon his leprosy. Richard petitioned the king, complaining of the actions of the monk of Abingdon whose letter to the Roman Curia had brought about the whole affair, while from the abbot and his Convent the Pope received the letter which follows (*Gesta*, ii. 290–1). In the words of the author of *Gesta Abbatum* (ii. 289), the letter is there rendered literally, 'sub tenore qui sequitur'. (In the context, this does not mean that the general sense of the letter was alone recorded. The word 'tenor' here has its legal meaning, and in fact the relevant sentence ends 'litteraliter porrexerunt'.) A similar letter is said to have been sent to the king. It is possible, of course, that the words of the letter owe more to the Convent than to Richard himself, but at several points there is more than a suspicion that he was abetting his monks in their declaration of trust in him. The Convent had considered and rejected the idea of having several Coadjutors. Nicholas of Flamstede was elected on Quinquagesima Sunday (14 February) 1332/3 (*Gesta*, ii. 289). Three weeks later the letter was sent to the Pope. It is worth noting that the letter singles out certain phrases from the Papal Bull addressed in the previous November to the Bishop of Lincoln, from which the following is an extract: 'Frater Ricardus, Abbas, . . . adeo est divino judicio lepre morbo respersus, quod nequit inter sanos absque scandalo conversari; propter quod prelibatum Monasterium varia patitur, et hactenus passum est in spiritualibus et temporalibus detrimenta' (*Gesta*, ii. 285). The Commissioners visited the abbey on 15 January 1332/3. The life of the Abingdon monk was threatened by the brethren of St. Albans on a later and less inappropriate occasion. Although not named in the letter, he was named elsewhere as 'Richard of Ildesle'. (Ilsley, Wallingford, and Abingdon are at the three vertices of a roughly equilateral triangle, eight or nine miles apart.)

Littera supplicatoria Domino Pape per ipsum Abbatem, et suum Conventum, directa, contra suggestionem predictam; et de Coadjutorio assignando Abbati memorato

Sanctissimo in Christo, patri et domino reverentissimo, Domino Johanni, divina Providencia, Sacrosancte Romane et universalis Ecclesie Summo Pontifici, sue precelse dominacionis servi et filii, supplices et devoti, Prior et Conventus Monasterii sui de Sancto Albano, Ordinis Sancti

A supplicatory letter to the Lord Pope by the same Abbot, and his Convent, against the aforesaid suggestion; and concerning the Coadjutor assigned to the Abbot

To the most Holy in Christ, to the most revered Father and Lord, to the Lord John, by divine Providence Supreme Pontiff of the Universal and Holy Roman Church, the servants and sons of his high dominion, the humble and devoted Prior and Convent of his Monastery of St. Albans, of the Order of Saint Benedict, of the

Benedicti, Lincolniensis Diocesis, sue Sanctitatis Sedi immediate subjecti, in sincera subjectione devota pedum oscula beatorum. Cum vos, Pater sanctissime, et acceptissime Deo Pontifex, in quo fons justicie consistit, et per quem injurie resolvuntur, pro subditorum quiete indefessa sollicitudine labores sustineatis continuos, et omnium juste postulantium preces sine accepcione personarum semper benigne exaudire dignemini;[a] nos vestri humillimi, Prior et Conventus predicti, pro damnis que quidam ambicionis filius nobis et dicto vestro Monasterio inferre molitur, vestre Sanctitatis remedium sitientes,[b] peticionem nostram Beatitudinis vestre celsitudini audemus, revelata facie, intimare. Sanctissime Pater, ad simplicitatis nostre audienciam dudum pervenit, pro eo quod Sanctitatis vestre Sedi suggestum extiterat, quod Dominus Ricardus, Abbas noster, quem nobis dederunt vestre manus Sacratissime in pastorem, est adeo divino judicio morbo lepre respersus, quod nequit inter sanos sine scandalo conversari; propter quod, dictum vestrum Monasterium in spiritualibus et temporalibus varia patitur detrimenta: Domino Episcopo Lincolniensi ad inquirendum super huiusmodi suggestiones sub certa forma commisit vestra Sanctitas potestatem. Ne igitur per huiusmodi suggestiones, que a fomite ambicionis primitus processerunt, vestra reddatur puritas circumventa, nos, huiusmodi rei noticiam plenarie obtinentes, veritatem plenam vestre Sanctitudini reseramus.

Devotus quidem filius vestre Sanctitatis, Abbas noster predictus, aliquali infirmitate laborat, quod moleste gerimus, ut debemus. Non

Diocese of Lincoln, and immediately subject to his Holy See, in true subjection faithfully kiss his blessed feet. When you, O most holy Father, and Pontiff most accepted by God, you in whom abides a fountain of justice, and through whom injustices are resolved, when you labour continually, concerned for the peace of mind of your subjects, and rightly deem the prayers of all worthy to be heard with kindness and without regard to persons; we your most humble Prior and Convent aforesaid make so bold as to bring yet again to the notice of your high Beatitude our petition, for the damages which a certain son of ambition undertook to do to us and your said Monastery. Most holy Father, it recently came to the ears of your simple subjects that a claim had been made, originating in your Holiness' See, that the Lord Richard, our Abbot, whom the hands of your Holiness gave to us as pastor, is by divine justice marked with leprosy, and that he is unable to live the monastic life with those in good health without giving offence; and that on account of this, your said Monastery suffers manifold temporal and spiritual hurts: and that your Holiness gave power to the Lord Bishop of Lincoln to investigate the claims *sub certa forma*. Lest your purity suffer by these claims, which were first set in motion by the fuel of ambition, we, completely apprised of the state of affairs, will reveal to your Holiness the complete truth of the matter.

One of your Holiness' devoted sons, our aforesaid Abbot, labours under some sort of infirmity, which with difficulty we bear as is our duty.

[a] MS. dignamini [b] MS. scitientes

tamen sic per eam inficitur, quin potest inter nos sine magno scandalo conversari; et cotidie conversatur, ac eciam in officio sibi commisso, tam in spiritualibus quam in temporalibus, a tempore sibi commissi regiminis laudabiliter se gessit, et gerit in presenti, prout opera manifestant. Ubi enim dictum vestrum Monasterium, propter frequentes vacaciones et alia adversa, que hactenus occurrerunt, diversis creditoribus in maximis pecuniarum summis oneratum et alias multipliciter depressum extiterat, iam per eius industriam et circumspectionem incipit relevari. In curandis animabus discretus extitit, injurias reformavit, dispersa ad statum debitum revocavit, et multa alia bona fecit, que enumerare infinitas non permittit. Veruntamen, quia idem Abbas noster, caute previdens, et discrete, quod licet in presenti sit habilis ad regendum, poterit tamen, infirmitate sua crescente, minus ydoneus effici in futurum, ut nobis et dicto vestro Monasterio uberius consulatur, et cura spiritualium et temporalium prosperius peragatur, religiosum et discretum virum, Fratrem Nicolaum de Flamstede, confratrem nostrum, triginta annis et amplius inter nos professum, presbyterum bone vite et conversacionis honeste, ac eciam in spiritualibus et temporalibus circumspectum. De consensu nostro expresso, tractatu super hoc in Capitulo nostro prius habito diligenti, sibi in Coadjutorem assumpsit, prout Sanctitudini vestre per instrumentum publicum, super hoc confectum, fieri poterit plena fides; quocirca, Sanctitatis vestre

He is not, however, so corrupted by it as to be unable to live the monastic life among us without causing great offence. He meets with us daily, and as his works make manifest, he has borne himself with honour from the time he undertook his office, and he continues to do so, in matters both spiritual and temporal. Whereas your aforesaid Monastery, by virtue of frequent vacancies and other misfortunes which have befallen it in the past, had been burdened with very large debts to a number of creditors, and had been on many occasions oppressed in other ways, by the Abbot's industry and circumspection its fortunes begin to rise once more. He is distinguished in the cure of souls, he puts injustice to rights, he restores order where there is disorder, and does many good acts—so many, indeed, that they are not to be numbered. Although our Abbot is at the present time perfectly capable of ruling, yet looking prudently to the future, and with great judgement, he sees that he will become less fitted to rule in the future, by virtue of his increasing infirmity. Therefore, in order that our affairs both spiritual and temporal and your aforesaid Monastery may prosper, he has chosen for himself a distinguished and religious man to be Coadjutor, Fr. Nicholas of Flamstede, a fellow of our confraternity for thirty years and more, an elder brother of good character and an honourable life in the monastery, and circumspect in matters spiritual and temporal. This he did with our express agreement, arrived at after a rigorous discussion in our Chapter, held so as to give your Holiness the full assurance of a public statement drawn up with this in view; wherefore we hasten with

pedibus provoluti, vobis humili precum instancia supplicamus, quatenus eundem Abbatem nostrum, et filium vestrum devotissimum, recommendatum habere, et huiusmodi Coadjutoris assumpcionem per vestre Sanctitatis Sedem litteras confirmatorias, de vestre Apostolice benignitatis clemencia, roborare dignemini; non permittentes eundem per suggestiones predictas, maxime cum non a nobis, quorum interesse vertitur principaliter in hac parte, sed a quodam nostri Ordinis, alterius tamen congregacionis, ad curam dicti vestri Monasterii, ut verum fateamur, ambiciose aspirante, primitus emanarunt, aliqualiter perturbari. Ad regimen Ecclesie sue sancte Altissimus vos conservet, in dierum longitudine Deo grata. Datum apud Sanctum Albanum, in Capitulo nostro, secundo Nonas Marcii, anno Domini MCCCXXXII°.

our humble prayers to your Holiness' presence to beg to recommend that we continue to have our same Abbot, and your most devoted son; and out of your benign Apostolic kindness we think it right that a letter be sent from your Holiness' See, confirming the Coadjutor in his office. The aforesaid claims do not originate with those who have most cause for concern, but with a certain member of our Order, although of another congregation. We do indeed confess that the claims spread abroad to the effect that the Monastery is in some way in a state of confusion were begun by a man of ambitious aspirations. May the Highest preserve you for the government of His Church; and thanks be to God to the end of time. Given at St. Albans, in our Chapter, 6 March 1332/3.

X. *Oraciones*

The four prayers (*Gesta*, ii. 294–9) speak for themselves. The first is an expression of gratitude for those, especially his parents, who have supported him. The second suggests that his early life was not of the utmost purity. The third, which is apparently unfinished, thanks God for the fact that, despite leprosy, he has not been deserted. It contains a passage ('ut et presentes . . . et reverenciam exhibere') which seems to take up, and deny, the charge mentioned in the Papal Bull, 'quod nequit inter sanos absque scandalo conversari' (see IX above). That the fourth and last prayer is a general confession for use by the laity is evident not only from its general tone, but from the confession of ingratitude towards 'venerabilem sponsam meam', a fact which Riley pointed out. From Riley, also, are taken the Biblical references. The translations are not meant to be imitative of the Restoration prayer book, or of any other; but, in the words of the preface to that version, they are meant to 'keep the mean between the two extremes', between the ancient and the modern. In particular, it seemed that too much would have been lost if the old form of second person singular had not been retained in addressing the deity. Finally, we observe that the word *oratio* in the Rule of Saint Benedict meant a *private* prayer (cap. xx).

De oracionibus per Ricardum Abbatem editis et usitatis

[1]

Domine Deus, Omnipotens Pater, et Fili, et Spiritus Sancte, Trinitas Sancta, unus Deus et unus Dominus omnium, Tu extraxisti me de utero matris mee, et providisti michi hactenus omnia necessaria ad hanc vitam; indidisti parentibus meis naturalem affectum, ut de sua inopia indigenciam meam sublevarent; et, decedentibus illis, michi amicos alios providisti, ut adhuc indignum peccatorem tuum tanto affectu diligerent, ut sibi de necessariis vite subtraherent, ut mihi ad habundanciam exhiberent. Da michi, Domine Deus meus, propter hec temporalia beneficia,—eciam quod irracionabile jumentum[a] faceret,—et devotum obsequium animo et corpore sine intermissione prestare, et illis, qui, Te ordinante, tam sedulo aluerunt, in regno tuo eterne retribucionis mercedem.

[2]

Alia Oratio eiusdem Abbatis

Domine Deus, Omnipotens Pater, et Fili, et Spiritus Sancte, Trinitas Sancta, unus Deus, et unus Dominus omnium, advertens quomodo laute, et non meo labore aut stipendiis, eleemosine titulo educatus, non cessavi per abrupta viciorum delectabilium, et ex electione, incedere; reputans me felicem, si, in seculo

Prayers written and used by Abbot Richard

[1]

Lord God, almighty Father, Son, and Holy Spirit, Holy Trinity, one God and alone Lord of all things, Thou didst take me from my mother's womb, and hast provided me to this time with all that is necessary for this life. Thou didst bestow on my parents a natural compassion, so that despite their want they did support me in my need; and when they were dead Thou didst provide me with other friends, so that until this present time they have shown such a great love for this thy unworthy sinner that they have deprived themselves of the necessities of life in order to give a greater abundance to me. Let me, O Lord my God, excel in devoted and unceasing service, both of mind and body, for these temporal blessings—and also because in the past I was left foolishly unmoved. And if it be Thy will, give to them who so zealously maintained me the reward of eternal recompense in Thy kingdom.

[2]

Lord God, almighty Father, Son, and Holy Spirit, Holy Trinity, one God and alone Lord of all things, how excellently hast Thou favoured me; and not for my work or tribute have I been sustained by the glory of alms, for I have not ceased to tread the dangerous paths of the delectable vices, and this by my own choice. I reckoned myself happy when, thrown into the life of the world, I

[a] MS. inmotum *altered to* inmetum (?).

positus, seculariter conversantibus in viciis carnis et spiritus non impar consodalibus invenirer; tantam longanimitatem provocate majestatis tue veneror et adoro, quia meo demerito reclamante, extraxisti me de seculo, ut residuum vite mee inter familiares et dedicatos[a] tuos ita ingererem,[b] et lugerem errores preteritos, et in futurum in scalam virtutum residuum vite extenderem;[c] ut essem disciplinatus et eruditus quomodo deberem regalibus mensis assistere, et in tam excelsi regis obsequio ministrare; nedum in inferiori ordine, sed et sacerdocio, coram Te fungi. Da michi, Domine Deus meus, obsecro, ex hoc nunc sic in regularibus institutis proficere, ut et iam tempus necgligenter perditum redimam, et sic in sancta tua disciplina regulariter erudiri, ut non frustra et indigne inter religiosos appaream.

[3]

Alia Oratio eiusdem Abbatis

Domine Deus, Omnipotens Pater, et Fili, et Spiritus Sancte, Trinitas Sancta, unus Deus, et unus Dominus omnium, recolo utcumque, quomodo cum essem obscurus genere, elevasti me de stercore, et sic honorasti me, ut sederem cum principibus, et solium gloriosum tenerem. Et quid sileam? Tu eduxisti me de tugurrio patris mei, et posuisti me pascere gregem populi tui, et fuisti mecum in omnibus viis meis; et requiem dedisti michi ab omnibus inimicis meis, et fecisti michi—'Nomen grande, juxta

was with those familiar with the sins of the flesh and of the spirit; and I was in no wise different from my comrades. I praise and revere that forbearance shown by thine offended majesty, because, although my demerits cried out against me, Thou didst take me out of the life of the world, that I might spend the remainder of my days among those dedicated to thy service, and that I might repent my former errors, and spend my life henceforth climbing the steps of virtue. Thou didst ensure that I became learned and able to serve obediently at the table of the Most High King, and that I might discharge a duty in thy presence also to lesser men, and to the priesthood.

I beseech Thee, O Lord my God, to enable me now to follow the established rules, lest by my negligence I return to my former dissipation. Let me become learned in accordance with the rule of thy holy teaching, so that I do not appear guilty of error and unworthy among men of religion.

[3]

Lord God, almighty Father, Son and Holy Spirit, Holy Trinity, one God and one Lord of all things, I recall to myself how, although I was of lowly birth, Thou didst raise me from the dung, and didst so honour me that I sit among princes and occupy a glorious throne. And wherefore shall I be silent? Thou didst lead me forth from the cottage of my father, and gavest me the task of pasturing the flock of thy people, and wentest with me in all my goings. Thou gavest me respite from all mine enemies, and made unto me 'a great

[a] MS. delicatos [b] MS. ingerem [c] MS. extonderem

nomen magnorum qui sunt in terris';[a] et, quod amplius est, cum sim homo extreme condicionis mala morphea, Te sic disponente, percussus ut non sim dignus inter homines conversari, sed, ex lege, ejici[b] extra castra, tamen, Domine Deus meus, nescio quo tuo occulto judicio, me sic contines in honorem; obsequia tanta procuras, ut non noverim aliquem de parentela mea vel genere, in optima sanitate corporisve, decorari tanto honore, quoad seculum predictum: et, quod non sine stupore immense bonitatis tue sepius recolo, sic inclinas corda magnorum in me, ut et presentes nedum non habominantur meum obsequium, colloquium, et faciei deformitatem, et manuum, sed delectantur mecum iungi colloquio, et reverenciam exhibere; absentes eciam, qui me nec visu noverint, delectantur benedicere, et obsequia michi prestare. Da igitur michi, Domine Deus meus, tuam gloriosam majestatem semper honorare et colere, et in hoc solummodo delectari, ut honores temporales michi exhibitos[c] possim in Te, omnium bonorum datorem, prout justum est, et debitum, retorquere. Videtur michi, Domine Deus meus, quod magne profectionis culmen attingerem, si tam sollicite, tam reverenter, Tibi obsequia et honores prestarem, quantum nedum familiares, sed et extranei michi, immo, Tibi in me, exhibent et impendunt. Sed scio, donec hausero de fonte gratie tue timorem et amorem . . .

[a] 2 Samuel 7: 9
[b] MS. eici (*allusion to* Numbers 5: 3)
[c] MS. exhibitas

name like unto the name of the great men that are in the Earth'. And moreover, though I be a man in an extreme state, and by thy providence smitten with a foul plague,[1] so that I am unworthy to have dealings with men, but by the law should be put out without the camp, yet Thou, O Lord my God, by what secret judgement I know not, dost yet hold me in honour. So great were thy services to me that I know not of any of my parents or kindred, although they were in sound health of body, who was before this generation graced with such honour. And often as I recall thy great goodness, and never without amazement, thou so inclinest the hearts of the great towards me that in my presence they do not abhor my service, my speech, and the deformity of my face and hands, but be pleased to partake of my company and to show their respect; and that those who are not present and who do not know me by sight be pleased to speak my praise and serve me. Let me, O Lord my God, ever honour and devote myself to thy glorious majesty, and in this alone take delight, so that I may return unto Thee the temporal honours bestowed upon me, this being what is just and due to Thee. It seemeth to me, O Lord my God, that I should have attained to the height of great excellence had I sworn to serve and honour Thee as not only my family but also strangers honour me, and Thee through me. But I know, until I draw reverence and love from the fountain of thy grace, . . .

[1] *Morphea* gave the word 'morphew' to the English language, but this name for a disease of the skin is little known, and was never precise. 'Leprosy' would here be perhaps too specific.

[4]

Alia Oracio eiusdem

Domine Deus, Omnipotens Pater, et Fili, et Spiritus Sancte, Trinitas Sancta, unus Deus, et unus Dominus omnium, cum gemitu an cum gaudio recolam quociens et quam misericorditer reconciliasti[a] me, post multas forinsecaciones[b] abs te, et offensas? Quociens elata cervice, et in contumacia cordis mei, sprevi mandata tua, restiti preceptis tuis, secundum totam latitudinem mandatorum, ita quod nullum preterivi quin inobedienter contempsi, aut insolenter neglexi. Quociens injuriam feci proximis meis, de quibus habui in mandatis eos tanquam me diligere, nec aliud velle illis facere, quam michimet ipsi optarem; et tamen, dolo, fraude, et machinacione mali, circumirem eos. Quociens creaturis Tui michi ad justum usum concessis abusus sum, magis in omnibus sequens appetitum commodi quam appetitum justi. Quociens, Domine Deus meus, abusus sum carne mea, in sui contumeliam Creatoris, abhominabiliter fornicatus in passionibus desiderii carnis mee; itaque, Domine, quod non sum expertus quod recolo me unquam ita intense delectatum spiritualiter in bono, utili, aut honesto, quantum delectabar[c] tunc explere desideria carnis mee, in omni illecebra et libidine illius fetide voluptatis. Quociens, Domine, ira, et odio, et mala voluntate, iratus sum proximis usque ad vehementem appetitum vindicte, quam vel intuli,

[4]

O Lord God, almighty Father, Son, and Holy Ghost, Holy Trinity, and one God alone Lord of all things, should I with lamentation or with rejoicing recall how often and with what compassion Thou hast restored me after many offences and many wanderings away from Thee? How often have I, proud of head and obstinate of heart, rejected thy commandment, and held out against thy laws and over the whole span of thy commandments, so that I came across nothing which I did not disobediently contemn or haughtily neglect. How often have I done injury to those nearest to me, whom I am enjoined by the commandments to love as I do myself, neither to do to them otherwise than I should choose to do to myself. But I did encompass them with grief, deceit, and evil machination. How often have I abused those of thy creatures given into my charge, following in all things my desire for that which was convenient rather than that which was just. How often, O Lord my God, have I abused my flesh, thus doing hurt to its Creator; and have I fornicated in the passionate yearning of my flesh, so that, O Lord, I have failed Thee inasmuch as I have never taken so intense a spiritual delight in the good, the useful, or the honourable as I have taken in sating the desires of my flesh, in every enticement and lust of that stinking and sensual pleasure. How often, O Lord, have I been roused by anger, by hatred, and by evil intent against those close at hand, so that I had an ardent desire for revenge, which I

[a] MS. reconsiliasti (*this spelling is followed throughout the MS.*).
[b] MS. forinaciones
[c] MS. delectabibar

vel contristabar inferre non posse; non deliberans inter justum et injuriam, sed solum vel libidine dominandi vel spiritu indignandi vel studio malignandi conatus sum explere libidinem iracundie quam concepi. Quociens, Domine Deus meus, crapula et ebrietate extinxi in me spiritum devocionis, quam michi impertiri sepe dignatus es; quociens huiusmodi execrabili voluptate[a] inhabilem me reddidi studio spirituali; et sepe partem sanitatis, quam michi dedisti, male superfluitatis desiderio suffocavi. Quociens, Domine Deus meus, ingratus sum Tibi; quum vidi manifesta beneficia tua in me, ita quod nulla tergiversacione possent negari, et tamen silui a laudibus, dissimulavi gracias agere, et tantum benefactorem diligere. Quociens ingratus sum proximis meis, quos tanquam me diligere teneor. Inprimis, venerabilem sponsam meam, domesticos et familiares, parentes, amicos, et omnes in fide conservos, quibus frequenter prodesse potui, sed neglexi; quibus, sine labore meo, potui bona temporalia et spiritualia procurasse, sed nolui; cum tamen ex lege gratitudinis hoc[b] deberem non parcere laboribus, ut saltem, quibus possem beneficiis, beneficia compensarem. Quociens, Domine Deus, negligens, desidiosus sum in utilibus et honestis, strenuus et alacer in otiosis, et vanus, lignum utique aridum et infructuosum; quod tamen adhuc non tam est horrendum, ut est esse lignum virulentas faciens labruscas, et vinum habominacionis distillans. Et quid dicam de multiloquio, magniloquio, vaniloquio, tur-

either took, or was sorry to be unable to take. I did not distinguish between justice and injury, but being governed by lust or a raging spirit or by scheming malice I did try to satisfy my wrathful passions which I nursed. How often, O Lord my God, have I by drunkenness and debauchery extinguished the spirit of devotion which Thou hast often thought fit to instil in me. Often the health which Thou gavest me have I stifled by a lust for an excess of evil. How often, O Lord my God, have I been ungrateful to Thee. When I saw thy manifest gifts to me, such that they could not by any pretence be denied, yet was I silent in Thy praise. I did but feign to give thanks, and to esteem so great a benefactor. How often have I been ungrateful to those near to me, whom I should so much value; and principally to my revered spouse, to the members of my house and family, to parents, friends, and all my fellow servants in the Faith, whom I could so often help, yet neglected to do so. For them I could have procured without effort temporal and spiritual good; but I did not wish it, when the law of gratitude required me to be not sparing in my labours, and to provide at least such benefit as was within my power. How often, O Lord my God, am I in my neglect indolent in useful and honourable matters, but prompt and eager in what is useless and empty. The wood which is both shrivelled and useless is yet less to be abhorred than the wood which makes poisonous wines, and distils the wine of abomination. And what shall I say of excessive speaking, of grandiloquence, of speaking vainly, of speak-

[a] MS. voluptati

[b] *Sic MS.*

piloquio, falsiloquio, et maliloquio? Et, ut innumerabilia mala mea brevi sermone concludam, videbatur mihi bonum elongari a Deo meo, recedere ab illo, forinsecari in regionem longinquam; et tamen clemens Pater meus cucurrit post me, revocavit me, et per Sacramentum penitentie reconciliavit me; non semel redeuntem, non bis, sed quociens volui ad vocem eius redire, tociens paratus est me reconciliare sibi, et ire sue ignoscere, sed insuper stolam primam et annulum amissum restituere et offerre.[a]

ing immodestly, of speaking false, and of speaking evil? And to bring quickly to an end the recounting of my numberless sins, it seemed to me good that I draw apart from my God, that I should turn away from him and take myself to a far country. But my forgiving Father hastened after me, called me back, and by the Sacrament of penitence reconciled me. I came back not once and not twice; but as often as I wished to return to his voice, so often was he prepared to reconcile me to himself and to overlook his wrath, but above all things to bring forth yet again and to offer to me the best robe and the lost ring.

[a] Cf. Luke 15: 13–24.

VIII

CANON SUPRA KALENDARIUM

SIGLA

A London, British Museum, MS. Harley 321.
B London, British Museum, MS. Harley 1612.
C Oxford, Bodleian Library, MS. Ashmole 191.

For full discussion of the manuscripts, Introduction to the text, and Commentary, see vol. ii, pp. 371–8.

DECLARACIONES ABBATIS SANCTI ALBANI SUPER KALENDARIUM REGINE[a]

[1]

Si fiat questio de nativitate viri et sint planete *in gradibus masculinis*, hec erit eis[b] fortis; et si fiat questio de nativitate mulieris et sint planete *in gradibus femineis*,[c] erit hec eis fortis. *Azamena gradus*. Azamena, id est gradus debilitatis corporis, sic expone: est enim azamena quedam corporis debilitatio, ut est cecitas, surditas membrorum amissio, et similia,[d] que in hac vita habet hoc. Cum ergo fuerit Luna in aliquo horum graduum in nativitate alicuius, accidet ei aliquis horum defectuum. *Lucidi*.[e] Et cum fuerint planete in gradibus lucidis, erit[f] eis fortius in significacione bonorum, et significat pulcritudinem, lucem, et fortunam. *Tenebrosi gradus*. Et cum fuerint in tenebrosis duriciem et tarditatem horribilem et rem tenebrosam[g] significabunt. *Vacui et fumosi*. Et cum fuerint in gradibus fumosis et vacuis, significant non modicum horribile. *Gradus putei*. Et[h] cum fuerint in puteo,[i] abibit[j] eius pulcritudo et aspectus debilitabitur a significacione[k] sua. *Augens fortunam*. Et cum fuerint planete in gradibus augentibus fortunam vel in gradibus sublimitatis,[l] augebunt fortunam nati et dominabitur terris, et habebit multas divicias; et si fuerit ascendens aliquis illorum in nativitate alicuius, habebit dominium ac multas divicias.[m]

[2]

Dignitates planetarum. Sunt autem in signis[n] quedam fortitudines que dicuntur dignitates planetarum, et sunt 5, scilicet domus, exaltacio, triplicitas, terminus, facies. Et dicitur[o] domus habere 5 fortitudines,

[a] From A. For rubric of B see introduction.

[b] Here and throughout A uses the spelling *eiis*.

[c] C omits one or two of the preceding words, but damage to the page makes it impossible to decide which.

[d] BC consimilia

[e] This type of rubricated heading tends to be omitted from B and C, and omissions are not here noted.

[f] B +hoc

[g] BC +et malam

[h] BC similiter

[i] B puteis

[j] C habibit (*altered to* habebit)

[k] A significaciones

[l] C sublimitatibus

[m] C follows with more than five sides of additional material, beginning 'Multum prodest scire in quo signo et in quo gradu signi sit Luna omni die secundum Ptholomeum in libro Almagesti . . .' Our text resumes at f. 9^{v}. It is not impossible that C is authentic in this respect, but A and B agree in omitting the passage.

[n] C signo

[o] B dicuntur

exaltacio 4, triplicitas 3, terminus 2, facies 1. Quidam tamen preposuerunt[a] terminos triplicitatibus, et volunt quod domus termini sit fortior domo triplicitatis. Iam autem[b] dicemus quantum quilibet eorum[c] valeat[d] assignantes talem comparacionem. Cum fuerit planeta in domo sua erit similis viro[e] in domo sua; et quando erit in exaltacione[f] sua erit similis viro in regno suo vel in gloria sua; et cum fuerit in triplicitate sua erit sicut vir in honore suo inter auxiliantes et ministros suos. Cum autem fuerit in termino suo erit similis viro existenti inter parentes[g] et cognatos suos et genus suum. Cum autem fuerit[h] in facie sua erit similis viro in magisterio suo existenti.[i]

[3]

Domus planetarum. Domus planete dicitur signum in quo planeta creatus fuit, aut primo apparuit homini super terram, aut signum quod maxime concordat ipsi planete in qualitatibus. Est ergo Leo domus Solis, Cancer Lune, Virgo principalis domus Mercurii, Libra principalis domus Veneris, Aries principalis domus Martis, Sagittarius principalis domus Iovis, Capricornus principalis domus Saturni. Habent itaque 5 planete[j] 5 domus accidentales, ut Saturnus Aquarium, Iupiter Pisces, Mars Scorpionem, Venus Taurum, Mercurius Geminos. Et est differencia[k] inter domum principalem et secundariam, quia principalis concordat cum planeta cuius erit domus in ambabus qualitatibus, accidentalis vero et secundaria in una sola scilicet[l] qualitate. Sol tamen[m] et Luna, quilibet illorum[n] unam solam domum habet, ut superius dicitur.

[4]

Exaltaciones planetarum. Exaltaciones planetarum sic patent: Sol exaltatur in 19 gradu Arietis, Luna in 3 gradu Tauri, Saturnus in 20 gradu Libre, Iupiter in 15 gradu Cancri, Mars in 28 gradu Capricorni, Venus in 27 gradu Piscium, Mercurius in 15 gradu Virginis. In 7 ergo signo ab exaltacione uniuscuiusque planete in gradu consimili sit eius descensio, et tunc incipit planeta declinare.

[5]

Triplicitates autem planetarum sic patent, et dicuntur eiusdem nature cum signo proprio eorum, ut si dividatur zodiacus in 4 partes, erunt in unaquaque parte tria signa. Eligantur igitur tria signa que sunt unius nature, non ut consequenter numerantur[o] in zodiaco, et dicuntur signa

[a] B preponunt; C ponunt [b] BC Nos autem iam [c] A λ [d] A habeant
[e] BC +cum fuerit [f] A exaltaciones [g] B +notos
[h] C que erit [i] AB λ [j] A planetas [k] BC differenciam [l] BC 2λ
[m] C uno [n] B eorum; C et eorum [o] (dubious reading)

illa facere triplicitatem,[a] eo quod in una natura conveniunt. Et si Sol erit in Sagittario, erit in triplicitate sua, et in Ariete similiter, quia sunt ignea sicut Leo, qui erit domus Solis; [et Luna cum sit in Scorpione, erit in triplicitate sua, et in Piscibus similiter],[b] quia sunt eiusdem nature cum Cancro, qui erit domus sua; et sic de ceteris, sicut in tabula hec patet.

Aries	Leo	Sagitt.	Ignea	Orientalia	Colerica
Cancer	Scorpio	Pisces	Aqua	Septentrionalia	Flematica
Taurus	Virgo	Capric.	Terrea	Meridionalia	Melancolia
Gemini	Libra	Aquar.	Aerea	Occidentalia	Sanguinea

[6]

Termini planetarum. De terminis vero diversi diversa senciunt, famosiores tamen sunt termini Egypciorum, qui sic patent. Quodlibet[c] signum habet 30 gradus. Divide ergo 30 in 5 partes equales, et erunt in qualibet 6. Incipe igitur ab Ariete cuius primos 6 gradus dabis Iovi, 6 Marti, 6 Soli, 6 Veneri, et ultimos 6 Mercurio, et[d] sunt 30 gradus Arietis. Incipe similiter a Tauro cuius primos 6 gradus dabis Lune, alios 6 Saturno, 6 Iovi, 6 Marti, 6 Soli, et sunt 30 gradus Tauri. Et sic facies de ceteris signis usque in fine Piscium per[e] ordinem, sicut patet per tabulam ad hoc factam in inferiore parte huius folii. [Vide p. 204 supra.]

[7]

Facies autem planetarum sic patent. Unumquodque signum dividitur in 30 gradus; divide ergo 30 in 3 partes equales, et erunt in unaquaque parte 10 gradus, qui dicuntur una facies. Incipe igitur ab Ariete cuius dabis primos 10 gradus Marti, reliquos 10 gradus Soli, ultimos 10 gradus Veneri, et sic dividitur in 3 facies Arietis. Incipe similiter a Tauro cuius dabis primos 10 gradus Mercurio, reliquos 10 Lune, et ultimos 10 Saturno. Et sic de ceteris gradibus signorum et planetis consequenter ordinabis quodlibet signum in 3 partes dividendo ordinem servando, tam in signis quam in planetis, sicut patet in tabulis ex parte alia.[f]

[8]

Ut igitur predictas 5 *fortitudines planetarum* sine labore possis invenire, describam tibi figuram 8 circulorum in quorum primo sunt 12 signa

[a] The damaged part of B ends here. Fragments of a torn page which follow are not from the next, but from the very last folio. It is interesting to note that this agrees with the end part of C, which is missing in A.

[b] All MSS. lack this passage, worded by analogy with the foregoing. It is required by the sense.

[c] C λ [d] A 5λ [e] C 2λ [f] A ends: in tabula sequenti

celi, in reliquis 7 planete per ordinem; et in unoquoque circulo ponam fortitudines quas habet planeta in signo ex directo signo sub quo invenitur. Ponam ergo pro domo 5 fortitudines, pro exaltacione 4, pro triplicitate 3, pro termino 2, pro facie 1, sicut patet hic in sequenti figura.[a] Et nota quod per istas 5 fortitudines sunt descripta fere omnia iudicia astrologie, et effectus in istis inferioribus.[b] [Vide fig., vol. iii, p. 17, *infra*.][c]

[9]

Nota quod *octava spera* stellarum fixarum in 100 annis movetur uno gradu versus orientem vel occidentem, sic spera Solis in 100 annis uno gradu movetur. Et nota quod octava spera ab inicio Arietis movetur versus orientem 8 gradibus usque ad 23 gradum Piscium, et inde revertitur et movetur iterum ab inicio Arietis versus occidentem totidem gradibus; et inde revertitur.

[10]

Corpus humanum iuxta 12 signa zodiaci ordinatur ita quod quodlibet signum habet membrum sibi appropriatum; sed quia Aries in prima mundi constitutione cepit vergi hinc est quod caput hominis habere dicitur, Taurus collum et guttur, [Gemini][d] spatulas et brachia ac manus, Cancer pectus et pulmonem, Leo cor et stomacum, Virgo epar et intestina, Libra renes, umbilicum, vesicam, et pancreas,[e] Scorpio putibias, Pisces pedes. Et ideo operari in membro cum ferro, Luna existente in signo significante membrum mortale est, ut dicit Ptholomeus.

[11]

Aries, Leo, Sagittarius faciunt *triplicitatem primam*, quia unumquodque istorum signorum est igneum, masculinum, diurnum, calidum, scilicet et fictum colericum sapore amaro. Hec triplicitas est orientalis, cuius domini sunt in die Sol, in nocte Iupiter, et eorum particeps in die et nocte est Saturnus. *Triplicitas secunda* est ex Tauro, Virgine, et Capricorno, quia hec signa sunt feminea, nocturna, terrea, frigida, sicca, malencolica, sapore aerea meridionalia. Huius quoque sunt domini in die Venus, in nocte Luna, quarum particeps est in die et nocte Mars.

[a] Although D has 'sicut patet in circulo superius descripto', the circle was actually placed at the beginning of the text on the first side of the roll, this ending being on the second side. Also in D is a diagram resembling a horoscope, but this is a table of exaltations.

[b] C ends in mid page, the figure occupying the remainder. An unrelated mnemonic verse follows on f. 11ʳ, together with a short table (*Tabula continuacionis motus Solis*) running from 1385 to 1469.

[c] The diagram was discussed in the commentary to *Exafrenon*, cap. 5. The surviving copies of the figure contain numerous mistakes.

[d] A Venus

[e] A hancras

Triplicitas tercia est ex Geminis, Libra, et Aquario. Hec sunt signa masculina, diurna, sanguinea, scilicet calida et humida, aerea, occidentalia, sapore dulcia, cuius triplicitatis domini sunt in die Saturnus et in nocte Mercurius; eorum particeps in die et nocte est Iupiter. *Quarta triplicitas* ex Cancro, Piscibus, et Scorpione. Et hec sunt signa feminea, nocturna, septentrionalia, aquatica, frigida scilicet et humida, sapore salsa; cuius triplicitatis domini sunt in die Venus et in nocte Mars, quorum particeps in die et nocte est Luna.

[12]

Quatuor quoque ex hiis signis dicuntur esse *mobilia*, scilicet Aries, Cancer, Libra, Capricornus, et quatuor *fixa*, scilicet Taurus, Leo, Scorpio, Aquarius. Reliqua vero, id est Gemini, Virgo, Sagittarius, et Pisces, sunt *communia*; dicuntur autem mobilia, quia quando ingreditur Sol, aliquod istorum movetur, id est mutatur tempus; vel figitur, id est in eodem statu perseverat; aut fit commune, id est, medietas illius erit unius temporis et medietas alterius vero similiter. Quando Sol signum Arietis ingreditur, mutatur tempus, scilicet hiemps in ver. Et quando intrat Taurum, figitur idem tempus vernale. Quando vero Sol ingreditur Geminos fit tempus commune, id est, erit [dimidium][a] veris et [dimidium][a] estatis; et sic de ceteris.

[13]

Aspectus planetarum. Sciendum quod 5 sunt aspectus planetarum, quorum primus dicitur coniunctio, secundus dicitur sextilis, tercius dicitur quartilis, quartus dicitur trinus, quintus dicitur opposicio.

Aspectus primus dicitur *coniunctio.* Aspectus primus, scilicet coniunctio, fit inter planetas cum approximant sibi invicem per minorem distanciam quam per 15 gradus; et [inimice][b] cadit sub radiis alterius, et tunc planeta dicitur combustus. Et iste primus aspectus est amicitie et manifeste benevolencie. Et quando bonus coniungitur bono, augetur amicabiliter bonitas eorum. Et si bonus coniungatur malo, separabitur amicabiliter malum illius, et debilitabitur mediocriter bonitas illius.

Aspectus secundus dicitur *sextilis*,[c] et hoc est quando planete distant ab inicio per sextam partem zodiaci, sed per duo signa. Et talis aspectus dicitur amicitie, non manifeste sed occulte.

Aspectus tercius dicitur *quartilis*, scilicet quando planeta aspicit planetam per distanciam trium signorum, que distancia est quarta pars zodiaci, ut si unus planeta esset in principio Arietis et alius in principio Cancri. Talis aspectus est occulte inimicitie, odii, et predictionis.

[a] A dividium [b] A minive (?)

[c] This, and all following opening phrases on the aspects, are given here only once, and not twice, as in the MS.

Aspectus quartus dicitur *trinus*, scilicet quando planete distant ab inicio per terciam partem zodiaci, scilicet per 4 signa, quia sunt semper similia. Et dicitur aspectus bonus, quia non est contrarietas in qualitate signorum, unde si fuerint planete concordantes, ut Iupiter et Venus, nulla reperietur in tali aspectu.

Aspectus quintus dicitur *opposicio*, scilicet quando unus planeta distat ab alio per medietatem zodiaci, scilicet per 12 signa, ut quando unus planeta fuerit in principio Arietis et alius in principio Libre. Et talis aspectus dicitur inimicitie maxime, ut belli et nocumenti evidentis, quia quilibet planeta manifeste impugnat virtutem planete existentis in suo opposito.[a]

[a] A ends here (f. 19^r). The text which follows (*De corporibus celestibus*, inc.: 'Corporalis mundi machina . . .') continues as though it were a part of the preceding, on account of the fact that the scribe has attempted to imitate the style of the foregoing, but in this he changes his mind after only one line.

IX

DOUBTFUL
AND
SPURIOUS
WORKS

I. DE COMPUTO

II. DE ECLIPSIBUS

III. DE DIAMETRIS and DE OPIMETRIS

IV. DE CHORDA ET ARCU

V. PARS ALBIONIS INSTRUMENTI *SAPHE* DICTA

For discussion of these works (not printed here) see vol. ii, pp. 379–87.

The manufacturer's authorised representative in the EU for product safety is Oxford University Press España S.A. of El Parque Empresarial San Fernando de Henares, Avenida de Castilla, 2 - 28830 Madrid (www.oup.es/en or product.safety@oup.com). OUP España S.A. also acts as importer into Spain of products made by the manufacturer.
Printed and bound by CPI Group (UK) Ltd, Croydon, CR0 4YY
14/08/2025
01936960-0018